Springer Texts in Statistics

Advisors:
George Casella   Stephen Fienberg   Ingram Olkin

Springer Texts in Statistics

(continued after index)

James E. Gentle

Matrix Algebra

Theory, Computations, and Applications in Statistics

 Springer

James E. Gentle
Department of Computational
  and Data Sciences
George Mason University
4400 University Drive
Fairfax, VA 22030-4444
jgentle@gmu.edu

e-ISBN 978-0-387-70873-7

Library of Congress Control Number: 2007930269

springer.com

To María

Preface

I began this book as an update of *Numerical Linear Algebra for Applications in Statistics*, published by Springer in 1998. There was a modest amount of new material to add, but I also wanted to supply more of the reasoning behind the facts about vectors and matrices. I had used material from that text in some courses, and I had spent a considerable amount of class time proving assertions made but not proved in that book. As I embarked on this project, the character of the book began to change markedly. In the previous book, I apologized for spending 30 pages on the theory and basic facts of linear algebra before getting on to the main interest: *numerical* linear algebra. In the present book, discussion of those basic facts takes up over half of the book.

The orientation and perspective of this book remains *numerical linear algebra for applications in statistics*. Computational considerations inform the narrative. There is an emphasis on the areas of matrix analysis that are important for statisticians, and the kinds of matrices encountered in statistical applications receive special attention.

This book is divided into three parts plus a set of appendices. The three parts correspond generally to the three areas of the book's subtitle — theory, computations, and applications — although the parts are in a different order, and there is no firm separation of the topics.

Part I, consisting of Chapters 1 through 7, covers most of the material in linear algebra needed by statisticians. (The word "matrix" in the title of the present book may suggest a somewhat more limited domain than "linear algebra"; but I use the former term only because it seems to be more commonly used by statisticians and is used more or less synonymously with the latter term.)

The first four chapters cover the basics of vectors and matrices, concentrating on topics that are particularly relevant for statistical applications. In Chapter 4, it is assumed that the reader is generally familiar with the basics of partial differentiation of scalar functions. Chapters 5 through 7 begin to take on more of an applications flavor, as well as beginning to give more consideration to computational methods. Although the details of the computations are not covered in those chapters, the topics addressed are oriented more toward computational algorithms. Chapter 5 covers methods for decomposing matrices into useful factors.

Chapter 6 addresses applications of matrices in setting up and solving linear systems, including overdetermined systems. We should not confuse statistical inference with fitting equations to data, although the latter task is a component of the former activity. In Chapter 6, we address the more mechanical aspects of the problem of fitting equations to data. Applications in statistical data analysis are discussed in Chapter 9. In those applications, we need to make statements (that is, assumptions) about relevant probability distributions.

Chapter 7 discusses methods for extracting eigenvalues and eigenvectors. There are many important details of algorithms for eigenanalysis, but they are beyond the scope of this book. As with other chapters in Part I, Chapter 7 makes some reference to statistical applications, but it focuses on the mathematical and mechanical aspects of the problem.

Although the first part is on "theory", the presentation is informal; neither definitions nor facts are highlighted by such words as "Definition", "Theorem", "Lemma", and so forth. It is assumed that the reader follows the natural development. Most of the facts have simple proofs, and most proofs are given naturally in the text. No "Proof" and "Q.E.D." or " ∎ " appear to indicate beginning and end; again, it is assumed that the reader is engaged in the development. For example, on page 270:

If A is nonsingular and symmetric, then A^{-1} is also symmetric because $(A^{-1})^{\mathrm{T}} = (A^{\mathrm{T}})^{-1} = A^{-1}$.

The first part of that sentence could have been stated as a theorem and given a number, and the last part of the sentence could have been introduced as the proof, with reference to some previous theorem that the inverse and transposition operations can be interchanged. (This had already been shown before page 270 — in an unnumbered theorem of course!)

None of the proofs are original (at least, I don't think they are), but in most cases I do not know the original source, or even the source where I first saw them. I would guess that many go back to C. F. Gauss. Most, whether they are as old as Gauss or not, have appeared somewhere in the work of C. R. Rao. Some lengthier proofs are only given in outline, but references are given for the details. Very useful sources of details of the proofs are Harville (1997), especially for facts relating to applications in linear models, and Horn and Johnson (1991) for more general topics, especially those relating to stochastic matrices. The older books by Gantmacher (1959) provide extensive coverage and often rather novel proofs. These two volumes have been brought back into print by the American Mathematical Society.

I also sometimes make simple assumptions without stating them explicitly. For example, I may write "for all i" when i is used as an index to a vector. I hope it is clear that "for all i" means only "for i that correspond to indices of the vector". Also, my use of an expression generally implies existence. For example, if "AB" is used to represent a matrix product, it implies that "A and B are conformable for the multiplication AB". Occasionally I remind the reader that I am taking such shortcuts.

The material in Part I, as in the entire book, was built up recursively. In the first pass, I began with some definitions and followed those with some facts that are useful in applications. In the second pass, I went back and added definitions and additional facts that lead to the results stated in the first pass. The supporting material was added as close to the point where it was needed as practical and as necessary to form a logical flow. Facts motivated by additional applications were also included in the second pass. In subsequent passes, I continued to add supporting material as necessary and to address the linear algebra for additional areas of application. I sought a bare-bones presentation that gets across what I considered to be the theory necessary for most applications in the data sciences. The material chosen for inclusion is motivated by applications.

Throughout the book, some attention is given to numerical methods for computing the various quantities discussed. This is in keeping with my belief that statistical computing should be dispersed throughout the statistics curriculum and statistical literature generally. Thus, unlike in other books on matrix "theory", I describe the "modified" Gram-Schmidt method, rather than just the "classical" GS. (I put "modified" and "classical" in quotes because, to me, GS *is* MGS. History is interesting, but in computational matters, I do not care to dwell on the methods of the past.) Also, condition numbers of matrices are introduced in the "theory" part of the book, rather than just in the "computational" part. Condition numbers also relate to fundamental properties of the model and the data.

The difference between an expression and a computing method is emphasized. For example, often we may write the solution to the linear system $Ax = b$ as $A^{-1}b$. Although this is the solution (so long as A is square and of full rank), solving the linear system does not involve computing A^{-1}. We may write $A^{-1}b$, but we know we can compute the solution without inverting the matrix.

> "This is an instance of a principle that we will encounter repeatedly:
> *the form of a mathematical expression and the way the expression*
> *should be evaluated in actual practice may be quite different."*

(The statement in quotes appears word for word in several places in the book.)

Standard textbooks on "matrices for statistical applications" emphasize their uses in the analysis of traditional linear models. This is a large and important field in which real matrices are of interest, and the important kinds of real matrices include symmetric, positive definite, projection, and generalized inverse matrices. This area of application also motivates much of the discussion in this book. In other areas of statistics, however, there are different matrices of interest, including similarity and dissimilarity matrices, stochastic matrices, rotation matrices, and matrices arising from graph-theoretic approaches to data analysis. These matrices have applications in clustering, data mining, stochastic processes, and graphics; therefore, I describe these matrices and their special properties. I also discuss the geometry of matrix algebra. This provides a better intuition of the operations. Homogeneous coordinates and special operations in $\mathrm{I\!R}^3$ are covered because of their geometrical applications in statistical graphics.

Part II addresses selected applications in data analysis. Applications are referred to frequently in Part I, and of course, the choice of topics for coverage was motivated by applications. The difference in Part II is in its orientation.

Only "selected" applications in data analysis are addressed; there are applications of matrix algebra in almost all areas of statistics, including the theory of estimation, which is touched upon in Chapter 4 of Part I. Certain types of matrices are more common in statistics, and Chapter 8 discusses in more detail some of the important types of matrices that arise in data analysis and statistical modeling. Chapter 9 addresses selected applications in data analysis. The material of Chapter 9 has no obvious definition that could be covered in a single chapter (or a single part, or even a single book), so I have chosen to discuss briefly a wide range of areas. Most of the sections and even subsections of Chapter 9 are on topics to which entire books are devoted; however, I do not believe that any single book addresses all of them.

Part III covers some of the important details of numerical computations, with an emphasis on those for linear algebra. I believe these topics constitute the most important material for an introductory course in numerical analysis for statisticians and should be covered in every such course.

Except for specific computational techniques for optimization, random number generation, and perhaps symbolic computation, Part III provides the basic material for a course in statistical computing. All statisticians should have a passing familiarity with the principles.

Chapter 10 provides some basic information on how data are stored and manipulated in a computer. Some of this material is rather tedious, but it is important to have a general understanding of computer arithmetic before considering computations for linear algebra. Some readers may skip or just skim Chapter 10, but the reader should be aware that the way the computer stores numbers and performs computations has far-reaching consequences. Computer arithmetic differs from ordinary arithmetic in many ways; for example, computer arithmetic lacks associativity of addition and multiplication, and series often converge even when they are not supposed to. (On the computer, a straightforward evaluation of $\sum_{x=1}^{\infty} x$ converges!)

I emphasize the differences between the abstract number system $\mathrm{I\!R}$, called the reals, and the computer number system $\mathrm{I\!F}$, the floating-point numbers unfortunately also often called "real". Table 10.3 on page 400 summarizes some of these differences. All statisticians should be aware of the effects of these differences. I also discuss the differences between $\mathrm{Z\!\!\!Z}$, the abstract number system called the integers, and the computer number system $\mathrm{I\!I}$, the fixed-point numbers. (Appendix A provides definitions for this and other notation that I use.)

Chapter 10 also covers some of the fundamentals of algorithms, such as iterations, recursion, and convergence. It also discusses software development. Software issues are revisited in Chapter 12.

While Chapter 10 deals with general issues in numerical analysis, Chapter 11 addresses specific issues in numerical methods for computations in linear algebra.

Chapter 12 provides a brief introduction to software available for computations with linear systems. Some specific systems mentioned include the IMSL^{TM} libraries for Fortran and C, Octave or MATLAB$^{\circledR}$ (or Matlab$^{\circledR}$), and R or S-PLUS$^{\circledR}$ (or S-Plus$^{\circledR}$). All of these systems are easy to use, and the best way to learn them is to begin using them for simple problems. I do not use any particular software system in the book, but in some exercises, and particularly in Part III, I do assume the ability to program in either Fortran or C and the availability of either R or S-Plus, Octave or Matlab, and Maple$^{\circledR}$ or Mathematica$^{\circledR}$. My own preferences for software systems are Fortran and R, and occasionally these preferences manifest themselves in the text.

Appendix A collects the notation used in this book. It is generally "standard" notation, but one thing the reader must become accustomed to is the lack of notational distinction between a vector and a scalar. All vectors are "column" vectors, although I usually write them as horizontal lists of their elements. (Whether vectors are "row" vectors or "column" vectors is generally only relevant for how we write expressions involving vector/matrix multiplication or partitions of matrices.)

I write algorithms in various ways, sometimes in a form that looks similar to Fortran or C and sometimes as a list of numbered steps. I believe all of the descriptions used are straightforward and unambiguous.

This book could serve as a basic reference either for courses in statistical computing or for courses in linear models or multivariate analysis. When the book is used as a reference, rather than looking for "Definition" or "Theorem", the user should look for items set off with bullets or look for numbered equations, or else should use the Index, beginning on page 519, or Appendix A, beginning on page 479.

The prerequisites for this text are minimal. Obviously some background in mathematics is necessary. Some background in statistics or data analysis and some level of scientific computer literacy are also required. References to rather advanced mathematical topics are made in a number of places in the text. To some extent this is because many sections evolved from class notes that I developed for various courses that I have taught. All of these courses were at the graduate level in the computational and statistical sciences, but they have had wide ranges in mathematical level. I have carefully reread the sections that refer to groups, fields, measure theory, and so on, and am convinced that if the reader does not know much about these topics, the material is still understandable, but if the reader is familiar with these topics, the references add to that reader's appreciation of the material. In many places, I refer to computer programming, and some of the exercises require some programming. A careful coverage of Part III requires background in numerical programming.

In regard to the use of the book as a text, most of the book evolved in one way or another for my own use in the classroom. I must quickly admit, however, that I have never used this whole book as a text for any single course. I have used Part III in the form of printed notes as the primary text for a course in the "foundations of computational science" taken by graduate students in the natural sciences (including a few statistics students, but dominated by physics students). I have provided several sections from Parts I and II in online PDF files as supplementary material for a two-semester course in mathematical statistics at the "baby measure theory" level (using Shao, 2003). Likewise, for my courses in computational statistics and statistical visualization, I have provided many sections, either as supplementary material or as the primary text, in online PDF files or printed notes. I have not taught a regular "applied statistics" course in almost 30 years, but if I did, I am sure that I would draw heavily from Parts I and II for courses in regression or multivariate analysis. If I ever taught a course in "matrices for statistics" (I don't even know if such courses exist), this book would be my primary text because I think it covers most of the things statisticians need to know about matrix theory and computations.

Some exercises are Monte Carlo studies. I do not discuss Monte Carlo methods in this text, so the reader lacking background in that area may need to consult another reference in order to work those exercises. The exercises should be considered an integral part of the book. For some exercises, the required software can be obtained from either `statlib` or `netlib` (see the bibliography). Exercises in any of the chapters, not just in Part III, may require computations or computer programming.

Penultimately, I must make some statement about the relationship of this book to some other books on similar topics. Much important statistical theory and many methods make use of matrix theory, and many statisticians have contributed to the advancement of matrix theory from its very early days. Widely used books with derivatives of the words "statistics" and "matrices/linear-algebra" in their titles include Basilevsky (1983), Graybill (1983), Harville (1997), Schott (2004), and Searle (1982). All of these are useful books. The computational orientation of this book is probably the main difference between it and these other books. Also, some of these other books only address topics of use in linear models, whereas this book also discusses matrices useful in graph theory, stochastic processes, and other areas of application. (If the applications are only in linear models, most matrices of interest are symmetric, and all eigenvalues can be considered to be real.) Other differences among all of these books, of course, involve the authors' choices of secondary topics and the ordering of the presentation.

Acknowledgments

I thank John Kimmel of Springer for his encouragement and advice on this book and other books on which he has worked with me. I especially thank Ken Berk for his extensive and insightful comments on a draft of this book. I thank my student Li Li for reading through various drafts of some of the chapters and pointing out typos or making helpful suggestions. I thank the anonymous reviewers of this edition for their comments and suggestions. I also thank the many readers of my previous book on numerical linear algebra who informed me of errors and who otherwise provided comments or suggestions for improving the exposition. Whatever strengths this book may have can be attributed in large part to these people, named or otherwise. The weaknesses can only be attributed to my own ignorance or hardheadedness.

I thank my wife, María, to whom this book is dedicated, for everything.

I used TeX via LaTeX 2_ε to write the book. I did all of the typing, programming, etc., myself, so all misteaks are mine. I would appreciate receiving suggestions for improvement and notification of errors. Notes on this book, including errata, are available at

```
http://mason.gmu.edu/~jgentle/books/matbk/
```

Fairfax County, Virginia

James E. Gentle
June 12, 2007

Contents

Part III   Numerical Methods and Software

Part I

Linear Algebra

1

Basic Vector/Matrix Structure and Notation

Vectors and matrices are useful in representing multivariate data, and they occur naturally in working with linear equations or when expressing linear relationships among objects. Numerical algorithms for a variety of tasks involve matrix and vector arithmetic. An optimization algorithm to find the minimum of a function, for example, may use a vector of first derivatives and a matrix of second derivatives; and a method to solve a differential equation may use a matrix with a few diagonals for computing differences.

There are various precise ways of defining vectors and matrices, but we will generally think of them merely as linear or rectangular arrays of numbers, or scalars, on which an algebra is defined. Unless otherwise stated, we will assume the scalars are real numbers. We denote both the set of real numbers and the field of real numbers as $\mathbb{R}$. (The *field* is the set together with the operators.) Occasionally we will take a geometrical perspective for vectors and will consider matrices to define geometrical transformations. In all contexts, however, the elements of vectors or matrices are real numbers (or, more generally, members of a field). When this is not the case, we will use more general phrases, such as "ordered lists" or "arrays".

Many of the operations covered in the first few chapters, especially the transformations and factorizations in Chapter 5, are important because of their use in solving systems of linear equations, which will be discussed in Chapter 6; in computing eigenvectors, eigenvalues, and singular values, which will be discussed in Chapter 7; and in the applications in Chapter 9.

Throughout the first few chapters, we emphasize the facts that are important in statistical applications. We also occasionally refer to relevant computational issues, although computational details are addressed specifically in Part III.

It is very important to understand that the form of a mathematical expression and the way the expression should be evaluated in actual practice may be quite different. We remind the reader of this fact from time to time. That there is a difference in mathematical expressions and computational methods is one of the main messages of Chapters 10 and 11. (An example of this, in notation that we will introduce later, is the expression $A^{-1}b$. If our goal is to solve a linear system $Ax = b$, we probably should never compute the matrix inverse A^{-1} and then multiply it times b. Nevertheless, it may be entirely appropriate to write the expression $A^{-1}b$.)

1.1 Vectors

For a positive integer n, a vector (or n-vector) is an n-tuple, ordered (multi)set, or array of n numbers, called *elements* or *scalars*. The number of elements is called the *order*, or sometimes the "length", of the vector. An n-vector can be thought of as representing a point in n-dimensional space. In this setting, the "length" of the vector may also mean the Euclidean distance from the origin to the point represented by the vector; that is, the square root of the sum of the squares of the elements of the vector. This Euclidean distance will generally be what we mean when we refer to the *length* of a vector (see page 17).

We usually use a lowercase letter to represent a vector, and we use the same letter with a single subscript to represent an element of the vector.

The first element of an n-vector is the first (1st) element and the last is the nth element. (This statement is not a tautology; in some computer systems, the first element of an object used to represent a vector is the 0th element of the object. This sometimes makes it difficult to preserve the relationship between the computer entity and the object that is of interest.) We will use paradigms and notation that maintain the priority of the object of interest rather than the computer entity representing it.

We may write the n-vector x as

$$x = \begin{pmatrix} x_1 \\ \vdots \\ x_n \end{pmatrix} \tag{1.1}$$

or

$$x = (x_1, \ldots, x_n). \tag{1.2}$$

We make no distinction between these two notations, although in some contexts we think of a vector as a "column", so the first notation may be more natural. The simplicity of the second notation recommends it for common use. (And this notation does not require the additional symbol for transposition that some people use when they write the elements of a vector horizontally.)

We use the notation

$$\mathrm{I\!R}^n$$

to denote the set of n-vectors with real elements.

1.2 Arrays

Arrays are structured collections of elements corresponding in shape to lines, rectangles, or rectangular solids. The number of dimensions of an array is often called the *rank* of the array. Thus, a vector is an array of rank 1, and a matrix is an array of rank 2. A scalar, which can be thought of as a degenerate array, has rank 0. When referring to computer software objects, "rank" is generally used in this sense. (This term comes from its use in describing a *tensor*. A rank 0 tensor is a scalar, a rank 1 tensor is a vector, a rank 2 tensor is a *square* matrix, and so on. In our usage referring to arrays, we do not require that the dimensions be equal, however.) When we refer to "rank of an array", we mean the number of dimensions. When we refer to "rank of a matrix", we mean something different, as we discuss in Section 3.3. In linear algebra, this latter usage is far more common than the former.

1.3 Matrices

A matrix is a rectangular or two-dimensional array. We speak of the *rows* and *columns* of a matrix. The rows or columns can be considered to be vectors, and we often use this equivalence. An $n \times m$ matrix is one with n rows and m columns. The number of rows and the number of columns determine the *shape* of the matrix. Note that the shape is the doubleton (n, m), not just a single number such as the ratio. If the number of rows is the same as the number of columns, the matrix is said to be square.

All matrices are two-dimensional in the sense of "dimension" used above. The word "dimension", however, when applied to matrices, often means something different, namely the number of columns. (This usage of "dimension" is common both in geometry and in traditional statistical applications.)

We usually use an uppercase letter to represent a matrix. To represent an element of the matrix, we usually use the corresponding lowercase letter with a subscript to denote the row and a second subscript to represent the column. If a nontrivial expression is used to denote the row or the column, we separate the row and column subscripts with a comma.

Although vectors and matrices are fundamentally quite different types of objects, we can bring some unity to our discussion and notation by occasionally considering a vector to be a "column vector" and in some ways to be the same as an $n \times 1$ matrix. (This has nothing to do with the way we may write the elements of a vector. The notation in equation (1.2) is more convenient than that in equation (1.1) and so will generally be used in this book, but its use should not change the nature of the vector. Likewise, this has nothing to do with the way the elements of a vector or a matrix are stored in the computer.) When we use vectors and matrices in the same expression, however, we use the symbol "T" (for "transpose") as a superscript to represent a vector that is being treated as a $1 \times n$ matrix.

We use the notation a_{*j} to correspond to the jth column of the matrix A and use a_{i*} to represent the (column) vector that corresponds to the ith row.

The first row is the 1st (first) row, and the first column is the 1st (first) column. (Again, we remark that computer entities used in some systems to represent matrices and to store elements of matrices as computer data sometimes index the elements beginning with 0. Furthermore, some systems use the first index to represent the column and the second index to indicate the row. We are not speaking here of the *storage order*— "row major" versus "column major" — we address that later, in Chapter 11. Rather, we are speaking of the mechanism of *referring to* the abstract entities. In image processing, for example, it is common practice to use the first index to represent the column and the second index to represent the row. In the software package PV-Wave, for example, there are two different kinds of two-dimensional objects: "arrays", in which the indexing is done as in image processing, and "matrices", in which the indexing is done as we have described.)

The $n \times m$ matrix A can be written

$$A = \begin{bmatrix} a_{11} & \dots & a_{1m} \\ \vdots & \vdots & \vdots \\ a_{n1} & \dots & a_{nm} \end{bmatrix}. \tag{1.3}$$

We also write the matrix A above as

$$A = (a_{ij}), \tag{1.4}$$

with the indices i and j ranging over $\{1, \dots, n\}$ and $\{1, \dots, m\}$, respectively. We use the notation $A_{n \times m}$ to refer to the matrix A and simultaneously to indicate that it is $n \times m$, and we use the notation

$$\mathbb{R}^{n \times m}$$

to refer to the set of all $n \times m$ matrices with real elements.

We use the notation $(A)_{ij}$ to refer to the element in the ith row and the jth column of the matrix A; that is, in equation (1.3), $(A)_{ij} = a_{ij}$.

Although vectors are column vectors and the notation in equations (1.1) and (1.2) represents the same entity, that would not be the same for matrices. If $x_1, \dots, x_n$ are scalars

$$X = \begin{bmatrix} x_1 \\ \vdots \\ x_n \end{bmatrix} \tag{1.5}$$

and

$$Y = [x_1, \dots, x_n], \tag{1.6}$$

then X is an $n \times 1$ matrix and Y is a $1 \times n$ matrix (and Y is the *transpose* of X). Although an $n \times 1$ matrix is a different type of object from a vector, we may treat X in equation (1.5) or Y^{T} in equation (1.6) as a vector when it is convenient to do so. Furthermore, although a 1×1 matrix, a 1-vector, and a scalar are all fundamentally different types of objects, we will treat a one by one matrix or a vector with only one element as a scalar whenever it is convenient.

One of the most important uses of matrices is as a transformation of a vector by vector/matrix multiplication. Such transformations are linear (a term that we define later). Although one can occasionally profitably distinguish matrices from linear transformations on vectors, for our present purposes there is no advantage in doing so. We will often treat matrices and linear transformations as equivalent.

Many of the properties of vectors and matrices we discuss hold for an infinite number of elements, but we will assume throughout this book that the number is finite.

Subvectors and Submatrices

We sometimes find it useful to work with only some of the elements of a vector or matrix. We refer to the respective arrays as "subvectors" or "submatrices". We also allow the rearrangement of the elements by row or column permutations and still consider the resulting object as a subvector or submatrix. In Chapter 3, we will consider special forms of submatrices formed by "partitions" of given matrices.

1.4 Representation of Data

Before we can do any serious analysis of data, the data must be represented in some structure that is amenable to the operations of the analysis. In simple cases, the data are represented by a list of scalar values. The ordering in the list may be unimportant, and the analysis may just consist of computation of simple summary statistics. In other cases, the list represents a time series of observations, and the relationships of observations to each other as a function of their distance apart in the list are of interest. Often, the data can be represented meaningfully in two lists that are related to each other by the positions in the lists. The generalization of this representation is a two-dimensional array in which each column corresponds to a particular type of data.

A major consideration, of course, is the nature of the individual items of data. The observational data may be in various forms: quantitative measures, colors, text strings, and so on. Prior to most analyses of data, they must be represented as real numbers. In some cases, they can be represented easily as real numbers, although there may be restrictions on the mapping into the reals. (For example, do the data naturally assume only integral values, or could any real number be mapped back to a possible observation?)

The most common way of representing data is by using a two-dimensional array in which the rows correspond to observational units ("instances") and the columns correspond to particular types of observations ("variables" or "features"). If the data correspond to real numbers, this representation is the familiar X data matrix. Much of this book is devoted to the matrix theory and computational methods for the analysis of data in this form. This type of matrix, perhaps with an adjoined vector, is the basic structure used in many familiar statistical methods, such as regression analysis, principal components analysis, analysis of variance, multidimensional scaling, and so on.

There are other types of structures that are useful in representing data based on graphs. A *graph* is a structure consisting of two components: a set of points, called *vertices* or *nodes* and a set of pairs of the points, called *edges*. (Note that this usage of the word "graph" is distinctly different from the more common one that refers to lines, curves, bars, and so on to represent data pictorially. The phrase "graph theory" is often used, or overused, to emphasize the present meaning of the word.) A graph $\mathcal{G} = (V, E)$ with vertices $V = \{v_1, \ldots, v_n\}$ is distinguished primarily by the nature of the edge elements (v_i, v_j) in E. Graphs are identified as complete graphs, directed graphs, trees, and so on, depending on E and its relationship with V. A tree may be used for data that are naturally aggregated in a hierarchy, such as political unit, subunit, household, and individual. Trees are also useful for representing clustering of data at different levels of association. In this type of representation, the individual data elements are the leaves of the tree.

In another type of graphical representation that is often useful in "data mining", where we seek to uncover relationships among objects, the vertices are the objects, either observational units or features, and the edges indicate some commonality between vertices. For example, the vertices may be text documents, and an edge between two documents may indicate that a certain number of specific words or phrases occur in both documents. Despite the differences in the basic ways of representing data, in graphical modeling of data, many of the standard matrix operations used in more traditional data analysis are applied to matrices that arise naturally from the graph.

However the data are represented, whether in an array or a network, the analysis of the data is often facilitated by using "association" matrices. The most familiar type of association matrix is perhaps a correlation matrix. We will encounter and use other types of association matrices in Chapter 8.

2

Vectors and Vector Spaces

In this chapter we discuss a wide range of basic topics related to vectors of real numbers. Some of the properties carry over to vectors over other fields, such as complex numbers, but the reader should not assume this. Occasionally, for emphasis, we will refer to "real" vectors or "real" vector spaces, but unless it is stated otherwise, we are assuming the vectors and vector spaces are real. The topics and the properties of vectors and vector spaces that we emphasize are motivated by applications in the data sciences.

2.1 Operations on Vectors

The elements of the vectors we will use in the following are real numbers, that is, elements of $\mathbb{R}$. We call elements of $\mathbb{R}$ *scalars*. Vector operations are defined in terms of operations on real numbers.

Two vectors can be added if they have the same number of elements. The sum of two vectors is the vector whose elements are the sums of the corresponding elements of the vectors being added. Vectors with the same number of elements are said to be *conformable* for addition. A vector all of whose elements are 0 is the *additive identity* for all conformable vectors.

We overload the usual symbols for the operations on the reals to signify the corresponding operations on vectors or matrices when the operations are defined. Hence, "+" can mean addition of scalars, addition of conformable vectors, or addition of a scalar to a vector. This last meaning of "+" may not be used in many mathematical treatments of vectors, but it is consistent with the semantics of modern computer languages such as Fortran 95, R, and Matlab. By the addition of a scalar to a vector, we mean the addition of the scalar to each element of the vector, resulting in a vector of the same number of elements.

A scalar multiple of a vector (that is, the product of a real number and a vector) is the vector whose elements are the multiples of the corresponding elements of the original vector. Juxtaposition of a symbol for a scalar and a symbol for a vector indicates the multiplication of the scalar with each element of the vector, resulting in a vector of the same number of elements.

A very common operation in working with vectors is the addition of a scalar multiple of one vector to another vector,

$$z = ax + y, \tag{2.1}$$

where a is a scalar and x and y are vectors conformable for addition. Viewed as a single operation with three operands, this is called an "axpy" for obvious reasons. (Because the Fortran versions of BLAS to perform this operation were called **saxpy** and **daxpy**, the operation is also sometimes called "saxpy" or "daxpy". See Section 12.1.4 on page 454, for a description of the BLAS.) The axpy operation is called a *linear combination*. Such linear combinations of vectors are the basic operations in most areas of linear algebra. The composition of axpy operations is also an axpy; that is, one linear combination followed by another linear combination is a linear combination. Furthermore, any linear combination can be decomposed into a sequence of axpy operations.

2.1.1 Linear Combinations and Linear Independence

If a given vector can be formed by a linear combination of one or more vectors, the set of vectors (including the given one) is said to be linearly dependent; conversely, if in a set of vectors no one vector can be represented as a linear combination of any of the others, the set of vectors is said to be *linearly independent*. In equation (2.1), for example, the vectors x, y, and z are not linearly independent. It is possible, however, that any two of these vectors are linearly independent. Linear independence is one of the most important concepts in linear algebra.

We can see that the definition of a linearly independent set of vectors $\{v_1, \ldots, v_k\}$ is equivalent to stating that if

$$a_1 v_1 + \cdots a_k v_k = 0, \tag{2.2}$$

then $a_1 = \cdots = a_k = 0$. If the set of vectors $\{v_1, \ldots, v_k\}$ is not linearly independent, then it is possible to select a *maximal linearly independent subset*; that is, a subset of $\{v_1, \ldots, v_k\}$ that is linearly independent and has maximum cardinality. We do this by selecting an arbitrary vector, v_{i_1}, and then seeking a vector that is independent of v_{i_1}. If there are none in the set that is linearly independent of v_{i_1}, then a maximum linearly independent subset is just the singleton, because all of the vectors must be a linear combination of just one vector (that is, a scalar multiple of that one vector). If there is a vector that is linearly independent of v_{i_1}, say v_{i_2}, we next seek a vector in the remaining set that is independent of v_{i_1} and v_{i_2}. If one does not exist, then $\{v_{i_1}, v_{i_2}\}$ is a maximal subset because any other vector can be represented in terms of these two and hence, within any subset of three vectors, one can be represented in terms of the two others. Thus, we see how to form a maximal linearly independent subset, and we see that the maximum cardinality of any subset of linearly independent vectors is unique.

It is easy to see that the maximum number of n-vectors that can form a set that is linearly independent is n. (We can see this by assuming n linearly independent vectors and then, for any $(n + 1)^{\text{th}}$ vector, showing that it is a linear combination of the others by building it up one by one from linear combinations of two of the given linearly independent vectors. In Exercise 2.1, you are asked to write out these steps.)

Properties of a set of vectors are usually invariant to a permutation of the elements of the vectors if the same permutation is applied to all vectors in the set. In particular, if a set of vectors is linearly independent, the set remains linearly independent if the elements of each vector are permuted in the same way.

If the elements of each vector in a set of vectors are separated into subvectors, linear independence of any set of corresponding subvectors implies linear independence of the full vectors. To state this more precisely for a set of three n-vectors, let $x = (x_1, \ldots, x_n)$, $y = (y_1, \ldots, y_n)$, and $z = (z_1, \ldots, z_n)$. Now let $\{i_1, \ldots, i_k\} \subset \{1, \ldots, n\}$, and form the k-vectors $\tilde{x} = (x_{i_1}, \ldots, x_{i_k})$, $\tilde{y} = (y_{i_1}, \ldots, y_{i_k})$, and $\tilde{z} = (z_{i_1}, \ldots, z_{i_k})$. Then linear independence of $\tilde{x}$, $\tilde{y}$, and $\tilde{z}$ implies linear independence of x, y, and z.

2.1.2 Vector Spaces and Spaces of Vectors

Let V be a set of n-vectors such that any linear combination of the vectors in V is also in V. Then the set together with the usual vector algebra is called a *vector space*. (Technically, the "usual algebra" is a *linear algebra* consisting of two operations: vector addition and scalar times vector multiplication, which are the two operations comprising an axpy. It has closure of the space under the combination of those operations, commutativity and associativity of addition, an additive identity and inverses, a multiplicative identity, distribution of multiplication over both vector addition and scalar addition, and associativity of scalar multiplication and scalar times vector multiplication. Vector spaces are *linear spaces*.) A vector space necessarily includes the additive identity. (In the axpy operation, let $a = -1$ and $y = x$.)

A vector space can also be made up of other objects, such as matrices. The key characteristic of a vector space is a linear algebra.

We generally use a calligraphic font to denote a vector space; $\mathcal{V}$, for example. Often, however, we think of the vector space merely in terms of the set of vectors on which it is built and denote it by an ordinary capital letter; V, for example.

The Order and the Dimension of a Vector Space

The maximum number of linearly independent vectors in a vector space is called the *dimension of the vector space*. We denote the dimension by

$$\dim(\cdot),$$

which is a mapping $\mathbb{R}^n \mapsto \mathbb{Z}_+$ (where $\mathbb{Z}_+$ denotes the positive integers). The length or order of the vectors in the space is the *order of the vector space*. The order is greater than or equal to the dimension, as we showed above.

The vector space consisting of all n-vectors with real elements is denoted $\mathbb{R}^n$. (As mentioned earlier, the notation $\mathbb{R}^n$ also refers to just the *set* of n-vectors with real elements; that is, to the set over which the vector space is defined.) Both the order and the dimension of $\mathbb{R}^n$ are n.

We also use the phrase *dimension of a vector* to mean the dimension of the vector space of which the vector is an element. This term is ambiguous, but its meaning is clear in certain applications, such as *dimension reduction*, that we will discuss later.

Many of the properties of vectors that we discuss hold for an infinite number of elements, but throughout this book we will assume the vector spaces have a finite number of dimensions.

Essentially Disjoint Vector Spaces

If the only element in common between two vector spaces $\mathcal{V}_1$ and $\mathcal{V}_2$ is the additive identity, the spaces are said to be *essentially disjoint*. If the vector spaces $\mathcal{V}_1$ and $\mathcal{V}_2$ are essentially disjoint, it is clear that any element in $\mathcal{V}_1$ (except the additive identity) is linearly independent of any set of elements in $\mathcal{V}_2$.

Some Special Vectors

We denote the additive identity in a vector space of order n by 0_n or sometimes by 0. This is the vector consisting of all zeros. We call this the *zero vector*. This vector by itself is sometimes called the *null vector space*. It is not a vector space in the usual sense; it would have dimension 0. (All linear combinations are the same.)

Likewise, we denote the vector consisting of all ones by 1_n or sometimes by 1. We call this the *one vector* and also the "summing vector" (see page 23). This vector and all scalar multiples of it are vector spaces with dimension 1. (This is true of any single nonzero vector; all linear combinations are just scalar multiples.) Whether 0 and 1 without a subscript represent vectors or scalars is usually clear from the context.

The ith *unit vector*, denoted by e_i, has a 1 in the ith position and 0s in all other positions:

$$e_i = (0, \ldots, 0, 1, 0, \ldots, 0). \tag{2.3}$$

Another useful vector is the *sign vector*, which is formed from signs of the elements of a given vector. It is denoted by "sign$(\cdot)$" and defined by

$$\begin{aligned}
\text{sign}(x)_i &= 1 & \text{if } x_i > 0, \\
&= 0 & \text{if } x_i = 0, \\
&= -1 & \text{if } x_i < 0.
\end{aligned} \tag{2.4}$$

Ordinal Relations among Vectors

There are several possible ways to form a rank ordering of vectors of the same order, but no complete ordering is entirely satisfactory. (Note the unfortunate overloading of the word "order" or "ordering" here.) If x and y are vectors of the same order and for corresponding elements $x_i > y_i$, we say x is *greater than y* and write

$$x > y. \tag{2.5}$$

In particular, if all of the elements of x are positive, we write $x > 0$.

If x and y are vectors of the same order and for corresponding elements $x_i \geq y_i$, we say x is *greater than or equal to y* and write

$$x \geq y. \tag{2.6}$$

This relationship is a *partial ordering* (see Exercise 8.1a). The expression $x \geq 0$ means that all of the elements of x are nonnegative.

Set Operations on Vector Spaces

Although a vector space is a set together with operations, we often speak of a vector space as if it were just the set, and we use some of the same notation to refer to vector spaces as we use to refer to sets. For example, if $\mathcal{V}$ is a vector space, the notation $\mathcal{W} \subseteq \mathcal{V}$ indicates that $\mathcal{W}$ is a vector space (that is, it has the properties listed above), that the set of vectors in the vector space $\mathcal{W}$ is a subset of the vectors in $\mathcal{V}$, and that the operations in the two objects are the same. A subset of a vector space $\mathcal{V}$ that is itself a vector space is called a *subspace* of $\mathcal{V}$.

The intersection of two vector spaces of the same order is a vector space. The union of two such vector spaces, however, is not necessarily a vector space (because for $v_1 \in \mathcal{V}_1$ and $v_2 \in \mathcal{V}_2$, $v_1 + v_2$ may not be in $\mathcal{V}_1 \cup \mathcal{V}_2$). We refer to a set of vectors of the same order together with the addition operator (whether or not the set is closed with respect to the operator) as a "space of vectors".

If $\mathcal{V}_1$ and $\mathcal{V}_2$ are spaces of vectors, the space of vectors

$$\mathcal{V} = \{v, \text{ s.t. } v = v_1 + v_2, \ v_1 \in \mathcal{V}_1, \ v_2 \in \mathcal{V}_2\}$$

is called the *sum* of the spaces $\mathcal{V}_1$ and $\mathcal{V}_2$ and is denoted by $\mathcal{V} = \mathcal{V}_1 + \mathcal{V}_2$. If the spaces $\mathcal{V}_1$ and $\mathcal{V}_2$ are vector spaces, then $\mathcal{V}_1 + \mathcal{V}_2$ is a vector space, as is easily verified.

If $\mathcal{V}_1$ and $\mathcal{V}_2$ are essentially disjoint vector spaces (not just spaces of vectors), the sum is called the *direct sum*. This relation is denoted by

$$\mathcal{V} = \mathcal{V}_1 \oplus \mathcal{V}_2. \tag{2.7}$$

Cones

A set of vectors that contains all positive scalar multiples of any vector in the set is called a *cone*. A cone always contains the zero vector. A set of vectors V is a *convex cone* if, for all $v_1, v_2 \in V$ and all $a, b \geq 0$, $av_1 + bv_2 \in V$. (Such a cone is called a homogeneous convex cone by some authors. Also, some authors require that $a + b = 1$ in the definition.) A convex cone is not necessarily a vector space because $v_1 - v_2$ may not be in V. An important convex cone in an n-dimensional vector space is the positive orthant together with the zero vector. This convex cone is not closed, in the sense that it does not contain some limits. The closure of the positive orthant (that is, the nonnegative orthant) is also a convex cone.

2.1.3 Basis Sets

If each vector in the vector space $\mathcal{V}$ can be expressed as a linear combination of the vectors in some set G, then G is said to be a *generating set* or *spanning set* of $\mathcal{V}$. If, in addition, all linear combinations of the elements of G are in $\mathcal{V}$, the vector space is the *space generated by* G and is denoted by $\mathcal{V}(G)$ or by $\mathrm{span}(G)$:

$$\mathcal{V}(G) \equiv \mathrm{span}(G).$$

A set of linearly independent vectors that generate or span a space is said to be a *basis* for the space.

- The representation of a given vector in terms of a basis set is unique.

To see this, let $\{v_1, \ldots, v_k\}$ be a basis for a vector space that includes the vector x, and let

$$x = c_1 v_1 + \cdots c_k v_k.$$

Now suppose

$$x = b_1 v_1 + \cdots b_k v_k,$$

so that we have

$$0 = (c_1 - b_1)v_1 + \cdots + (c_k - b_k)v_k.$$

Since $\{v_1, \ldots, v_k\}$ are independent, the only way this is possible is if $c_i = b_i$ for each i.

A related fact is that if $\{v_1, \ldots, v_k\}$ is a basis for a vector space of order n that includes the vector x and $x = c_1 v_1 + \cdots c_k v_k$, then $x = 0_n$ if and only if $c_i = 0$ for each i.

If B_1 is a basis set for $\mathcal{V}_1$, B_2 is a basis set for $\mathcal{V}_2$, and $\mathcal{V}_1 \oplus \mathcal{V}_2 = \mathcal{V}$, then $B_1 \cup B_2$ is a generating set for $\mathcal{V}$ because from the definition of $\oplus$ we see that any vector in $\mathcal{V}$ can be represented as a linear combination of vectors in B_1 plus a linear combination of vectors in B_2.

The number of vectors in a generating set is at least as great as the dimension of the vector space. Because the vectors in a basis set are independent, the number of vectors in a basis set is exactly the same as the dimension of the vector space; that is, if B is a basis set of the vector space $\mathcal{V}$, then

$$\dim(\mathcal{V}) = \#(B). \tag{2.8}$$

A *generating set* or *spanning set* of a cone C is a set of vectors $S = \{v_i\}$ such that for any vector v in C there exists scalars $a_i \geq 0$ so that $v = \sum a_i v_i$, and if for scalars $b_i \geq 0$ and $\sum b_i v_i = 0$, then $b_i = 0$ for all i. If a generating set of a cone has a finite number of elements, the cone is a *polyhedron*. A generating set consisting of the minimum number of vectors of any generating set for that cone is a basis set for the cone.

2.1.4 Inner Products

A useful operation on two vectors x and y of the same order is the *dot product*, which we denote by $\langle x, y \rangle$ and define as

$$\langle x, y \rangle = \sum_i x_i y_i. \tag{2.9}$$

The dot product is also called the *inner product* or the *scalar product*. The dot product is actually a special type of inner product, but it is the most commonly used inner product, and so we will use the terms synonymously. A vector space together with an inner product is called an *inner product space*.

The dot product is also sometimes written as $x \cdot y$, hence the name. Yet another notation for the dot product is $x^{\mathrm{T}} y$, and we will see later that this notation is natural in the context of matrix multiplication. We have the equivalent notations

$$\langle x, y \rangle \equiv x \cdot y \equiv x^{\mathrm{T}} y.$$

The dot product is a mapping from a vector space $\mathcal{V}$ to $\mathbb{R}$ that has the following properties:

1. Nonnegativity and mapping of the identity:
   if $x \neq 0$, then $\langle x, x \rangle > 0$ and $\langle 0, x \rangle = \langle x, 0 \rangle = \langle 0, 0 \rangle = 0$.
2. Commutativity:
   $\langle x, y \rangle = \langle y, x \rangle$.
3. Factoring of scalar multiplication in dot products:
   $\langle ax, y \rangle = a \langle x, y \rangle$ for real a.
4. Relation of vector addition to addition of dot products:
   $\langle x + y, z \rangle = \langle x, z \rangle + \langle y, z \rangle$.

These properties in fact define a more general inner product for other kinds of mathematical objects for which an addition, an additive identity, and a multiplication by a scalar are defined. (We should restate here that we assume the vectors have real elements. The dot product of vectors over the complex field is not an inner product because, if x is complex, we can have $x^{\mathrm{T}} x = 0$ when

$x \neq 0$. An alternative definition of a dot product using complex conjugates is an inner product, however.) Inner products are also defined for matrices, as we will discuss on page 74. We should note in passing that there are two different kinds of multiplication used in property 3. The first multiplication is scalar multiplication, which we have defined above, and the second multiplication is ordinary multiplication in $\mathrm{I\!R}$. There are also two different kinds of addition used in property 4. The first addition is vector addition, defined above, and the second addition is ordinary addition in $\mathrm{I\!R}$. The dot product can reveal fundamental relationships between the two vectors, as we will see later.

A useful property of inner products is the *Cauchy-Schwarz inequality*:

$$\langle x, y \rangle \leq \langle x, x \rangle^{\frac{1}{2}} \langle y, y \rangle^{\frac{1}{2}}. \tag{2.10}$$

This relationship is also sometimes called the Cauchy-Bunyakovskii-Schwarz inequality. (Augustin-Louis Cauchy gave the inequality for the kind of discrete inner products we are considering here, and Viktor Bunyakovskii and Hermann Schwarz independently extended it to more general inner products, defined on functions, for example.) The inequality is easy to see, by first observing that for every real number t,

$$\begin{aligned}
0 &\leq \langle (tx + y), (tx + y) \rangle^2 \\
&= \langle x, x \rangle t^2 + 2 \langle x, y \rangle t + \langle y, y \rangle \\
&= at^2 + bt + c,
\end{aligned}$$

where the constants a, b, and c correspond to the dot products in the preceding equation. This quadratic in t cannot have two distinct real roots. Hence the discriminant, $b^2 - 4ac$, must be less than or equal to zero; that is,

$$\left(\frac{1}{2} b \right)^2 \leq ac.$$

By substituting and taking square roots, we get the Cauchy-Schwarz inequality. It is also clear from this proof that equality holds only if $x = 0$ or if $y = rx$, for some scalar r.

2.1.5 Norms

We consider a set of objects S that has an addition-type operator, $+$, a corresponding additive identity, 0, and a scalar multiplication; that is, a multiplication of the objects by a real (or complex) number. On such a set, a *norm* is a function, $\| \cdot \|$, from S to $\mathrm{I\!R}$ that satisfies the following three conditions:

1. Nonnegativity and mapping of the identity:
   if $x \neq 0$, then $\|x\| > 0$, and $\|0\| = 0$.
2. Relation of scalar multiplication to real multiplication:
   $\|ax\| = |a| \, \|x\|$ for real a.

3. Triangle inequality:
$$\|x + y\| \le \|x\| + \|y\|.$$

(If property 1 is relaxed to require only $\|x\| \ge 0$ for $x \ne 0$, the function is called a *seminorm*.) Because a norm is a function whose argument is a vector, we also often use a functional notation such as $\rho(x)$ to represent a norm.

Sets of various types of objects (functions, for example) can have norms, but our interest in the present context is in norms for vectors and (later) for matrices. (The three properties above in fact define a more general norm for other kinds of mathematical objects for which an addition, an additive identity, and multiplication by a scalar are defined. Norms are defined for matrices, as we will discuss later. Note that there are two different kinds of multiplication used in property 2 and two different kinds of addition used in property 3.)

A vector space together with a norm is called a *normed space*.

For some types of objects, a norm of an object may be called its "length" or its "size". (Recall the ambiguity of "length" of a vector that we mentioned at the beginning of this chapter.)

L_p Norms

There are many norms that could be defined for vectors. One type of norm is called an L_p norm, often denoted as $\| \cdot \|_p$. For $p \ge 1$, it is defined as

$$\|x\|_p = \left(\sum_i |x_i|^p \right)^{\frac{1}{p}}. \tag{2.11}$$

This is also sometimes called the *Minkowski norm* and also the *Hölder norm*.

It is easy to see that the L_p norm satisfies the first two conditions above. For general $p \ge 1$ it is somewhat more difficult to prove the triangular inequality (which for the L_p norms is also called the Minkowski inequality), but for some special cases it is straightforward, as we will see below.

The most common L_p norms, and in fact the most commonly used vector norms, are:

- $\|x\|_1 = \sum_i |x_i|$, also called the *Manhattan norm* because it corresponds to sums of distances along coordinate axes, as one would travel along the rectangular street plan of Manhattan.
- $\|x\|_2 = \sqrt{\sum_i x_i^2}$, also called the *Euclidean norm*, the *Euclidean length*, or just the *length* of the vector. The L_p norm is the square root of the inner product of the vector with itself: $\|x\|_2 = \sqrt{\langle x, x \rangle}$.
- $\|x\|_\infty = \max_i |x_i|$, also called the *max norm* or the *Chebyshev norm*. The L_∞ norm is defined by taking the limit in an L_p norm, and we see that it is indeed $\max_i |x_i|$ by expressing it as

$$\|x\|_\infty = \lim_{p\to\infty} \|x\|_p = \lim_{p\to\infty} \left(\sum_i |x_i|^p\right)^{\frac{1}{p}} = m \lim_{p\to\infty} \left(\sum_i \left|\frac{x_i}{m}\right|^p\right)^{\frac{1}{p}}$$

with $m = \max_i |x_i|$. Because the quantity of which we are taking the pth root is bounded above by the number of elements in x and below by 1, that factor goes to 1 as p goes to ∞.

An L_p norm is also called a p-norm, or 1-norm, 2-norm, or ∞-norm in those special cases.

It is easy to see that, for any n-vector x, the L_p norms have the relationships

$$\|x\|_\infty \le \|x\|_2 \le \|x\|_1. \tag{2.12}$$

More generally, for given x and for $p \ge 1$, we see that $\|x\|_p$ is a nonincreasing function of p.

We also have bounds that involve the number of elements in the vector:

$$\|x\|_\infty \le \|x\|_2 \le \sqrt{n}\|x\|_\infty, \tag{2.13}$$

and

$$\|x\|_2 \le \|x\|_1 \le \sqrt{n}\|x\|_2. \tag{2.14}$$

The triangle inequality obviously holds for the L_1 and L_∞ norms. For the L_2 norm it can be seen by expanding $\sum(x_i + y_i)^2$ and then using the Cauchy-Schwarz inequality (2.10) on page 16. Rather than approaching it that way, however, we will show below that the L_2 norm can be defined in terms of an inner product, and then we will establish the triangle inequality for any norm defined similarly by an inner product; see inequality (2.19). Showing that the triangle inequality holds for other L_p norms is more difficult; see Exercise 2.6.

A generalization of the L_p vector norm is the *weighted L_p vector norm* defined by

$$\|x\|_{wp} = \left(\sum_i w_i|x_i|^p\right)^{\frac{1}{p}}, \tag{2.15}$$

where $w_i \ge 0$.

Basis Norms

If $\{v_1, \ldots, v_k\}$ is a basis for a vector space that includes a vector x with $x = c_1 v_1 + \cdots + c_k v_k$, then

$$\rho(x) = \left(\sum_i c_i^2\right)^{\frac{1}{2}} \tag{2.16}$$

is a norm. It is straightforward to see that $\rho(x)$ is a norm by checking the following three conditions:

- $\rho(x) \geq 0$ and $\rho(x) = 0$ if and only if $x = 0$ because $x = 0$ if and only if $c_i = 0$ for all i.
- $\rho(ax) = \left(\sum_i a^2 c_i^2\right)^{\frac{1}{2}} = |a| \left(\sum_i a^2 c_i^2\right)^{\frac{1}{2}} = |a|\rho(x)$.
- If $y = b_1 v_1 + \cdots + b_k v_k$, then

$$\rho(x + y) = \left(\sum_i (c_i + b_i)^2\right)^{\frac{1}{2}} \leq \left(\sum_i c_i^2\right)^{\frac{1}{2}} \left(\sum_i b_i^2\right)^{\frac{1}{2}} = \rho(x)\rho(y).$$

The last inequality is just the triangle inequality for the L_2 norm for the vectors $(c_1, \cdots, c_k)$ and $(b_1, \cdots, b_k)$.

In Section 2.2.5, we will consider special forms of basis sets in which the norm in equation (2.16) is identically the L_2 norm. (This is called Parseval's identity, equation (2.38).)

Equivalence of Norms

There is an equivalence among any two norms over a normed linear space in the sense that if $\| \cdot \|_a$ and $\| \cdot \|_b$ are norms, then there are positive numbers r and s such that for any x in the space,

$$r\|x\|_b \leq \|x\|_a \leq s\|x\|_b. \tag{2.17}$$

Expressions (2.13) and (2.14) are examples of this general equivalence for three L_p norms.

We can prove inequality (2.17) by using the norm defined in equation (2.16). We need only consider the case $x \neq 0$, because the inequality is obviously true if $x = 0$. Let $\| \cdot \|_a$ be any norm over a given normed linear space and let $\{v_1, \ldots, v_k\}$ be a basis for the space. Any x in the space has a representation in terms of the basis, $x = c_1 v_1 + \cdots + c_k v_k$. Then

$$\|x\|_a = \left\| \sum_{1=i}^{k} c_i v_i \right\|_a \leq \sum_{1=i}^{k} |c_i| \, \|v_i\|_a.$$

Applying the Cauchy-Schwarz inequality to the two vectors $(c_1, \cdots, c_k)$ and $(\|v_1\|_a, \cdots, \|v_k\|_a)$, we have

$$\sum_{1=i}^{k} |c_i| \, \|v_i\|_a \leq \left(\sum_{1=i}^{k} c_i^2\right)^{\frac{1}{2}} \left(\sum_{1=i}^{k} \|v_i\|_a^2\right)^{\frac{1}{2}}.$$

Hence, with $\tilde{s} = (\sum_i \|v_i\|_a^2)^{\frac{1}{2}}$, which must be positive, we have

$$\|x\|_a \leq \tilde{s}\rho(x).$$

Now, to establish a lower bound for $\|x\|_a$, let us define a subset C of the linear space consisting of all vectors $(u_1, \ldots, u_k)$ such that $\sum |u_i|^2 = 1$. This set is obviously closed. Next, we define a function $f(\cdot)$ over this closed subset by

$$f(u) = \left\| \sum_{1=i}^{k} u_i v_i \right\|_a.$$

Because f is continuous, it attains a minimum in this closed subset, say for the vector u_*; that is, $f(u_*) \le f(u)$ for any u such that $\sum |u_i|^2 = 1$. Let

$$\tilde{r} = f(u_*),$$

which must be positive, and again consider any x in the normed linear space and express it in terms of the basis, $x = c_1 v_1 + \cdots c_k v_k$. If $x \ne 0$, we have

$$\|x\|_a = \left\| \sum_{1=i}^{k} c_i v_i \right\|_a$$

$$= \left(\sum_{1=i}^{k} c_i^2 \right)^{\frac{1}{2}} \left\| \sum_{1=i}^{k} \left(\frac{c_i}{\left(\sum_{1=i}^{k} c_i^2 \right)^{\frac{1}{2}}} \right) v_i \right\|_a$$

$$= \rho(x) f(\tilde{c}),$$

where $\tilde{c} = (c_1, \cdots, c_k)/(\sum_{1=i}^{k} c_i^2)^{1/2}$. Because $\tilde{c}$ is in the set C, $f(\tilde{c}) \ge r$; hence, combining this with the inequality above, we have

$$\tilde{r}\rho(x) \le \|x\|_a \le \tilde{s}\rho(x).$$

This expression holds for any norm $\|\cdot\|_a$ and so, after obtaining similar bounds for any other norm $\|\cdot\|_b$ and then combining the inequalities for $\|\cdot\|_a$ and $\|\cdot\|_b$, we have the bounds in the equivalence relation (2.17). (This is an equivalence relation because it is reflexive, symmetric, and transitive. Its transitivity is seen by the same argument that allowed us to go from the inequalities involving $\rho(\cdot)$ to ones involving $\|\cdot\|_b$.)

Convergence of Sequences of Vectors

A sequence of real numbers $a_1, a_2, \ldots$ is said to converge to a finite number a if for any given $\epsilon > 0$ there is an integer M such that, for $k > M$, $|a_k - a| < \epsilon$, and we write $\lim_{k \to \infty} a_k = a$, or we write $a_k \to a$ as $k \to \infty$. If M does not depend on ϵ, the convergence is said to be uniform.

We define convergence of a sequence of vectors in terms of the convergence of a sequence of their norms, which is a sequence of real numbers. We say that a sequence of vectors $x_1, x_2, \ldots$ (of the same order) converges to the vector x with respect to the norm $\|\cdot\|$ if the sequence of real numbers $\|x_1 - x\|, \|x_2 - x\|, \ldots$ converges to 0. Because of the bounds (2.17), the choice of the norm is irrelevant, and so convergence of a sequence of vectors is well-defined without reference to a specific norm. (This is the reason equivalence of norms is an important property.)

Norms Induced by Inner Products

There is a close relationship between a norm and an inner product. For any inner product space with inner product $\langle \cdot, \cdot \rangle$, a norm of an element of the space can be defined in terms of the square root of the inner product of the element with itself:

$$\|x\| = \sqrt{\langle x, x \rangle}. \tag{2.18}$$

Any function $\| \cdot \|$ defined in this way satisfies the properties of a norm. It is easy to see that $\|x\|$ satisfies the first two properties of a norm, nonnegativity and scalar equivariance. Now, consider the square of the right-hand side of the triangle inequality, $\|x\| + \|y\|$:

$$\begin{aligned}
(\|x\| + \|y\|)^2 &= \langle x, x \rangle + 2\sqrt{\langle x, x \rangle \langle y, y \rangle} + \langle y, y \rangle \\
&\geq \langle x, x \rangle + 2\langle x, y \rangle + \langle y, y \rangle \\
&= \langle x + y, \, x + y \rangle \\
&= \|x + y\|^2;
\end{aligned} \tag{2.19}$$

hence, the triangle inequality holds. Therefore, given an inner product, $\langle x, y \rangle$, then $\sqrt{\langle x, x \rangle}$ is a norm.

Equation (2.18) defines a norm given any inner product. It is called the *norm induced by the inner product*. In the case of vectors and the inner product we defined for vectors in equation (2.9), the induced norm is the L_2 norm, $\| \cdot \|_2$, defined above.

In the following, when we use the unqualified symbol $\| \cdot \|$ for a vector norm, we will mean the L_2 norm; that is, the Euclidean norm, the induced norm.

In the sequence of equations above for an induced norm of the sum of two vectors, one equation (expressed differently) stands out as particularly useful in later applications:

$$\|x + y\|^2 = \|x\|^2 + \|y\|^2 + 2\langle x, y \rangle. \tag{2.20}$$

2.1.6 Normalized Vectors

The Euclidean norm of a vector corresponds to the length of the vector x in a natural way; that is, it agrees with our intuition regarding "length". Although, as we have seen, this is just one of many vector norms, in most applications it is the most useful one. (I must warn you, however, that occasionally I will carelessly but naturally use "length" to refer to the order of a vector; that is, the number of elements. This usage is common in computer software packages such as R and SAS IML, and software necessarily shapes our vocabulary.)

Dividing a given vector by its length *normalizes* the vector, and the resulting vector with length 1 is said to be *normalized*; thus

$$\tilde{x} = \frac{1}{\|x\|} x \tag{2.21}$$

is a normalized vector. Normalized vectors are sometimes referred to as "unit vectors", although we will generally reserve this term for a special kind of normalized vector (see page 12). A normalized vector is also sometimes referred to as a "normal vector". I use "normalized vector" for a vector such as $\tilde{x}$ in equation (2.21) and use the latter phrase to denote a vector that is orthogonal to a subspace.

2.1.7 Metrics and Distances

It is often useful to consider how far apart two vectors are; that is, the "distance" between them. A reasonable distance measure would have to satisfy certain requirements, such as being a nonnegative real number. A function Δ that maps any two objects in a set S to $\mathrm{I\!R}$ is called a *metric* on S if, for all x, y, and z in S, it satisfies the following three conditions:

1. $\Delta(x, y) > 0$ if $x \neq y$ and $\Delta(x, y) = 0$ if $x = y$;
2. $\Delta(x, y) = \Delta(y, x)$;
3. $\Delta(x, y) \leq \Delta(x, z) + \Delta(z, y)$.

These conditions correspond in an intuitive manner to the properties we expect of a distance between objects.

Metrics Induced by Norms

If subtraction and a norm are defined for the elements of S, the most common way of forming a metric is by using the norm. If $\| \cdot \|$ is a norm, we can verify that

$$\Delta(x, y) = \|x - y\| \tag{2.22}$$

is a metric by using the properties of a norm to establish the three properties of a metric above (Exercise 2.7).

The general inner products, norms, and metrics defined above are relevant in a wide range of applications. The sets on which they are defined can consist of various types of objects. In the context of real vectors, the most common inner product is the dot product; the most common norm is the Euclidean norm that arises from the dot product; and the most common metric is the one defined by the Euclidean norm, called the Euclidean distance.

2.1.8 Orthogonal Vectors and Orthogonal Vector Spaces

Two vectors v_1 and v_2 such that

$$\langle v_1, v_2 \rangle = 0 \tag{2.23}$$

are said to be *orthogonal*, and this condition is denoted by $v_1 \perp v_2$. (Sometimes we exclude the zero vector from this definition, but it is not important to do so.) Normalized vectors that are all orthogonal to each other are called *orthonormal* vectors. (If the elements of the vectors are from the field of complex numbers, orthogonality and normality are defined in terms of the dot products of a vector with a complex conjugate of a vector.)

A set of nonzero vectors that are mutually orthogonal are necessarily linearly independent. To see this, we show it for any two orthogonal vectors and then indicate the pattern that extends to three or more vectors. Suppose v_1 and v_2 are nonzero and are orthogonal; that is, $\langle v_1, v_2 \rangle = 0$. We see immediately that if there is a scalar a such that $v_1 = av_2$, then a must be nonzero and we have a contradiction because $\langle v_1, v_2 \rangle = a\langle v_1, v_1 \rangle \neq 0$. For three mutually orthogonal vectors, v_1, v_2, and v_3, we consider $v_1 = av_2 + bv_3$ for a or b nonzero, and arrive at the same contradiction.

Two vector spaces $\mathcal{V}_1$ and $\mathcal{V}_2$ are said to be *orthogonal*, written $\mathcal{V}_1 \perp \mathcal{V}_2$, if each vector in one is orthogonal to every vector in the other. If $\mathcal{V}_1 \perp \mathcal{V}_2$ and $\mathcal{V}_1 \oplus \mathcal{V}_2 = \mathrm{I\!R}^n$, then $\mathcal{V}_2$ is called the *orthogonal complement* of $\mathcal{V}_1$, and this is written as $\mathcal{V}_2 = \mathcal{V}_1^\perp$. More generally, if $\mathcal{V}_1 \perp \mathcal{V}_2$ and $\mathcal{V}_1 \oplus \mathcal{V}_2 = \mathcal{V}$, then $\mathcal{V}_2$ is called the orthogonal complement of $\mathcal{V}_1$ with respect to $\mathcal{V}$. This is obviously a symmetric relationship; if $\mathcal{V}_2$ is the orthogonal complement of $\mathcal{V}_1$, then $\mathcal{V}_1$ is the orthogonal complement of $\mathcal{V}_2$.

If B_1 is a basis set for $\mathcal{V}_1$, B_2 is a basis set for $\mathcal{V}_2$, and $\mathcal{V}_2$ is the orthogonal complement of $\mathcal{V}_1$ with respect to $\mathcal{V}$, then $B_1 \cup B_2$ is a basis set for $\mathcal{V}$. It is a basis set because since $\mathcal{V}_1$ and $\mathcal{V}_2$ are orthogonal, it must be the case that $B_1 \cap B_2 = \emptyset$.

If $\mathcal{V}_1 \subset \mathcal{V}$, $\mathcal{V}_2 \subset \mathcal{V}$, $\mathcal{V}_1 \perp \mathcal{V}_2$, and $\dim(\mathcal{V}_1) + \dim(\mathcal{V}_2) = \dim(\mathcal{V})$, then

$$\mathcal{V}_1 \oplus \mathcal{V}_2 = \mathcal{V}; \qquad (2.24)$$

that is, $\mathcal{V}_2$ is the orthogonal complement of $\mathcal{V}_1$. We see this by first letting B_1 and B_2 be bases for $\mathcal{V}_1$ and $\mathcal{V}_2$. Now $\mathcal{V}_1 \perp \mathcal{V}_2$ implies that $B_1 \cap B_2 = \emptyset$ and $\dim(\mathcal{V}_1) + \dim(\mathcal{V}_2) = \dim(\mathcal{V})$ implies $\#(B_1) + \#(B_2) = \#(B)$, for any basis set B for $\mathcal{V}$; hence, $B_1 \cup B_2$ is a basis set for $\mathcal{V}$.

The intersection of two orthogonal vector spaces consists only of the zero vector (Exercise 2.9).

A set of linearly independent vectors can be mapped to a set of mutually orthogonal (and orthonormal) vectors by means of the Gram-Schmidt transformations (see equation (2.34) below).

2.1.9 The "One Vector"

Another often useful vector is the vector with all elements equal to 1. We call this the "one vector" and denote it by 1 or by 1_n. The one vector can be used in the representation of the sum of the elements in a vector:

$$1^{\mathrm{T}} x = \sum x_i. \qquad (2.25)$$

The one vector is also called the "summing vector".

The Mean and the Mean Vector

Because the elements of x are real, they can be summed; however, in applications it may or may not make sense to add the elements in a vector, depending on what is represented by those elements. If the elements have some kind of essential commonality, it may make sense to compute their sum as well as their *arithmetic mean*, which for the n-vector x is denoted by $\bar{x}$ and defined by

$$\bar{x} = 1_n^{\mathrm{T}} x / n. \tag{2.26}$$

We also refer to the arithmetic mean as just the "mean" because it is the most commonly used mean.

It is often useful to think of the mean as an n-vector all of whose elements are $\bar{x}$. The symbol $\bar{x}$ is also used to denote this vector; hence, we have

$$\bar{x} = \bar{x} 1_n, \tag{2.27}$$

in which $\bar{x}$ on the left-hand side is a vector and $\bar{x}$ on the right-hand side is a scalar. We also have, for the two different objects,

$$\|\bar{x}\|^2 = n\bar{x}^2. \tag{2.28}$$

The meaning, whether a scalar or a vector, is usually clear from the context. In any event, an expression such as $x - \bar{x}$ is unambiguous; the addition (subtraction) has the same meaning whether $\bar{x}$ is interpreted as a vector or a scalar. (In some mathematical treatments of vectors, addition of a scalar to a vector is not defined, but here we are following the conventions of modern computer languages.)

2.2 Cartesian Coordinates and Geometrical Properties of Vectors

Points in a Cartesian geometry can be identified with vectors. Several definitions and properties of vectors can be motivated by this geometric interpretation. In this interpretation, vectors are directed line segments with a common origin. The geometrical properties can be seen most easily in terms of a Cartesian coordinate system, but the properties of vectors defined in terms of a Cartesian geometry have analogues in Euclidean geometry without a coordinate system. In such a system, only length and direction are defined, and two vectors are considered to be the same vector if they have the same length and direction. Generally, we will not assume that there is a "position" associated with a vector.

2.2.1 Cartesian Geometry

A Cartesian coordinate system in d dimensions is defined by d unit vectors, e_i in equation (2.3), each with d elements. A unit vector is also called a *principal axis* of the coordinate system. The set of unit vectors is orthonormal. (There is an implied number of elements of a unit vector that is inferred from the context. Also parenthetically, we remark that the phrase "unit vector" is sometimes used to refer to a vector the sum of whose squared elements is 1, that is, whose length, in the Euclidean distance sense, is 1. As we mentioned above, we refer to this latter type of vector as a "normalized vector".)

The sum of all of the unit vectors is the one vector:

$$\sum_{1=1}^{d} e_i = 1_d.$$

A point x with Cartesian coordinates $(x_1, \ldots, x_d)$ is associated with a vector from the origin to the point, that is, the vector $(x_1, \ldots, x_d)$. The vector can be written as the linear combination

$$x = x_1 e_1 + \ldots + x_d e_d$$

or, equivalently, as

$$x = \langle x, e_1 \rangle e_1 + \ldots + \langle x, e_d \rangle e_n.$$

(This is a Fourier expansion, equation (2.36) below.)

2.2.2 Projections

The *projection* of the vector y onto the vector x is the vector

$$\hat{y} = \frac{\langle x, y \rangle}{\|x\|^2} x. \tag{2.29}$$

This definition is consistent with a geometrical interpretation of vectors as directed line segments with a common origin. The projection of y onto x is the inner product of the normalized x and y times the normalized x; that is, $\langle \tilde{x}, y \rangle \tilde{x}$, where $\tilde{x} = x/\|x\|$. Notice that the order of y and x is the same.

An important property of a projection is that when it is subtracted from the vector that was projected, the resulting vector, called the "residual", is orthogonal to the projection; that is, if

$$r = y - \frac{\langle x, y \rangle}{\|x\|^2} x$$
$$= y - \hat{y} \tag{2.30}$$

then r and $\hat{y}$ are orthogonal, as we can easily see by taking their inner product (see Figure 2.1). Notice also that the Pythagorean relationship holds:

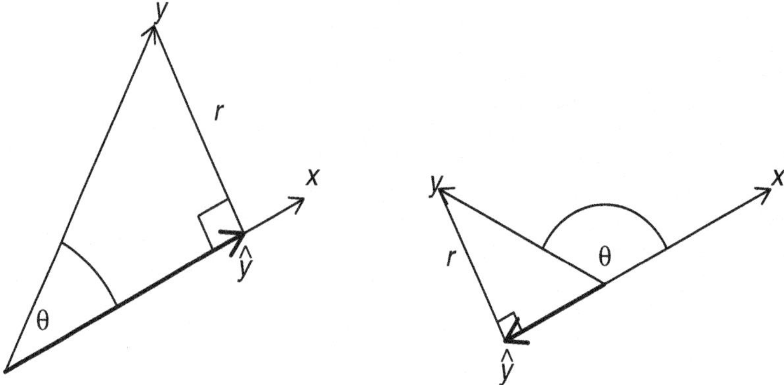

Fig. 2.1. Projections and Angles

$$\|y\|^2 = \|\hat{y}\|^2 + \|r\|^2. \tag{2.31}$$

As we mentioned on page 24, the mean $\bar{y}$ can be interpreted either as a scalar or as a vector all of whose elements are $\bar{y}$. As a vector, it is the projection of y onto the one vector 1_n,

$$\frac{\langle 1_n, y\rangle}{\|1_n\|^2}\, 1_n = \frac{1_n^{\mathrm{T}} y}{n}\, 1_n$$
$$= \bar{y}\, 1_n,$$

from equations (2.26) and (2.29).

We will consider more general projections (that is, projections onto planes or other subspaces) on page 280, and on page 331 we will view linear regression fitting as a projection onto the space spanned by the independent variables.

2.2.3 Angles between Vectors

The *angle* between the vectors x and y is determined by its cosine, which we can compute from the length of the projection of one vector onto the other. Hence, denoting the angle between x and y as angle(x, y), we define

$$\text{angle}(x, y) = \cos^{-1}\left(\frac{\langle x, y\rangle}{\|x\|\|y\|}\right), \tag{2.32}$$

with $\cos^{-1}(\cdot)$ being taken in the interval $[0, \pi]$. The cosine is $\pm\|\hat{y}\|/\|y\|$, with the sign chosen appropriately; see Figure 2.1. Because of this choice of $\cos^{-1}(\cdot)$, we have that angle$(y, x) =$ angle(x, y) — but see Exercise 2.13e on page 39.

The word "orthogonal" is appropriately defined by equation (2.23) on page 22 because orthogonality in that sense is equivalent to the corresponding geometric property. (The cosine is 0.)

Notice that the angle between two vectors is invariant to scaling of the vectors; that is, for any nonzero scalar a, $\text{angle}(ax, y) = \text{angle}(x, y)$.

A given vector can be defined in terms of its length and the angles θ_i that it makes with the unit vectors. The cosines of these angles are just the scaled coordinates of the vector:

$$
\begin{aligned}
\cos(\theta_i) &= \frac{\langle x, e_i \rangle}{\|x\|\|e_i\|} \\
&= \frac{1}{\|x\|}\, x_i.
\end{aligned}
\tag{2.33}
$$

These quantities are called the *direction cosines* of the vector.

Although geometrical intuition often helps us in understanding properties of vectors, sometimes it may lead us astray in high dimensions. Consider the direction cosines of an arbitrary vector in a vector space with large dimensions. If the elements of the arbitrary vector are nearly equal (that is, if the vector is a diagonal through an orthant of the coordinate system), the direction cosine goes to 0 as the dimension increases. In high dimensions, any two vectors are "almost orthogonal" to each other; see Exercise 2.11.

The geometric property of the angle between vectors has important implications for certain operations both because it may indicate that rounding in computations will have deleterious effects and because it may indicate a deficiency in the understanding of the application.

We will consider more general projections and angles between vectors and other subspaces on page 287. In Section 5.2.1, we will consider rotations of vectors onto other vectors or subspaces. Rotations are similar to projections, except that the length of the vector being rotated is preserved.

2.2.4 Orthogonalization Transformations

Given m nonnull, linearly independent vectors, $x_1, \ldots, x_m$, it is easy to form m orthonormal vectors, $\tilde{x}_1, \ldots, \tilde{x}_m$, that span the same space. A simple way to do this is sequentially. First normalize x_1 and call this $\tilde{x}_1$. Next, project x_2 onto $\tilde{x}_1$ and subtract this projection from x_2. The result is orthogonal to $\tilde{x}_1$; hence, normalize this and call it $\tilde{x}_2$. These first two steps, which are illustrated in Figure 2.2, are

$$
\begin{aligned}
\tilde{x}_1 &= \frac{1}{\|x_1\|}\, x_1, \\[2ex]
\tilde{x}_2 &= \frac{1}{\|x_2 - \langle \tilde{x}_1, x_2 \rangle \tilde{x}_1\|}\, (x_2 - \langle \tilde{x}_1, x_2 \rangle \tilde{x}_1).
\end{aligned}
\tag{2.34}
$$

These are called *Gram-Schmidt transformations*.

The Gram-Schmidt transformations can be continued with all of the vectors in the linearly independent set. There are two straightforward ways equations (2.34) can be extended. One method generalizes the second equation in

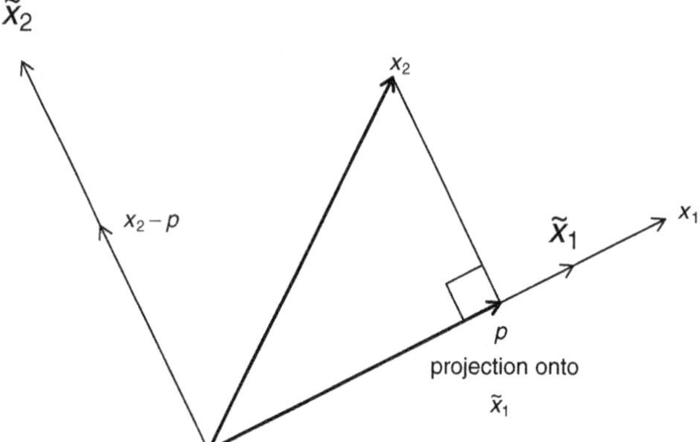

Fig. 2.2. Orthogonalization of x_1 and x_2

an obvious way:

for $k = 2, 3\ldots,$

$$\tilde{x}_k = \left(x_k - \sum_{i=1}^{k-1} \langle \tilde{x}_i, x_k \rangle \tilde{x}_i \right) \Big/ \left\| x_k - \sum_{i=1}^{k-1} \langle \tilde{x}_i, x_k \rangle \tilde{x}_i \right\|.$$

$$(2.35)$$

In this method, at the kth step, we orthogonalize the kth vector by computing its residual with respect to the plane formed by all the previous $k-1$ orthonormal vectors.

Another way of extending the transformation of equations (2.34) is, at the kth step, to compute the residuals of all remaining vectors with respect just to the kth normalized vector. We describe this method explicitly in Algorithm 2.1.

Algorithm 2.1 Gram-Schmidt Orthonormalization of a Set of Linearly Independent Vectors, $x_1, \ldots, x_m$

0. For $k = 1, \ldots, m,$
   {
   set $\tilde{x}_i = x_i.$
   }
1. Ensure that $\tilde{x}_1 \neq 0;$
   set $\tilde{x}_1 = \tilde{x}_1 / \|\tilde{x}_1\|.$
2. If $m > 1,$ for $k = 2, \ldots, m,$
   {
       for $j = k, \ldots, m,$
       {
           set $\tilde{x}_j = \tilde{x}_j - \langle \tilde{x}_{k-1}, \tilde{x}_j \rangle \tilde{x}_{k-1}.$
       }

    ensure that $\tilde{x}_k \neq 0$;
     set $\tilde{x}_k = \tilde{x}_k / \|\tilde{x}_k\|$.
}

Although the method indicated in equation (2.35) is mathematically equivalent to this method, the use of Algorithm 2.1 is to be preferred for computations because it is less subject to rounding errors. (This may not be immediately obvious, although a simple numerical example can illustrate the fact — see Exercise 11.1c on page 441. We will not digress here to consider this further, but the difference in the two methods has to do with the relative magnitudes of the quantities in the subtraction. The method of Algorithm 2.1 is sometimes called the "modified Gram-Schmidt method". We will discuss this method again in Section 11.2.1.) This is an instance of a principle that we will encounter repeatedly: *the form of a mathematical expression and the way the expression should be evaluated in actual practice may be quite different.*

These orthogonalizing transformations result in a set of orthogonal vectors that span the same space as the original set. They are not unique; if the order in which the vectors are processed is changed, a different set of orthogonal vectors will result.

Orthogonal vectors are useful for many reasons: perhaps to improve the stability of computations; or in data analysis to capture the variability most efficiently; or for dimension reduction as in principal components analysis; or in order to form more meaningful quantities as in a vegetative index in remote sensing. We will discuss various specific orthogonalizing transformations later.

2.2.5 Orthonormal Basis Sets

A basis for a vector space is often chosen to be an orthonormal set because it is easy to work with the vectors in such a set.

If $u_1, \ldots, u_n$ is an orthonormal basis set for a space, then a vector x in that space can be expressed as

$$x = c_1 u_1 + \cdots + c_n u_n, \tag{2.36}$$

and because of orthonormality, we have

$$c_i = \langle x, u_i \rangle. \tag{2.37}$$

(We see this by taking the inner product of both sides with u_i.) A representation of a vector as a linear combination of orthonormal basis vectors, as in equation (2.36), is called a *Fourier expansion*, and the c_i are called *Fourier coefficients*.

By taking the inner product of each side of equation (2.36) with itself, we have *Parseval's identity*:

$$\|x\|^2 = \sum c_i^2. \tag{2.38}$$

This shows that the L_2 norm is the same as the norm in equation (2.16) (on page 18) for the case of an orthogonal basis.

Although the Fourier expansion is not unique because a different orthogonal basis set could be chosen, Parseval's identity removes some of the arbitrariness in the choice; no matter what basis is used, the sum of the squares of the Fourier coefficients is equal to the square of the norm that arises from the inner product. ("The" inner product means the inner product used in defining the orthogonality.)

Another useful expression of Parseval's identity in the Fourier expansion is

$$\left\| x - \sum_{i=1}^{k} c_i u_i \right\|^2 = \langle x,\, x \rangle - \sum_{i=1}^{k} c_i^2 \qquad (2.39)$$

(because the term on the left-hand side is 0).

The expansion (2.36) is a special case of a very useful expansion in an orthogonal basis set. In the finite-dimensional vector spaces we consider here, the series is finite. In function spaces, the series is generally infinite, and so issues of convergence are important. For different types of functions, different orthogonal basis sets may be appropriate. Polynomials are often used, and there are some standard sets of orthogonal polynomials, such as Jacobi, Hermite, and so on. For periodic functions especially, orthogonal trigonometric functions are useful.

2.2.6 Approximation of Vectors

In high-dimensional vector spaces, it is often useful to approximate a given vector in terms of vectors from a lower dimensional space. Suppose, for example, that $\mathcal{V} \subset \mathbb{R}^n$ is a vector space of dimension k (necessarily, $k \leq n$) and x is a given n-vector. We wish to determine a vector $\tilde{x}$ in $\mathcal{V}$ that approximates x.

Optimality of the Fourier Coefficients

The first question, of course, is what constitutes a "good" approximation. One obvious criterion would be based on a norm of the difference of the given vector and the approximating vector. So now, choosing the norm as the Euclidean norm, we may pose the problem as one of finding $\tilde{x} \in \mathcal{V}$ such that

$$\| x - \tilde{x} \| \leq \| x - v \| \quad \forall\, v \in \mathcal{V}. \qquad (2.40)$$

This difference is a *truncation error*. Let $u_1, \ldots, u_k$ be an orthonormal basis set for $\mathcal{V}$, and let

$$\tilde{x} = c_1 u_1 + \cdots + c_k u_k, \qquad (2.41)$$

where the c_i are the Fourier coefficients of x, $\langle x,\, u_i \rangle$. Now let $v = a_1 u_1 + \cdots + a_k u_k$ be any other vector in $\mathcal{V}$, and consider

$$
\begin{aligned}
\|x - v\|^2 &= \left\| x - \sum_{i=1}^{k} a_i u_i \right\|^2 \\
&= \left\langle x - \sum_{i=1}^{k} a_i u_i,\ x - \sum_{i=1}^{k} a_i u_i \right\rangle \\
&= \langle x,\, x \rangle - 2 \sum_{i=1}^{k} a_i \langle x,\, u_i \rangle + \sum_{i=1}^{k} a_i^2 \\
&= \langle x,\, x \rangle - 2 \sum_{i=1}^{k} a_i c_i + \sum_{i=1}^{k} a_i^2 + \sum_{i=1}^{k} c_i^2 - \sum_{i=1}^{k} c_i^2 \\
&= \langle x,\, x \rangle + \sum_{i=1}^{k} (a_i - c_i)^2 - \sum_{i=1}^{k} c_i^2 \\
&= \left\| x - \sum_{i=1}^{k} c_i u_i \right\|^2 + \sum_{i=1}^{k} (a_i - c_i)^2 \\
&\geq \left\| x - \sum_{i=1}^{k} c_i u_i \right\|^2 .
\end{aligned}
\tag{2.42}
$$

Therefore we have $\|x - \tilde{x}\| \leq \|x - v\|$, and so $\tilde{x}$ is the best approximation of x with respect to the Euclidean norm in the k-dimensional vector space $\mathcal{V}$.

Choice of the Best Basis Subset

Now, posing the problem another way, we may seek the best k-dimensional subspace of $\mathrm{I\!R}^n$ from which to choose an approximating vector. This question is not well-posed (because the one-dimensional vector space determined by x is the solution), but we can pose a related interesting question: suppose we have a Fourier expansion of x in terms of a set of n orthogonal basis vectors, $u_1, \ldots, u_n$, and we want to choose the "best" k basis vectors from this set and use them to form an approximation of x. (This restriction of the problem is equivalent to choosing a coordinate system.) We see the solution immediately from inequality (2.42): we choose the k u_is corresponding to the k largest c_is in absolute value, and we take

$$
\tilde{x} = c_{i_1} u_{i_1} + \cdots + c_{i_k} u_{i_k},
\tag{2.43}
$$

where $\min(\{|c_{i_j}| \ : \ j = 1, \ldots, k\}) \geq \max(\{|c_{i_j}| \ : \ j = k+1, \ldots, n\})$.

2.2.7 Flats, Affine Spaces, and Hyperplanes

Given an n-dimensional vector space of order n, $\mathrm{I\!R}^n$ for example, consider a system of m linear equations in the n-vector variable x,

$$c_1^{\mathrm{T}} x = b_1$$

$$\vdots \quad \vdots$$

$$c_m^{\mathrm{T}} x = b_m,$$

where $c_1, \ldots, c_m$ are linearly independent n-vectors (and hence $m \leq n$). The set of points defined by these linear equations is called a *flat*. Although it is not necessarily a vector space, a flat is also called an *affine space*. An intersection of two flats is a flat.

If the equations are *homogeneous* (that is, if $b_1 = \cdots = b_m = 0$), then the point $(0, \ldots, 0)$ is included, and the flat is an $(n - m)$-dimensional subspace (also a vector space, of course). Stating this another way, a flat through the origin is a vector space, but other flats are not vector spaces.

If $m = 1$, the flat is called a *hyperplane*. A hyperplane through the origin is an $(n - 1)$-dimensional vector space.

If $m = n-1$, the flat is a line. A line through the origin is a one-dimensional vector space.

2.2.8 Cones

A cone is an important type of vector set (see page 14 for definitions). The most important type of cone is a convex cone, which corresponds to a solid geometric object with a single finite vertex.

Given a set of vectors V (usually but not necessarily a cone), the *dual cone* of V, denoted V^*, is defined as

$$V^* = \{y^* \text{ s.t. } y^{*\mathrm{T}} y \geq 0 \text{ for all } y \in V\},$$

and the *polar cone* of V, denoted V^0, is defined as

$$V^0 = \{y^0 \text{ s.t. } y^{0\mathrm{T}} y \leq 0 \text{ for all } y \in V\}.$$

Obviously, V^0 can be formed by multiplying all of the vectors in V^* by -1, and so we write $V^0 = -V^*$, and we also have $(-V)^* = -V^*$.

Although the definitions can apply to any set of vectors, dual cones and polar cones are of the most interest in the case in which the underlying set of vectors is a cone in the nonnegative orthant (the set of all vectors all of whose elements are nonnegative). In that case, the dual cone is just the full nonnegative orthant, and the polar cone is just the nonpositive orthant (the set of all vectors all of whose elements are nonpositive).

Although a convex cone is not necessarily a vector space, the union of the dual cone and the polar cone of a convex cone is a vector space. (You are asked to prove this in Exercise 2.12.) The nonnegative orthant, which is an important convex cone, is its own dual.

Geometrically, the dual cone V^* of V consists of all vectors that form nonobtuse angles with the vectors in V. Convex cones, dual cones, and polar cones play important roles in optimization.

2.2.9 Cross Products in $\mathbb{R}^3$

For the special case of the vector space $\mathbb{R}^3$, another useful vector product is the cross product, which is a mapping from $\mathbb{R}^3 \times \mathbb{R}^3$ to $\mathbb{R}^3$. Before proceeding, we note an overloading of the term "cross product" and of the symbol "$\times$" used to denote it. If A and B are sets, the *set cross product* or the *set Cartesian product* of A and B is the set consisting of all doubletons (a, b) where a ranges over all elements of A, and b ranges independently over all elements of B. Thus, $\mathbb{R}^3 \times \mathbb{R}^3$ is the set of all pairs of all real 3-vectors.

The *vector cross product* of the vectors

$$x = (x_1, x_2, x_3),$$
$$y = (y_1, y_2, y_3),$$

written $x \times y$, is defined as

$$x \times y = (x_2 y_3 - x_3 y_2,\ x_3 y_1 - x_1 y_3,\ x_1 y_2 - x_2 y_1). \tag{2.44}$$

(We also use the term "cross products" in a different way to refer to another type of product formed by several inner products; see page 287.) The cross product has the following properties, which are immediately obvious from the definition:

1. Self-nilpotency:
   $x \times x = 0$, for all x.
2. Anti-commutativity:
   $x \times y = -y \times x$.
3. Factoring of scalar multiplication;
   $ax \times y = a(x \times y)$ for real a.
4. Relation of vector addition to addition of cross products:
   $(x + y) \times z = (x \times z) + (y \times z)$.

The cross product is useful in modeling phenomena in nature, which are often represented as vectors in $\mathbb{R}^3$. The cross product is also useful in "three-dimensional" computer graphics for determining whether a given surface is visible from a given perspective and for simulating the effect of lighting on a surface.

2.3 Centered Vectors and Variances and Covariances of Vectors

In this section, we define some scalar-valued functions of vectors that are analogous to functions of random variables averaged over their probabilities or probability density. The functions of vectors discussed here are the same as the ones that define sample statistics. This short section illustrates the properties of norms, inner products, and angles in terms that should be familiar to the reader.

These functions, and transformations using them, are useful for applications in the data sciences. It is important to know the effects of various transformations of data on data analysis.

2.3.1 The Mean and Centered Vectors

When the elements of a vector have some kind of common interpretation, the sum of the elements or the mean (equation (2.26)) of the vector may have meaning. In this case, it may make sense to *center* the vector; that is, to subtract the mean from each element. For a given vector x, we denote its centered counterpart as x_c:

$$x_\mathrm{c} = x - \bar{x}. \tag{2.45}$$

We refer to any vector whose sum of elements is 0 as a centered vector.

From the definitions, it is easy to see that

$$(x + y)_\mathrm{c} = x_\mathrm{c} + y_\mathrm{c} \tag{2.46}$$

(see Exercise 2.14). Interpreting $\bar{x}$ as a vector, and recalling that it is the projection of x onto the one vector, we see that x_c is the residual in the sense of equation (2.30). Hence, we see that x_c and x are orthogonal, and the Pythagorean relationship holds:

$$\|x\|^2 = \|\bar{x}\|^2 + \|x_\mathrm{c}\|^2. \tag{2.47}$$

From this we see that the length of a centered vector is less than or equal to the length of the original vector. (Notice that equation (2.47) is just the formula familiar to data analysts, which with some rearrangement is $\sum(x_i - \bar{x})^2 = \sum x_i^2 - n\bar{x}^2$.)

For any scalar a and n-vector x, expanding the terms, we see that

$$\|x - a\|^2 = \|x_\mathrm{c}\|^2 + n(a - \bar{x})^2, \tag{2.48}$$

where we interpret $\bar{x}$ as a scalar here.

Notice that a nonzero vector when centered may be the zero vector. This leads us to suspect that some properties that depend on a dot product are not invariant to centering. This is indeed the case. The angle between two vectors, for example, is not invariant to centering; that is, in general,

$$\mathrm{angle}(x_\mathrm{c}, y_\mathrm{c}) \neq \mathrm{angle}(x, y) \tag{2.49}$$

(see Exercise 2.15).

2.3.2 The Standard Deviation, the Variance, and Scaled Vectors

We also sometimes find it useful to scale a vector by both its length (normalize the vector) and by a function of its number of elements. We denote this *scaled* vector as x_s and define it as

$$x_s = \sqrt{n-1} \, \frac{x}{\|x_c\|}. \tag{2.50}$$

For comparing vectors, it is usually better to center the vectors prior to any scaling. We denote this *centered and scaled* vector as x_{cs} and define it as

$$x_{cs} = \sqrt{n-1} \, \frac{x_c}{\|x_c\|}. \tag{2.51}$$

Centering and scaling is also called *standardizing*. Note that the vector is centered before being scaled. The angle between two vectors is not changed by scaling (but, of course, it may be changed by centering).

The multiplicative inverse of the scaling factor,

$$s_x = \|x_c\|/\sqrt{n-1}, \tag{2.52}$$

is called the *standard deviation* of the vector x. The standard deviation of x_c is the same as that of x; in fact, the standard deviation is invariant to the addition of any constant. The standard deviation is a measure of how much the elements of the vector vary. If all of the elements of the vector are the same, the standard deviation is 0 because in that case $x_c = 0$.

The square of the standard deviation is called the *variance*, denoted by V:

$$\begin{aligned}
\mathrm{V}(x) &= s_x^2 \\
&= \frac{\|x_c\|^2}{n-1}. \tag{2.53}
\end{aligned}$$

(In perhaps more familiar notation, equation (2.53) is just $\mathrm{V}(x) = \sum(x_i - \bar{x})^2/(n-1)$.) From equation (2.45), we see that

$$\mathrm{V}(x) = \frac{1}{n-1} \left(\|x\|^2 - \|\bar{x}\|^2 \right).$$

(The terms "mean", "standard deviation", "variance", and other terms we will mention below are also used in an analogous, but slightly different, manner to refer to properties of *random variables*. In that context, the terms to refer to the quantities we are discussing here would be preceded by the word "sample", and often for clarity I will use the phrases "sample standard deviation" and "sample variance" to refer to what is defined above, especially if the elements of x are interpreted as independent realizations of a random variable. Also, recall the two possible meanings of "mean", or $\bar{x}$; one is a vector, and one is a scalar, as in equation (2.27).)

If a and b are scalars (or b is a vector with all elements the same), the definition, together with equation (2.48), immediately gives

$$V(ax + b) = a^2 V(x).$$

This implies that for the scaled vector x_s,

$$V(x_s) = 1.$$

If a is a scalar and x and y are vectors with the same number of elements, from the equation above, and using equation (2.20) on page 21, we see that the variance following an axpy operation is given by

$$V(ax + y) = a^2 V(x) + V(y) + 2a \frac{\langle x_c,\, y_c \rangle}{n-1}. \tag{2.54}$$

While equation (2.53) appears to be relatively simple, evaluating the expression for a given x may not be straightforward. We discuss computational issues for this expression on page 410. This is an instance of a principle that we will encounter repeatedly: *the form of a mathematical expression and the way the expression should be evaluated in actual practice may be quite different.*

2.3.3 Covariances and Correlations between Vectors

If x and y are n-vectors, the *covariance* between x and y is

$$\text{Cov}(x, y) = \frac{\langle x - \bar{x},\, y - \bar{y} \rangle}{n-1}. \tag{2.55}$$

By representing $x - \bar{x}$ as $x - \bar{x}1$ and $y - \bar{y}$ similarly, and expanding, we see that $\text{Cov}(x, y) = (\langle x,\, y \rangle - n\bar{x}\bar{y})/(n-1)$. Also, we see from the definition of covariance that $\text{Cov}(x, x)$ is the variance of the vector x, as defined above.

From the definition and the properties of an inner product given on page 15, if x, y, and z are conformable vectors, we see immediately that

- $\text{Cov}(a1, y) = 0$
  for any scalar a (where 1 is the one vector);
- $\text{Cov}(ax, y) = a\text{Cov}(x, y)$
  for any scalar a;
- $\text{Cov}(y, x) = \text{Cov}(x, y)$;
- $\text{Cov}(y, y) = V(y)$; and
- $\text{Cov}(x + z, y) = \text{Cov}(x, y) + \text{Cov}(z, y)$,
  in particular,
  - $\text{Cov}(x + y, y) = \text{Cov}(x, y) + V(y), and$
  - $\text{Cov}(x + a, y) = \text{Cov}(x, y)$
    for any scalar a.

Using the definition of the covariance, we can rewrite equation (2.54) as

$$V(ax + y) = a^2 V(x) + V(y) + 2a\mathrm{Cov}(x, y). \qquad (2.56)$$

The covariance is a measure of the extent to which the vectors point in the same direction. A more meaningful measure of this is obtained by the covariance of the centered and scaled vectors. This is the *correlation* between the vectors,

$$\begin{aligned}
\mathrm{Corr}(x, y) &= \mathrm{Cov}(x_{\mathrm{cs}}, y_{\mathrm{cs}}) \\
&= \left\langle \frac{x_{\mathrm{c}}}{\|x_{\mathrm{c}}\|}, \frac{y_{\mathrm{c}}}{\|y_{\mathrm{c}}\|} \right\rangle \\
&= \frac{\langle x_{\mathrm{c}}, y_{\mathrm{c}} \rangle}{\|x_{\mathrm{c}}\| \|y_{\mathrm{c}}\|},
\end{aligned} \qquad (2.57)$$

which we see immediately from equation (2.32) is the cosine of the angle between x_{c} and y_{c}:

$$\mathrm{Corr}(x, y) = \cos(\mathrm{angle}(x_{\mathrm{c}}, y_{\mathrm{c}})). \qquad (2.58)$$

(Recall that this is not the same as the angle between x and y.)

An equivalent expression for the correlation is

$$\mathrm{Corr}(x, y) = \frac{\mathrm{Cov}(x, y)}{\sqrt{V(x)V(y)}}. \qquad (2.59)$$

It is clear that the correlation is in the interval $[-1, 1]$ (from the Cauchy-Schwarz inequality). A correlation of -1 indicates that the vectors point in opposite directions, a correlation of 1 indicates that the vectors point in the same direction, and a correlation of 0 indicates that the vectors are orthogonal.

While the covariance is equivariant to scalar multiplication, the absolute value of the correlation is invariant to it; that is, the correlation changes only as the sign of the scalar multiplier,

$$\mathrm{Corr}(ax, y) = \mathrm{sign}(a)\mathrm{Corr}(x, y), \qquad (2.60)$$

for any scalar a.

Exercises

2.1. Write out the step-by-step proof that the maximum number of n-vectors that can form a set that is linearly independent is n, as stated on page 11.

2.2. Give an example of two vector spaces whose union is not a vector space.

2.3. Let $\{v_i\}_{i=1}^n$ be an orthonormal basis for the n-dimensional vector space $\mathcal{V}$. Let $x \in \mathcal{V}$ have the representation

$$x = \sum b_i v_i.$$

Show that the Fourier coefficients b_i can be computed as

$$b_i = \langle x, v_i \rangle.$$

2.4. Let $p = \frac{1}{2}$ in equation (2.11); that is, let $\rho(x)$ be defined for the n-vector x as

$$\rho(x) = \left(\sum_{i=1}^n |x_i|^{1/2} \right)^2.$$

Show that $\rho(\cdot)$ is not a norm.

2.5. Prove equation (2.12) and show that the bounds are sharp by exhibiting instances of equality. (Use the fact that $\|x\|_\infty = \max_i |x_i|$.)

2.6. Prove the following inequalities.

a) Prove Hölder's inequality: for any p and q such that $p \geq 1$ and $p + q = pq$, and for vectors x and y of the same order,

$$\langle x, y \rangle \leq \|x\|_p \|y\|_q.$$

b) Prove the triangle inequality for any L_p norm. (This is sometimes called Minkowski's inequality.)

Hint: Use Hölder's inequality.

2.7. Show that the expression defined in equation (2.22) on page 22 is a metric.

2.8. Show that equation (2.31) on page 26 is correct.

2.9. Show that the intersection of two orthogonal vector spaces consists only of the zero vector.

2.10. From the definition of direction cosines in equation (2.33), it is easy to see that the sum of the squares of the direction cosines is 1. For the special case of $\mathbb{R}^3$, draw a sketch and use properties of right triangles to show this geometrically.

2.11. In $\mathbb{R}^2$ with a Cartesian coordinate system, the diagonal directed line segment through the positive quadrant (orthant) makes a 45° angle with each of the positive axes. In 3 dimensions, what is the angle between the diagonal and each of the positive axes? In 10 dimensions? In 100 dimensions? In 1000 dimensions? We see that in higher dimensions any two lines are almost orthogonal. (That is, the angle between them approaches 90°.) What are some of the implications of this for data analysis?

2.12. Show that if C is a convex cone, then $C^* \cup C^0$ together with the usual operations is a vector space, where C^* is the dual of C and C^0 is the polar cone of C.

Hint: Just apply the definitions of the individual terms.

2.13. $\mathbb{R}^3$ and the cross product.

    a) Is the cross product associative? Prove or disprove.

    b) For $x, y \in \mathbb{R}^3$, show that the area of the triangle with vertices $(0, 0, 0)$, x, and y is $\|x \times y\|/2$.

    c) For $x, y, z \in \mathbb{R}^3$, show that

$$\langle x, \ y \times z \rangle = \langle x \times y, \ z \rangle.$$

    This is called the "triple scalar product".

    d) For $x, y, z \in \mathbb{R}^3$, show that

$$x \times (y \times z) = \langle x, \ z \rangle y - \langle x, \ y \rangle z.$$

    This is called the "triple vector product". It is in the plane determined by y and z.

    e) The magnitude of the angle between two vectors is determined by the cosine, formed from the inner product. Show that in the special case of $\mathbb{R}^3$, the angle is also determined by the sine and the cross product, and show that this method can determine both the magnitude and the *direction* of the angle; that is, the way a particular vector is rotated into the other.

2.14. Using equations (2.26) and (2.45), establish equation (2.46).

2.15. Show that the angle between the centered vectors x_c and y_c is not the same in general as the angle between the uncentered vectors x and y of the same order.

2.16. Formally prove equation (2.54) (and hence equation (2.56)).

2.17. Prove that for any vectors x and y of the same order,

$$(\mathrm{Cov}(x, y))^2 \le \mathrm{V}(x)\mathrm{V}(y).$$

3

Basic Properties of Matrices

In this chapter, we build on the notation introduced on page 5, and discuss a wide range of basic topics related to matrices with real elements. Some of the properties carry over to matrices with complex elements, but the reader should not assume this. Occasionally, for emphasis, we will refer to "real" matrices, but unless it is stated otherwise, we are assuming the matrices are real.

The topics and the properties of matrices that we choose to discuss are motivated by applications in the data sciences. In Chapter 8, we will consider in more detail some special types of matrices that arise in regression analysis and multivariate data analysis, and then in Chapter 9 we will discuss some specific applications in statistics.

3.1 Basic Definitions and Notation

It is often useful to treat the rows or columns of a matrix as vectors. Terms such as linear independence that we have defined for vectors also apply to rows and/or columns of a matrix. The vector space generated by the columns of the $n \times m$ matrix A is of order n and of dimension m or less, and is called the *column space* of A, the *range* of A, or the *manifold* of A. This vector space is denoted by

$$\mathcal{V}(A)$$

or

$$\text{span}(A).$$

(The argument of $\mathcal{V}(\cdot)$ or $\text{span}(\cdot)$ can be either a matrix or a set of vectors. Recall from Section 2.1.3 that if G is a set of vectors, the symbol $\text{span}(G)$ denotes the vector space generated by the vectors in G.) We also define the *row space* of A to be the vector space of order m (and of dimension n or less) generated by the rows of A; notice, however, the preference given to the column space.

Many of the properties of matrices that we discuss hold for matrices with an infinite number of elements, but throughout this book we will assume that the matrices have a finite number of elements, and hence the vector spaces are of finite order and have a finite number of dimensions.

Similar to our definition of multiplication of a vector by a scalar, we define the multiplication of a matrix A by a scalar c as

$$cA = (ca_{ij}).$$

The a_{ii} elements of a matrix are called *diagonal elements*; an element a_{ij} with $i < j$ is said to be "above the diagonal", and one with $i > j$ is said to be "below the diagonal". The vector consisting of all of the a_{ii}'s is called the *principal diagonal* or just the diagonal. The elements $a_{i,i+c_k}$ are called "codiagonals" or "minor diagonals". If the matrix has m columns, the $a_{i,m+1-i}$ elements of the matrix are called *skew diagonal elements*. We use terms similar to those for diagonal elements for elements above and below the skew diagonal elements. These phrases are used with both square and nonsquare matrices.

If, in the matrix A with elements a_{ij} for all i and j, $a_{ij} = a_{ji}$, A is said to be *symmetric*. A symmetric matrix is necessarily square. A matrix A such that $a_{ij} = -a_{ji}$ is said to be *skew symmetric*. The diagonal entries of a skew symmetric matrix must be 0. If $a_{ij} = \bar{a}_{ji}$ (where $\bar{a}$ represents the conjugate of the complex number a), A is said to be *Hermitian*. A Hermitian matrix is also necessarily square, and, of course, a real symmetric matrix is Hermitian. A Hermitian matrix is also called a *self-adjoint* matrix.

Many matrices of interest are *sparse*; that is, they have a large proportion of elements that are 0. ("A large proportion" is subjective, but generally means more than 75%, and in many interesting cases is well over 95%.) Efficient and accurate computations often require that the sparsity of a matrix be accommodated explicitly.

If all except the principal diagonal elements of a matrix are 0, the matrix is called a *diagonal matrix*. A diagonal matrix is the most common and most important type of sparse matrix. If all of the principal diagonal elements of a matrix are 0, the matrix is called a *hollow matrix*. A skew symmetric matrix is hollow, for example. If all except the principal skew diagonal elements of a matrix are 0, the matrix is called a *skew diagonal matrix*.

An $n \times m$ matrix A for which

$$|a_{ii}| > \sum_{j \neq i}^{m} |a_{ij}| \quad \text{for each } i = 1, \ldots, n \tag{3.1}$$

is said to be row *diagonally dominant*; one for which $|a_{jj}| > \sum_{i \neq j}^{n} |a_{ij}|$ for each $j = 1, \ldots, m$ is said to be column diagonally dominant. (Some authors refer to this as *strict* diagonal dominance and use "diagonal dominance" without qualification to allow the possibility that the inequalities in the definitions are not strict.) Most interesting properties of such matrices hold whether the dominance is by row or by column. If A is symmetric, row and column diagonal dominances are equivalent, so we refer to row or column diagonally dominant symmetric matrices without the qualification; that is, as just diagonally dominant.

If all elements below the diagonal are 0, the matrix is called an *upper triangular matrix*; and a *lower triangular matrix* is defined similarly. If all elements of a column or row of a triangular matrix are zero, we still refer to the matrix as triangular, although sometimes we speak of its form as *trapezoidal*. Another form called trapezoidal is one in which there are more columns than rows, and the additional columns are possibly nonzero. The four general forms of triangular or trapezoidal matrices are shown below.

$$
\begin{bmatrix} X & X & X \\ 0 & X & X \\ 0 & 0 & X \end{bmatrix}
\quad
\begin{bmatrix} X & X & X \\ 0 & X & X \\ 0 & 0 & 0 \end{bmatrix}
\quad
\begin{bmatrix} X & X & X \\ 0 & X & X \\ 0 & 0 & X \\ 0 & 0 & 0 \end{bmatrix}
\quad
\begin{bmatrix} X & X & X & X \\ 0 & X & X & X \\ 0 & 0 & X & X \end{bmatrix}
$$

In this notation, X indicates that the element is possibly not zero. It does not mean each element is the same. In other cases, X and 0 may indicate "submatrices", which we discuss in the section on partitioned matrices.

If all elements are 0 except $a_{i,i+c_k}$ for some small number of integers c_k, the matrix is called a *band matrix* (or *banded matrix*). In many applications, $c_k \in \{-w_l, -w_l + 1, \ldots, -1, 0, 1, \ldots, w_u - 1, w_u\}$. In such a case, w_l is called the *lower band width* and w_u is called the *upper band width*. These patterned matrices arise in time series and other stochastic process models as well as in solutions of differential equations, and so they are very important in certain applications. Although it is often the case that interesting band matrices are symmetric, or at least have the same number of codiagonals that are nonzero, neither of these conditions always occurs in applications of band matrices. If all elements below the principal skew diagonal elements of a matrix are 0, the matrix is called a *skew upper triangular matrix*. A common form of Hankel matrix, for example, is the skew upper triangular matrix (see page 312). Notice that the various terms defined here, such as triangular and band, also apply to nonsquare matrices.

Band matrices occur often in numerical solutions of partial differential equations. A band matrix with lower and upper band widths of 1 is a *tridiagonal matrix*. If all diagonal elements and all elements $a_{i,i\pm1}$ are nonzero, a tridiagonal matrix is called a "matrix of type 2". The inverse of a covariance matrix that occurs in common stationary time series models is a matrix of type 2 (see page 312).

Because the matrices with special patterns are usually characterized by the locations of zeros and nonzeros, we often use an intuitive notation with X and 0 to indicate the pattern. Thus, a band matrix may be written as

$$
\begin{bmatrix}
X & X & 0 & \cdots & 0 & 0 \\
X & X & X & \cdots & 0 & 0 \\
0 & X & X & \cdots & 0 & 0 \\
& & & \ddots & \ddots & \\
0 & 0 & 0 & \cdots & X & X
\end{bmatrix}.
$$

Computational methods for matrices may be more efficient if the patterns are taken into account.

A matrix is in upper *Hessenberg form*, and is called a *Hessenberg matrix*, if it is upper triangular except for the first subdiagonal, which may be nonzero. That is, $a_{ij} = 0$ for $i > j + 1$:

$$
\begin{bmatrix}
X & X & X & \cdots & X & X \\
X & X & X & \cdots & X & X \\
0 & X & X & \cdots & X & X \\
0 & 0 & X & \cdots & X & X \\
\vdots & \vdots & & \ddots & \vdots & \vdots \\
0 & 0 & 0 & \cdots & X & X
\end{bmatrix}.
$$

A symmetric matrix that is in Hessenberg form is necessarily *tridiagonal*.

Hessenberg matrices arise in some methods for computing eigenvalues (see Chapter 7).

3.1.1 Matrix Shaping Operators

In order to perform certain operations on matrices and vectors, it is often useful first to reshape a matrix. The most common reshaping operation is the transpose, which we define in this section. Sometimes we may need to rearrange the elements of a matrix or form a vector into a special matrix. In this section, we define three operators for doing this.

Transpose

The *transpose* of a matrix is the matrix whose i^{th} row is the ith column of the original matrix and whose j^{th} column is the jth row of the original matrix. We use a superscript "T" to denote the transpose of a matrix; thus, if $A = (a_{ij})$, then

$$
A^{\text{T}} = (a_{ji}). \tag{3.2}
$$

(In other literature, the transpose is often denoted by a prime, as in $A' = (a_{ji}) = A^{\text{T}}$.)

If the elements of the matrix are from the field of complex numbers, the *conjugate transpose*, also called the *adjoint*, is more useful than the transpose. ("Adjoint" is also used to denote another type of matrix, so we will generally avoid using that term. This meaning of the word is the origin of the other term for a Hermitian matrix, a "self-adjoint matrix".) We use a superscript "H" to denote the conjugate transpose of a matrix; thus, if $A = (a_{ij})$, then $A^{\mathrm{H}} = (\bar{a}_{ji})$. We also use a similar notation for vectors. If the elements of A are all real, then $A^{\mathrm{H}} = A^{\mathrm{T}}$. (The conjugate transpose is often denoted by an asterisk, as in $A^* = (\bar{a}_{ji}) = A^{\mathrm{H}}$. This notation is more common if a prime is used to denote the transpose. We sometimes use the notation A^* to denote a g_2 inverse of the matrix A; see page 102.)

If (and only if) A is symmetric, $A = A^{\mathrm{T}}$; if (and only if) A is skew symmetric, $A^{\mathrm{T}} = -A$; and if (and only if) A is Hermitian, $A = A^{\mathrm{H}}$.

Diagonal Matrices and Diagonal Vectors: diag($\cdot$) and vecdiag($\cdot$)

A square diagonal matrix can be specified by the diag($\cdot$) constructor function that operates on a vector and forms a diagonal matrix with the elements of the vector along the diagonal:

$$\mathrm{diag}\big((d_1, d_2, \ldots, d_n)\big) = \begin{bmatrix} d_1 & 0 & \cdots & 0 \\ 0 & d_2 & \cdots & 0 \\ & & \ddots & \\ 0 & 0 & \cdots & d_n \end{bmatrix}. \tag{3.3}$$

(Notice that the argument of diag is a vector; that is why there are two sets of parentheses in the expression above, although sometimes we omit one set without loss of clarity.) The diag function defined here is a mapping $\mathrm{I\!R}^n \mapsto \mathrm{I\!R}^{n \times n}$. Later we will extend this definition slightly.

The vecdiag($\cdot$) function forms a vector from the principal diagonal elements of a matrix. If A is an $n \times m$ matrix, and $k = \min(n, m)$,

$$\mathrm{vecdiag}(A) = (a_{11}, \ldots, a_{kk}). \tag{3.4}$$

The vecdiag function defined here is a mapping $\mathrm{I\!R}^{n \times m} \mapsto \mathrm{I\!R}^{\min(n,m)}$.

Sometimes we overload diag($\cdot$) to allow its argument to be a matrix, and in that case, it is the same as vecdiag($\cdot$). The R system, for example, uses this overloading.

Forming a Vector from the Elements of a Matrix: vec($\cdot$) and vech($\cdot$)

It is sometimes useful to consider the elements of a matrix to be elements of a single vector. The most common way this is done is to string the columns of the matrix end-to-end into a vector. The vec($\cdot$) function does this:

$$\mathrm{vec}(A) = (a_1^{\mathrm{T}}, a_2^{\mathrm{T}}, \ldots, a_m^{\mathrm{T}}), \tag{3.5}$$

where $a_1, a_2, \ldots, a_m$ are the column vectors of the matrix A. The vec function is also sometimes called the "pack" function. (A note on the notation: the right side of equation (3.5) is the notation for a column vector with elements a_i^{T}; see Chapter 1.) The vec function is a mapping $\mathrm{I\!R}^{n \times m} \mapsto \mathrm{I\!R}^{nm}$.

For a symmetric matrix A with elements a_{ij}, the "vech" function stacks the unique elements into a vector:

$$\mathrm{vech}(A) = (a_{11}, a_{21}, \ldots, a_{m1}, a_{22}, \ldots, a_{m2}, \ldots, a_{mm}). \tag{3.6}$$

There are other ways that the unique elements could be stacked that would be simpler and perhaps more useful (see the discussion of symmetric storage mode on page 451), but equation (3.6) is the standard definition of $\mathrm{vech}(\cdot)$. The vech function is a mapping $\mathrm{I\!R}^{n \times n} \mapsto \mathrm{I\!R}^{n(n+1)/2}$.

3.1.2 Partitioned Matrices

We often find it useful to partition a matrix into submatrices; for example, in many applications in data analysis, it is often convenient to work with submatrices of various types representing different subsets of the data.

We usually denote the submatrices with capital letters with subscripts indicating the relative positions of the submatrices. Hence, we may write

$$A = \begin{bmatrix} A_{11} & A_{12} \\ A_{21} & A_{22} \end{bmatrix}, \tag{3.7}$$

where the matrices A_{11} and A_{12} have the same number of rows, A_{21} and A_{22} have the same number of rows, A_{11} and A_{21} have the same number of columns, and A_{12} and A_{22} have the same number of columns. Of course, the submatrices in a partitioned matrix may be denoted by different letters. Also, for clarity, sometimes we use a vertical bar to indicate a partition:

$$A = [\,B\,|\,C\,].$$

The vertical bar is used just for clarity and has no special meaning in this representation.

The term "submatrix" is also used to refer to a matrix formed from a given matrix by deleting various rows and columns of the given matrix. In this terminology, B is a submatrix of A if for each element b_{ij} there is an a_{kl} with $k \geq i$ and $l \geq j$ such that $b_{ij} = a_{kl}$; that is, the rows and/or columns of the submatrix are not necessarily contiguous in the original matrix. This kind of subsetting is often done in data analysis, for example, in variable selection in linear regression analysis.

A square submatrix whose principal diagonal elements are elements of the principal diagonal of the given matrix is called a *principal submatrix*. If A_{11} in the example above is square, it is a principal submatrix, and if A_{22} is square, it is also a principal submatrix. Sometimes the term "principal submatrix" is restricted to square submatrices. If a matrix is diagonally dominant, then it is clear that any principal submatrix of it is also diagonally dominant.

A principal submatrix that contains the $(1,1)$ elements and whose rows and columns are contiguous in the original matrix is called a *leading principal submatrix*. If A_{11} is square, it is a leading principal submatrix in the example above.

Partitioned matrices may have useful patterns. A "block diagonal" matrix is one of the form

$$\begin{bmatrix} X & 0 & \cdots & 0 \\ 0 & X & \cdots & 0 \\ & & \ddots & \\ 0 & 0 & \cdots & X \end{bmatrix},$$

where 0 represents a submatrix with all zeros and X represents a general submatrix with at least some nonzeros.

The $\mathrm{diag}(\cdot)$ function previously introduced for a vector is also defined for a list of matrices:

$$\mathrm{diag}(A_1, A_2, \ldots, A_k)$$

denotes the block diagonal matrix with submatrices $A_1, A_2, \ldots, A_k$ along the diagonal and zeros elsewhere. A matrix formed in this way is sometimes called a *direct sum* of $A_1, A_2, \ldots, A_k$, and the operation is denoted by $\oplus$:

$$A_1 \oplus \cdots \oplus A_k = \mathrm{diag}(A_1, \ldots, A_k).$$

Although the direct sum is a binary operation, we are justified in defining it for a list of matrices because the operation is clearly associative.

The A_i may be of different sizes and they may not be square, although in most applications the matrices are square (and some authors define the direct sum only for square matrices).

We will define vector spaces of matrices below and then recall the definition of a direct sum of vector spaces (page 13), which is different from the direct sum defined above in terms of $\mathrm{diag}(\cdot)$.

Transposes of Partitioned Matrices

The transpose of a partitioned matrix is formed in the obvious way; for example,

$$\begin{bmatrix} A_{11} & A_{12} & A_{13} \\ A_{21} & A_{22} & A_{23} \end{bmatrix}^{\mathrm{T}} = \begin{bmatrix} A_{11}^{\mathrm{T}} & A_{21}^{\mathrm{T}} \\ A_{12}^{\mathrm{T}} & A_{22}^{\mathrm{T}} \\ A_{13}^{\mathrm{T}} & A_{23}^{\mathrm{T}} \end{bmatrix}. \tag{3.8}$$

3.1.3 Matrix Addition

The sum of two matrices of the same shape is the matrix whose elements are the sums of the corresponding elements of the addends. As in the case of vector addition, we overload the usual symbols for the operations on the reals to signify the corresponding operations on matrices when the operations are defined; hence, addition of matrices is also indicated by "+", as with scalar addition and vector addition. We assume throughout that writing a sum of matrices $A + B$ implies that they are of the same shape; that is, that they are *conformable for addition.*

The "+" operator can also mean addition of a scalar to a matrix, as in $A + a$, where A is a matrix and a is a scalar. Although this meaning of "+" is generally not used in mathematical treatments of matrices, in this book we use it to mean the addition of the scalar to each element of the matrix, resulting in a matrix of the same shape. This meaning is consistent with the semantics of modern computer languages such as Fortran 90/95 and R.

The addition of two $n \times m$ matrices or the addition of a scalar to an $n \times m$ matrix requires nm scalar additions.

The *matrix additive identity* is a matrix with all elements zero. We sometimes denote such a matrix with n rows and m columns as $0_{n \times m}$, or just as 0. We may denote a square additive identity as 0_n.

There are several possible ways to form a rank ordering of matrices of the same shape, but no complete ordering is entirely satisfactory. If all of the elements of the matrix A are positive, we write

$$A > 0; \tag{3.9}$$

if all of the elements are nonnegative, we write

$$A \geq 0. \tag{3.10}$$

The terms "positive" and "nonnegative" and these symbols are not to be confused with the terms "positive definite" and "nonnegative definite" and similar symbols for important classes of matrices having different properties (which we will introduce in equation (3.62) and discuss further in Section 8.3.)

The transpose of the sum of two matrices is the sum of the transposes:

$$(A + B)^{\mathrm{T}} = A^{\mathrm{T}} + B^{\mathrm{T}}.$$

The sum of two symmetric matrices is therefore symmetric.

Vector Spaces of Matrices

Having defined scalar multiplication, matrix addition (for conformable matrices), and a matrix additive identity, we can define a vector space of $n \times m$ matrices as any set that is closed with respect to those operations (which necessarily would contain the additive identity; see page 11). As with any vector space, we have the concepts of linear independence, generating set or spanning set, basis set, essentially disjoint spaces, and direct sums of matrix vector spaces (as in equation (2.7), which is different from the direct sum of matrices defined in terms of $\mathrm{diag}(\cdot)$).

With scalar multiplication, matrix addition, and a matrix additive identity, we see that $\mathrm{I\!R}^{n \times m}$ is a vector space. If $n \geq m$, a set of nm $n \times m$ matrices whose columns consist of all combinations of a set of n n-vectors that span $\mathrm{I\!R}^n$ is a basis set for $\mathrm{I\!R}^{n \times m}$. If $n < m$, we can likewise form a basis set for $\mathrm{I\!R}^{n \times m}$ or for subspaces of $\mathrm{I\!R}^{n \times m}$ in a similar way. If $\{B_1, \ldots, B_k\}$ is a basis set for $\mathrm{I\!R}^{n \times m}$, then any $n \times m$ matrix can be represented as $\sum_{i=1}^{k} c_i B_i$. Subsets of a basis set generate subspaces of $\mathrm{I\!R}^{n \times m}$.

Because the sum of two symmetric matrices is symmetric, and a scalar multiple of a symmetric matrix is likewise symmetric, we have a vector space of the $n \times n$ symmetric matrices. This is clearly a subspace of the vector space $\mathrm{I\!R}^{n \times n}$. All vectors in any basis for this vector space must be symmetric. Using a process similar to our development of a basis for a general vector space of matrices, we see that there are $n(n+1)/2$ matrices in the basis (see Exercise 3.1).

3.1.4 Scalar-Valued Operators on Square Matrices: The Trace

There are several useful mappings from matrices to real numbers; that is, from $\mathrm{I\!R}^{n \times m}$ to $\mathrm{I\!R}$. Some important ones are norms, which are similar to vector norms and which we will consider later. In this section and the next, we define two scalar-valued operators, the trace and the determinant, that apply to square matrices.

The Trace: $\mathrm{tr}(\cdot)$

The sum of the diagonal elements of a square matrix is called the *trace* of the matrix. We use the notation "$\mathrm{tr}(A)$" to denote the trace of the matrix A:

$$\mathrm{tr}(A) = \sum_i a_{ii}. \tag{3.11}$$

The Trace of the Transpose of Square Matrices

From the definition, we see

$$\mathrm{tr}(A) = \mathrm{tr}(A^{\mathrm{T}}). \tag{3.12}$$

The Trace of Scalar Products of Square Matrices

For a scalar c and an $n \times n$ matrix A,

$$\mathrm{tr}(cA) = c\,\mathrm{tr}(A).$$

This follows immediately from the definition because for $\mathrm{tr}(cA)$ each diagonal element is multiplied by c.

The Trace of Partitioned Square Matrices

If the square matrix A is partitioned such that the diagonal blocks are square submatrices, that is,

$$A = \begin{bmatrix} A_{11} & A_{12} \\ A_{21} & A_{22} \end{bmatrix}, \tag{3.13}$$

where A_{11} and A_{22} are square, then from the definition, we see that

$$\text{tr}(A) = \text{tr}(A_{11}) + \text{tr}(A_{22}). \tag{3.14}$$

The Trace of the Sum of Square Matrices

If A and B are square matrices of the same order, a useful (and obvious) property of the trace is

$$\text{tr}(A + B) = \text{tr}(A) + \text{tr}(B). \tag{3.15}$$

3.1.5 Scalar-Valued Operators on Square Matrices: The Determinant

The determinant, like the trace, is a mapping from $\mathbb{R}^{n \times n}$ to $\mathbb{R}$. Although it may not be obvious from the definition below, the determinant has far-reaching applications in matrix theory.

The Determinant: $|\cdot|$ or $\det(\cdot)$

For an $n \times n$ (square) matrix A, consider the product $a_{1j_1} a_{2j_2} \cdots a_{nj_n}$, where $\pi_j = (j_1, j_2, \ldots, j_n)$ is one of the $n!$ permutations of the integers from 1 to n. Define a permutation to be *even* or *odd* according to the number of times that a smaller element follows a larger one in the permutation. (For example, 1, 3, 2 is an odd permutation, and 3, 1, 2 is an even permutation.) Let $\sigma(\pi_j) = 1$ if $\pi_j = (j_1, \ldots, j_n)$ is an even permutation, and let $\sigma(\pi_j) = -1$ otherwise. Then the *determinant* of A, denoted by $|A|$, is defined by

$$|A| = \sum_{\text{all permutations}} \sigma(\pi_j) a_{1j_1} \cdots a_{nj_n}. \tag{3.16}$$

The determinant is also sometimes written as $\det(A)$, especially, for example, when we wish to refer to the absolute value of the determinant. (The determinant of a matrix may be negative.)

The definition is not as daunting as it may appear at first glance. Many properties become obvious when we realize that $\sigma(\cdot)$ is always ± 1, and it can be built up by elementary exchanges of adjacent elements. For example, consider $\sigma(3, 2, 1)$. There are three elementary exchanges beginning with the natural ordering:

$$(1,2,3) \to (2,1,3) \to (2,3,1) \to (3,2,1);$$

hence, $\sigma(3,2,1) = (-1)^3 = -1$.

If π_j consists of the interchange of exactly two elements in $(1,\ldots,n)$, say elements p and q with $p < q$, then there are $q - p$ elements before p that are larger than p, and there are $q - p + 1$ elements between q and p in the permutation each with exactly one larger element preceding it. The total number is $2q - 2p + 1$, which is an odd number. Therefore, if π_j consists of the interchange of exactly two elements, then $\sigma(\pi_j) = -1$.

If the integers $1,\ldots,m$ and $m+1,\ldots,n$ are together in a given permutation, they can be considered separately:

$$\sigma(j_1,\ldots,j_n) = \sigma(j_1,\ldots,j_m)\sigma(j_{m+1},\ldots,j_n). \tag{3.17}$$

Furthermore, we see that the product $a_{1j_1} \cdots a_{nj_n}$ has exactly one factor from each unique row-column pair. These observations facilitate the derivation of various properties of the determinant (although the details are sometimes quite tedious).

We see immediately from the definition that the determinant of an upper or lower triangular matrix (or a diagonal matrix) is merely the product of the diagonal elements (because in each term of equation (3.16) there is a 0, except in the term in which the subscripts on each factor are the same).

Minors, Cofactors, and Adjugate Matrices

Consider the 2×2 matrix

$$A = \begin{bmatrix} a_{11} & a_{12} \\ a_{21} & a_{22} \end{bmatrix}.$$

From the definition, we see $|A| = a_{11}a_{22} + (-1)a_{21}a_{12}$.

Now let A be a 3×3 matrix:

$$A = \begin{bmatrix} a_{11} & a_{12} & a_{13} \\ a_{21} & a_{22} & a_{23} \\ a_{31} & a_{32} & a_{33} \end{bmatrix}.$$

In the definition of the determinant, consider all of the terms in which the elements of the first row of A appear. With some manipulation of those terms, we can express the determinant in terms of determinants of submatrices as

$$\begin{aligned}
|A| = {} & a_{11}(-1)^{1+1} \left| \begin{bmatrix} a_{22} & a_{32} \\ a_{32} & a_{33} \end{bmatrix} \right| \\[2ex]
& + a_{12}(-1)^{1+2} \left| \begin{bmatrix} a_{21} & a_{32} \\ a_{31} & a_{33} \end{bmatrix} \right| \\[2ex]
& + a_{13}(-1)^{1+3} \left| \begin{bmatrix} a_{21} & a_{22} \\ a_{31} & a_{32} \end{bmatrix} \right|.
\end{aligned} \tag{3.18}$$

This exercise in manipulation of the terms in the determinant could be carried out with other rows of A.

The determinants of the 2×2 submatrices in equation (3.18) are called *minors* or *complementary minors* of the associated element. The definition can be extended to $(n-1) \times (n-1)$ submatrices of an $n \times n$ matrix. We denote the minor associated with the a_{ij} element as

$$|A_{-(i)(j)}|, \tag{3.19}$$

in which $A_{-(i)(j)}$ denotes the submatrix that is formed from A by removing the ith row and the jth column. The sign associated with the minor corresponding to a_{ij} is $(-1)^{i+j}$. The minor together with its appropriate sign is called the *cofactor* of the associated element; that is, the cofactor of a_{ij} is $(-1)^{i+j}|A_{-(i)(j)}|$. We denote the cofactor of a_{ij} as $a_{(ij)}$:

$$a_{(ij)} = (-1)^{i+j}|A_{-(i)(j)}|. \tag{3.20}$$

Notice that both minors and cofactors are scalars.

The manipulations leading to equation (3.18), though somewhat tedious, can be carried out for a square matrix of any size, and minors and cofactors are defined as above. An expression such as in equation (3.18) is called an expansion in minors or an expansion in cofactors.

The extension of the expansion (3.18) to an expression involving a sum of signed products of complementary minors arising from $(n-1) \times (n-1)$ submatrices of an $n \times n$ matrix A is

$$
\begin{aligned}
|A| &= \sum_{j=1}^{n} a_{ij}(-1)^{i+j}|A_{-(i)(j)}| \\
&= \sum_{j=1}^{n} a_{ij} a_{(ij)},
\end{aligned}
\tag{3.21}
$$

or, over the rows,

$$|A| = \sum_{i=1}^{n} a_{ij} a_{(ij)}. \tag{3.22}$$

These expressions are called *Laplace expansions*. Each determinant $|A_{-(i)(j)}|$ can likewise be expressed recursively in a similar expansion.

Expressions (3.21) and (3.22) are special cases of a more general Laplace expansion based on an extension of the concept of a complementary minor of an element to that of a complementary minor of a minor. The derivation of the general Laplace expansion is straightforward but rather tedious (see Harville, 1997, for example, for the details).

Laplace expansions could be used to compute the determinant, but the main value of these expansions is in proving properties of determinants. For example, from the special Laplace expansion (3.21) or (3.22), we can quickly see that the determinant of a matrix with two rows that are the same is zero. We see this by recursively expanding all of the minors until we have only 2×2 matrices consisting of a duplicated row. The determinant of such a matrix is 0, so the expansion is 0.

The expansion in equation (3.21) has an interesting property: if instead of the elements a_{ij} from the ith row we use elements from a different row, say the kth row, the sum is zero. That is, for $k \neq i$,

$$\sum_{j=1}^{n} a_{kj}(-1)^{i+j}|A_{-(i)(j)}| = \sum_{j=1}^{n} a_{kj} a_{(ij)}$$
$$= 0. \tag{3.23}$$

This is true because such an expansion is exactly the same as an expansion for the determinant of a matrix whose k^{th} row has been replaced by its ith row; that is, a matrix with two identical rows. The determinant of such a matrix is 0, as we saw above.

A certain matrix formed from the cofactors has some interesting properties. We define the matrix here but defer further discussion. The *adjugate* of the $n \times n$ matrix A is defined as

$$\text{adj}(A) = (a_{(ji)}), \tag{3.24}$$

which is an $n \times n$ matrix of the cofactors of the elements of the transposed matrix. (The adjugate is also called the *adjoint*, but as we noted above, the term adjoint may also mean the conjugate transpose. To distinguish it from the conjugate transpose, the adjugate is also sometimes called the "classical adjoint". We will generally avoid using the term "adjoint".) Note the reversal of the subscripts; that is,

$$\text{adj}(A) = (a_{(ij)})^{\text{T}}.$$

The adjugate has an interesting property:

$$A\,\text{adj}(A) = \text{adj}(A)A = |A|I. \tag{3.25}$$

To see this, consider the $(ij)^{\text{T}}$ element of $A\,\text{adj}(A)$, $\sum_k a_{ik}(\text{adj}(A))_{kj}$. Now, noting the reversal of the subscripts in $\text{adj}(A)$ in equation (3.24), and using equations (3.21) and (3.23), we have

$$\sum_{k} a_{ik}(\text{adj}(A))_{kj} = \begin{cases} |A| & \text{if } i = j \\ 0 & \text{if } i \neq j; \end{cases}$$

that is, $A\,\text{adj}(A) = |A|I$.

The adjugate has a number of useful properties, some of which we will encounter later, as in equation (3.131).

The Determinant of the Transpose of Square Matrices

One important property we see immediately from a manipulation of the definition of the determinant is

$$|A| = |A^{\mathrm{T}}|. \tag{3.26}$$

The Determinant of Scalar Products of Square Matrices

For a scalar c and an $n \times n$ matrix A,

$$|cA| = c^n |A|. \tag{3.27}$$

This follows immediately from the definition because, for $|cA|$, each factor in each term of equation (3.16) is multiplied by c.

The Determinant of an Upper (or Lower) Triangular Matrix

If A is an $n \times n$ upper (or lower) triangular matrix, then

$$|A| = \prod_{i=1}^{n} a_{ii}. \tag{3.28}$$

This follows immediately from the definition. It can be generalized, as in the next section.

The Determinant of Certain Partitioned Square Matrices

Determinants of square partitioned matrices that are block diagonal or upper or lower block triangular depend only on the diagonal partitions:

$$|A| = \left| \begin{bmatrix} A_{11} & 0 \\ 0 & A_{22} \end{bmatrix} \right| = \left| \begin{bmatrix} A_{11} & 0 \\ A_{21} & A_{22} \end{bmatrix} \right| = \left| \begin{bmatrix} A_{11} & A_{12} \\ 0 & A_{22} \end{bmatrix} \right|$$

$$= |A_{11}|\,|A_{22}|. \tag{3.29}$$

We can see this by considering the individual terms in the determinant, equation (3.16). Suppose the full matrix is $n \times n$, and A_{11} is $m \times m$. Then A_{22} is $(n-m) \times (n-m)$, A_{21} is $(n-m) \times m$, and A_{12} is $m \times (n-m)$. In equation (3.16), any addend for which $(j_1, \ldots, j_m)$ is not a permutation of the integers $1, \ldots, m$ contains a factor a_{ij} that is in a 0 diagonal block, and hence the addend is 0. The determinant consists only of those addends for which $(j_1, \ldots, j_m)$ is a permutation of the integers $1, \ldots, m$, and hence $(j_{m+1}, \ldots, j_n)$ is a permutation of the integers $m+1, \ldots, n$,

$$|A| = \sum \sum \sigma(j_1, \ldots, j_m, j_{m+1}, \ldots, j_n) a_{1j_1} \cdots a_{mj_m} a_{m+1,j_n} \cdots a_{nj_n},$$

where the first sum is taken over all permutations that keep the first m integers together while maintaining a fixed ordering for the integers $m+1$ through n, and the second sum is taken over all permutations of the integers from $m+1$ through n while maintaining a fixed ordering of the integers from 1 to m. Now, using equation (3.17), we therefore have for A of this special form

$$
\begin{aligned}
|A| &= \sum \sum \sigma(j_1, \ldots, j_m, j_{m+1}, \ldots, j_n) a_{1j_1} \cdots a_{mj_m} a_{m+1,j_{m+1}} \cdots a_{nj_n} \\
&= \sum \sigma(j_1, \ldots, j_m) a_{1j_1} \cdots a_{mj_m} \sum \sigma(j_{m+1}, \ldots, j_n) a_{m+1,j_{m+1}} \cdots a_{nj_n} \\
&= |A_{11}||A_{22}|,
\end{aligned}
$$

which is equation (3.29). We use this result to give an expression for the determinant of more general partitioned matrices in Section 3.4.2.

Another useful partitioned matrix of the form of equation (3.13) has $A_{11} = 0$ and $A_{21} = -I$:

$$
A = \begin{bmatrix} 0 & A_{12} \\ -I & A_{22} \end{bmatrix}.
$$

In this case, using equation (3.21), we get

$$
\begin{aligned}
|A| &= ((-1)^{n+1+1}(-1))^n |A_{12}| \\
&= (-1)^{n(n+3)} |A_{12}| \\
&= |A_{12}|.
\end{aligned}
\tag{3.30}
$$

The Determinant of the Sum of Square Matrices

Occasionally it is of interest to consider the determinant of the sum of square matrices. We note in general that

$$
|A + B| \neq |A| + |B|,
$$

which we can see easily by an example. (Consider matrices in $\mathbb{R}^{2 \times 2}$, for example, and let $A = I$ and $B = \begin{bmatrix} -1 & 0 \\ 0 & 0 \end{bmatrix}$.)

In some cases, however, simplified expressions for the determinant of a sum can be developed. We consider one in the next section.

A Diagonal Expansion of the Determinant

A particular sum of matrices whose determinant is of interest is one in which a diagonal matrix D is added to a square matrix A, that is, $|A + D|$. (Such a determinant arises in eigenanalysis, for example, as we see in Section 3.8.2.)

For evaluating the determinant $|A + D|$, we can develop another expansion of the determinant by restricting our choice of minors to determinants of matrices formed by deleting the same rows and columns and then continuing to delete rows and columns recursively from the resulting matrices. The expansion is a polynomial in the elements of D; and for our purposes later, that is the most useful form.

Before considering the details, let us develop some additional notation. The matrix formed by deleting the same row and column of A is denoted $A_{-(i)(i)}$ as above (following equation (3.19)). In the current context, however, it is more convenient to adopt the notation $A_{(i_1,\ldots,i_k)}$ to represent the matrix formed from rows $i_1,\ldots,i_k$ and columns $i_1,\ldots,i_k$ from a given matrix A. That is, the notation $A_{(i_1,\ldots,i_k)}$ indicates the rows and columns *kept* rather than those deleted; and furthermore, in this notation, the indexes of the rows and columns are the same. We denote the determinant of this $k \times k$ matrix in the obvious way, $|A_{(i_1,\ldots,i_k)}|$. Because the principal diagonal elements of this matrix are principal diagonal elements of A, we call $|A_{(i_1,\ldots,i_k)}|$ a *principal minor* of A.

Now consider $|A + D|$ for the 2×2 case:

$$\left| \begin{bmatrix} a_{11} + d_1 & a_{12} \\ a_{21} & a_{22} + d_2 \end{bmatrix} \right|.$$

Expanding this, we have

$$|A + D| = (a_{11} + d_1)(a_{22} + d_2) - a_{12}a_{21}$$

$$= \left| \begin{bmatrix} a_{11} & a_{12} \\ a_{21} & a_{22} \end{bmatrix} \right| + d_1 d_2 + a_{22}d_1 + a_{11}d_2$$

$$= |A_{(1,2)}| + d_1 d_2 + a_{22}d_1 + a_{11}d_2.$$

Of course, $|A_{(1,2)}| = |A|$, but we are writing it this way to develop the pattern. Now, for the 3×3 case, we have

$$\begin{aligned} |A + D| = {} & |A_{(1,2,3)}| \\ & + |A_{(2,3)}|d_1 + |A_{(1,3)}|d_2 + |A_{(1,2)}|d_3 \\ & + a_{33}d_1 d_2 + a_{22}d_1 d_3 + a_{11}d_2 d_3 \\ & + d_1 d_2 d_3. \end{aligned} \tag{3.31}$$

In the applications of interest, the elements of the diagonal matrix D may be a single variable: d, say. In this case, the expression simplifies to

$$|A + D| = |A_{(1,2,3)}| + \sum_{i \neq j} |A_{(i,j)}|d + \sum_i a_{i,i}d^2 + d^3. \tag{3.32}$$

Carefully continuing in this way for an $n \times n$ matrix, either as in equation (3.31) for n variables or as in equation (3.32) for a single variable, we can make use of a Laplace expansion to evaluate the determinant.

Consider the expansion in a single variable because that will prove most useful. The pattern persists; the constant term is $|A|$, the coefficient of the first-degree term is the sum of the $(n-1)$-order principal minors, and, at the other end, the coefficient of the $(n-1)^{\text{th}}$-degree term is the sum of the first-order principal minors (that is, just the diagonal elements), and finally the coefficient of the nth-degree term is 1.

This kind of representation is called a *diagonal expansion* of the determinant because the coefficients are principal minors. It has occasional use for matrices with large patterns of zeros, but its main application is in analysis of eigenvalues, which we consider in Section 3.8.2.

Computing the Determinant

For an arbitrary matrix, the determinant is rather difficult to compute. The method for computing a determinant is not the one that would arise directly from the definition or even from a Laplace expansion. The more efficient methods involve first factoring the matrix, as we discuss in later sections.

The determinant is not very often directly useful, but although it may not be obvious from its definition, the determinant, along with minors, cofactors, and adjoint matrices, is very useful in discovering and proving properties of matrices. The determinant is used extensively in eigenanalysis (see Section 3.8).

A Geometrical Perspective of the Determinant

In Section 2.2, we discussed a useful geometric interpretation of vectors in a linear space with a Cartesian coordinate system. The elements of a vector correspond to measurements along the respective axes of the coordinate system. When working with several vectors, or with a matrix in which the columns (or rows) are associated with vectors, we may designate a vector x_i as $x_i = (x_{i1}, \ldots, x_{id})$. A set of d linearly independent d-vectors define a parallelotope in d dimensions. For example, in a two-dimensional space, the linearly independent 2-vectors x_1 and x_2 define a parallelogram, as shown in Figure 3.1.

The area of this parallelogram is the base times the height, bh, where, in this case, b is the length of the vector x_1, and h is the length of x_2 times the sine of the angle θ. Thus, making use of equation (2.32) on page 26 for the cosine of the angle, we have

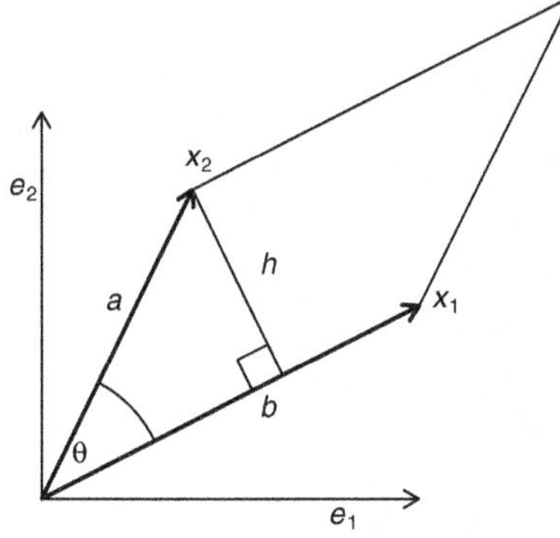

Fig. 3.1. Volume (Area) of Region Determined by x_1 and x_2

$$
\begin{aligned}
\text{area} &= bh \\
&= \|x_1\| \|x_2\| \sin(\theta) \\
&= \|x_1\| \|x_2\| \sqrt{1 - \left(\frac{\langle x_1, x_2 \rangle}{\|x_1\| \|x_2\|} \right)^2} \\
&= \sqrt{\|x_1\|^2 \|x_2\|^2 - (\langle x_1, x_2 \rangle)^2} \\
&= \sqrt{(x_{11}^2 + x_{12}^2)(x_{21}^2 + x_{22}^2) - (x_{11}x_{21} - x_{12}x_{22})^2} \\
&= |x_{11}x_{22} - x_{12}x_{21}| \\
&= |\det(X)|,
\end{aligned}
\tag{3.33}
$$

where $x_1 = (x_{11}, x_{12})$, $x_2 = (x_{21}, x_{22})$, and

$$
\begin{aligned}
X &= [x_1 \mid x_2] \\
&= \begin{bmatrix} x_{11} & x_{21} \\ x_{12} & x_{22} \end{bmatrix}.
\end{aligned}
$$

Although we will not go through the details here, this equivalence of a volume of a parallelotope that has a vertex at the origin and the absolute value of the determinant of a square matrix whose columns correspond to the vectors that form the sides of the parallelotope extends to higher dimensions.

In making a change of variables in integrals, as in equation (4.37) on page 165, we use the absolute value of the determinant of the Jacobian as a volume element. Another instance of the interpretation of the determinant as a volume is in the generalized variance, discussed on page 296.

3.2 Multiplication of Matrices and Multiplication of Vectors and Matrices

The elements of a vector or matrix are elements of a field, and most matrix and vector operations are defined in terms of the two operations of the field. Of course, in this book, the field of most interest is the field of real numbers.

3.2.1 Matrix Multiplication (Cayley)

There are various kinds of multiplication of matrices that may be useful. The most common kind of multiplication is *Cayley multiplication*. If the number of columns of the matrix A, with elements a_{ij}, and the number of rows of the matrix B, with elements b_{ij}, are equal, then the *(Cayley) product* of A and B is defined as the matrix C with elements

$$c_{ij} = \sum_k a_{ik}b_{kj}. \tag{3.34}$$

This is the most common type of matrix product, and we refer to it by the unqualified phrase "matrix multiplication".

Cayley matrix multiplication is indicated by juxtaposition, with no intervening symbol for the operation: $C = AB$.

If the matrix A is $n \times m$ and the matrix B is $m \times p$, the product $C = AB$ is $n \times p$:

$$
\begin{array}{ccc}
C & = A & B
\end{array}
$$

$$
\left[\quad\right]_{n\times p} = \left[\quad\right]_{n\times m} \left[\quad\right]_{m\times p}
$$

Cayley matrix multiplication is a mapping,

$$\mathrm{I\!R}^{n\times m} \times \mathrm{I\!R}^{m\times p} \mapsto \mathrm{I\!R}^{n\times p}.$$

The multiplication of an $n \times m$ matrix and an $m \times p$ matrix requires nmp scalar multiplications and $np(m - 1)$ scalar additions. Here, as always in numerical analysis, we must remember that the definition of an operation, such as matrix multiplication, does not necessarily define a good algorithm for evaluating the operation.

It is obvious that while the product AB may be well-defined, the product BA is defined only if $n = p$; that is, if the matrices AB and BA are square. We assume throughout that writing a product of matrices AB implies that the number of columns of the first matrix is the same as the number of rows of the second; that is, they are *conformable for multiplication* in the order given.

It is easy to see from the definition of matrix multiplication (3.34) that in general, even for square matrices, $AB \neq BA$. It is also obvious that if AB exists, then $B^{\mathrm{T}}A^{\mathrm{T}}$ exists and, in fact,

$$B^{\mathrm{T}}A^{\mathrm{T}} = (AB)^{\mathrm{T}}. \tag{3.35}$$

The product of symmetric matrices is not, in general, symmetric. If (but not only if) A and B are symmetric, then $AB = (BA)^{\mathrm{T}}$.

Various matrix shapes are preserved under matrix multiplication. Assume A and B are square matrices of the same number of rows. If A and B are diagonal, AB is diagonal; if A and B are upper triangular, AB is upper triangular; and if A and B are lower triangular, AB is lower triangular.

Because matrix multiplication is not commutative, we often use the terms "premultiply" and "postmultiply" and the corresponding nominal forms of these terms. Thus, in the product AB, we may say B is premultiplied by A, or, equivalently, A is postmultiplied by B.

Although matrix multiplication is *not commutative*, it is *associative*; that is, if the matrices are conformable,

$$A(BC) = (AB)C. \tag{3.36}$$

It is also *distributive* over addition; that is,

$$A(B + C) = AB + AC \tag{3.37}$$

and

$$(B + C)A = BA + CA. \tag{3.38}$$

These properties are obvious from the definition of matrix multiplication. (Note that left-sided distribution is not the same as right-sided distribution because the multiplication is not commutative.)

An $n \times n$ matrix consisting of 1s along the diagonal and 0s everywhere else is a *multiplicative identity* for the set of $n \times n$ matrices and Cayley multiplication. Such a matrix is called the *identity matrix of order n*, and is denoted by I_n, or just by I. The columns of the identity matrix are unit vectors.

The identity matrix is a multiplicative identity for any matrix so long as the matrices are conformable for the multiplication. If A is $n \times m$, then

$$I_n A = A I_m = A.$$

Powers of Square Matrices

For a square matrix A, its product with itself is defined, and so we will use the notation A^2 to mean the Cayley product AA, with similar meanings for A^k for a positive integer k. As with the analogous scalar case, A^k for a negative integer may or may not exist, and when it exists, it has a meaning for Cayley multiplication similar to the meaning in ordinary scalar multiplication. We will consider these issues later (in Section 3.3.3).

For an $n \times n$ matrix A, if A^k exists for negative integers, we define A^0 by

$$A^0 = I_n. \tag{3.39}$$

For a diagonal matrix $D = \mathrm{diag}\,((d_1, \ldots, d_n))$, we have

$$D^k = \mathrm{diag}\left((d_1^k, \ldots, d_n^k)\right). \tag{3.40}$$

Matrix Polynomials

Polynomials in square matrices are similar to the more familiar polynomials in scalars. We may consider

$$p(A) = b_0 I + b_1 A + \cdots b_k A^k.$$

The value of this polynomial is a matrix.

The theory of polynomials in general holds, and in particular, we have the useful factorizations of monomials: for any positive integer k,

$$I - A^k = (I - A)(I + A + \cdots A^{k-1}), \tag{3.41}$$

and for an odd positive integer k,

$$I + A^k = (I + A)(I - A + \cdots A^{k-1}). \tag{3.42}$$

3.2.2 Multiplication of Partitioned Matrices

Multiplication and other operations with partitioned matrices are carried out with their submatrices in the obvious way. Thus, assuming the submatrices are conformable for multiplication,

$$\begin{bmatrix} A_{11} & A_{12} \\ A_{21} & A_{22} \end{bmatrix} \begin{bmatrix} B_{11} & B_{12} \\ B_{21} & B_{22} \end{bmatrix} = \begin{bmatrix} A_{11}B_{11} + A_{12}B_{21} & A_{11}B_{12} + A_{12}B_{22} \\ A_{21}B_{11} + A_{22}B_{21} & A_{21}B_{12} + A_{22}B_{22} \end{bmatrix}.$$

Sometimes a matrix may be partitioned such that one partition is just a single column or row, that is, a vector or the transpose of a vector. In that case, we may use a notation such as

$$[X \; y]$$

or

$$[X \mid y],$$

where X is a matrix and y is a vector. We develop the notation in the obvious fashion; for example,

$$[X \; y]^{\mathrm{T}} [X \; y] = \begin{bmatrix} X^{\mathrm{T}}X & X^{\mathrm{T}}y \\ y^{\mathrm{T}}X & y^{\mathrm{T}}y \end{bmatrix}. \tag{3.43}$$

3.2.3 Elementary Operations on Matrices

Many common computations involving matrices can be performed as a sequence of three simple types of operations on either the rows or the columns of the matrix:

- the interchange of two rows (columns),

- a scalar multiplication of a given row (column), and
- the replacement of a given row (column) by the sum of that row (columns) and a scalar multiple of another row (column); that is, an axpy operation.

Such an operation on the rows of a matrix can be performed by premultiplication by a matrix in a standard form, and an operation on the columns of a matrix can be performed by postmultiplication by a matrix in a standard form. To repeat:

- premultiplication: operation on rows;
- postmultiplication: operation on columns.

The matrix used to perform the operation is called an *elementary transformation matrix* or *elementary operator matrix*. Such a matrix is the identity matrix transformed by the corresponding operation performed on its unit rows, e_p^{T}, or columns, e_p.

In actual computations, we do not form the elementary transformation matrices explicitly, but their formulation allows us to discuss the operations in a systematic way and better understand the properties of the operations. Products of any of these elementary operator matrices can be used to effect more complicated transformations.

Operations on the rows are more common, and that is what we will discuss here, although operations on columns are completely analogous. These transformations of rows are called *elementary row operations*.

Interchange of Rows or Columns; Permutation Matrices

By first interchanging the rows or columns of a matrix, it may be possible to partition the matrix in such a way that the partitions have interesting or desirable properties. Also, in the course of performing computations on a matrix, it is often desirable to interchange the rows or columns of the matrix. (This is an instance of "pivoting", which will be discussed later, especially in Chapter 6.) In matrix computations, we almost never actually move data from one row or column to another; rather, the interchanges are effected by changing the indexes to the data.

Interchanging two rows of a matrix can be accomplished by premultiplying the matrix by a matrix that is the identity with those same two rows interchanged; for example,

$$
\begin{bmatrix} 1 & 0 & 0 & 0 \\ 0 & 0 & 1 & 0 \\ 0 & 1 & 0 & 0 \\ 0 & 0 & 0 & 1 \end{bmatrix}
\begin{bmatrix} a_{11} & a_{12} & a_{13} \\ a_{21} & a_{22} & a_{23} \\ a_{31} & a_{32} & a_{33} \\ a_{41} & a_{42} & a_{43} \end{bmatrix}
=
\begin{bmatrix} a_{11} & a_{12} & a_{13} \\ a_{31} & a_{32} & a_{33} \\ a_{21} & a_{22} & a_{23} \\ a_{41} & a_{42} & a_{43} \end{bmatrix} .
$$

The first matrix in the expression above is called an *elementary permutation matrix*. It is the identity matrix with its second and third rows (or columns) interchanged. An elementary permutation matrix, which is the identity with the pth and q^{th} rows interchanged, is denoted by E_{pq}. That is, E_{pq} is the identity, except the pth row is e_q^{T} and the qth row is e_p^{T}. Note that $E_{pq} = E_{qp}$. Thus, for example, if the given matrix is $4 \times m$, to interchange the second and third rows, we use

$$E_{23} = E_{32} = \begin{bmatrix} 1 & 0 & 0 & 0 \\ 0 & 0 & 1 & 0 \\ 0 & 1 & 0 & 0 \\ 0 & 0 & 0 & 1 \end{bmatrix}.$$

It is easy to see from the definition that an elementary permutation matrix is symmetric. Note that the notation E_{pq} does not indicate the order of the elementary permutation matrix; that must be specified in the context.

Premultiplying a matrix A by a (conformable) E_{pq} results in an interchange of the pth and q^{th} rows of A as we see above. Any permutation of rows of A can be accomplished by successive premultiplications by elementary permutation matrices. Note that the order of multiplication matters. Although a given permutation can be accomplished by different elementary permutations, the number of elementary permutations that effect a given permutation is always either even or odd; that is, if an odd number of elementary permutations results in a given permutation, any other sequence of elementary permutations to yield the given permutation is also odd in number. Any given permutation can be effected by successive interchanges of adjacent rows.

Postmultiplying a matrix A by a (conformable) E_{pq} results in an interchange of the pth and q^{th} columns of A:

$$\begin{bmatrix} a_{11} & a_{12} & a_{13} \\ a_{21} & a_{22} & a_{23} \\ a_{31} & a_{32} & a_{33} \\ a_{41} & a_{42} & a_{43} \end{bmatrix} \begin{bmatrix} 1 & 0 & 0 \\ 0 & 0 & 1 \\ 0 & 1 & 0 \end{bmatrix} = \begin{bmatrix} a_{11} & a_{13} & a_{12} \\ a_{21} & a_{23} & a_{22} \\ a_{31} & a_{33} & a_{32} \\ a_{41} & a_{43} & a_{42} \end{bmatrix}.$$

Note that

$$A = E_{pq} E_{pq} A = A E_{pq} E_{pq}; \tag{3.44}$$

that is, as an operator, an elementary permutation matrix is its own inverse operator: $E_{pq} E_{pq} = I$.

Because all of the elements of a permutation matrix are 0 or 1, the trace of an $n \times n$ elementary permutation matrix is $n - 2$.

The product of elementary permutation matrices is also a *permutation matrix* in the sense that it permutes several rows or columns. For example, premultiplying A by the matrix $Q = E_{pq} E_{qr}$ will yield a matrix whose p^{th} row is the rth row of the original A, whose q^{th} row is the pth row of A, and whose r^{th} row is the qth row of A. We often use the notation E_π to denote a more general permutation matrix. This expression will usually be used generically, but sometimes we will specify the permutation, π.

A general permutation matrix (that is, a product of elementary permutation matrices) is not necessarily symmetric, but its transpose is also a permutation matrix. It is not necessarily its own inverse, but its permutations can be reversed by a permutation matrix formed by products of elementary permutation matrices in the opposite order; that is,

$$E_\pi^{\mathrm{T}} E_\pi = I.$$

As a prelude to other matrix operations, we often permute both rows and columns, so we often have a representation such as

$$B = E_{\pi_1} A E_{\pi_2}, \tag{3.45}$$

where E_{π_1} is a permutation matrix to permute the rows and E_{π_2} is a permutation matrix to permute the columns. We use these kinds of operations to arrive at the important equation (3.99) on page 80, and combine these operations with others to yield equation (3.113) on page 86. These equations are used to determine the number of linearly independent rows and columns and to represent the matrix in a form with a maximal set of linearly independent rows and columns clearly identified.

The Vec-Permutation Matrix

A special permutation matrix is the matrix that transforms the vector $\mathrm{vec}(A)$ into $\mathrm{vec}(A^{\mathrm{T}})$. If A is $n \times m$, the matrix K_{nm} that does this is $nm \times nm$. We have

$$\mathrm{vec}(A^{\mathrm{T}}) = K_{nm}\mathrm{vec}(A). \tag{3.46}$$

The matrix K_{nm} is called the *nm vec-permutation matrix*.

Scalar Row or Column Multiplication

Often, numerical computations with matrices are more accurate if the rows have roughly equal norms. For this and other reasons, we often transform a matrix by multiplying one of its rows by a scalar. This transformation can also be performed by premultiplication by an elementary transformation matrix. For multiplication of the pth row by the scalar, the elementary transformation matrix, which is denoted by $E_p(a)$, is the identity matrix in which the pth diagonal element has been replaced by a. Thus, for example, if the given matrix is $4 \times m$, to multiply the second row by a, we use

$$E_2(a) = \begin{bmatrix} 1 & 0 & 0 & 0 \\ 0 & a & 0 & 0 \\ 0 & 0 & 1 & 0 \\ 0 & 0 & 0 & 1 \end{bmatrix}.$$

Postmultiplication of a given matrix by the multiplier matrix $E_p(a)$ results in the multiplication of the pth column by the scalar. For this, $E_p(a)$ is a square matrix of order equal to the number of columns of the given matrix.

Note that the notation $E_p(a)$ does not indicate the number of rows and columns. This must be specified in the context.

Note that, if $a \neq 0$,

$$A = E_p(1/a)E_p(a)A, \tag{3.47}$$

that is, as an operator, the inverse operator is a row multiplication matrix on the same row and with the reciprocal as the multiplier.

Axpy Row or Column Transformations

The other elementary operation is an axpy on two rows and a replacement of one of those rows with the result

$$a_p \leftarrow aa_q + a_p.$$

This operation also can be effected by premultiplication by a matrix formed from the identity matrix by inserting the scalar in the (p, q) position. Such a matrix is denoted by $E_{pq}(a)$. Thus, for example, if the given matrix is $4 \times m$, to add a times the third row to the second row, we use

$$E_{23}(a) = \begin{bmatrix} 1 & 0 & 0 & 0 \\ 0 & 1 & a & 0 \\ 0 & 0 & 1 & 0 \\ 0 & 0 & 0 & 1 \end{bmatrix}.$$

Premultiplication of a matrix A by such a matrix,

$$E_{pq}(a)A, \tag{3.48}$$

yields a matrix whose p^{th} row is a times the qth row plus the original row.

Given the 4×3 matrix $A = (a_{ij})$, we have

$$E_{23}(a)A = \begin{bmatrix} a_{11} & a_{12} & a_{13} \\ a_{21} + aa_{31} & a_{22} + aa_{32} & a_{23} + aa_{33} \\ a_{31} & a_{32} & a_{33} \\ a_{41} & a_{42} & a_{43} \end{bmatrix}.$$

Postmultiplication of a matrix A by an axpy operator matrix,

$$AE_{pq}(a),$$

yields a matrix whose q^{th} column is a times the pth column plus the original column. For this, $E_{pq}(a)$ is a square matrix of order equal to the number of columns of the given matrix. Note that the column that is changed corresponds to the *second* subscript in $E_{pq}(a)$.

Note that

$$A = E_{pq}(-a)E_{pq}(a)A; \tag{3.49}$$

that is, as an operator, the inverse operator is the same axpy elementary operator matrix with the negative of the multiplier.

A common use of axpy operator matrices is to form a matrix with zeros in all positions of a given column below a given position in the column. These operations usually follow an operation by a scalar row multiplier matrix that puts a 1 in the position of interest. For example, given an $n \times m$ matrix A with $a_{ij} \neq 0$, to put a 1 in the (i, j) position and 0s in all positions of the jth column below the ith row, we form the product

$$E_{mi}(-a_{mj}) \cdots E_{i+1,i}(-a_{i+1,j})E_i(1/a_{ij})A. \tag{3.50}$$

This process is called *Gaussian elimination.*

Gaussian elimination is often performed sequentially down the diagonal elements of a matrix. If at some point $a_{ii} = 0$, the operations of equation (3.50) cannot be performed. In that case, we may first interchange the ith row with the kth row, where $k > i$ and $a_{ki} \neq 0$. Such an interchange is called *pivoting.* We will discuss pivoting in more detail on page 209 in Chapter 6.

To form a matrix with zeros in all positions of a given column except one, we use additional matrices for the rows above the given element:

$$E_{mi}(-a_{mj}) \cdots E_{i+1,i}(-a_{i+1,j}) \cdots E_{i-1,i}(-a_{i-1,j}) \cdots E_{1i}(-a_{1j})E_i(1/a_{ij})A.$$

We can likewise zero out all elements in the ith row except the one in the $(ij)^{\text{th}}$ position by similar postmultiplications.

These elementary transformations are the basic operations in Gaussian elimination, which is discussed in Sections 5.6 and 6.2.1.

Determinants of Elementary Operator Matrices

The determinant of an elementary permutation matrix E_{pq} has only one term in the sum that defines the determinant (equation (3.16), page 50), and that term is 1 times σ evaluated at the permutation that exchanges p and q. As we have seen (page 51), this is an odd permutation; hence, for an elementary permutation matrix E_{pq},

$$|E_{pq}| = -1. \tag{3.51}$$

Because all terms in $|E_{pq}A|$ are exactly the same terms as in $|A|$ but with one different permutation in each term, we have

$$|E_{pq}A| = -|A|.$$

More generally, if A and E_π are $n \times n$ matrices, and E_π is any permutation matrix (that is, any product of E_{pq} matrices), then $|E_\pi A|$ is either $|A|$ or $-|A|$ because all terms in $|E_\pi A|$ are exactly the same as the terms in $|A|$ but possibly with different signs because the permutations are different. In fact, the differences in the permutations are exactly the same as the permutation of $1, \ldots, n$ in E_π; hence,

$$|E_\pi A| = |E_\pi|\,|A|.$$

(In equation (3.57) below, we will see that this equation holds more generally.)

The determinant of an elementary row multiplication matrix $E_p(a)$ is

$$|E_p(a)| = a. \tag{3.52}$$

If A and $E_p(a)$ are $n \times n$ matrices, then

$$|E_p(a)A| = a|A|,$$

as we see from the definition of the determinant, equation (3.16).

The determinant of an elementary axpy matrix $E_{pq}(a)$ is 1,

$$|E_{pq}(a)| = 1, \tag{3.53}$$

because the term consisting of the product of the diagonals is the only term in the determinant.

Now consider $|E_{pq}(a)A|$ for an $n \times n$ matrix A. Expansion in the minors (equation (3.21)) along the pth row yields

$$
\begin{aligned}
|E_{pq}(a)A| &= \sum_{j=1}^{n} (a_{pj} + a a_{qj})(-1)^{p+j}|A_{(ij)}| \\
&= \sum_{j=1}^{n} a_{pj}(-1)^{p+j}|A_{(ij)}| + a \sum_{j=1}^{n} a_{qj}(-1)^{p+j}|A_{(ij)}|.
\end{aligned}
$$

From equation (3.23) on page 53, we see that the second term is 0, and since the first term is just the determinant of A, we have

$$|E_{pq}(a)A| = |A|. \tag{3.54}$$

3.2.4 Traces and Determinants of Square Cayley Products

The Trace

A useful property of the trace for the matrices A and B that are conformable for the multiplications AB and BA is

$$\text{tr}(AB) = \text{tr}(BA). \tag{3.55}$$

This is obvious from the definitions of matrix multiplication and the trace.

Because of the associativity of matrix multiplication, this relation can be extended as

$$\text{tr}(ABC) = \text{tr}(BCA) = \text{tr}(CAB) \tag{3.56}$$

for matrices A, B, and C that are conformable for the multiplications indicated. Notice that the individual matrices need not be square.

The Determinant

An important property of the determinant is

$$|AB| = |A|\,|B| \tag{3.57}$$

if A and B are square matrices conformable for multiplication. We see this by first forming

$$\left| \begin{bmatrix} I & A \\ 0 & I \end{bmatrix} \begin{bmatrix} A & 0 \\ -I & B \end{bmatrix} \right| = \left| \begin{bmatrix} 0 & AB \\ -I & B \end{bmatrix} \right| \tag{3.58}$$

and then observing from equation (3.30) that the right-hand side is $|AB|$. Now consider the left-hand side. The matrix that is the first factor is a product of elementary axpy transformation matrices; that is, it is a matrix that when postmultiplied by another matrix merely adds multiples of rows in the lower part of the matrix to rows in the upper part of the matrix. If A and B are $n \times n$ (and so the identities are likewise $n \times n$), the full matrix is the product:

$$\begin{bmatrix} I & A \\ 0 & I \end{bmatrix} = E_{1,n+1}(a_{11}) \cdots E_{1,2n}(a_{1n})E_{2,n+1}(a_{21}) \cdots E_{2,2n}(a_{2,n}) \cdots E_{n,2n}(a_{nn}).$$

Hence, applying equation (3.54) recursively, we have

$$\left| \begin{bmatrix} I & A \\ 0 & I \end{bmatrix} \begin{bmatrix} A & 0 \\ -I & B \end{bmatrix} \right| = \left| \begin{bmatrix} A & 0 \\ -I & B \end{bmatrix} \right|,$$

and from equation (3.29) we have

$$\left| \begin{bmatrix} A & 0 \\ -I & B \end{bmatrix} \right| = |A||B|,$$

and so finally we have equation (3.57).

3.2.5 Multiplication of Matrices and Vectors

It is often convenient to think of a vector as a matrix with only one element in one of its dimensions. This provides for an immediate extension of the definitions of transpose and matrix multiplication to include vectors as either or both factors. In this scheme, we follow the convention that a vector corresponds to a *column*; that is, if x is a vector and A is a matrix, Ax or $x^{\mathrm{T}}A$ may be well-defined, but neither xA nor Ax^{T} would represent anything, except in the case when all dimensions are 1. (In some computer systems for matrix algebra, these conventions are not enforced; see, for example the R code in Figure 12.4 on page 468.) The alternative notation $x^{\mathrm{T}}y$ we introduced earlier for the dot product or inner product, $\langle x, y \rangle$, of the vectors x and y is consistent with this paradigm. We will continue to write vectors as $x = (x_1, \ldots, x_n)$, however. This does not imply that the vector is a row vector. We would represent a matrix with one row as $Y = [y_{11} \ldots y_{1n}]$ and a matrix with one column as $Z = [z_{11} \ldots z_{m1}]^{\mathrm{T}}$.

The Matrix/Vector Product as a Linear Combination

If we represent the vectors formed by the columns of an $n \times m$ matrix A as $a_1, \ldots, a_m$, the matrix/vector product Ax is a linear combination of these columns of A:

$$Ax = \sum_{i=1}^{m} x_i a_i. \tag{3.59}$$

(Here, each x_i is a scalar, and each a_i is a vector.)

Given the equation $Ax = b$, we have $b \in \mathrm{span}(A)$; that is, the n-vector b is in the k-dimensional column space of A, where $k \leq m$.

3.2.6 Outer Products

The *outer product* of the vectors x and y is the matrix

$$xy^{\mathrm{T}}. \tag{3.60}$$

Note that the definition of the outer product does not require the vectors to be of equal length. Note also that while the inner product is commutative, the outer product is not commutative (although it does have the property $xy^{\mathrm{T}} = (yx^{\mathrm{T}})^{\mathrm{T}}$).

A very common outer product is of a vector with itself:

$$xx^{\mathrm{T}}.$$

The outer product of a vector with itself is obviously a symmetric matrix.

We should again note some subtleties of differences in the types of objects that result from operations. If A and B are matrices conformable for the operation, the product $A^{\mathrm{T}}B$ is a *matrix* even if both A and B are $n \times 1$ and so the result is 1×1. For the vectors x and y and matrix C, however, $x^{\mathrm{T}}y$ and $x^{\mathrm{T}}Cy$ are *scalars*; hence, the dot product and a quadratic form are *not* the same as the result of a matrix multiplication. The dot product is a scalar, and the result of a matrix multiplication is a matrix. The outer product of vectors is a matrix, even if both vectors have only one element. Nevertheless, as we have mentioned before, in the following, *we will treat a one by one matrix or a vector with only one element as a scalar whenever it is convenient to do so.*

3.2.7 Bilinear and Quadratic Forms; Definiteness

A variation of the vector dot product, $x^{\mathrm{T}}Ay$, is called a *bilinear form*, and the special bilinear form $x^{\mathrm{T}}Ax$ is called a *quadratic form*. Although in the definition of quadratic form we do not require A to be symmetric — because for a given value of x and a given value of the quadratic form $x^{\mathrm{T}}Ax$ there is a unique symmetric matrix A_s such that $x^{\mathrm{T}}A_s x = x^{\mathrm{T}}Ax$ — we generally work only with symmetric matrices in dealing with quadratic forms. (The matrix A_s is $\frac{1}{2}(A + A^{\mathrm{T}})$; see Exercise 3.3.) Quadratic forms correspond to sums of squares and hence play an important role in statistical applications.

Nonnegative Definite and Positive Definite Matrices

A symmetric matrix A such that for any (conformable and real) vector x the quadratic form $x^{\mathrm{T}}Ax$ is nonnegative, that is,

$$x^{\mathrm{T}}Ax \geq 0, \tag{3.61}$$

is called a *nonnegative definite matrix*. We denote the fact that A is nonnegative definite by

$$A \succeq 0.$$

(Note that we consider $0_{n \times n}$ to be nonnegative definite.)

A symmetric matrix A such that for any (conformable) vector $x \neq 0$ the quadratic form

$$x^{\mathrm{T}}Ax > 0 \tag{3.62}$$

is called a *positive definite matrix*. We denote the fact that A is positive definite by

$$A \succ 0.$$

(Recall that $A \geq 0$ and $A > 0$ mean, respectively, that all elements of A are nonnegative and positive.) When A and B are symmetric matrices of the same order, we write $A \succeq B$ to mean $A - B \succeq 0$ and $A \succ B$ to mean $A - B \succ 0$. Nonnegative and positive definite matrices are very important in applications. We will encounter them from time to time in this chapter, and then we will discuss more of their properties in Section 8.3.

In this book we use the terms "nonnegative definite" and "positive definite" only for symmetric matrices. In other literature, these terms may be used more generally; that is, for any (square) matrix that satisfies (3.61) or (3.62).

The Trace of Inner and Outer Products

The invariance of the trace to permutations of the factors in a product (equation (3.55)) is particularly useful in working with quadratic forms. Because the quadratic form itself is a scalar (or a 1×1 matrix), and because of the invariance, we have the very useful fact

$$\begin{aligned} x^{\mathrm{T}}Ax &= \mathrm{tr}(x^{\mathrm{T}}Ax) \\ &= \mathrm{tr}(Axx^{\mathrm{T}}). \end{aligned} \tag{3.63}$$

Furthermore, for any scalar a, n-vector x, and $n \times n$ matrix A, we have

$$(x - a)^{\mathrm{T}}A(x - a) = \mathrm{tr}(Ax_{\mathrm{c}}x_{\mathrm{c}}^{\mathrm{T}}) + n(a - \bar{x})^2\mathrm{tr}(A). \tag{3.64}$$

(Compare this with equation (2.48) on page 34.)

3.2.8 Anisometric Spaces

In Section 2.1, we considered various properties of vectors that depend on the inner product, such as orthogonality of two vectors, norms of a vector, angles between two vectors, and distances between two vectors. All of these properties and measures are invariant to the orientation of the vectors; the space is *isometric* with respect to a Cartesian coordinate system. Noting that the inner product is the bilinear form $x^{\mathrm{T}}Iy$, we have a heuristic generalization to an anisometric space. Suppose, for example, that the scales of the coordinates differ; say, a given distance along one axis in the natural units of the axis is equivalent (in some sense depending on the application) to twice that distance along another axis, again measured in the natural units of the axis. The properties derived from the inner product, such as a norm and a metric, may correspond to the application better if we use a bilinear form in which the matrix reflects the different effective distances along the coordinate axes. A diagonal matrix whose entries have relative values corresponding to the inverses of the relative scales of the axes may be more useful. Instead of $x^{\mathrm{T}}y$, we may use $x^{\mathrm{T}}Dy$, where D is this diagonal matrix.

Rather than differences in scales being just in the directions of the coordinate axes, more generally we may think of anisometries being measured by general (but perhaps symmetric) matrices. (The covariance and correlation matrices defined on page 294 come to mind. Any such matrix to be used in this context should be positive definite because we will generalize the dot product, which is necessarily nonnegative, in terms of a quadratic form.) A bilinear form $x^{\mathrm{T}}Ay$ may correspond more closely to the properties of the application than the standard inner product.

We define orthogonality of two vectors x and y with respect to A by

$$x^{\mathrm{T}}Ay = 0. \tag{3.65}$$

In this case, we say x and y are *A-conjugate*.

The L_2 norm of a vector is the square root of the quadratic form of the vector with respect to the identity matrix. A generalization of the L_2 vector norm, called an *elliptic norm* or a *conjugate norm*, is defined for the vector x as the square root of the quadratic form $x^{\mathrm{T}}Ax$ for any symmetric positive definite matrix A. It is sometimes denoted by $\|x\|_A$:

$$\|x\|_A = \sqrt{x^{\mathrm{T}}Ax}. \tag{3.66}$$

It is easy to see that

$$\|x\|_A$$

satisfies the definition of a norm given on page 16. If A is a diagonal matrix with elements $w_i \geq 0$, the elliptic norm is the weighted L_2 norm of equation (2.15).

The elliptic norm yields an *elliptic metric* in the usual way of defining a metric in terms of a norm. The distance between the vectors x and y with respect to A is $\sqrt{(x-y)^{\mathrm{T}}A(x-y)}$. It is easy to see that this satisfies the definition of a metric given on page 22.

A metric that is widely useful in statistical applications is the Mahalanobis distance, which uses a covariance matrix as the scale for a given space. (The sample covariance matrix is defined in equation (8.70) on page 294.) If S is the covariance matrix, the Mahalanobis distance, with respect to that matrix, between the vectors x and y is

$$\sqrt{(x-y)^{\mathrm{T}}S^{-1}(x-y)}. \tag{3.67}$$

3.2.9 Other Kinds of Matrix Multiplication

The most common kind of product of two matrices is the Cayley product, and when we speak of matrix multiplication without qualification, we mean the Cayley product. Three other types of matrix multiplication that are useful are *Hadamard multiplication*, *Kronecker multiplication*, and *dot product multiplication*.

The Hadamard Product

Hadamard multiplication is defined for matrices of the same shape as the multiplication of each element of one matrix by the corresponding element of the other matrix. Hadamard multiplication immediately inherits the commutativity, associativity, and distribution over addition of the ordinary multiplication of the underlying field of scalars. Hadamard multiplication is also called array multiplication and element-wise multiplication. Hadamard matrix multiplication is a mapping

$$\mathbb{R}^{n \times m} \times \mathbb{R}^{n \times m} \mapsto \mathbb{R}^{n \times m}.$$

The identity for Hadamard multiplication is the matrix of appropriate shape whose elements are all 1s.

The Kronecker Product

Kronecker multiplication, denoted by $\otimes$, is defined for any two matrices $A_{n \times m}$ and $B_{p \times q}$ as

$$A \otimes B = \begin{bmatrix} a_{11}B & \dots & a_{1m}B \\ \vdots & \dots & \vdots \\ a_{n1}B & \dots & a_{nm}B \end{bmatrix}.$$

The Kronecker product of A and B is $np \times mq$; that is, Kronecker matrix multiplication is a mapping

$$\mathbb{R}^{n \times m} \times \mathbb{R}^{p \times q} \mapsto \mathbb{R}^{np \times mq}.$$

The Kronecker product is also called the "right direct product" or just *direct product*. (A left direct product is a Kronecker product with the factors reversed.)

Kronecker multiplication is not commutative, but it is associative and it is distributive over addition, as we will see below.

The identity for Kronecker multiplication is the 1×1 matrix with the element 1; that is, it is the same as the scalar 1.

The determinant of the Kronecker product of two square matrices $A_{n \times n}$ and $B_{m \times m}$ has a simple relationship to the determinants of the individual matrices:

$$|A \otimes B| = |A|^m |B|^n. \tag{3.68}$$

The proof of this, like many facts about determinants, is straightforward but involves tedious manipulation of cofactors. The manipulations in this case can be facilitated by using the vec-permutation matrix. See Harville (1997) for a detailed formal proof.

We can understand the properties of the Kronecker product by expressing the (i, j) element of $A \otimes B$ in terms of the elements of A and B,

$$(A \otimes B)_{i,j} = A_{[i/p]+1,\ [j/q]+1} B_{i-p[i/p],\ j-q[i/q]}, \tag{3.69}$$

where $[\cdot]$ is the greatest integer function.

Some additional properties of Kronecker products that are immediate results of the definition are, assuming the matrices are conformable for the indicated operations,

$$\begin{aligned}
(aA) \otimes (bB) &= ab(A \otimes B) \\
&= (abA) \otimes B \\
&= A \otimes (abB), \text{ for scalars } a, b,
\end{aligned} \tag{3.70}$$

$$(A + B) \otimes (C) = A \otimes C + B \otimes C, \tag{3.71}$$

$$(A \otimes B) \otimes C = A \otimes (B \otimes C), \tag{3.72}$$

$$(A \otimes B)^{\mathrm{T}} = A^{\mathrm{T}} \otimes B^{\mathrm{T}}, \tag{3.73}$$

$$(A \otimes B)(C \otimes D) = AC \otimes BD. \tag{3.74}$$

These properties are all easy to see by using equation (3.69) to express the (i, j) element of the matrix on either side of the equation, taking into account the size of the matrices involved. For example, in the first equation, if A is $n \times m$ and B is $p \times q$, the (i, j) element on the left-hand side is

$$aA_{[i/p]+1,\ [j/q]+1} bB_{i-p[i/p],\ j-q[i/q]}$$

and that on the right-hand side is

$$ab A_{[i/p]+1,\ [j/q]+1} B_{i-p[i/p],\ j-q[i/q]}.$$

They are all this easy! Hence, they are Exercise 3.6.

Another property of the Kronecker product of square matrices is

$$\mathrm{tr}(A \otimes B) = \mathrm{tr}(A)\mathrm{tr}(B). \tag{3.75}$$

This is true because the trace of the product is merely the sum of all possible products of the diagonal elements of the individual matrices.

The Kronecker product and the vec function often find uses in the same application. For example, an $n \times m$ normal random matrix X with parameters M, Σ, and Ψ can be expressed in terms of an ordinary np-variate normal random variable $Y = \mathrm{vec}(X)$ with parameters $\mathrm{vec}(M)$ and $\Sigma \otimes \Psi$. (We discuss matrix random variables briefly on page 168. For a fuller discussion, the reader is referred to a text on matrix random variables such as Carmeli, 1983.)

A relationship between the vec function and Kronecker multiplication is

$$\mathrm{vec}(ABC) = (C^{\mathrm{T}} \otimes A)\mathrm{vec}(B) \tag{3.76}$$

for matrices A, B, and C that are conformable for the multiplication indicated.

The Dot Product or the Inner Product of Matrices

Another product of two matrices of the same shape is defined as the sum of the dot products of the vectors formed from the columns of one matrix with vectors formed from the corresponding columns of the other matrix; that is, if $a_1, \ldots, a_m$ are the columns of A and $b_1, \ldots, b_m$ are the columns of B, then the dot product of A and B, denoted $\langle A, B \rangle$, is

$$\langle A, B \rangle = \sum_{j=1}^{m} a_j^{\mathrm{T}} b_j. \tag{3.77}$$

For conformable matrices A, B, and C, we can easily confirm that this product satisfies the general properties of an inner product listed on page 15:

- If $A \neq 0$, $\langle A, A \rangle > 0$, and $\langle 0, A \rangle = \langle A, 0 \rangle = \langle 0, 0 \rangle = 0$.
- $\langle A, B \rangle = \langle B, A \rangle$.
- $\langle sA, B \rangle = s\langle A, B \rangle$, for a scalar s.
- $\langle (A + B), C \rangle = \langle A, C \rangle + \langle B, C \rangle$.

We also call this inner product of matrices the dot product of the matrices. (As in the case of the dot product of vectors, the dot product of matrices defined over the complex field is not an inner product because the first property listed above does not hold.)

As with any inner product (restricted to objects in the field of the reals), its value is a real number. Thus the matrix dot product is a mapping

$$\mathbb{R}^{n \times m} \times \mathbb{R}^{n \times m} \mapsto \mathbb{R}.$$

The dot product of the matrices A and B with the same shape is denoted by $A \cdot B$, or $\langle A, B \rangle$, just like the dot product of vectors.

We see from the definition above that the dot product of matrices satisfies

$$\langle A, B \rangle = \mathrm{tr}(A^{\mathrm{T}} B), \tag{3.78}$$

which could alternatively be taken as the definition.

Rewriting the definition of $\langle A, B \rangle$ as $\sum_{j=1}^{m} \sum_{i=1}^{n} a_{ij} b_{ij}$, we see that

$$\langle A, B \rangle = \langle A^{\mathrm{T}}, B^{\mathrm{T}} \rangle. \tag{3.79}$$

Like any inner product, dot products of matrices obey the Cauchy-Schwarz inequality (see inequality (2.10), page 16),

$$\langle A, B \rangle \leq \langle A, A \rangle^{\frac{1}{2}} \langle B, B \rangle^{\frac{1}{2}}, \tag{3.80}$$

with equality holding only if $A = 0$ or $B = sA$ for some scalar s.

In Section 2.1.8, we defined orthogonality and orthonormality of two or more vectors in terms of dot products. We can likewise define an orthogonal binary relationship between two matrices in terms of dot products of matrices. We say the matrices A and B of the same shape are *orthogonal to each other* if

$$\langle A, B \rangle = 0. \tag{3.81}$$

From equations (3.78) and (3.79) we see that the matrices A and B are orthogonal to each other if and only if $A^{\mathrm{T}} B$ and $B^{\mathrm{T}} A$ are hollow (that is, they have 0s in all diagonal positions). We also use the term "orthonormal" to refer to matrices that are orthogonal to each other and for which each has a dot product with itself of 1. In Section 3.7, we will define orthogonality as a unary property of matrices. The term "orthogonal", when applied to matrices, generally refers to that property rather than the binary property we have defined here.

On page 48 we identified a vector space of matrices and defined a basis for the space $\mathbb{R}^{n \times m}$. If $\{U_1, \ldots, U_k\}$ is a basis set for $\mathcal{M} \subset \mathbb{R}^{n \times m}$, with the property that $\langle U_i, U_j \rangle = 0$ for $i \neq j$ and $\langle U_i, U_i \rangle = 1$, and A is an $n \times m$ matrix, with the Fourier expansion

$$A = \sum_{i=1}^{k} c_i U_i, \tag{3.82}$$

we have, analogous to equation (2.37) on page 29,

$$c_i = \langle A, U_i \rangle. \tag{3.83}$$

The c_i have the same properties (such as the Parseval identity, equation (2.38), for example) as the Fourier coefficients in any orthonormal expansion. Best approximations within $\mathcal{M}$ can also be expressed as truncations of the sum in equation (3.82) as in equation (2.41). The objective of course is to reduce the truncation error. (The norms in Parseval's identity and in measuring the goodness of an approximation are matrix norms in this case. We discuss matrix norms in Section 3.9 beginning on page 128.)

3.3 Matrix Rank and the Inverse of a Full Rank Matrix

The linear dependence or independence of the vectors forming the rows or columns of a matrix is an important characteristic of the matrix.

The maximum number of linearly independent vectors (those forming either the rows or the columns) is called the *rank* of the matrix. We use the notation

$$\mathrm{rank}(A)$$

to denote the rank of the matrix A. (We have used the term "rank" before to denote dimensionality of an array. "Rank" as we have just defined it applies only to a matrix or to a set of vectors, and this is by far the more common meaning of the word. The meaning is clear from the context, however.)

Because multiplication by a nonzero scalar does not change the linear independence of vectors, for the scalar a with $a \neq 0$, we have

$$\mathrm{rank}(aA) = \mathrm{rank}(A). \tag{3.84}$$

From results developed in Section 2.1, we see that for the $n \times m$ matrix A,

$$\mathrm{rank}(A) \leq \min(n, m). \tag{3.85}$$

Row Rank and Column Rank

We have defined matrix rank in terms of numbers of linearly independent rows *or* columns. This is because the number of linearly independent rows is the same as the number of linearly independent columns. Although we may use the terms "row rank" or "column rank", the single word "rank" is sufficient because they are the same. To see this, assume we have an $n \times m$ matrix A and that there are exactly p linearly independent rows and exactly q linearly independent columns. We can permute the rows and columns of the matrix so that the first p rows are linearly independent rows and the first q columns are linearly independent and the remaining rows or columns are linearly dependent on the first ones. (Recall that applying the same permutation to all of the elements of each vector in a set of vectors does not change the linear dependencies over the set.) After these permutations, we have a matrix B with submatrices W, X, Y, and Z,

$$B = \begin{bmatrix} W_{p \times q} & X_{p \times m-q} \\ Y_{n-p \times q} & Z_{n-p \times m-q} \end{bmatrix}, \tag{3.86}$$

where the rows of $R = [W|X]$ correspond to p linearly independent m-vectors and the columns of $C = \begin{bmatrix} W \\ Y \end{bmatrix}$ correspond to q linearly independent n-vectors. Without loss of generality, we can assume $p \leq q$. Now, if $p < q$, it must be the case that the columns of W are linearly dependent because there are q of them, but they have only p elements. Therefore, there is some q-vector a such that $Wa = 0$. Now, since the rows of R are the full set of linearly independent rows, any row in $[Y|Z]$ can be expressed as a linear combination of the rows of R, and any row in Y can be expressed as a linear combination of the rows of W. This means, for some $n-p \times p$ matrix T, that $Y = TW$. In this case, however, $Ca = 0$. But this contradicts the assumption that the columns of C are linearly independent; therefore it cannot be the case that $p < q$. We conclude therefore that $p = q$; that is, that the maximum number of linearly independent rows is the same as the maximum number of linearly independent columns.

Because the row rank, the column rank, and the rank of A are all the same, we have

$$\text{rank}(A) = \dim(\mathcal{V}(A)), \tag{3.87}$$

$$\text{rank}(A^{\text{T}}) = \text{rank}(A), \tag{3.88}$$

$$\dim(\mathcal{V}(A^{\text{T}})) = \dim(\mathcal{V}(A)). \tag{3.89}$$

(Note, of course, that in general $\mathcal{V}(A^{\text{T}}) \neq \mathcal{V}(A)$; the orders of the vector spaces are possibly different.)

Full Rank Matrices

If the rank of a matrix is the same as its smaller dimension, we say the matrix is of *full rank*. In the case of a nonsquare matrix, we may say the matrix is of full row rank or full column rank just to emphasize which is the smaller number.

If a matrix is not of full rank, we say it is *rank deficient* and define the *rank deficiency* as the difference between its smaller dimension and its rank.

A full rank matrix that is square is called *nonsingular*, and one that is not nonsingular is called *singular*.

A square matrix that is either row or column diagonally dominant is nonsingular. The proof of this is Exercise 3.8. (It's easy!)

A positive definite matrix is nonsingular. The proof of this is Exercise 3.9.

Later in this section, we will identify additional properties of square full rank matrices. (For example, they have inverses and their determinants are nonzero.)

Rank of Elementary Operator Matrices and Matrix Products Involving Them

Because within any set of rows of an elementary operator matrix (see Section 3.2.3), for some given column, only one of those rows contains a nonzero element, the elementary operator matrices are all obviously of full rank (with the proviso that $a \neq 0$ in $E_p(a)$).

Furthermore, the rank of the product of any given matrix with an elementary operator matrix is the same as the rank of the given matrix. To see this, consider each type of elementary operator matrix in turn. For a given matrix A, the set of rows of $E_{pq}A$ is the same as the set of rows of A; hence, the rank of $E_{pq}A$ is the same as the rank of A. Likewise, the set of columns of AE_{pq} is the same as the set of columns of A; hence, again, the rank of AE_{pq} is the same as the rank of A.

The set of rows of $E_p(a)A$ for $a \neq 0$ is the same as the set of rows of A, except for one, which is a nonzero scalar multiple of the corresponding row of A; therefore, the rank of $E_p(a)A$ is the same as the rank of A. Likewise, the set of columns of $AE_p(a)$ is the same as the set of columns of A, except for one, which is a nonzero scalar multiple of the corresponding row of A; therefore, again, the rank of $AE_p(a)$ is the same as the rank of A.

Finally, the set of rows of $E_{pq}(a)A$ for $a \neq 0$ is the same as the set of rows of A, except for one, which is a nonzero scalar multiple of some row of A added to the corresponding row of A; therefore, the rank of $E_{pq}(a)A$ is the same as the rank of A. Likewise, we conclude that the rank of $AE_{pq}(a)$ is the same as the rank of A.

We therefore have that if P and Q are the products of elementary operator matrices,

$$\text{rank}(PAQ) = \text{rank}(A). \tag{3.90}$$

On page 88, we will extend this result to products by any full rank matrices.

3.3.1 The Rank of Partitioned Matrices, Products of Matrices, and Sums of Matrices

The partitioning in equation (3.86) leads us to consider partitioned matrices in more detail.

Rank of Partitioned Matrices and Submatrices

Let the matrix A be partitioned as

$$A = \begin{bmatrix} A_{11} & A_{12} \\ A_{21} & A_{22} \end{bmatrix},$$

where any pair of submatrices in a column or row may be null (that is, where for example, it may be the case that $A = [A_{11}|A_{12}]$). Then the number of linearly independent rows of A must be at least as great as the number of linearly independent rows of $[A_{11}|A_{12}]$ and the number of linearly independent rows of $[A_{21}|A_{22}]$. By the properties of subvectors in Section 2.1.1, the number of linearly independent rows of $[A_{11}|A_{12}]$ must be at least as great as the number of linearly independent rows of A_{11} or A_{21}. We could go through a similar argument relating to the number of linearly independent columns and arrive at the inequality

$$\operatorname{rank}(A_{ij}) \le \operatorname{rank}(A). \tag{3.91}$$

Furthermore, we see that

$$\operatorname{rank}(A) \le \operatorname{rank}([A_{11}|A_{12}]) + \operatorname{rank}([A_{21}|A_{22}]) \tag{3.92}$$

because $\operatorname{rank}(A)$ is the number of linearly independent columns of A, which is less than or equal to the number of linearly independent rows of $[A_{11}|A_{12}]$ plus the number of linearly independent rows of $[A_{12}|A_{22}]$. Likewise, we have

$$\operatorname{rank}(A) \le \operatorname{rank}\left(\begin{bmatrix} A_{11} \\ A_{21} \end{bmatrix}\right) + \operatorname{rank}\left(\begin{bmatrix} A_{12} \\ A_{22} \end{bmatrix}\right). \tag{3.93}$$

In a similar manner, by merely counting the number of independent rows, we see that, if

$$\mathcal{V}\left([A_{11}|A_{12}]^{\mathrm{T}}\right) \perp \mathcal{V}\left([A_{21}|A_{22}]^{\mathrm{T}}\right),$$

then

$$\operatorname{rank}(A) = \operatorname{rank}([A_{11}|A_{12}]) + \operatorname{rank}([A_{21}|A_{22}]); \tag{3.94}$$

and, if

$$\mathcal{V}\left(\begin{bmatrix} A_{11} \\ A_{21} \end{bmatrix}\right) \perp \mathcal{V}\left(\begin{bmatrix} A_{12} \\ A_{22} \end{bmatrix}\right),$$

then

$$\operatorname{rank}(A) = \operatorname{rank}\left(\begin{bmatrix} A_{11} \\ A_{21} \end{bmatrix}\right) + \operatorname{rank}\left(\begin{bmatrix} A_{12} \\ A_{22} \end{bmatrix}\right). \tag{3.95}$$

An Upper Bound on the Rank of Products of Matrices

The rank of the product of two matrices is less than or equal to the lesser of
the ranks of the two:

$$\operatorname{rank}(AB) \leq \min(\operatorname{rank}(A), \operatorname{rank}(B)). \qquad (3.96)$$

We can show this by separately considering two cases for the $n \times k$ matrix A
and the $k \times m$ matrix B. In one case, we assume k is at least as large as n
and $n \leq m$, and in the other case we assume $k < n \leq m$. In both cases, we
represent the rows of AB as k linear combinations of the rows of B.

From equation (3.96), we see that the rank of an outer product matrix
(that is, a matrix formed as the outer product of two vectors) is 1.

Equation (3.96) provides a useful upper bound on $\operatorname{rank}(AB)$. In Section 3.3.8, we will develop a lower bound on $\operatorname{rank}(AB)$.

An Upper and a Lower Bound on the Rank of Sums of Matrices

The rank of the sum of two matrices is less than or equal to the sum of their
ranks; that is,

$$\operatorname{rank}(A + B) \leq \operatorname{rank}(A) + \operatorname{rank}(B). \qquad (3.97)$$

We can see this by observing that

$$A + B = [A|B] \begin{bmatrix} I \\ I \end{bmatrix},$$

and so $\operatorname{rank}(A + B) \leq \operatorname{rank}([A|B])$ by equation (3.96), which in turn is $\leq$
$\operatorname{rank}(A) + \operatorname{rank}(B)$ by equation (3.92).

Using inequality (3.97) and the fact that $\operatorname{rank}(-B) = \operatorname{rank}(B)$, we write
$\operatorname{rank}(A - B) \leq \operatorname{rank}(A) + \operatorname{rank}(B)$, and so, replacing A in (3.97) by $A + B$, we
have $\operatorname{rank}(A) \leq \operatorname{rank}(A + B) + \operatorname{rank}(B)$, or $\operatorname{rank}(A + B) \geq \operatorname{rank}(A) - \operatorname{rank}(B)$.
By a similar procedure, we get $\operatorname{rank}(A + B) \geq \operatorname{rank}(B) - \operatorname{rank}(A)$, or

$$\operatorname{rank}(A + B) \geq |\operatorname{rank}(A) - \operatorname{rank}(B)|. \qquad (3.98)$$

3.3.2 Full Rank Partitioning

As we saw above, the matrix W in the partitioned B in equation (3.86) is
square; in fact, it is $r \times r$, where r is the rank of B:

$$B = \begin{bmatrix} W_{r \times r} & X_{r \times m-r} \\ Y_{n-r \times r} & Z_{n-r \times m-r} \end{bmatrix}. \qquad (3.99)$$

This is called a *full rank partitioning* of B.

The matrix B in equation (3.99) has a very special property: the full set
of linearly independent rows are the first r rows, and the full set of linearly
independent columns are the first r columns.

Any rank r matrix can be put in the form of equation (3.99) by using permutation matrices as in equation (3.45), assuming that $r \geq 1$. That is, if A is a nonzero matrix, there is a matrix of the form of B above that has the same rank. For some permutation matrices E_{π_1} and E_{π_2},

$$B = E_{\pi_1} A E_{\pi_2}. \tag{3.100}$$

The inverses of these permutations coupled with the full rank partitioning of B form a full rank partitioning of the original matrix A.

For a square matrix of rank r, this kind of partitioning implies that there is a full rank $r \times r$ principal submatrix, and the principal submatrix formed by including any of the remaining diagonal elements is singular. The principal minor formed from the full rank principal submatrix is nonzero, but if the order of the matrix is greater than r, a principal minor formed from a submatrix larger than $r \times r$ is zero.

The partitioning in equation (3.99) is of general interest, and we will use this type of partitioning often. We express an equivalent partitioning of a transformed matrix in equation (3.113) below.

The same methods as above can be used to form a full rank square submatrix of any order less than or equal to the rank. That is, if the $n \times m$ matrix A is of rank r and $q \leq r$, we can form

$$E_{\pi_r} A E_{\pi_c} = \begin{bmatrix} S_{q \times q} & T_{q \times m-q} \\ U_{n-q \times r} & V_{n-q \times m-q} \end{bmatrix}, \tag{3.101}$$

where S is of rank q.

It is obvious that the rank of a matrix can never exceed its smaller dimension (see the discussion of linear independence on page 10). Whether or not a matrix has more rows than columns, the rank of the matrix is the same as the dimension of the column space of the matrix. (As we have just seen, the dimension of the column space is necessarily the same as the dimension of the row space, but the order of the column space is different from the order of the row space unless the matrix is square.)

3.3.3 Full Rank Matrices and Matrix Inverses

We have already seen that full rank matrices have some important properties. In this section, we consider full rank matrices and matrices that are their Cayley multiplicative inverses.

Solutions of Linear Equations

Important applications of vectors and matrices involve systems of linear equations:

$$a_{11}x_1 + \cdots + a_{1m}x_m \overset{?}{=} b_1$$
$$\vdots \qquad\qquad \vdots \qquad \vdots \qquad\qquad (3.102)$$
$$a_{n1}x_1 + \cdots + a_{nm}x_m \overset{?}{=} b_n$$

or

$$Ax \overset{?}{=} b. \qquad\qquad (3.103)$$

In this system, A is called the coefficient matrix. An x that satisfies this system of equations is called a *solution* to the system. For given A and b, a solution may or may not exist. From equation (3.59), a solution exists if and only if the n-vector b is in the k-dimensional column space of A, where $k \le m$. A system for which a solution exists is said to be *consistent*; otherwise, it is *inconsistent*.

We note that if $Ax = b$, for any conformable y,

$$y^{\mathrm{T}}Ax = 0 \iff y^{\mathrm{T}}b = 0. \qquad\qquad (3.104)$$

Consistent Systems

A linear system $A_{n \times m}x = b$ is consistent if and only if

$$\mathrm{rank}([A \,|\, b]) = \mathrm{rank}(A). \qquad\qquad (3.105)$$

We can see this by recognizing that the space spanned by the columns of A is the same as that spanned by the columns of A and the vector b; therefore b must be a linear combination of the columns of A. Furthermore, the linear combination is the solution to the system $Ax = b$. (Note, of course, that it is not necessary that it be a unique linear combination.)

Equation (3.105) is equivalent to the condition

$$[A \,|\, b]y = 0 \Leftrightarrow Ay = 0. \qquad\qquad (3.106)$$

A special case that yields equation (3.105) for any b is

$$\mathrm{rank}(A_{n \times m}) = n, \qquad\qquad (3.107)$$

and so if A is of full row rank, the system is consistent regardless of the value of b. In this case, of course, the number of rows of A must be no greater than the number of columns (by inequality (3.85)). A square system in which A is nonsingular is clearly consistent.

A generalization of the linear system $Ax = b$ is $AX = B$, where B is an $n \times k$ matrix. This is the same as k systems $Ax_1 = b_1, \ldots, Ax_k = b_k$, where the x_1 and the b_i are the columns of the respective matrices. Such a system is consistent if each of the $Ax_i = b_i$ systems is consistent. Consistency of $AX = B$, as above, is the condition for a solution in X to exist.

We discuss methods for solving linear systems in Section 3.5 and in Chapter 6. In the next section, we consider a special case of $n \times n$ (square) A when equation (3.107) is satisfied (that is, when A is nonsingular).

Matrix Inverses

Let A be an $n \times n$ nonsingular matrix, and consider the linear systems

$$Ax_i = e_i,$$

where e_i is the ith unit vector. For each e_i, this is a consistent system by equation (3.105).

We can represent all n such systems as

$$A \begin{bmatrix} x_1 | \cdots | x_n \end{bmatrix} = \begin{bmatrix} e_1 | \cdots | e_n \end{bmatrix}$$

or

$$AX = I_n,$$

and this full system must have a solution; that is, there must be an X such that $AX = I_n$. Because $AX = I$, we call X a "right inverse" of A. The matrix X must be $n \times n$ and nonsingular (because I is); hence, it also has a right inverse, say Y, and $XY = I$. From $AX = I$, we have $AXY = Y$, so $A = Y$, and so finally $XA = I$; that is, the right inverse of A is also the "left inverse". We will therefore just call it the *inverse* of A and denote it as A^{-1}. This is the Cayley multiplicative inverse. Hence, for an $n \times n$ nonsingular matrix A, we have a matrix A^{-1} such that

$$A^{-1}A = AA^{-1} = I_n. \tag{3.108}$$

We have already encountered the idea of a matrix inverse in our discussions of elementary transformation matrices. The matrix that performs the inverse of the elementary operation is the inverse matrix.

From the definitions of the inverse and the transpose, we see that

$$(A^{-1})^{\text{T}} = (A^{\text{T}})^{-1}, \tag{3.109}$$

and because in applications we often encounter the inverse of a transpose of a matrix, we adopt the notation

$$A^{-\text{T}}$$

to denote the inverse of the transpose.

In the linear system (3.103), if $n = m$ and A is nonsingular, the solution is

$$x = A^{-1}b. \tag{3.110}$$

For scalars, the combined operations of inversion and multiplication are equivalent to the single operation of division. From the analogy with scalar operations, we sometimes denote AB^{-1} by A/B. Because matrix multiplication is not commutative, we often use the notation "\" to indicate the combined operations of inversion and multiplication on the left; that is, $B \backslash A$ is the same as $B^{-1}A$. The solution given in equation (3.110) is also sometimes represented as $A\backslash b$.

We discuss the solution of systems of equations in Chapter 6, but here we will point out that when we write an expression that involves computations to evaluate it, such as $A^{-1}b$ or $A\backslash b$, the form of the expression does not specify how to do the computations. This is an instance of a principle that we will encounter repeatedly: *the form of a mathematical expression and the way the expression should be evaluated in actual practice may be quite different.*

Nonsquare Full Rank Matrices; Right and Left Inverses

Suppose A is $n \times m$ and $\text{rank}(A) = n$; that is, $n \leq m$ and A is of full row rank. Then $\text{rank}([A \,|\, e_i]) = \text{rank}(A)$, where e_i is the ith unit vector of length n; hence the system

$$Ax_i = e_i$$

is consistent for each e_i, and ,as before, we can represent all n such systems as

$$A\left[x_1|\cdots|x_n\right] = \left[e_1|\cdots|e_n\right]$$

or

$$AX = I_n.$$

As above, there must be an X such that $AX = I_n$, and we call X a *right inverse* of A. The matrix X must be $m \times n$ and it must be of rank n (because I is). This matrix is not necessarily the inverse of A, however, because A and X may not be square. We denote the right inverse of A as

$$A^{-\text{R}}.$$

Furthermore, we could only have solved the system AX if A was of full row rank because $n \leq m$ and $n = \text{rank}(I) = \text{rank}(AX) \leq \text{rank}(A)$. To summarize, A has a right inverse if and only if A is of full row rank.

Now, suppose A is $n \times m$ and $\text{rank}(A) = m$; that is, $m \leq n$ and A is of full column row rank. Writing $YA = I_m$ and reversing the roles of the coefficient matrix and the solution matrix in the argument above, we have that Y exists and is a *left inverse* of A. We denote the left inverse of A as

$$A^{-\text{L}}.$$

Also, using a similar argument as above, we see that the matrix A has a left inverse if and only if A is of full column rank.

We also note that if AA^{T} is of full rank, the right inverse of A is $A^{\text{T}}(AA^{\text{T}})^{-1}$. Likewise, if $A^{\text{T}}A$ is of full rank, the left inverse of A is $(A^{\text{T}}A)^{-1}A^{\text{T}}$.

3.3.4 Full Rank Factorization

The partitioning of an $n \times m$ matrix as in equation (3.99) on page 80 leads to an interesting factorization of a matrix. Recall that we had an $n \times m$ matrix B partitioned as

$$B = \begin{bmatrix} W_{r \times r} & X_{r \times m-r} \\ Y_{n-r \times r} & Z_{n-r \times m-r} \end{bmatrix},$$

where r is the rank of B, W is of full rank, the rows of $R = [W|X]$ span the full row space of B, and the columns of $C = \begin{bmatrix} W \\ Y \end{bmatrix}$ span the full column space of B.

Therefore, for some T, we have $[Y|Z] = TR$, and for some S, we have $\begin{bmatrix} X \\ Z \end{bmatrix} = CS$. From this, we have $Y = TW$, $Z = TX$, $X = WS$, and $Z = YS$, so $Z = TWS$. Since W is nonsingular, we have $T = YW^{-1}$ and $S = W^{-1}X$, so $Z = YW^{-1}X$.

We can therefore write the partitions as

$$B = \begin{bmatrix} W & X \\ Y & YW^{-1}X \end{bmatrix}$$

$$= \begin{bmatrix} I \\ YW^{-1} \end{bmatrix} W \left[I \mid W^{-1}X \right]. \tag{3.111}$$

From this, we can form two equivalent factorizations of B:

$$B = \begin{bmatrix} W \\ Y \end{bmatrix} [I \mid W^{-1}X] = \begin{bmatrix} I \\ YW^{-1} \end{bmatrix} [W \mid X].$$

The matrix B has a very special property: the full set of linearly independent rows are the first r rows, and the full set of linearly independent columns are the first r columns. We have seen, however, that any matrix A of rank r can be put in this form, and $A = E_{\pi_2} B E_{\pi_1}$ for an $n \times n$ permutation matrix E_{π_2} and an $m \times m$ permutation matrix E_{π_1}.

We therefore have, for the $n \times m$ matrix A with rank r, two equivalent factorizations,

$$A = \begin{bmatrix} QW \\ QY \end{bmatrix} [P \mid W^{-1}XP]$$

$$= \begin{bmatrix} Q \\ QYW^{-1} \end{bmatrix} [WP \mid XP],$$

both of which are in the general form

$$A_{n \times m} = L_{n \times r}\, R_{r \times m}, \tag{3.112}$$

where L is of full column rank and R is of row column rank. This is called a *full rank factorization* of the matrix A. We will use a full rank factorization in proving various properties of matrices. We will consider other factorizations in Chapter 5 that have more practical uses in computations.

3.3.5 Equivalent Matrices

Matrices of the same order that have the same rank are said to be *equivalent matrices*.

Equivalent Canonical Forms

For any $n \times m$ matrix A with $\text{rank}(A) = r > 0$, by combining the permutations that yield equation (3.99) with other operations, we have, for some matrices P and Q that are products of various elementary operator matrices,

$$PAQ = \begin{bmatrix} I_r & 0 \\ 0 & 0 \end{bmatrix}. \tag{3.113}$$

This is called an *equivalent canonical form* of A, and it exists for any matrix A that has at least one nonzero element (which is the same as requiring $\text{rank}(A) > 0$).

We can see by construction that an equivalent canonical form exists for any $n \times m$ matrix A that has a nonzero element. First, assume $a_{ij} \neq 0$. By two successive permutations, we move a_{ij} to the $(1, 1)$ position; specifically, $(E_{i1}AE_{1j})_{11} = a_{ij}$. We then divide the first row by a_{ij}; that is, we form $E_1(1/a_{ij})E_{i1}AE_{1j}$. We then proceed with a sequence of $n - 1$ premultiplications by axpy matrices to zero out the first column of the matrix, as in expression (3.50), followed by a sequence of $(m - 1)$ postmultiplications by axpy matrices to zero out the first row. We then have a matrix of the form

$$\begin{bmatrix} 1 & 0 & \cdots & 0 \\ 0 & & & \\ \vdots & & [\ X \ ] & \\ 0 & & & \end{bmatrix}. \tag{3.114}$$

If $X = 0$, we are finished; otherwise, we perform the same kinds of operations on the $(n - 1) \times (m - 1)$ matrix X and continue until we have the form of equation (3.113).

The matrices P and Q in equation (3.113) are not unique. The order in which they are built from elementary operator matrices can be very important in preserving the accuracy of the computations.

Although the matrices P and Q in equation (3.113) are not unique, the equivalent canonical form itself (the right-hand side) is obviously unique because the only thing that determines it, aside from the shape, is the r in I_r, and that is just the rank of the matrix. There are two other, more general, equivalent forms that are often of interest. These equivalent forms, row echelon form and Hermite form, are not unique. A matrix R is said to be in *row echelon form*, or just *echelon form*, if

- $r_{ij} = 0$ for $i > j$, and

-     if k is such that $r_{ik} \neq 0$ and $r_{il} = 0$ for $l < k$, then $r_{i+1,j} = 0$ for $j \leq k$.

A matrix in echelon form is upper triangular. An upper triangular matrix H is said to be in *Hermite form* if

-     $h_{ii} = 0$ or 1,
-     if $h_{ii} = 0$, then $h_{ij} = 0$ for all j, and
-     if $h_{ii} = 1$, then $h_{ki} = 0$ for all $k \neq i$.

If H is in Hermite form, then $H^2 = H$, as is easily verified. (A matrix H such that $H^2 = H$ is said to be *idempotent*. We discuss idempotent matrices beginning on page 280.) Another, more specific, equivalent form, called the *Jordan form*, is a special row echelon form based on eigenvalues.

Any of these equivalent forms is useful in determining the rank of a matrix. Each form may have special uses in proving properties of matrices. We will often make use of the equivalent canonical form in other sections of this chapter.

Products with a Nonsingular Matrix

It is easy to see that if A is a square full rank matrix (that is, A is nonsingular), and if B and C are conformable matrices for the multiplications AB and CA, respectively, then

$$\text{rank}(AB) = \text{rank}(B) \tag{3.115}$$

and

$$\text{rank}(CA) = \text{rank}(C). \tag{3.116}$$

This is true because, for a given conformable matrix B, by the inequality (3.96), we have $\text{rank}(AB) \leq \text{rank}(B)$. Forming $B = A^{-1}AB$, and again applying the inequality, we have $\text{rank}(B) \leq \text{rank}(AB)$; hence, $\text{rank}(AB) = \text{rank}(B)$. Likewise, for a square full rank matrix A, we have $\text{rank}(CA) = \text{rank}(C)$. (Here, we should recall that all matrices are real.)

On page 88, we give a more general result for products with general full rank matrices.

A Factorization Based on an Equivalent Canonical Form

Elementary operator matrices and products of them are of full rank and thus have inverses. When we introduced the matrix operations that led to the definitions of the elementary operator matrices in Section 3.2.3, we mentioned the inverse operations, which would then define the inverses of the matrices.

The matrices P and Q in the equivalent canonical form of the matrix A, PAQ in equation (3.113), have inverses. From an equivalent canonical form of a matrix A with rank r, we therefore have the equivalent canonical factorization of A:

$$A = P^{-1} \begin{bmatrix} I_r & 0 \\ 0 & 0 \end{bmatrix} Q^{-1}. \tag{3.117}$$

A factorization based on an equivalent canonical form is also a full rank factorization and could be written in the same form as equation (3.112).

Equivalent Forms of Symmetric Matrices

If A is symmetric, the equivalent form in equation (3.113) can be written as $PAP^{\mathrm{T}} = \mathrm{diag}(I_r, 0)$ and the equivalent canonical factorization of A in equation (3.117) can be written as

$$A = P^{-1} \begin{bmatrix} I_r & 0 \\ 0 & 0 \end{bmatrix} P^{-\mathrm{T}}. \tag{3.118}$$

These facts follow from the same process that yielded equation (3.113) for a general matrix.

Also a full rank factorization for a symmetric matrix, as in equation (3.112), can be given as

$$A = LL^{\mathrm{T}}. \tag{3.119}$$

3.3.6 Multiplication by Full Rank Matrices

We have seen that a matrix has an inverse if it is square and of full rank. Conversely, it has an inverse only if it is square and of full rank. We see that a matrix that has an inverse must be square because $A^{-1}A = AA^{-1}$, and we see that it must be full rank by the inequality (3.96). In this section, we consider other properties of full rank matrices. In some cases, we require the matrices to be square, but in other cases, these properties hold whether or not they are square.

Using matrix inverses allows us to establish important properties of products of matrices in which at least one factor is a full rank matrix.

Products with a General Full Rank Matrix

If A is a full column rank matrix and if B is a matrix conformable for the multiplication AB, then

$$\mathrm{rank}(AB) = \mathrm{rank}(B). \tag{3.120}$$

If A is a full row rank matrix and if C is a matrix conformable for the multiplication CA, then

$$\mathrm{rank}(CA) = \mathrm{rank}(C). \tag{3.121}$$

Consider a full rank $n \times m$ matrix A with $\mathrm{rank}(A) = m$ (that is, $m \leq n$) and let B be conformable for the multiplication AB. Because A is of full column rank, it has a left inverse (see page 84); call it $A^{-\mathrm{L}}$, and so $A^{-\mathrm{L}}A = I_m$. From inequality (3.96), we have $\mathrm{rank}(AB) \leq \mathrm{rank}(B)$, and applying the inequality again, we have $\text{rank}(B) = \text{rank}(A^{-\text{L}}AB) \leq \text{rank}(AB)$; hence $\text{rank}(AB) = \text{rank}(B)$.

Now consider a full rank $n \times m$ matrix A with $\text{rank}(A) = n$ (that is, $n \leq m$) and let C be conformable for the multiplication CA. Because A is of full row rank, it has a right inverse; call it $A^{-\text{R}}$, and so $AA^{-\text{R}} = I_n$. From inequality (3.96), we have $\text{rank}(CA) \leq \text{rank}(C)$, and applying the inequality again, we have $\text{rank}(C) = \text{rank}(CAA^{-\text{L}}) \leq \text{rank}(CA)$; hence $\text{rank}(CA) = \text{rank}(C)$.

To state this more simply:

- Premultiplication of a given matrix by a full column rank matrix does not change the rank of the given matrix, and postmultiplication of a given matrix by a full row rank matrix does not change the rank of the given matrix.

From this we see that $A^{\text{T}}A$ is of full rank if (and only if) A is of full column rank, and AA^{T} is of full rank if (and only if) A is of full row rank. We will develop a stronger form of these statements in Section 3.3.7.

Preservation of Positive Definiteness

A certain type of product of a full rank matrix and a positive definite matrix preserves not only the rank, but also the positive definiteness: if C is $n \times n$ and positive definite, and A is $n \times m$ and of rank m (hence, $m \leq n$), then $A^{\text{T}}CA$ is positive definite. (Recall from inequality (3.62) that a matrix C is positive definite if it is symmetric and for any $x \neq 0$, $x^{\text{T}}Cx > 0$.)

To see this, assume matrices C and A as described. Let x be any m-vector such that $x \neq 0$, and let $y = Ax$. Because A is of full column rank, $y \neq 0$. We have

$$\begin{aligned}
x^{\text{T}}(A^{\text{T}}CA)x &= (xA)^{\text{T}}C(Ax) \\
&= y^{\text{T}}Cy \\
&> 0.
\end{aligned} \tag{3.122}$$

Therefore, since $A^{\text{T}}CA$ is symmetric,

- if C is positive definite and A is of full column rank, then $A^{\text{T}}CA$ is positive definite.

Furthermore, we have the converse:

- if $A^{\text{T}}CA$ is positive definite, then A is of full column rank,

for otherwise there exists an $x \neq 0$ such that $Ax = 0$, and so $x^{\text{T}}(A^{\text{T}}CA)x = 0$.

The General Linear Group

Consider the set of all square $n \times n$ full rank matrices together with the usual (Cayley) multiplication. As we have seen, this set is closed under multiplication. (The product of two square matrices of full rank is of full rank, and of course the product is also square.) Furthermore, the (multiplicative) identity is a member of this set, and each matrix in the set has a (multiplicative) inverse in the set; therefore, the set together with the usual multiplication is a mathematical structure called a *group*. (See any text on modern algebra.) This group is called the *general linear group* and is denoted by $\mathcal{G}L(n)$. General group-theoretic properties can be used in the derivation of properties of these full-rank matrices. Note that this group is not commutative.

As we mentioned earlier (before we had considered inverses in general), if A is an $n \times n$ matrix and if A^{-1} exists, we define A^0 to be I_n.

The $n \times n$ elementary operator matrices are members of the general linear group $\mathcal{G}L(n)$.

The elements in the general linear group are matrices and, hence, can be viewed as transformations or operators on n-vectors. Another set of linear operators on n-vectors are the doubletons (A, v), where A is an $n \times n$ full-rank matrix and v is an n-vector. As an operator on $x \in \mathbb{R}^n$, (A, v) is the transformation $Ax + v$, which preserves affine spaces. Two such operators, (A, v) and (B, w), are combined by composition: $(A, v)((B, w)(x)) = ABx + Aw + v$. The set of such doubletons together with composition forms a group, called the affine group. It is denoted by $\mathcal{A}L(n)$.

3.3.7 Products of the Form $A^{\mathrm{T}}A$

Given a real matrix A, an important matrix product is $A^{\mathrm{T}}A$. (This is called a Gramian matrix. We will discuss this kind of matrix in more detail beginning on page 287.)

Matrices of this form have several interesting properties. First, for any $n \times m$ matrix A, we have the fact that $A^{\mathrm{T}}A = 0$ if and only if $A = 0$. We see this by noting that if $A = 0$, then $\mathrm{tr}(A^{\mathrm{T}}A) = 0$. Conversely, if $\mathrm{tr}(A^{\mathrm{T}}A) = 0$, then $a_{ij}^2 = 0$ for all i, j, and so $a_{ij} = 0$, that is, $A = 0$. Summarizing, we have

$$\mathrm{tr}(A^{\mathrm{T}}A) = 0 \;\Leftrightarrow\; A = 0 \tag{3.123}$$

and

$$A^{\mathrm{T}}A = 0 \;\Leftrightarrow\; A = 0. \tag{3.124}$$

Another useful fact about $A^{\mathrm{T}}A$ is that it is nonnegative definite. This is because for any y, $y^{\mathrm{T}}(A^{\mathrm{T}}A)y = (yA)^{\mathrm{T}}(Ay) \geq 0$. In addition, we see that $A^{\mathrm{T}}A$ is positive definite if and only if A is of full column rank. This follows from (3.124), and if A is of full column rank, $Ay = 0 \Rightarrow y = 0$.

Now consider a generalization of the equation $A^{\mathrm{T}}A = 0$:

$$A^{\mathrm{T}}A(B - C) = 0.$$

Multiplying by $B^{\mathrm{T}} - C^{\mathrm{T}}$ and factoring $(B^{\mathrm{T}} - C^{\mathrm{T}})A^{\mathrm{T}}A(B - C)$, we have

$$(AB - AC)^{\mathrm{T}}(AB - AC) = 0;$$

hence, from (3.124), we have $AB - AC = 0$. Furthermore, if $AB - AC = 0$, then clearly $A^{\mathrm{T}}A(B - C) = 0$. We therefore conclude that

$$A^{\mathrm{T}}AB = A^{\mathrm{T}}AC \ \Leftrightarrow\ AB = AC. \tag{3.125}$$

By the same argument, we have

$$BA^{\mathrm{T}}A = CA^{\mathrm{T}}A \ \Leftrightarrow\ BA^{\mathrm{T}} = CA^{\mathrm{T}}.$$

Now, let us consider $\operatorname{rank}(A^{\mathrm{T}}A)$. We have seen that $(A^{\mathrm{T}}A)$ is of full rank if and only if A is of full column rank. Next, preparatory to our main objective, we note from above that

$$\operatorname{rank}(A^{\mathrm{T}}A) = \operatorname{rank}(AA^{\mathrm{T}}). \tag{3.126}$$

Let A be an $n \times m$ matrix, and let $r = \operatorname{rank}(A)$. If $r = 0$, then $A = 0$ (hence, $A^{\mathrm{T}}A = 0$) and $\operatorname{rank}(A^{\mathrm{T}}A) = 0$. If $r > 0$, interchange columns of A if necessary to obtain a partitioning similar to equation (3.99),

$$A = [A_1 A_2],$$

where A_1 is an $n \times r$ matrix of rank r. (Here, we are ignoring the fact that the columns might have been permuted. All properties of the rank are unaffected by these interchanges.) Now, because A_1 is of full column rank, there is an $r \times m - r$ matrix B such that $A_2 = A_1 B$; hence we have $A = A_1[I_r B]$ and

$$A^{\mathrm{T}}A = \begin{bmatrix} I_r \\ B^{\mathrm{T}} \end{bmatrix} A_1^{\mathrm{T}}A_1[I_r B].$$

Because A_1 is of full rank, $\operatorname{rank}(A_1^{\mathrm{T}}A_1) = r$. Now let

$$T = \begin{bmatrix} I_r & 0 \\ -B^{\mathrm{T}} & I_{m-r} \end{bmatrix}.$$

It is clear that T is of full rank, and so

$$\operatorname{rank}(A^{\mathrm{T}}A) = \operatorname{rank}(TA^{\mathrm{T}}AT^{\mathrm{T}})$$
$$= \operatorname{rank}\left(\begin{bmatrix} A_1^{\mathrm{T}}A_1 & 0 \\ 0 & 0 \end{bmatrix}\right)$$
$$= \operatorname{rank}(A_1^{\mathrm{T}}A_1)$$
$$= r;$$

that is,

$$\text{rank}(A^{\text{T}}A) = \text{rank}(A). \tag{3.127}$$

From this equation, we have a useful fact for Gramian matrices. The system

$$A^{\text{T}}Ax = A^{\text{T}}b \tag{3.128}$$

is consistent for any A and b.

3.3.8 A Lower Bound on the Rank of a Matrix Product

Equation (3.96) gives an upper bound on the rank of the product of two matrices; the rank cannot be greater than the rank of either of the factors. Now, using equation (3.117), we develop a lower bound on the rank of the product of two matrices if one of them is square.

If A is $n \times n$ (that is, square) and B is a matrix with n rows, then

$$\text{rank}(AB) \geq \text{rank}(A) + \text{rank}(B) - n. \tag{3.129}$$

We see this by first letting $r = \text{rank}(A)$, letting P and Q be matrices that form an equivalent canonical form of A (see equation (3.117)), and then forming

$$C = P^{-1} \begin{bmatrix} 0 & 0 \\ 0 & I_{n-r} \end{bmatrix} Q^{-1},$$

so that $A + C = P^{-1}Q^{-1}$. Because P^{-1} and Q^{-1} are of full rank, $\text{rank}(C) = \text{rank}(I_{n-r}) = n - \text{rank}(A)$. We now develop an upper bound on $\text{rank}(B)$,

$$\begin{aligned}
\text{rank}(B) &= \text{rank}(P^{-1}Q^{-1}B) \\
&= \text{rank}(AB + CB) \\
&\leq \text{rank}(AB) + \text{rank}(CB), \text{ by equation (3.97)} \\
&\leq \text{rank}(AB) + \text{rank}(C), \text{ by equation (3.96)} \\
&= \text{rank}(AB) + n - \text{rank}(A),
\end{aligned}$$

yielding (3.129), a lower bound on $\text{rank}(AB)$.

The inequality (3.129) is called *Sylvester's law of nullity*. It provides a lower bound on $\text{rank}(AB)$ to go with the upper bound of inequality (3.96), $\min(\text{rank}(A), \text{rank}(B))$.

3.3.9 Determinants of Inverses

From the relationship $|AB| = |A|\,|B|$ for square matrices mentioned earlier, it is easy to see that for nonsingular square A,

$$|A| = 1/|A^{-1}|, \tag{3.130}$$

and so

- $|A| = 0$ if and only if A is singular.

(From the definition of the determinant in equation (3.16), we see that the determinant of any finite-dimensional matrix with finite elements is finite, and we implicitly assume that the elements are finite.)

For a matrix whose determinant is nonzero, from equation (3.25) we have

$$A^{-1} = \frac{1}{|A|}\mathrm{adj}(A). \tag{3.131}$$

3.3.10 Inverses of Products and Sums of Matrices

The inverse of the Cayley product of two nonsingular matrices of the same size is particularly easy to form. If A and B are square full rank matrices of the same size,

$$(AB)^{-1} = B^{-1}A^{-1}. \tag{3.132}$$

We can see this by multiplying $B^{-1}A^{-1}$ and (AB).

Often in linear regression analysis we need inverses of various sums of matrices. This may be because we wish to update regression estimates based on additional data or because we wish to delete some observations. If A and B are full rank matrices of the same size, the following relationships are easy to show (and are easily proven if taken in the order given; see Exercise 3.12):

$$A(I + A)^{-1} = (I + A^{-1})^{-1}, \tag{3.133}$$

$$(A + BB^{\mathrm{T}})^{-1}B = A^{-1}B(I + B^{\mathrm{T}}A^{-1}B)^{-1}, \tag{3.134}$$

$$(A^{-1} + B^{-1})^{-1} = A(A + B)^{-1}B, \tag{3.135}$$

$$A - A(A + B)^{-1}A = B - B(A + B)^{-1}B, \tag{3.136}$$

$$A^{-1} + B^{-1} = A^{-1}(A + B)B^{-1}, \tag{3.137}$$

$$(I + AB)^{-1} = I - A(I + BA)^{-1}B, \tag{3.138}$$

$$(I + AB)^{-1}A = A(I + BA)^{-1}. \tag{3.139}$$

When A and/or B are not of full rank, the inverses may not exist, but in that case these equations hold for a generalized inverse, which we will discuss in Section 3.6.

There is also an analogue to the expansion of the inverse of $(1 - a)$ for a scalar a:

$$(1 - a)^{-1} = 1 + a + a^2 + a^3 + \cdots, \quad \text{if } |a| < 1.$$

This comes from a factorization of the binomial $1 - a^k$, similar to equation (3.41), and the fact that $a^k \rightarrow 0$ if $|a| < 1$. In Section 3.9 on page 128, we will discuss conditions that ensure the convergence of A^k for a square matrix A. We will define a norm $\|A\|$ on A and show that if $\|A\| < 1$, then $A^k \rightarrow 0$. Then, analogous to the scalar series, using equation (3.41) for a square matrix A, we have

$$(I - A)^{-1} = I + A + A^2 + A^3 + \cdots, \quad \text{if } \|A\| < 1. \tag{3.140}$$

We include this equation here because of its relation to equations (3.133) through (3.139). We will discuss it further on page 134, after we have introduced and discussed $\|A\|$ and other conditions that ensure convergence. This expression and the condition that determines it are very important in the analysis of time series and other stochastic processes.

Also, looking ahead, we have another expression similar to equations (3.133) through (3.139) and (3.140) for a special type of matrix. If $A^2 = A$, for any $a \neq -1$,

$$(I + aA)^{-1} = I - \frac{a}{a+1}A$$

(see page 282).

3.3.11 Inverses of Matrices with Special Forms

Matrices with various special patterns may have relatively simple inverses. For example, the inverse of a diagonal matrix with nonzero entries is a diagonal matrix consisting of the reciprocals of those elements. Likewise, a block diagonal matrix consisting of full-rank submatrices along the diagonal has an inverse that is merely the block diagonal matrix consisting of the inverses of the submatrices. We discuss inverses of various special matrices in Chapter 8.

Inverses of Kronecker Products of Matrices

If A and B are square full rank matrices, then

$$(A \otimes B)^{-1} = A^{-1} \otimes B^{-1}. \tag{3.141}$$

We can see this by multiplying $A^{-1} \otimes B^{-1}$ and $A \otimes B$.

3.3.12 Determining the Rank of a Matrix

Although the equivalent canonical form (3.113) immediately gives the rank of a matrix, in practice the numerical determination of the rank of a matrix is not an easy task. The problem is that rank is a mapping $\mathbb{R}^{n \times m} \mapsto \mathbb{Z}_+$, where $\mathbb{Z}_+$ represents the positive integers. Such a function is often difficult to compute because the domain is relatively dense and the range is sparse.

Small changes in the domain may result in large discontinuous changes in the function value.

It is not even always clear whether a matrix is nonsingular. Because of rounding on the computer, a matrix that is mathematically nonsingular may appear to be singular. We sometimes use the phrase "nearly singular" or "algorithmically singular" to describe such a matrix. In Sections 6.1 and 11.4, we consider this kind of problem in more detail.

3.4 More on Partitioned Square Matrices: The Schur Complement

A square matrix A that can be partitioned as

$$A = \begin{bmatrix} A_{11} & A_{12} \\ A_{21} & A_{22} \end{bmatrix}, \tag{3.142}$$

where A_{11} is nonsingular, has interesting properties that depend on the matrix

$$Z = A_{22} - A_{21} A_{11}^{-1} A_{12}, \tag{3.143}$$

which is called the *Schur complement* of A_{11} in A.

We first observe from equation (3.111) that if equation (3.142) represents a full rank partitioning (that is, if the rank of A_{11} is the same as the rank of A), then

$$A = \begin{bmatrix} A_{11} & A_{12} \\ A_{21} & A_{21} A_{11}^{-1} A_{12} \end{bmatrix}, \tag{3.144}$$

and $Z = 0$.

There are other useful properties, which we mention below. There are also some interesting properties of certain important random matrices partitioned in this way. For example, suppose A_{22} is $k \times k$ and A is an $m \times m$ Wishart matrix with parameters n and Σ partitioned like A in equation (3.142). (This of course means A is symmetrical, and so $A_{12} = A_{21}^{\mathrm{T}}$.) Then Z has a Wishart distribution with parameters $n - m + k$ and $\Sigma_{22} - \Sigma_{21} \Sigma_{11}^{-1} \Sigma_{12}$, and is independent of A_{21} and A_{11}. (See Exercise 4.8 on page 171 for the probability density function for a Wishart distribution.)

3.4.1 Inverses of Partitioned Matrices

Suppose A is nonsingular and can be partitioned as above with both A_{11} and A_{22} nonsingular. It is easy to see (Exercise 3.13, page 141) that the inverse of A is given by

$$A^{-1} = \begin{bmatrix} A_{11}^{-1} + A_{11}^{-1} A_{12} Z^{-1} A_{21} A_{11}^{-1} & -A_{11}^{-1} A_{12} Z^{-1} \\ -Z^{-1} A_{21} A_{11}^{-1} & Z^{-1} \end{bmatrix}, \tag{3.145}$$

where Z is the Schur complement of A_{11}.

If

$$A = [X\,y]^{\mathrm{T}}\,[X\,y]$$

and is partitioned as in equation (3.43) on page 61 and X is of full column rank, then the Schur complement of $X^{\mathrm{T}}X$ in $[X\,y]^{\mathrm{T}}\,[X\,y]$ is

$$y^{\mathrm{T}}y - y^{\mathrm{T}}X(X^{\mathrm{T}}X)^{-1}X^{\mathrm{T}}y. \tag{3.146}$$

This particular partitioning is useful in linear regression analysis, where this Schur complement is the residual sum of squares and the more general Wishart distribution mentioned above reduces to a chi-squared one. (Although the expression is useful, this is an instance of a principle that we will encounter repeatedly: *the form of a mathematical expression and the way the expression should be evaluated in actual practice may be quite different.*)

3.4.2 Determinants of Partitioned Matrices

If the square matrix A is partitioned as

$$A = \begin{bmatrix} A_{11} & A_{12} \\ A_{21} & A_{22} \end{bmatrix},$$

and A_{11} is square and nonsingular, then

$$|A| = |A_{11}|\,\left|A_{22} - A_{21}A_{11}^{-1}A_{12}\right|; \tag{3.147}$$

that is, the determinant is the product of the determinant of the principal submatrix and the determinant of its Schur complement.

This result is obtained by using equation (3.29) on page 54 and the factorization

$$\begin{bmatrix} A_{11} & A_{12} \\ A_{21} & A_{22} \end{bmatrix} = \begin{bmatrix} A_{11} & 0 \\ A_{21} & A_{22} - A_{21}A_{11}^{-1}A_{12} \end{bmatrix} \begin{bmatrix} I & A_{11}^{-1}A_{12} \\ 0 & I \end{bmatrix}. \tag{3.148}$$

The factorization in equation (3.148) is often useful in other contexts as well.

3.5 Linear Systems of Equations

Some of the most important applications of matrices are in representing and solving systems of n linear equations in m unknowns,

$$Ax = b,$$

where A is an $n \times m$ matrix, x is an m-vector, and b is an n-vector. As we observed in equation (3.59), the product Ax in the linear system is a linear combination of the columns of A; that is, if a_j is the jth column of A, $Ax = \sum_{j=1}^{m} x_j a_j$.

If $b = 0$, the system is said to be *homogeneous*. In this case, unless $x = 0$, the columns of A must be linearly dependent.

3.5.1 Solutions of Linear Systems

When in the linear system $Ax = b$, A is square and nonsingular, the solution is obviously $x = A^{-1}b$. We will not discuss this simple but common case further here. Rather, we will discuss it in detail in Chapter 6 after we have discussed matrix factorizations later in this chapter and in Chapter 5.

When A is not square or is singular, the system may not have a solution or may have more than one solution. A consistent system (see equation (3.105)) has a solution. For consistent systems that are singular or not square, the *generalized inverse* is an important concept. We introduce it in this section but defer its discussion to Section 3.6.

Underdetermined Systems

A consistent system in which $\mathrm{rank}(A) < m$ is said to be *underdetermined*. An underdetermined system may have fewer equations than variables, or the coefficient matrix may just not be of full rank. For such a system there is more than one solution. In fact, there are infinitely many solutions because if the vectors x_1 and x_2 are solutions, the vector $wx_1 + (1 - w)x_2$ is likewise a solution for any scalar w.

Underdetermined systems arise in analysis of variance in statistics, and it is useful to have a compact method of representing the solution to the system. It is also desirable to identify a unique solution that has some kind of optimal properties. Below, we will discuss types of solutions and the number of linearly independent solutions and then describe a unique solution of a particular type.

Overdetermined Systems

Often in mathematical modeling applications, the number of equations in the system $Ax = b$ is not equal to the number of variables; that is the coefficient matrix A is $n \times m$ and $n \neq m$. If $n > m$ and $\mathrm{rank}([A \mid b]) > \mathrm{rank}(A)$, the system is said to be *overdetermined*. There is no x that satisfies such a system, but approximate solutions are useful. We discuss approximate solutions of such systems in Section 6.7 on page 222 and in Section 9.2.2 on page 330.

Generalized Inverses

A matrix G such that $AGA = A$ is called a *generalized inverse* and is denoted by A^-:

$$AA^- A = A. \tag{3.149}$$

Note that if A is $n \times m$, then A^- is $m \times n$. If A is nonsingular (square and of full rank), then obviously $A^- = A^{-1}$.

Without additional restrictions on A, the generalized inverse is not unique. Various types of generalized inverses can be defined by adding restrictions to the definition of the inverse. In Section 3.6, we will discuss various types of generalized inverses and show that A^- exists for any $n \times m$ matrix A. Here we will consider some properties of any generalized inverse.

From equation (3.149), we see that

$$A^{\mathrm{T}}(A^-)^{\mathrm{T}}A^{\mathrm{T}} = A^{\mathrm{T}};$$

thus, if A^- is a generalized inverse of A, then $(A^-)^{\mathrm{T}}$ is a generalized inverse of A^{T}.

The $m \times m$ square matrices A^-A and $(I - A^-A)$ are often of interest. By using the definition (3.149), we see that

$$(A^-A)(A^-A) = A^-A. \tag{3.150}$$

(Such a matrix is said to be *idempotent*. We discuss idempotent matrices beginning on page 280.) From equation (3.96) together with the fact that $AA^-A = A$, we see that

$$\mathrm{rank}(A^-A) = \mathrm{rank}(A). \tag{3.151}$$

By multiplication as above, we see that

$$A(I - A^-A) = 0, \tag{3.152}$$

that

$$(I - A^-A)(A^-A) = 0, \tag{3.153}$$

and that $(I - A^-A)$ is also idempotent:

$$(I - A^-A)(I - A^-A) = (I - A^-A). \tag{3.154}$$

The fact that $(A^-A)(A^-A) = A^-A$ yields the useful fact that

$$\mathrm{rank}(I - A^-A) = m - \mathrm{rank}(A). \tag{3.155}$$

This follows from equations (3.153), (3.129), and (3.151), which yield $0 \geq \mathrm{rank}(I - A^-A) + \mathrm{rank}(A) - m$, and from equation (3.97), which gives $m = \mathrm{rank}(I) \leq \mathrm{rank}(I - A^-A) + \mathrm{rank}(A)$. The two inequalities result in the equality of equation (3.155).

Multiple Solutions in Consistent Systems

Suppose the system $Ax = b$ is consistent and A^- is a generalized inverse of A; that is, it is any matrix such that $AA^-A = A$. Then

$$x = A^-b \tag{3.156}$$

is a solution to the system because if $AA^-A = A$, then $AA^-Ax = Ax$ and since $Ax = b$,

$$AA^-b = b; \tag{3.157}$$

that is, A^-b is a solution. Furthermore, if $x = Gb$ is any solution, then $AGA = A$; that is, G is a generalized inverse of A. This can be seen by the following argument. Let a_j be the jth column of A. The m systems of n equations, $Ax = a_j$, $j = 1, \ldots, m$, all have solutions. (Each solution is a vector with 0s in all positions except the jth position, which is a 1.) Now, if Gb is a solution to the original system, then Ga_j is a solution to the system $Ax = a_j$. So $AGa_j = a_j$ for all j; hence $AGA = A$.

If $Ax = b$ is consistent, not only is A^-b a solution but also, for any z,

$$A^-b + (I - A^-A)z \tag{3.158}$$

is a solution because $A(A^-b + (I - A^-A)z) = AA^-b + (A - AA^-A)z = b$. Furthermore, any solution to $Ax = b$ can be represented as $A^-b + (I - A^-A)z$ for some z. This is because if y is any solution (that is, if $Ay = b$), we have

$$y = A^-b - A^-Ay + y = A^-b - (A^-A - I)y = A^-b + (I - A^-A)z.$$

The number of linearly independent solutions arising from $(I - A^-A)z$ is just the rank of $(I - A^-A)$, which from equation (3.155) is $\text{rank}(I - A^-A) = m - \text{rank}(A)$.

3.5.2 Null Space: The Orthogonal Complement

The solutions of a consistent system $Ax = b$, which we characterized in equation (3.158) as $A^-b + (I - A^-A)z$ for any z, are formed as a given solution to $Ax = b$ plus all solutions to $Az = 0$.

For an $n \times m$ matrix A, the set of vectors generated by all solutions, z, of the homogeneous system

$$Az = 0 \tag{3.159}$$

is called the *null space* of A. We denote the null space of A by

$$\mathcal{N}(A).$$

The null space is either the single 0 vector (in which case we say the null space is empty or null) or it is a vector space.

We see that the null space of A is a vector space if it is not empty because the zero vector is in $\mathcal{N}(A)$, and if x and y are in $\mathcal{N}(A)$ and a is any scalar, $ax + y$ is also a solution of $Az = 0$. We call the dimension of $\mathcal{N}(A)$ the *nullity* of A. The nullity of A is

$$\dim(\mathcal{N}(A)) = \text{rank}(I - A^-A)$$
$$= m - \text{rank}(A) \tag{3.160}$$

from equation (3.155).

The order of $\mathcal{N}(A)$ is m. (Recall that the order of $\mathcal{V}(A)$ is n. The order of $\mathcal{V}(A^{\mathrm{T}})$ is m.)

If A is square, we have

$$\mathcal{N}(A) \subset \mathcal{N}(A^2) \subset \mathcal{N}(A^3) \subset \cdots \tag{3.161}$$

and

$$\mathcal{V}(A) \supset \mathcal{V}(A^2) \supset \mathcal{V}(A^3) \supset \cdots . \tag{3.162}$$

(We see this easily from the inequality (3.96).)

If $Ax = b$ is consistent, any solution can be represented as $A^-b + z$, for some z in the null space of A, because if y is some solution, $Ay = b = AA^-b$ from equation (3.157), and so $A(y - A^-b) = 0$; that is, $z = y - A^-b$ is in the null space of A. If A is nonsingular, then there is no such z, and the solution is unique. The number of linearly independent solutions to $Az = 0$, is the same as the nullity of A.

If a is in $\mathcal{V}(A^{\mathrm{T}})$ and b is in $\mathcal{N}(A)$, we have $b^{\mathrm{T}}a = b^{\mathrm{T}}A^{\mathrm{T}}x = 0$. In other words, the null space of A is orthogonal to the row space of A; that is, $\mathcal{N}(A) \perp \mathcal{V}(A^{\mathrm{T}})$. This is because $A^{\mathrm{T}}x = a$ for some x, and $Ab = 0$ or $b^{\mathrm{T}}A^{\mathrm{T}} = 0$. For any matrix B whose columns are in $\mathcal{N}(A)$, $A^{\mathrm{T}}B = 0$, and $B^{\mathrm{T}}A = 0$.

Because $\dim(\mathcal{N}(A)) + \dim(\mathcal{V}(A^{\mathrm{T}})) = m$ and $\mathcal{N}(A) \perp \mathcal{V}(A^{\mathrm{T}})$, by equation (2.24) we have

$$\mathcal{N}(A) \oplus \mathcal{V}(A^{\mathrm{T}}) = \mathbb{R}^m; \tag{3.163}$$

that is, the null space of A is the *orthogonal complement* of $\mathcal{V}(A^{\mathrm{T}})$. All vectors in the null space of the matrix A are orthogonal to all vectors in the column space of A.

3.6 Generalized Inverses

On page 97, we defined a generalized inverse of a matrix A as a matrix A^- such that $AA^-A = A$, and we observed several interesting properties of generalized inverses.

Immediate Properties of Generalized Inverses

The properties of a generalized inverse A^- derived in equations (3.150) through (3.158) include:

- $(A^-)^{\mathrm{T}}$ is a generalized inverse of A^{T}.
- $\mathrm{rank}(A^-A) = \mathrm{rank}(A)$.
- A^-A is idempotent.
- $I - A^-A$ is idempotent.
- $\mathrm{rank}(I - A^-A) = m - \mathrm{rank}(A)$.

In this section, we will first consider some more properties of "general" generalized inverses, which are analogous to properties of inverses, and then we will discuss some additional requirements on the generalized inverse that make it unique.

3.6.1 Generalized Inverses of Sums of Matrices

Often we need generalized inverses of various sums of matrices. On page 93, we gave a number of relationships that hold for inverses of sums of matrices. All of the equations (3.133) through (3.139) hold for generalized inverses. For example,

$$A(I + A)^- = (I + A^-)^-.$$

(Again, these relationships are easily proven if taken in the order given on page 93.)

3.6.2 Generalized Inverses of Partitioned Matrices

If A is partitioned as

$$A = \begin{bmatrix} A_{11} & A_{12} \\ A_{21} & A_{22} \end{bmatrix}, \tag{3.164}$$

then, similar to equation (3.145), a generalized inverse of A is given by

$$A^- = \begin{bmatrix} A_{11}^- + A_{11}^- A_{12} Z^- A_{21} A_{11}^- & -A_{11}^- A_{12} Z^- \\ -Z^- A_{21} A_{11}^- & Z^- \end{bmatrix}, \tag{3.165}$$

where $Z = A_{22} - A_{21} A_{11}^- A_{12}$ (see Exercise 3.14, page 141).

If the partitioning in (3.164) happens to be such that A_{11} is of full rank and of the same rank as A, a generalized inverse of A is given by

$$A^- = \begin{bmatrix} A_{11}^{-1} & 0 \\ 0 & 0 \end{bmatrix}, \tag{3.166}$$

where 0 represents matrices of the appropriate shapes. This is not necessarily the same generalized inverse as in equation (3.165). The fact that it is a generalized inverse is easy to establish by using the definition of generalized inverse and equation (3.144).

3.6.3 Pseudoinverse or Moore-Penrose Inverse

A generalized inverse is not unique in general. As we have seen, a generalized inverse determines a set of linearly independent solutions to a linear system $Ax = b$. We may impose other conditions on the generalized inverse to arrive at a unique matrix that yields a solution that has some desirable properties. If we impose three more conditions, we have a unique matrix, denoted by A^+, that yields a solution A^+b that has the minimum length of any solution to $Ax = b$. We define this matrix and discuss some of its properties below, and in Section 6.7 we discuss properties of the solution A^+b.

Definition and Terminology

To the general requirement $AA^-A = A$, we successively add three requirements that define special generalized inverses, sometimes called respectively g_2 or g_{12}, g_3 or g_{123}, and g_4 or g_{1234} inverses. The "general" generalized inverse is sometimes called a g_1 inverse. The g_4 inverse is called the Moore-Penrose inverse. As we will see below, it is unique. The terminology distinguishing the various types of generalized inverses is not used consistently in the literature. I will indicate some alternative terms in the definition below.

For a matrix A, a *Moore-Penrose inverse*, denoted by A^+, is a matrix that has the following four properties.

1. $AA^+A = A$. Any matrix that satisfies this condition is called a generalized inverse, and as we have seen above is denoted by A^-. For many applications, this is the only condition necessary. Such a matrix is also called a g_1 *inverse*, an *inner pseudoinverse*, or a *conditional inverse*.
2. $A^+AA^+ = A^+$. A matrix A^+ that satisfies this condition is called an *outer pseudoinverse*. A g_1 inverse that also satisfies this condition is called a g_2 *inverse* or *reflexive generalized inverse*, and is denoted by A^*.
3. A^+A is symmetric.
4. AA^+ is symmetric.

The Moore-Penrose inverse is also called the *pseudoinverse*, the *p-inverse*, and the *normalized generalized inverse*. (My current preferred term is "Moore-Penrose inverse", but out of habit, I often use the term "pseudoinverse" for this special generalized inverse. I generally avoid using any of the other alternative terms introduced above. I use the term "generalized inverse" to mean the "general generalized inverse", the g_1.) The name Moore-Penrose derives from the preliminary work of Moore (1920) and the more thorough later work of Penrose (1955), who laid out the conditions above and proved existence and uniqueness.

Existence

We can see by construction that the Moore-Penrose inverse exists for any matrix A. First, if $A = 0$, note that $A^+ = 0$. If $A \neq 0$, it has a full rank factorization, $A = LR$, as in equation (3.112), so $L^{\mathrm{T}}AR^{\mathrm{T}} = L^{\mathrm{T}}LRR^{\mathrm{T}}$. Because the $n \times r$ matrix L is of full column rank and the $r \times m$ matrix R is of row column rank, $L^{\mathrm{T}}L$ and RR^{T} are both of full rank, and hence $L^{\mathrm{T}}LRR^{\mathrm{T}}$ is of full rank. Furthermore, $L^{\mathrm{T}}AR^{\mathrm{T}} = L^{\mathrm{T}}LRR^{\mathrm{T}}$, so it is of full rank, and $(L^{\mathrm{T}}AR^{\mathrm{T}})^{-1}$ exists. Now, form $R^{\mathrm{T}}(L^{\mathrm{T}}AR^{\mathrm{T}})^{-1}L^{\mathrm{T}}$. By checking properties 1 through 4 above, we see that

$$A^+ = R^{\mathrm{T}}(L^{\mathrm{T}}AR^{\mathrm{T}})^{-1}L^{\mathrm{T}} \tag{3.167}$$

is a Moore-Penrose inverse of A. This expression for the Moore-Penrose inverse based on a full rank decomposition of A is not as useful as another expression we will consider later, based on QR decomposition (equation (5.38) on page 190).

Uniqueness

We can see that the Moore-Penrose inverse is unique by considering any matrix G that satisfies the properties 1 through 4 for $A \neq 0$. (The Moore-Penrose inverse of $A = 0$ (that is, $A^+ = 0$) is clearly unique, as there could be no other matrix satisfying property 2.) By applying the properties and using A^+ given above, we have the following sequence of equations:

$$G =$$
$$GAG = (GA)^{\mathrm{T}}G = A^{\mathrm{T}}G^{\mathrm{T}}G = (AA^+A)^{\mathrm{T}}G^{\mathrm{T}}G = (A^+A)^{\mathrm{T}}A^{\mathrm{T}}G^{\mathrm{T}}G =$$
$$A^+AA^{\mathrm{T}}G^{\mathrm{T}}G = A^+A(GA)^{\mathrm{T}}G = A^+AGAG = A^+AG = A^+AA^+AG =$$
$$A^+(AA^+)^{\mathrm{T}}(AG)^{\mathrm{T}} = A^+(A^+)^{\mathrm{T}}A^{\mathrm{T}}G^{\mathrm{T}}A^{\mathrm{T}} = A^+(A^+)^{\mathrm{T}}(AGA)^{\mathrm{T}} =$$
$$A^+(A^+)^{\mathrm{T}}A^{\mathrm{T}} = A^+(AA^+)^{\mathrm{T}} = A^+AA^+$$
$$= A^+.$$

Other Properties

If A is nonsingular, then obviously $A^+ = A^{-1}$, just as for any generalized inverse.

Because A^+ is a generalized inverse, all of the properties for a generalized inverse A^- discussed above hold; in particular, A^+b is a solution to the linear system $Ax = b$ (see equation (3.156)). In Section 6.7, we will show that this unique solution has a kind of optimality.

If the inverses on the right-hand side of equation (3.165) are pseudoinverses, then the result is the pseudoinverse of A.

The generalized inverse given in equation (3.166) is the same as the pseudoinverse given in equation (3.167).

Pseudoinverses also have a few additional interesting properties not shared by generalized inverses; for example

$$(I - A^+A)A^+ = 0. \tag{3.168}$$

3.7 Orthogonality

In Section 2.1.8, we defined orthogonality and orthonormality of two or more vectors in terms of dot products. On page 75, in equation (3.81), we also defined the orthogonal binary relationship between two matrices. Now we define the orthogonal unary property of a matrix. This is the more important property and is what is commonly meant when we speak of orthogonality of matrices. We use the orthonormality property of vectors, which is a binary relationship, to define orthogonality of a single matrix.

Orthogonal Matrices; Definition and Simple Properties

A matrix whose rows or columns constitute a set of orthonormal vectors is said to be an *orthogonal* matrix. If Q is an $n \times m$ orthogonal matrix, then $QQ^{\mathrm{T}} = I_n$ if $n \leq m$, and $Q^{\mathrm{T}}Q = I_m$ if $n \geq m$. If Q is a square orthogonal matrix, then $QQ^{\mathrm{T}} = Q^{\mathrm{T}}Q = I$. An orthogonal matrix is also called a *unitary matrix*. (For matrices whose elements are complex numbers, a matrix is said to be *unitary* if the matrix times its conjugate transpose is the identity; that is, if $QQ^{\mathrm{H}} = I$.)

The determinant of a square orthogonal matrix is ± 1 (because the determinant of the product is the product of the determinants and the determinant of I is 1).

The matrix dot product of an $n \times m$ orthogonal matrix Q with itself is its number of columns:

$$\langle Q, Q \rangle = m. \tag{3.169}$$

This is because $Q^{\mathrm{T}}Q = I_m$. Recalling the definition of the orthogonal binary relationship from page 75, we note that if Q is an orthogonal matrix, then Q is not orthogonal to itself.

A permutation matrix (see page 62) is orthogonal. We can see this by building the permutation matrix as a product of elementary permutation matrices, and it is easy to see that they are all orthogonal.

One further property we see by simple multiplication is that if A and B are orthogonal, then $A \otimes B$ is orthogonal.

The definition of orthogonality is sometimes made more restrictive to require that the matrix be square.

Orthogonal and Orthonormal Columns

The definition given above for orthogonal matrices is sometimes relaxed to require only that the columns or rows be orthogonal (rather than orthonormal). If orthonormality is not required, the determinant is not necessarily 1. If Q is a matrix that is "orthogonal" in this weaker sense of the definition, and Q has more rows than columns, then

$$Q^{\mathrm{T}}Q = \begin{bmatrix} \mathrm{X} & 0 & \cdots & 0 \\ 0 & \mathrm{X} & \cdots & 0 \\ & & \ddots & \\ 0 & 0 & \cdots & \mathrm{X} \end{bmatrix}.$$

Unless stated otherwise, I use the term "orthogonal matrix" to refer to a matrix whose columns are orthonormal; that is, for which $Q^{\mathrm{T}}Q = I$.

The Orthogonal Group

The set of $n \times m$ orthogonal matrices for which $n \geq m$ is called an (n, m) Stiefel manifold, and an (n, n) Stiefel manifold together with Cayley multiplication is a group, sometimes called the *orthogonal group* and denoted as $\mathcal{O}(n)$. The orthogonal group $\mathcal{O}(n)$ is a subgroup of the general linear group $\mathcal{G}L(n)$, defined on page 90. The orthogonal group is useful in multivariate analysis because of the invariance of the so-called Haar measure over this group (see Section 4.5.1).

Because the Euclidean norm of any column of an orthogonal matrix is 1, no element in the matrix can be greater than 1 in absolute value. We therefore have an analogue of the Bolzano-Weierstrass theorem for sequences of orthogonal matrices. The standard Bolzano-Weierstrass theorem for real numbers states that if a sequence a_i is bounded, then there exists a subsequence a_{i_j} that converges. (See any text on real analysis.) From this, we conclude that if $Q_1, Q_2, \ldots$ is a sequence of $n \times n$ orthogonal matrices, then there exists a subsequence $Q_{i_1}, Q_{i_2}, \ldots$, such that

$$\lim_{j \to \infty} Q_{i_j} = Q, \tag{3.170}$$

where Q is some fixed matrix. The limiting matrix Q must also be orthogonal because $Q_{i_j}^{\mathrm{T}} Q_{i_j} = I$, and so, taking limits, we have $Q^{\mathrm{T}} Q = I$. The set of $n \times n$ orthogonal matrices is therefore compact.

Conjugate Vectors

Instead of defining orthogonality of vectors in terms of dot products as in Section 2.1.8, we could define it more generally in terms of a bilinear form as in Section 3.2.8. If the bilinear form $x^{\mathrm{T}} A y = 0$, we say x and y are orthogonal with respect to the matrix A. We also often use a different term and say that the vectors are *conjugate* with respect to A, as in equation (3.65). The usual definition of orthogonality in terms of a dot product is equivalent to the definition in terms of a bilinear form in the identity matrix.

Likewise, but less often, orthogonality of matrices is generalized to conjugacy of two matrices with respect to a third matrix: $Q^{\mathrm{T}} A Q = I$.

3.8 Eigenanalysis; Canonical Factorizations

Multiplication of a given vector by a square matrix may result in a scalar multiple of the vector. Stating this more formally, and giving names to such a special vector and scalar, if A is an $n \times n$ (square) matrix, v is a vector not equal to 0, and c is a scalar such that

$$Av = cv, \tag{3.171}$$

we say v is an *eigenvector* of the matrix A, and c is an *eigenvalue* of the matrix A. We refer to the pair c and v as an associated eigenvector and eigenvalue or as an *eigenpair*. While we restrict an eigenvector to be nonzero (or else we would have 0 as an eigenvector associated with any number being an eigenvalue), an eigenvalue can be 0; in that case, of course, the matrix must be singular. (Some authors restrict the definition of an eigenvalue to real values that satisfy (3.171), and there is an important class of matrices for which it is known that all eigenvalues are real. In this book, we do not want to restrict ourselves to that class; hence, we do not require c or v in equation (3.171) to be real.)

We use the term "eigenanalysis" or "eigenproblem" to refer to the general theory, applications, or computations related to either eigenvectors or eigenvalues.

There are various other terms used for eigenvalues and eigenvectors. An eigenvalue is also called a *characteristic value* (that is why I use a "c" to represent an eigenvalue), a *latent root*, or a *proper value*, and similar synonyms exist for an eigenvector. An eigenvalue is also sometimes called a *singular value*, but the latter term has a different meaning that we will use in this book (see page 127; the absolute value of an eigenvalue is a singular value, and singular values are also defined for nonsquare matrices).

Although generally throughout this chapter we have assumed that vectors and matrices are real, in eigenanalysis, even if A is real, it may be the case that c and v are complex. Therefore, in this section, we must be careful about the nature of the eigenpairs, even though we will continue to assume the basic matrices are real.

Before proceeding to consider properties of eigenvalues and eigenvectors, we should note how remarkable the relationship $Av = cv$ is: the effect of a matrix multiplication of an eigenvector is the same as a scalar multiplication of the eigenvector. The eigenvector is an *invariant* of the transformation in the sense that its direction does not change. This would seem to indicate that the eigenvalue and eigenvector depend on some kind of deep properties of the matrix, and indeed this is the case, as we will see. Of course, the first question is whether such special vectors and scalars exist. The answer is yes, but before considering that and other more complicated issues, we will state some simple properties of any scalar and vector that satisfy $Av = cv$ and introduce some additional terminology.

Left Eigenvectors

In the following, when we speak of an eigenvector or eigenpair without qualification, we will mean the objects defined by equation (3.171). There is another type of eigenvector for A, however, a *left eigenvector*, defined as a nonzero w in

$$w^{\mathrm{T}} A = cw^{\mathrm{T}}. \tag{3.172}$$

For emphasis, we sometimes refer to the eigenvector of equation (3.171), $Av = cv$, as a *right eigenvector*.

We see from the definition of a left eigenvector, that if a matrix is symmetric, each left eigenvector is an eigenvector (a *right* eigenvector).

If v is an eigenvector of A and w is a left eigenvector of A with a different associated eigenvalue, then v and w are orthogonal; that is, if $Av = c_1v$, $w^{\mathrm{T}}A = c_2w^{\mathrm{T}}$, and $c_1 \neq c_2$, then $w^{\mathrm{T}}v = 0$. We see this by multiplying both sides of $w^{\mathrm{T}}A = c_2w^{\mathrm{T}}$ by v to get $w^{\mathrm{T}}Av = c_2w^{\mathrm{T}}v$ and multiplying both sides of $Av = c_1v$ by w^{T} to get $w^{\mathrm{T}}Av = c_1w^{\mathrm{T}}v$. Hence, we have $c_1w^{\mathrm{T}}v = c_2w^{\mathrm{T}}v$, and because $c_1 \neq c_2$, we have $w^{\mathrm{T}}v = 0$.

3.8.1 Basic Properties of Eigenvalues and Eigenvectors

If c is an eigenvalue and v is a corresponding eigenvector for a real matrix A, we see immediately from the definition of eigenvector and eigenvalue in equation (3.171) the following properties. (In Exercise 3.16, you are asked to supply the simple proofs for these properties, or you can see a text such as Harville, 1997, for example.)

Assume that $Av = cv$ and that all elements of A are real.

1. bv is an eigenvector of A, where b is any nonzero scalar.

    It is often desirable to scale an eigenvector v so that $v^{\mathrm{T}}v = 1$. Such a normalized eigenvector is also called a *unit eigenvector*.

    For a given eigenvector, there is always a particular eigenvalue associated with it, but for a given eigenvalue there is a space of associated eigenvectors. (The space is a vector space if we consider the zero vector to be a member.) It is therefore not appropriate to speak of "the" eigenvector associated with a given eigenvalue — although we do use this term occasionally. (We could interpret it as referring to the normalized eigenvector.) There is, however, another sense in which an eigenvalue does not determine a unique eigenvector, as we discuss below.

2. bc is an eigenvalue of bA, where b is any nonzero scalar.

3. $1/c$ and v are an eigenpair of A^{-1} (if A is nonsingular).

4. $1/c$ and v are an eigenpair of A^{+} if A (and hence A^{+}) is square and c is nonzero.

5. If A is diagonal or triangular with elements a_{ii}, the eigenvalues are the a_{ii} with corresponding eigenvectors e_i (the unit vectors).

6. c^2 and v are an eigenpair of A^2. More generally, c^k and v are an eigenpair of A^k for $k = 1, 2, \ldots$.

7. If A and B are conformable for the multiplications AB and BA, the nonzero eigenvalues of AB are the same as the nonzero eigenvalues of BA. (Note that A and B are not necessarily square.) The set of eigenvalues is the same if A and B are square. (Note, however, that if A and B are square and d is an eigenvalue of B, cd is not necessarily an eigenvalue of AB.)

8. If A and B are square and of the same order and if B^{-1} exists, then the eigenvalues of BAB^{-1} are the same as the eigenvalues of A. (This is called a similarity transformation; see page 114.)

3.8.2 The Characteristic Polynomial

From the equation $(A - cI)v = 0$ that defines eigenvalues and eigenvectors, we see that in order for v to be nonnull, $(A - cI)$ must be singular, and hence

$$|A - cI| = |cI - A| = 0. \tag{3.173}$$

Equation (3.173) is sometimes taken as the definition of an eigenvalue c. It is definitely a fundamental relation, and, as we will see, allows us to identify a number of useful properties.

The determinant is a polynomial of degree n in c, $p_A(c)$, called the *characteristic polynomial*, and when it is equated to 0, it is called the *characteristic equation*:

$$p_A(c) = s_0 + s_1 c + \cdots + s_n c^n = 0. \tag{3.174}$$

From the expansion of the determinant $|cI - A|$, as in equation (3.32) on page 56, we see that $s_0 = (-1)^n |A|$ and $s_n = 1$, and, in general, $s_k = (-1)^{n-k}$ times the sums of all principal minors of A of order $n - k$. (Often, we equivalently define the characteristic polynomial as the determinant of $(A - cI)$. The difference would just be changes of signs of the coefficients in the polynomial.)

An eigenvalue of A is a root of the characteristic polynomial. The existence of n roots of the polynomial (by the Fundamental Theorem of Algebra) establishes the existence of n eigenvalues, some of which may be complex and some may be zero. We can write the characteristic polynomial in factored form as

$$p_A(c) = (-1)^n (c - c_1) \cdots (c - c_n). \tag{3.175}$$

The "number of eigenvalues" must be distinguished from the cardinality of the spectrum, which is the number of unique values.

A real matrix may have complex eigenvalues (and, hence, eigenvectors), just as a polynomial with real coefficients can have complex roots. Clearly, the eigenvalues of a real matrix must occur in conjugate pairs just as in the case of roots of polynomials. (As mentioned above, some authors restrict the definition of an eigenvalue to real values that satisfy (3.171). As we have seen, the eigenvalues of a symmetric matrix are always real, and this is a case that we will emphasize, but in this book we do not restrict the definition.)

The characteristic polynomial has many interesting properties that we will not discuss here. One, stated by the Cayley-Hamilton theorem, is that the matrix itself is a root of the matrix polynomial formed by the characteristic polynomial; that is,

$$p_A(A) = s_0 I + s_1 A + \cdots + s_n A^n = 0_n.$$

We see this by using equation (3.25) to write the matrix in equation (3.173) as

$$(A - cI)\text{adj}(A - cI) = p_A(c)I. \tag{3.176}$$

Hence $\text{adj}(A - cI)$ is a polynomial in c of degree less than or equal to $n - 1$, so we can write it as

$$\text{adj}(A - cI) = B_0 + B_1 c + \cdots + B_{n-1} c^{n-1},$$

where the B_i are $n \times n$ matrices. Now, equating the coefficients of c on the two sides of equation (3.176), we have

$$AB_0 = s_0 I$$
$$AB_1 - B_0 = s_1 I$$
$$\vdots$$
$$AB_{n-1} - B_{n-2} = s_{n-1} I$$
$$B_{n-1} = s_n I.$$

Now, multiply the second equation by A, the third equation by A^2, and the ith equation by A^{i-1}, and add all equations. We get the desired result: $p_A(A) = 0$. See also Exercise 3.17.

Another interesting fact is that any given n^{th}-degree polynomial, p, is the characteristic polynomial of an $n \times n$ matrix, A, of particularly simple form. Consider the polynomial

$$p(c) = s_0 + s_1 c + \cdots + s_{n-1} c^{n-1} + c^n$$

and the matrix

$$A = \begin{bmatrix} 0 & 1 & 0 & \cdots & 0 \\ 0 & 0 & 1 & \cdots & 0 \\ & & & \ddots & \\ 0 & 0 & 0 & \cdots & 1 \\ -s_0 & -s_1 & -s_2 & \cdots & -s_{n-1} \end{bmatrix}. \tag{3.177}$$

The matrix A is called the *companion matrix* of the polynomial p, and it is easy to see (by a tedious expansion) that the characteristic polynomial of A is p. This, of course, shows that a characteristic polynomial does not uniquely determine a matrix, although the converse is true (within signs).

Eigenvalues and the Trace and Determinant

If the eigenvalues of the matrix A are $c_1, \ldots, c_n$, because they are the roots of the characteristic polynomial, we can readily form that polynomial as

$$p_A(c) = (c - c_1) \cdots (c - c_n)$$
$$= (-1)^n \prod c_i + \cdots + (-1)^{n-1} \sum c_i c^{n-1} + c^n. \tag{3.178}$$

Because this is the same polynomial as obtained by the expansion of the determinant in equation (3.174), the coefficients must be equal. In particular, by simply equating the corresponding coefficients of the constant terms and $(n-1)^{\text{th}}$-degree terms, we have the two very important facts:

$$|A| = \prod c_i \tag{3.179}$$

and

$$\text{tr}(A) = \sum c_i. \tag{3.180}$$

Additional Properties of Eigenvalues and Eigenvectors

Using the characteristic polynomial yields the following properties. This is a continuation of the list that began on page 107. We assume A is a real matrix with eigenpair (c, v).

10. c is an eigenvalue of A^{T} (because $|A^{\text{T}} - cI| = |A - cI|$ for any c). The eigenvectors of A^{T}, which are left eigenvectors of A, are not necessarily the same as the eigenvectors of A, however.
11. There is a left eigenvector such that c is the associated eigenvalue.
12. $(\bar{c}, \bar{v})$ is an eigenpair of A, where $\bar{c}$ and $\bar{v}$ are the complex conjugates and A, as usual, consists of real elements. (If c and v are real, this is a tautology.)
13. $c\bar{c}$ is an eigenvalue of $A^{\text{T}}A$.
14. c is real if A is symmetric.

In Exercise 3.18, you are asked to supply the simple proofs for these properties, or you can see a text such as Harville (1997), for example.

3.8.3 The Spectrum

Although, for an $n \times n$ matrix, from the characteristic polynomial we have n roots, and hence n eigenvalues, some of these roots may be the same. It may also be the case that more than one eigenvector corresponds to a given eigenvalue. The set of all the distinct eigenvalues of a matrix is often of interest. This set is called the *spectrum* of the matrix.

Notation

Sometimes it is convenient to refer to the distinct eigenvalues and sometimes we wish to refer to all eigenvalues, as in referring to the number of roots of the characteristic polynomial. To refer to the distinct eigenvalues in a way that allows us to be consistent in the subscripts, we will call the distinct eigenvalues $\lambda_1, \ldots, \lambda_k$. The set of these constitutes the spectrum.

We denote the spectrum of the matrix A by $\sigma(A)$:

$$\sigma(A) = \{\lambda_1, \ldots, \lambda_k\}. \tag{3.181}$$

In terms of the spectrum, equation (3.175) becomes

$$p_A(c) = (-1)^n (c - \lambda_1)^{m_1} \cdots (c - \lambda_k)^{m_k}, \tag{3.182}$$

for $m_i \geq 1$.

We label the c_i and v_i so that

$$|c_1| \geq \cdots \geq |c_n|. \tag{3.183}$$

We likewise label the λ_i so that

$$|\lambda_1| > \cdots > |\lambda_k|. \tag{3.184}$$

With this notation, we have

$$|\lambda_1| = |c_1|$$

and

$$|\lambda_k| = |c_n|,$$

but we cannot say anything about the other λs and cs.

The Spectral Radius

For the matrix A with these eigenvalues, $|c_1|$ is called the *spectral radius* and is denoted by $\rho(A)$:

$$\rho(A) = \max |c_i|. \tag{3.185}$$

The set of complex numbers

$$\{x \,:\, |x| = \rho(A)\} \tag{3.186}$$

is called the *spectral circle* of A.

An eigenvalue corresponding to $\max |c_i|$ (that is, c_1) is called a *dominant eigenvalue*. We are more often interested in the absolute value (or modulus) of a dominant eigenvalue rather than the eigenvalue itself; that is, $\rho(A)$ (that is, $|c_1|$) is more often of interest than just c_1.)

Interestingly, we have for all i

$$|c_i| \leq \max_j \sum_k |a_{kj}| \tag{3.187}$$

and

$$|c_i| \leq \max_k \sum_j |a_{kj}|. \tag{3.188}$$

The inequalities of course also hold for $\rho(A)$ on the left-hand side. Rather than proving this here, we show this fact in a more general setting relating to matrix norms in inequality (3.243) on page 134. (These bounds relate to the L_1 and L_∞ matrix norms, respectively.)

A matrix may have all eigenvalues equal to 0 but yet the matrix itself may not be 0. Any upper triangular matrix with all 0s on the diagonal is an example.

Because, as we saw on page 107, if c is an eigenvalue of A, then bc is an eigenvalue of bA where b is any nonzero scalar, we can scale a matrix with a nonzero eigenvalue so that its spectral radius is 1. The scaled matrix is simply $A/|c_1|$.

Linear Independence of Eigenvectors Associated with Distinct Eigenvalues

Suppose that $\{\lambda_1, \ldots, \lambda_k\}$ is a set of distinct eigenvalues of the matrix A and $\{x_1, \ldots, x_k\}$ is a set of eigenvectors such that (λ_i, x_i) is an eigenpair. Then $x_1, \ldots, x_k$ are linearly independent; that is, eigenvectors associated with distinct eigenvalues are linearly independent.

We can see that this must be the case by assuming that the eigenvectors are not linearly independent. In that case, let $\{y_1, \ldots, y_j\} \subset \{x_1, \ldots, x_k\}$, for some $j < k$, be a maximal linearly independent subset. Let the corresponding eigenvalues be $\{\mu_1, \ldots, \mu_j\} \subset \{\lambda_1, \ldots, \lambda_k\}$. Then, for some eigenvector y_{j+1}, we have

$$y_{j+1} = \sum_{i=1}^{j} t_i y_i$$

for some t_i. Now, multiplying both sides of the equation by $A - \mu_{j+1}I$, where μ_{j+1} is the eigenvalue corresponding to y_{j+1}, we have

$$0 = \sum_{i=1}^{j} t_i(\mu_i - \mu_{j+1})y_i.$$

If the eigenvalues are unique (that is, for each $i \le j$), we have $\mu_i \neq \mu_{j+1}$, then the assumption that the eigenvalues are not linearly independent is contradicted because otherwise we would have a linear combination with nonzero coefficients equal to zero.

The Eigenspace and Geometric Multiplicity

Rewriting the definition (3.171) for the ith eigenvalue and associated eigenvector of the $n \times n$ matrix A as

$$(A - c_i I)v_i = 0, \tag{3.189}$$

we see that the eigenvector v_i is in $\mathcal{N}(A - c_i I)$, the null space of $(A - c_i I)$. For such a nonnull vector to exist, of course, $(A - c_i I)$ must be singular; that is, $\text{rank}(A - c_i I)$ must be less than n. This null space is called the *eigenspace* of the eigenvalue c_i.

It is possible that a given eigenvalue may have more than one associated eigenvector that are linearly independent of each other. For example, we easily see that the identity matrix has only one unique eigenvalue, namely 1, but any vector is an eigenvector, and so the number of linearly independent eigenvectors is equal to the number of rows or columns of the identity. If u and v are eigenvectors corresponding to the same eigenvalue c, then any linear combination of u and v is an eigenvector corresponding to c; that is, if $Au = cu$ and $Av = cv$, for any scalars a and b,

$$A(au + bv) = c(au + bv).$$

The dimension of the eigenspace corresponding to the eigenvalue c_i is called the *geometric multiplicity* of c_i; that is, the geometric multiplicity of c_i is the nullity of $A - c_i I$. If g_i is the geometric multiplicity of c_i, an eigenvalue of the $n \times n$ matrix A, then we can see from equation (3.160) that $\text{rank}(A - c_i I) + g_i = n$.

The multiplicity of 0 as an eigenvalue is just the nullity of A. If A is of full rank, the multiplicity of 0 will be 0, but, in this case, we do not consider 0 to be an eigenvalue. If A is singular, however, we consider 0 to be an eigenvalue, and the multiplicity of the 0 eigenvalue is the rank deficiency of A.

Multiple linearly independent eigenvectors corresponding to the same eigenvalue can be chosen to be orthogonal to each other using, for example, the Gram-Schmidt transformations, as in equation (2.34) on page 27. These orthogonal eigenvectors span the same eigenspace. They are not unique, of course, as any sequence of Gram-Schmidt transformations could be applied.

Algebraic Multiplicity

A single value that occurs as a root of the characteristic equation m times is said to have *algebraic multiplicity* m. Although we sometimes refer to this as just the multiplicity, algebraic multiplicity should be distinguished from geometric multiplicity, defined above. These are not the same, as we will see in an example later. An eigenvalue whose algebraic multiplicity and geometric multiplicity are the same is called a *semisimple* eigenvalue. An eigenvalue with algebraic multiplicity 1 is called a *simple* eigenvalue.

Because the determinant that defines the eigenvalues of an $n \times n$ matrix is an nth-degree polynomial, we see that the sum of the multiplicities of distinct eigenvalues is n.

Because most of the matrices in statistical applications are real, in the following we will generally restrict our attention to real matrices. It is important to note that the eigenvalues and eigenvectors of a real matrix are not necessarily real, but as we have observed, the eigenvalues of a symmetric real matrix are real. (The proof, which was stated as an exercise, follows by noting that if A is symmetric, the eigenvalues of $A^{\mathrm{T}}A$ are the eigenvalues of A^2, which from the definition are obviously nonnegative.)

3.8.4 Similarity Transformations

Two $n \times n$ matrices, A and B, are said to be *similar* if there exists a nonsingular matrix P such that

$$B = P^{-1}AP. \tag{3.190}$$

The transformation in equation (3.190) is called a *similarity transformation*. (Compare this with *equivalent matrices* on page 86. The matrices A and B in equation (3.190) are equivalent, as we see using equations (3.115) and (3.116).)

It is clear from the definition that the similarity relationship is both commutative and transitive.

If A and B are similar, as in equation (3.190), then for any scalar c

$$
\begin{aligned}
|A - cI| &= |P^{-1}||A - cI||P| \\
&= |P^{-1}AP - cP^{-1}IP| \\
&= |B - cI|,
\end{aligned}
$$

and, hence, A and B have the same eigenvalues. (This simple fact was stated as property 8 on page 108.)

Orthogonally Similar Transformations

An important type of similarity transformation is based on an orthogonal matrix in equation (3.190). If Q is orthogonal and

$$B = Q^{\mathrm{T}}AQ, \tag{3.191}$$

A and B are said to be *orthogonally similar*.

If B in the equation $B = Q^{\mathrm{T}}AQ$ is a diagonal matrix, A is said to be *orthogonally diagonalizable*, and QBQ^{T} is called the *orthogonally diagonal factorization* or *orthogonally similar factorization* of A. We will discuss characteristics of orthogonally diagonalizable matrices in Sections 3.8.5 and 3.8.6 below.

Schur Factorization

If B in equation (3.191) is an upper triangular matrix, QBQ^{T} is called the *Schur factorization* of A.

For any square matrix, the Schur factorization exists; hence, it is one of the most useful similarity transformations. The Schur factorization clearly exists in the degenerate case of a 1×1 matrix.

To see that it exists for any $n \times n$ matrix A, let (c, v) be an arbitrary eigenpair of A with v normalized, and form an orthogonal matrix U with v as its first column. Let U_2 be the matrix consisting of the remaining columns; that is, U is partitioned as $[v \,|\, U_2]$.

$$U^{\mathrm{T}}AU = \begin{bmatrix} v^{\mathrm{T}}Av & v^{\mathrm{T}}AU_2 \\ U_2^{\mathrm{T}}Av & U_2^{\mathrm{T}}AU_2 \end{bmatrix}$$

$$= \begin{bmatrix} c & v^{\mathrm{T}}AU_2 \\ 0 & U_2^{\mathrm{T}}AU_2 \end{bmatrix}$$

$$= B,$$

where $U_2^{\mathrm{T}}AU_2$ is an $(n-1) \times (n-1)$ matrix. Now the eigenvalues of $U^{\mathrm{T}}AU$ are the same as those of A; hence, if $n = 2$, then $U_2^{\mathrm{T}}AU_2$ is a scalar and must equal the other eigenvalue, and so the statement is proven.

We now use induction on n to establish the general case. Assume that the factorization exists for any $(n-1) \times (n-1)$ matrix, and let A be any $n \times n$ matrix. We let (c, v) be an arbitrary eigenpair of A (with v normalized), follow the same procedure as in the preceding paragraph, and get

$$U^{\mathrm{T}}AU = \begin{bmatrix} c & v^{\mathrm{T}}AU_2 \\ 0 & U_2^{\mathrm{T}}AU_2 \end{bmatrix}.$$

Now, since $U_2^{\mathrm{T}}AU_2$ is an $(n-1) \times (n-1)$ matrix, by the induction hypothesis there exists an $(n-1) \times (n-1)$ orthogonal matrix V such that $V^{\mathrm{T}}(U_2^{\mathrm{T}}AU_2)V = T$, where T is upper triangular. Now let

$$Q = U \begin{bmatrix} 1 & 0 \\ 0 & V \end{bmatrix}.$$

By multiplication, we see that $Q^{\mathrm{T}}Q = I$ (that is, Q is orthogonal). Now form

$$Q^{\mathrm{T}}AQ = \begin{bmatrix} c & v^{\mathrm{T}}AU_2V \\ 0 & V^{\mathrm{T}}U_2^{\mathrm{T}}AU_2V \end{bmatrix} = \begin{bmatrix} c & v^{\mathrm{T}}AU_2V \\ 0 & T \end{bmatrix} = B.$$

We see that B is upper triangular because T is, and so by induction the Schur factorization exists for any $n \times n$ matrix.

Note that the Schur factorization is also based on orthogonally similar transformations, but the term "orthogonally similar factorization" is generally used only to refer to the diagonal factorization.

Uses of Similarity Transformations

Similarity transformations are very useful in establishing properties of matrices, such as convergence properties of sequences (see, for example, Section 3.9.5). Similarity transformations are also used in algorithms for computing eigenvalues (see, for example, Section 7.3). In an orthogonally similar factorization, the elements of the diagonal matrix are the eigenvalues. Although the diagonals in the upper triangular matrix of the Schur factorization are the eigenvalues, that particular factorization is rarely used in computations.

Although similar matrices have the same eigenvalues, they do not necessarily have the same eigenvectors. If A and B are similar, for some nonzero vector v and some scalar c, $Av = cv$ implies that there exists a nonzero vector u such that $Bu = cu$, but it does not imply that $u = v$ (see Exercise 3.19b).

3.8.5 Similar Canonical Factorization; Diagonalizable Matrices

If V is a matrix whose columns correspond to the eigenvectors of A, and C is a diagonal matrix whose entries are the eigenvalues corresponding to the columns of V, using the definition (equation (3.171)) we can write

$$AV = VC. \tag{3.192}$$

Now, if V is nonsingular, we have

$$A = VCV^{-1}. \tag{3.193}$$

Expression (3.193) represents a *diagonal factorization* of the matrix A. We see that a matrix A with eigenvalues $c_1, \ldots, c_n$ that can be factorized this way is similar to the matrix $\mathrm{diag}(c_1, \ldots, c_n)$, and this representation is sometimes called the *similar canonical form* of A or the *similar canonical factorization* of A.

Not all matrices can be factored as in equation (3.193). It obviously depends on V being nonsingular; that is, that the eigenvectors are linearly independent. If a matrix can be factored as in (3.193), it is called a *diagonalizable matrix*, a *simple matrix*, or a *regular matrix* (the terms are synonymous, and we will generally use the term "diagonalizable"); a matrix that cannot be factored in that way is called a *deficient matrix* or a *defective matrix* (the terms are synonymous).

Any matrix all of whose eigenvalues are unique is diagonalizable (because, as we saw on page 112, in that case the eigenvectors are linearly independent), but uniqueness of the eigenvalues is not a necessary condition. A necessary and sufficient condition for a matrix to be diagonalizable can be stated in terms of the unique eigenvalues and their multiplicities: suppose for the $n \times n$ matrix A that the distinct eigenvalues $\lambda_1, \ldots, \lambda_k$ have algebraic multiplicities $m_1, \ldots, m_k$. If, for $l = 1, \ldots, k$,

$$\mathrm{rank}(A - \lambda_l I) = n - m_l \tag{3.194}$$

(that is, if all eigenvalues are semisimple), then A is diagonalizable, and this condition is also necessary for A to be diagonalizable. This fact is called the "diagonalizability theorem". Recall that A being diagonalizable is equivalent to V in $AV = VC$ (equation (3.192)) being nonsingular.

To see that the condition is sufficient, assume, for each i, $\mathrm{rank}(A - c_i I) = n - m_i$, and so the equation $(A - c_i I)x = 0$ has exactly $n - (n - m_i)$ linearly independent solutions, which are by definition eigenvectors of A associated with c_i. (Note the somewhat complicated notation. Each c_i is the same as some λ_l, and for each λ_l, we have $\lambda_l = c_{l_1} = c_{l_{m_l}}$ for $1 \le l_1 < \cdots < l_{m_l} \le n$.) Let $w_1, \ldots, w_{m_i}$ be a set of linearly independent eigenvectors associated with c_i, and let u be an eigenvector associated with c_j and $c_j \ne c_i$. (The vectors $w_1, \ldots, w_{m_i}$ and u are columns of V.) Now if u is not linearly independent of $w_1, \ldots, w_{m_i}$, we write $u = \sum b_k w_k$, and so $Au = A \sum b_k w_k = c_i \sum b_k w_k = c_i u$, contradicting the assumption that u is not an eigenvector associated with c_i. Therefore, the eigenvectors associated with different eigenvalues are linearly independent, and so V is nonsingular.

Now, to see that the condition is necessary, assume V is nonsingular; that is, V^{-1} exists. Because C is a diagonal matrix of all n eigenvalues, the matrix $(C - c_i I)$ has exactly m_i zeros on the diagonal, and hence, $\operatorname{rank}(C - c_i I) = n - m_i$. Because $V(C - c_i I)V^{-1} = (A - c_i I)$, and multiplication by a full rank matrix does not change the rank (see page 88), we have $\operatorname{rank}(A - c_i I) = n - m_i$.

Symmetric Matrices

A symmetric matrix is a diagonalizable matrix. We see this by first letting A be any $n \times n$ symmetric matrix with eigenvalue c of multiplicity m. We need to show that $\operatorname{rank}(A - cI) = n - m$. Let $B = A - cI$, which is symmetric because A and I are. First, we note that c is real, and therefore B is real. Let $r = \operatorname{rank}(B)$. From equation (3.127), we have

$$\operatorname{rank}\left(B^2\right) = \operatorname{rank}\left(B^{\mathrm{T}} B\right) = \operatorname{rank}(B) = r.$$

In the full rank partitioning of B, there is at least one $r \times r$ principal submatrix of full rank. The r-order principal minor in B^2 corresponding to any full rank $r \times r$ principal submatrix of B is therefore positive. Furthermore, any j-order principal minor in B^2 for $j > r$ is zero. Now, rewriting the characteristic polynomial in equation (3.174) slightly by attaching the sign to the variable w, we have

$$p_{B^2}(w) = t_{n-r}(-w)^{n-r} + \cdots + t_{n-1}(-w)^{n-1} + (-w)^n = 0,$$

where t_{n-j} is the sum of all j-order principal minors. Because $t_{n-r} \ne 0$, $w = 0$ is a root of multiplicity $n - r$. It is likewise an eigenvalue of B with multiplicity $n - r$. Because $A = B + cI$, $0 + c$ is an eigenvalue of A with multiplicity $n - r$; hence, $m = n - r$. Therefore $n - m = r = \operatorname{rank}(A - cI)$.

A Defective Matrix

Although most matrices encountered in statistics applications are diagonalizable, it may be of interest to consider an example of a matrix that is not diagonalizable. Searle (1982) gives an example of a small matrix:

$$A = \begin{bmatrix} 0 & 1 & 2 \\ 2 & 3 & 0 \\ 0 & 4 & 5 \end{bmatrix}.$$

The three strategically placed 0s make this matrix easy to work with, and the determinant of $(cI - A)$ yields the characteristic polynomial equation

$$c^3 - 8c^2 + 13c - 6 = 0.$$

This can be factored as $(c-6)(c-1)^2$, hence, we have eigenvalues $c_1 = 6$ with algebraic multiplicity $m_1 = 1$, and $c_2 = 1$ with algebraic multiplicity $m_2 = 2$. Now, consider $A - c_2 I$:

$$A - I = \begin{bmatrix} -1 & 1 & 2 \\ 2 & 2 & 0 \\ 0 & 4 & 4 \end{bmatrix}.$$

This is clearly of rank 2; hence the rank of the null space of $A - c_2 I$ (that is, the geometric multiplicity of c_2) is $3 - 2 = 1$. The matrix A is not diagonalizable.

3.8.6 Properties of Diagonalizable Matrices

If the matrix A has the similar canonical factorization VCV^{-1} of equation (3.193), some important properties are immediately apparent. First of all, this factorization implies that the eigenvectors of a diagonalizable matrix are linearly independent.

Other properties are easy to derive or to show because of this factorization. For example, the general equations (3.179) and (3.180) concerning the product and the sum of eigenvalues follow easily from

$$|A| = |VCV^{-1}| = |V|\,|C|\,|V^{-1}| = |C|$$

and

$$\mathrm{tr}(A) = \mathrm{tr}(VCV^{-1}) = \mathrm{tr}(V^{-1}VC) = \mathrm{tr}(C).$$

One important fact is that the number of nonzero eigenvalues of a diagonalizable matrix A is equal to the rank of A. This must be the case because the rank of the diagonal matrix C is its number of nonzero elements and the rank of A must be the same as the rank of C. Another way of saying this is that the sum of the multiplicities of the unique nonzero eigenvalues is equal to the rank of the matrix; that is, $\sum_{i=1}^{k} m_i = \mathrm{rank}(A)$, for the matrix A with k distinct eigenvalues with multiplicities m_i.

Matrix Functions

We use the diagonal factorization (3.193) of the matrix $A = VCV^{-1}$ to define a function of the matrix that corresponds to a function of a scalar, $f(x)$,

$$f(A) = V\mathrm{diag}(f(c_1),\ldots,f(c_n))V^{-1}, \tag{3.195}$$

if $f(\cdot)$ is defined for each eigenvalue c_i. (Notice the relationship of this definition to the Cayley-Hamilton theorem and to Exercise 3.17.)

Another useful feature of the diagonal factorization of the matrix A in equation (3.193) is that it allows us to study functions of powers of A because $A^k = VC^kV^{-1}$. In particular, we may assess the convergence of a function of a power of A,

$$\lim_{k\to\infty} g(k,A).$$

Functions of scalars that have power series expansions may be defined for matrices in terms of power series expansions in A, which are effectively power series in the diagonal elements of C. For example, using the power series expansion of $\mathrm{e}^x = \sum_{k=0}^{\infty} \frac{x^k}{k!}$, we can define the *matrix exponential* for the square matrix A as the matrix

$$\mathrm{e}^A = \sum_{k=0}^{\infty} \frac{A^k}{k!}, \tag{3.196}$$

where $A^0/0!$ is defined as I. (Recall that we did not define A^0 if A is singular.) If A is represented as VCV^{-1}, this expansion becomes

$$\mathrm{e}^A = V\sum_{k=0}^{\infty} \frac{C^k}{k!} V^{-1}$$
$$= V\mathrm{diag}\left((\mathrm{e}^{c_1},\ldots,\mathrm{e}^{c_n})\right)V^{-1}.$$

3.8.7 Eigenanalysis of Symmetric Matrices

The eigenvalues and eigenvectors of symmetric matrices have some interesting properties. First of all, as we have already observed, for a real symmetric matrix, the eigenvalues are all real. We have also seen that symmetric matrices are diagonalizable; therefore all of the properties of diagonalizable matrices carry over to symmetric matrices.

Orthogonality of Eigenvectors

In the case of a symmetric matrix A, any eigenvectors corresponding to distinct eigenvalues are orthogonal. This is easily seen by assuming that c_1 and c_2 are unequal eigenvalues with corresponding eigenvectors v_1 and v_2. Now consider $v_1^{\mathrm{T}}v_2$. Multiplying this by c_2, we get

$$c_2 v_1^{\mathrm{T}} v_2 = v_1^{\mathrm{T}} A v_2 = v_2^{\mathrm{T}} A v_1 = c_1 v_2^{\mathrm{T}} v_1 = c_1 v_1^{\mathrm{T}} v_2.$$

Because $c_1 \neq c_2$, we have $v_1^{\mathrm{T}} v_2 = 0$.

Now, consider two eigenvalues $c_i = c_j$, that is, an eigenvalue of multiplicity greater than 1 and distinct associated eigenvectors v_i and v_j. By what we just saw, an eigenvector associated with $c_k \neq c_i$ is orthogonal to the space spanned by v_i and v_j. Assume v_i is normalized and apply a Gram-Schmidt transformation to form

$$\tilde{v}_j = \frac{1}{\|v_j - \langle v_i, v_j \rangle v_i\|} \, (v_j - \langle v_i, v_j \rangle v_i),$$

as in equation (2.34) on page 27, yielding a vector orthogonal to v_i. Now, we have

$$
\begin{aligned}
A\tilde{v}_j &= \frac{1}{\|v_j - \langle v_i, v_j \rangle v_i\|} \, (Av_j - \langle v_i, v_j \rangle Av_i) \\
&= \frac{1}{\|v_j - \langle v_i, v_j \rangle v_i\|} \, (c_j v_j - \langle v_i, v_j \rangle c_i v_i) \\
&= c_j \frac{1}{\|v_j - \langle v_i, v_j \rangle v_i\|} \, (v_j - \langle v_i, v_j \rangle v_i) \\
&= c_j \tilde{v}_j;
\end{aligned}
$$

hence, $\tilde{v}_j$ is an eigenvector of A associated with c_j. We conclude therefore that the eigenvectors of a symmetric matrix can be chosen to be orthogonal.

A symmetric matrix is orthogonally diagonalizable, because the V in equation (3.193) can be chosen to be orthogonal, and can be written as

$$A = VCV^{\mathrm{T}}, \tag{3.197}$$

where $VV^{\mathrm{T}} = V^{\mathrm{T}}V = I$, and so we also have

$$V^{\mathrm{T}}AV = C. \tag{3.198}$$

Such a matrix is orthogonally similar to a diagonal matrix formed from its eigenvalues.

Spectral Decomposition

When A is symmetric and the eigenvectors v_i are chosen to be orthonormal,

$$I = \sum_i v_i v_i^{\mathrm{T}}, \tag{3.199}$$

so

$$
\begin{aligned}
A &= A \sum_i v_i v_i^{\mathrm{T}} \\
&= \sum_i A v_i v_i^{\mathrm{T}} \\
&= \sum_i c_i v_i v_i^{\mathrm{T}}. \tag{3.200}
\end{aligned}
$$

This representation is called the *spectral decomposition* of the symmetric matrix A. It is essentially the same as equation (3.197), so $A = VCV^{\mathrm{T}}$ is also called the spectral decomposition.

The representation is unique except for the ordering and the choice of eigenvectors for eigenvalues with multiplicities greater than 1. If the rank of the matrix is r, we have $|c_1| \geq \cdots \geq |c_r| > 0$, and if $r < n$, then $c_{r+1} = \cdots = c_n = 0$.

Note that the matrices in the spectral decomposition are projection matrices that are orthogonal to each other (but they are not orthogonal matrices) and they sum to the identity. Let

$$P_i = v_i v_i^{\mathrm{T}}. \tag{3.201}$$

Then we have

$$P_i P_i = P_i, \tag{3.202}$$
$$P_i P_j = 0 \text{ for } i \neq j, \tag{3.203}$$
$$\sum_i P_i = I, \tag{3.204}$$

and the spectral decomposition,

$$A = \sum_i c_i P_i. \tag{3.205}$$

The P_i are called *spectral projectors*.

The spectral decomposition also applies to powers of A,

$$A^k = \sum_i c_i^k v_i v_i^{\mathrm{T}}, \tag{3.206}$$

where k is an integer. If A is nonsingular, k can be negative in the expression above.

The spectral decomposition is one of the most important tools in working with symmetric matrices.

Although we will not prove it here, all diagonalizable matrices have a spectral decomposition in the form of equation (3.205) with projection matrices that satisfy properties (3.202) through (3.204). These projection matrices cannot necessarily be expressed as outer products of eigenvectors, however. The eigenvalues and eigenvectors of a nonsymmetric matrix might not be real, the left and right eigenvectors might not be the same, and two eigenvectors might not be mutually orthogonal. In the spectral representation $A = \sum_i c_i P_i$, however, if c_j is a simple eigenvalue with associated left and right eigenvectors y_j and x_j, respectively, then the projection matrix P_j is $x_j y_j^{\mathrm{H}} / y_j^{\mathrm{H}} x_j$. (Note that because the eigenvectors may not be real, we take the conjugate transpose.) This is Exercise 3.20.

Quadratic Forms and the Rayleigh Quotient

Equation (3.200) yields important facts about quadratic forms in A. Because V is of full rank, an arbitrary vector x can be written as Vb for some vector b. Therefore, for the quadratic form $x^{\mathrm{T}}Ax$ we have

$$
\begin{aligned}
x^{\mathrm{T}}Ax &= x^{\mathrm{T}} \sum_i c_i v_i v_i^{\mathrm{T}} x \\
&= \sum_i b^{\mathrm{T}} V^{\mathrm{T}} v_i v_i^{\mathrm{T}} V b c_i \\
&= \sum_i b_i^2 c_i.
\end{aligned}
$$

This immediately gives the inequality

$$
x^{\mathrm{T}}Ax \le \max\{c_i\} b^{\mathrm{T}} b.
$$

(Notice that $\max\{c_i\}$ here is not necessarily c_1; in the important case when all of the eigenvalues are nonnegative, it is, however.) Furthermore, if $x \neq 0$, $b^{\mathrm{T}}b = x^{\mathrm{T}}x$, and we have the important inequality

$$
\frac{x^{\mathrm{T}}Ax}{x^{\mathrm{T}}x} \le \max\{c_i\}. \tag{3.207}
$$

Equality is achieved if x is the eigenvector corresponding to $\max\{c_i\}$, so we have

$$
\max_{x \neq 0} \frac{x^{\mathrm{T}}Ax}{x^{\mathrm{T}}x} = \max\{c_i\}. \tag{3.208}
$$

If $c_1 > 0$, this is the spectral radius, $\rho(A)$.

The expression on the left-hand side in (3.207) as a function of x is called the *Rayleigh quotient* of the symmetric matrix A and is denoted by $R_A(x)$:

$$
\begin{aligned}
R_A(x) &= \frac{x^{\mathrm{T}}Ax}{x^{\mathrm{T}}x} \\
&= \frac{\langle x, Ax \rangle}{\langle x, x \rangle}.
\end{aligned} \tag{3.209}
$$

Because if $x \neq 0$, $x^{\mathrm{T}}x > 0$, it is clear that the Rayleigh quotient is nonnegative for all x if and only if A is nonnegative definite and is positive for all x if and only if A is positive definite.

The Fourier Expansion

The $v_i v_i^{\mathrm{T}}$ matrices in equation (3.200) have the property that $\langle v_i v_i^{\mathrm{T}}, v_j v_j^{\mathrm{T}} \rangle = 0$ for $i \neq j$ and $\langle v_i v_i^{\mathrm{T}}, v_i v_i^{\mathrm{T}} \rangle = 1$, and so the spectral decomposition is a Fourier expansion as in equation (3.82) and the eigenvalues are Fourier coefficients.

From equation (3.83), we see that the eigenvalues can be represented as the dot product

$$c_i = \langle A,\, v_i v_i^{\mathrm{T}} \rangle. \tag{3.210}$$

The eigenvalues c_i have the same properties as the Fourier coefficients in any orthonormal expansion. In particular, the best approximating matrices within the subspace of $n \times n$ symmetric matrices spanned by $\{v_1 v_1^{\mathrm{T}}, \ldots, v_n v_n^{\mathrm{T}}\}$ are partial sums of the form of equation (3.200). In Section 3.10, however, we will develop a stronger result for approximation of matrices that does not rely on the restriction to this subspace and which applies to general, nonsquare matrices.

Powers of a Symmetric Matrix

If (c, v) is an eigenpair of the symmetric matrix A with $v^{\mathrm{T}} v = 1$, then for any $k = 1, 2, \ldots,$

$$\left(A - cvv^{\mathrm{T}}\right)^k = A^k - c^k vv^{\mathrm{T}}. \tag{3.211}$$

This follows from induction on k, for it clearly is true for $k = 1$, and if for a given k it is true that for $k - 1$

$$\left(A - cvv^{\mathrm{T}}\right)^{k-1} = A^{k-1} - c^{k-1} vv^{\mathrm{T}},$$

then by multiplying both sides by $(A - cvv^{\mathrm{T}})$, we see it is true for k:

$$\begin{aligned}
\left(A - cvv^{\mathrm{T}}\right)^k &= \left(A^{k-1} - c^{k-1} vv^{\mathrm{T}}\right)\left(A - cvv^{\mathrm{T}}\right) \\
&= A^k - c^{k-1} vv^{\mathrm{T}} A - c A^{k-1} vv^{\mathrm{T}} + c^k vv^{\mathrm{T}} \\
&= A^k - c^k vv^{\mathrm{T}} - c^k vv^{\mathrm{T}} + c^k vv^{\mathrm{T}} \\
&= A^k - c^k vv^{\mathrm{T}}.
\end{aligned}$$

There is a similar result for nonsymmetric square matrices, where w and v are left and right eigenvectors, respectively, associated with the same eigenvalue c that can be scaled so that $w^{\mathrm{T}} v = 1$. (Recall that an eigenvalue of A is also an eigenvalue of A^{T}, and if w is a left eigenvector associated with the eigenvalue c, then $A^{\mathrm{T}} w = cw$.) The only property of symmetry used above was that we could scale $v^{\mathrm{T}} v$ to be 1; hence, we just need $w^{\mathrm{T}} v \neq 0$. This is clearly true for a diagonalizable matrix (from the definition). It is also true if c is simple (which is somewhat harder to prove). It is thus true for the dominant eigenvalue, which is simple, in two important classes of matrices we will consider in Sections 8.7.1 and 8.7.2, positive matrices and irreducible nonnegative matrices.

If w and v are left and right eigenvectors of A associated with the same eigenvalue c and $w^{\mathrm{T}} v = 1$, then for $k = 1, 2, \ldots,$

$$\left(A - cvw^{\mathrm{T}}\right)^k = A^k - c^k vw^{\mathrm{T}}. \tag{3.212}$$

We can prove this by induction as above.

The Trace and Sums of Eigenvalues

For a general $n \times n$ matrix A with eigenvalues $c_1, \ldots, c_n$, we have $\mathrm{tr}(A) = \sum_{i=1}^{n} c_i$. (This is equation (3.180).) This is particularly easy to see for symmetric matrices because of equation (3.197), rewritten as $V^{\mathrm{T}}AV = C$, the diagonal matrix of the eigenvalues. For a symmetric matrix, however, we have a stronger result.

If A is an $n \times n$ symmetric matrix with eigenvalues $c_1 \geq \cdots \geq c_n$, and U is an $n \times k$ orthogonal matrix, with $k \leq n$, then

$$\mathrm{tr}(U^{\mathrm{T}}AU) \leq \sum_{i=1}^{k} c_i. \tag{3.213}$$

To see this, we represent U in terms of the columns of V, which span $\mathbb{R}^n$, as $U = VX$. Hence,

$$\begin{aligned}
\mathrm{tr}(U^{\mathrm{T}}AU) &= \mathrm{tr}(X^{\mathrm{T}}V^{\mathrm{T}}AVX) \\
&= \mathrm{tr}(X^{\mathrm{T}}CX) \\
&= \sum_{i=1}^{n} x_i^{\mathrm{T}} x_i \, c_i,
\end{aligned} \tag{3.214}$$

where x_i^{T} is the ith row of X.

Now $X^{\mathrm{T}}X = X^{\mathrm{T}}V^{\mathrm{T}}VX = U^{\mathrm{T}}U = I_k$, so either $x_i^{\mathrm{T}}x_i = 0$ or $x_i^{\mathrm{T}}x_i = 1$, and $\sum_{i=1}^{n} x_i^{\mathrm{T}}x_i = k$. Because $c_1 \geq \cdots \geq c_n$, therefore $\sum_{i=1}^{n} x_i^{\mathrm{T}}x_i \, c_i \leq \sum_{i=1}^{k} c_i$, and so from equation (3.214) we have $\mathrm{tr}(U^{\mathrm{T}}AU) \leq \sum_{i=1}^{k} c_i$.

3.8.8 Positive Definite and Nonnegative Definite Matrices

The factorization of symmetric matrices in equation (3.197) yields some useful properties of positive definite and nonnegative definite matrices (introduced on page 70). We will briefly discuss these properties here and then return to the subject in Section 8.3 and discuss more properties of positive definite and nonnegative definite matrices.

Eigenvalues of Positive and Nonnegative Definite Matrices

In this book, we use the terms "nonnegative definite" and "positive definite" only for real symmetric matrices, so the eigenvalues of nonnegative definite or positive definite matrices are real.

Any real symmetric matrix is positive (nonnegative) definite if and only if all of its eigenvalues are positive (nonnegative). We can see this using the factorization (3.197) of a symmetric matrix. One factor is the diagonal matrix

C of the eigenvalues, and the other factors are orthogonal. Hence, for any x, we have $x^{\mathrm{T}}Ax = x^{\mathrm{T}}VCV^{\mathrm{T}}x = y^{\mathrm{T}}Cy$, where $y = V^{\mathrm{T}}x$, and so

$$x^{\mathrm{T}}Ax > (\geq)\, 0$$

if and only if

$$y^{\mathrm{T}}Cy > (\geq)\, 0.$$

This, together with the resulting inequality (3.122) on page 89, implies that if P is a nonsingular matrix and D is a diagonal matrix, $P^{\mathrm{T}}DP$ is positive (nonnegative) if and only if the elements of D are positive (nonnegative).

A matrix (whether symmetric or not and whether real or not) all of whose eigenvalues have positive real parts is said to be *positive stable*. Positive stability is an important property in some applications, such as numerical solution of systems of nonlinear differential equations. Clearly, a positive definite matrix is positive stable.

Inverse of Positive Definite Matrices

If A is positive definite and $A = VCV^{\mathrm{T}}$ as in equation (3.197), then $A^{-1} = VC^{-1}V^{\mathrm{T}}$ and A^{-1} is positive definite because the elements of C^{-1} are positive.

Diagonalization of Positive Definite Matrices

If A is positive definite, the elements of the diagonal matrix C in equation (3.197) are positive, and so their square roots can be absorbed into V to form a nonsingular matrix P. The diagonalization in equation (3.198), $V^{\mathrm{T}}AV = C$, can therefore be reexpressed as

$$P^{\mathrm{T}}AP = I. \tag{3.215}$$

Square Roots of Positive and Nonnegative Definite Matrices

The factorization (3.197) together with the nonnegativity of the eigenvalues of positive and nonnegative definite matrices allows us to define a square root of such a matrix.

Let A be a nonnegative definite matrix and let V and C be as in equation (3.197): $A = VCV^{\mathrm{T}}$. Now, let S be a diagonal matrix whose elements are the square roots of the corresponding elements of C. Then $(VSV^{\mathrm{T}})^2 = A$; hence, we write

$$A^{\frac{1}{2}} = VSV^{\mathrm{T}} \tag{3.216}$$

and call this matrix the *square root* of A. This definition of the square root of a matrix is an instance of equation (3.195) with $f(x) = \sqrt{x}$. We also can similarly define $A^{\frac{1}{r}}$ for $r > 0$.

We see immediately that $A^{\frac{1}{2}}$ is symmetric because A is symmetric.

If A is positive definite, A^{-1} exists and is positive definite. It therefore has a square root, which we denote as $A^{-\frac{1}{2}}$.

The square roots are nonnegative, and so $A^{\frac{1}{2}}$ is nonnegative definite. Furthermore, $A^{\frac{1}{2}}$ and $A^{-\frac{1}{2}}$ are positive definite if A is positive definite.

In Section 5.9.1, we will show that this $A^{\frac{1}{2}}$ is unique, so our reference to it as the square root is appropriate. (There is occasionally some ambiguity in the terms "square root" and "second root" and the symbols used to denote them. If x is a nonnegative scalar, the usual meaning of its square root, denoted by $\sqrt{x}$, is a nonnegative number, while its second roots, which may be denoted by $x^{\frac{1}{2}}$, are usually considered to be either of the numbers $\pm\sqrt{x}$. In our notation $A^{\frac{1}{2}}$, we mean *the* square root; that is, the nonnegative matrix, if it exists. Otherwise, we say the square root of the matrix does not exist. For example, $I_2^{\frac{1}{2}} = I_2$, and while if $J = \begin{bmatrix} 0 & 1 \\ 1 & 0 \end{bmatrix}$, $J^2 = I_2$, we do not consider J to be a square root of I_2.)

3.8.9 The Generalized Eigenvalue Problem

The characterization of an eigenvalue as a root of the determinant equation (3.173) can be extended to define a *generalized eigenvalue* of the square matrices A and B to be a root in c of the equation

$$|A - cB| = 0 \tag{3.217}$$

if a root exists.

Equation (3.217) is equivalent to $A - cB$ being singular; that is, for some c and some nonzero, finite v,

$$Av = cBv.$$

Such a v (if it exists) is called the *generalized eigenvector*. In contrast to the existence of eigenvalues of any square matrix with finite elements, the generalized eigenvalues may not exist; that is, they may be infinite.

If B is nonsingular and A and B are $n \times n$, all n eigenvalues of A and B exist (and are finite). These generalized eigenvalues are the eigenvalues of AB^{-1} or $B^{-1}A$. We see this because $|B| \neq 0$, and so if c_0 is any of the n (finite) eigenvalues of AB^{-1} or $B^{-1}A$, then $0 = |AB^{-1} - c_0I| = |B^{-1}A - c_0I| = |A - c_0B| = 0$. Likewise, we see that any eigenvector of AB^{-1} or $B^{-1}A$ is a generalized eigenvector of A and B.

In the case of ordinary eigenvalues, we have seen that symmetry of the matrix induces some simplifications. In the case of generalized eigenvalues, symmetry together with positive definiteness yields some useful properties, which we will discuss in Section 7.6.

Generalized eigenvalue problems often arise in multivariate statistical applications. Roy's maximum root statistic, for example, is the largest generalized eigenvalue of two matrices that result from operations on a partitioned matrix of sums of squares.

Matrix Pencils

As c ranges over the reals (or, more generally, the complex numbers), the set of matrices of the form $A - cB$ is called the *matrix pencil*, or just the *pencil*, generated by A and B, denoted as

$$(A, B).$$

(In this definition, A and B do not need to be square.) A generalized eigenvalue of the square matrices A and B is called an eigenvalue of the pencil.

A pencil is said to be *regular* if $|A - cB|$ is not identically 0 (and, of course, if $|A - cB|$ is defined, meaning A and B are square). An interesting special case of a regular pencil is when B is nonsingular. As we have seen, in that case, eigenvalues of the pencil (A, B) exist (and are finite) and are the same as the ordinary eigenvalues of AB^{-1} or $B^{-1}A$, and the ordinary eigenvectors of AB^{-1} or $B^{-1}A$ are eigenvectors of the pencil (A, B).

3.8.10 Singular Values and the Singular Value Decomposition

An $n \times m$ matrix A can be factored as

$$A = UDV^{\mathrm{T}}, \tag{3.218}$$

where U is an $n \times n$ orthogonal matrix, V is an $m \times m$ orthogonal matrix, and D is an $n \times m$ diagonal matrix with nonnegative entries. (An $n \times m$ diagonal matrix has $\min(n, m)$ elements on the diagonal, and all other entries are zero.)

The number of positive entries in D is the same as the rank of A. (We see this by first recognizing that the number of nonzero entries of D is obviously the rank of D, and multiplication by the full rank matrices U and V^{T} yields a product with the same rank from equations (3.120) and (3.121).) The factorization (3.218) is called the *singular value decomposition* (SVD) or the *canonical singular value factorization* of A. The elements on the diagonal of D, d_i, are called the *singular values* of A.

If the rank of the matrix is r, we have $d_1 \geq \cdots \geq d_r > 0$, and if $r < \min(n, m)$, then $d_{r+1} = \cdots = d_{\min(n,m)} = 0$. In this case

$$D = \begin{bmatrix} D_r & 0 \\ 0 & 0 \end{bmatrix},$$

where $D_r = \mathrm{diag}(d_1, \ldots, d_r)$.

From the factorization (3.218) defining the singular values, we see that the singular values of A^{T} are the same as those of A.

For a matrix with more rows than columns, in an alternate definition of the singular value decomposition, the matrix U is $n \times m$ with orthogonal columns, and D is an $m \times m$ diagonal matrix with nonnegative entries. Likewise, for a matrix with more columns than rows, the singular value decomposition can be defined as above but with the matrix V being $m \times n$ with orthogonal columns and D being $m \times m$ and diagonal with nonnegative entries.

If A is symmetric, we see from equations (3.197) and (3.218) that the singular values are the absolute values of the eigenvalues.

The Fourier Expansion in Terms of the Singular Value Decomposition

From equation (3.218), we see that the general matrix A with rank r also has a Fourier expansion, similar to equation (3.200), in terms of the singular values and outer products of the columns of the U and V matrices:

$$A = \sum_{i=1}^{r} d_i u_i v_i^{\mathrm{T}}. \tag{3.219}$$

This is also called a spectral decomposition. The $u_i v_i^{\mathrm{T}}$ matrices in equation (3.219) have the property that $\langle u_i v_i^{\mathrm{T}}, u_j v_j^{\mathrm{T}} \rangle = 0$ for $i \neq j$ and $\langle u_i v_i^{\mathrm{T}}, u_i v_i^{\mathrm{T}} \rangle = 1$, and so the spectral decomposition is a Fourier expansion as in equation (3.82), and the singular values are Fourier coefficients.

The singular values d_i have the same properties as the Fourier coefficients in any orthonormal expansion. For example, from equation (3.83), we see that the singular values can be represented as the dot product

$$d_i = \langle A, u_i v_i^{\mathrm{T}} \rangle.$$

After we have discussed matrix norms in the next section, we will formulate Parseval's identity for this Fourier expansion.

3.9 Matrix Norms

Norms on matrices are scalar functions of matrices with the three properties on page 16 that define a norm in general. Matrix norms are often required to have another property, called the *consistency property*, in addition to the properties listed on page 16, which we repeat here for convenience. Assume A and B are matrices conformable for the operations shown.

1. Nonnegativity and mapping of the identity:
   if $A \neq 0$, then $\|A\| > 0$, and $\|0\| = 0$.
2. Relation of scalar multiplication to real multiplication:
   $\|aA\| = |a|\,\|A\|$ for real a.
3. Triangle inequality:
   $\|A + B\| \leq \|A\| + \|B\|$.
4. Consistency property:
   $\|AB\| \leq \|A\|\,\|B\|$.

Some people do not require the consistency property for a matrix norm. Most useful matrix norms have the property, however, and we will consider it to be a requirement in the definition. The consistency property for multiplication is similar to the triangular inequality for addition.

Any function from $\mathrm{I\!R}^{n \times m}$ to $\mathrm{I\!R}$ that satisfies these four properties is a matrix norm.

We note that the four properties of a matrix norm do not imply that it is invariant to transposition of a matrix, and in general, $\|A^{\mathrm{T}}\| \neq \|A\|$. Some matrix norms are the same for the transpose of a matrix as for the original matrix. For instance, because of the property of the matrix dot product given in equation (3.79), we see that a norm defined by that inner product would be invariant to transposition.

For a square matrix A, the consistency property for a matrix norm yields

$$\|A^k\| \leq \|A\|^k \tag{3.220}$$

for any positive integer k.

A matrix norm $\|\cdot\|$ is *orthogonally invariant* if A and B being orthogonally similar implies $\|A\| = \|B\|$.

3.9.1 Matrix Norms Induced from Vector Norms

Some matrix norms are defined in terms of vector norms. For clarity, we will denote a vector norm as $\|\cdot\|_{\mathrm{v}}$ and a matrix norm as $\|\cdot\|_{\mathrm{M}}$. (This notation is meant to be generic; that is, $\|\cdot\|_{\mathrm{v}}$ represents any vector norm.) The matrix norm $\|\cdot\|_{\mathrm{M}}$ *induced* by $\|\cdot\|_{\mathrm{v}}$ is defined by

$$\|A\|_{\mathrm{M}} = \max_{x \neq 0} \frac{\|Ax\|_{\mathrm{v}}}{\|x\|_{\mathrm{v}}}. \tag{3.221}$$

It is easy to see that an induced norm is indeed a matrix norm. The first three properties of a norm are immediate, and the consistency property can be verified by applying the definition (3.221) to AB and replacing Bx with y; that is, using Ay.

We usually drop the v or M subscript, and the notation $\|\cdot\|$ is overloaded to mean either a vector or matrix norm. (Overloading of symbols occurs in many contexts, and we usually do not even recognize that the meaning is context-dependent. In computer language design, overloading must be recognized explicitly because the language specifications must be explicit.)

The induced norm of A given in equation (3.221) is sometimes called the *maximum magnification* by A. The expression looks very similar to the maximum eigenvalue, and indeed it is in some cases.

For any vector norm and its induced matrix norm, we see from equation (3.221) that

$$\|Ax\| \leq \|A\|\,\|x\| \tag{3.222}$$

because $\|x\| \geq 0$.

L_p Matrix Norms

The matrix norms that correspond to the L_p vector norms are defined for the $n \times m$ matrix A as

$$\|A\|_p = \max_{\|x\|_p = 1} \|Ax\|_p. \tag{3.223}$$

(Notice that the restriction on $\|x\|_p$ makes this an induced norm as defined in equation (3.221). Notice also the overloading of the symbols; the norm on the left that is being defined is a matrix norm, whereas those on the right of the equation are vector norms.) It is clear that the L_p matrix norms satisfy the consistency property, because they are induced norms.

The L_1 and L_∞ norms have interesting simplifications of equation (3.221):

$$\|A\|_1 = \max_j \sum_i |a_{ij}|, \tag{3.224}$$

so the L_1 is also called the *column-sum norm*; and

$$\|A\|_\infty = \max_i \sum_j |a_{ij}|, \tag{3.225}$$

so the L_∞ is also called the *row-sum norm*. We see these relationships by considering the L_p norm of the vector

$$v = (a_{1*}^{\mathrm{T}} x, \ldots, a_{n*}^{\mathrm{T}} x),$$

where a_{i*} is the ith row of A, with the restriction that $\|x\|_p = 1$. The L_p norm of this vector is based on the absolute values of the elements; that is, $|\sum_j a_{ij} x_j|$ for $i = 1, \ldots, n$. Because we are free to choose x (subject to the restriction that $\|x\|_p = 1$), for a given i, we can choose the sign of each x_j to maximize the overall expression. For example, for a fixed i, we can choose each x_j to have the same sign as a_{ij}, and so $|\sum_j a_{ij} x_j|$ is the same as $\sum_j |a_{ij}| |x_j|$.

For the column-sum norm, the L_1 norm of v is $\sum_i |a_{i*}^{\mathrm{T}} x|$. The elements of x are chosen to maximize this under the restriction that $\sum |x_j| = 1$. The maximum of the expression is attained by setting $x_k = \text{sign}(\sum_i a_{ik})$, where k is such that $|\sum_i a_{ik}| \geq |\sum_i a_{ij}|$, for $j = 1, \ldots, m$, and $x_q = 0$ for $q = 1, \ldots m$ and $q \neq k$. (If there is no unique k, any choice will yield the same result.) This yields equation (3.224).

For the row-sum norm, the L_∞ norm of v is

$$\max_i |a_{i*}^{\mathrm{T}} x| = \max_i \sum_j |a_{ij}| |x_j|$$

when the sign of x_j is chosen appropriately (for a given i). The elements of x must be chosen so that $\max |x_j| = 1$; hence, each x_j is chosen as ± 1. The maximum $|a_{i*}^{\mathrm{T}} x|$ is attained by setting $x_j = \text{sign}(a_{kj})$, for $j = 1, \ldots m$, where k is such that $\sum_j |a_{kj}| \geq \sum_j |a_{ij}|$, for $i = 1, \ldots, n$. This yields equation (3.225).

From equations (3.224) and (3.225), we see that

$$\|A^{\mathrm{T}}\|_\infty = \|A\|_1. \tag{3.226}$$

Alternative formulations of the L_2 norm of a matrix are not so obvious from equation (3.223). It is related to the eigenvalues (or the singular values) of the matrix. The L_2 matrix norm is related to the spectral radius (page 111):

$$\|A\|_2 = \sqrt{\rho(A^{\mathrm{T}}A)}, \tag{3.227}$$

(see Exercise 3.24, page 142). Because of this relationship, the L_2 matrix norm is also called the *spectral norm*.

From the invariance of the singular values to matrix transposition, we see that positive eigenvalues of $A^{\mathrm{T}}A$ are the same as those of AA^{T}; hence, $\|A^{\mathrm{T}}\|_2 = \|A\|_2$.

For Q orthogonal, the L_2 vector norm has the important property

$$\|Qx\|_2 = \|x\|_2 \tag{3.228}$$

(see Exercise 3.25a, page 142). For this reason, an orthogonal matrix is sometimes called an *isometric matrix*. By the proper choice of x, it is easy to see from equation (3.228) that

$$\|Q\|_2 = 1. \tag{3.229}$$

Also from this we see that if A and B are orthogonally similar, then $\|A\|_2 = \|B\|_2$; hence, the spectral matrix norm is orthogonally invariant.

The L_2 matrix norm is a Euclidean-type norm since it is induced by the Euclidean vector norm (but it is not called the Euclidean matrix norm; see below).

L_1, L_2, and L_∞ Norms of Symmetric Matrices

For a symmetric matrix A, we have the obvious relationships

$$\|A\|_1 = \|A\|_\infty \tag{3.230}$$

and, from equation (3.227),

$$\|A\|_2 = \rho(A). \tag{3.231}$$

3.9.2 The Frobenius Norm — The "Usual" Norm

The *Frobenius norm* is defined as

$$\|A\|_{\mathrm{F}} = \sqrt{\sum_{i,j} a_{ij}^2}. \tag{3.232}$$

It is easy to see that this measure has the consistency property (Exercise 3.27), as a norm must. The Frobenius norm is sometimes called the *Euclidean matrix norm* and denoted by $\|\cdot\|_E$, although the L_2 matrix norm is more directly based on the Euclidean vector norm, as we mentioned above. We will usually use the notation $\|\cdot\|_F$ to denote the Frobenius norm. Occasionally we use $\|\cdot\|$ without the subscript to denote the Frobenius norm, but usually the symbol without the subscript indicates that any norm could be used in the expression. The Frobenius norm is also often called the "usual norm", which emphasizes the fact that it is one of the most useful matrix norms. Other names sometimes used to refer to the Frobenius norm are *Hilbert-Schmidt norm* and *Schur norm*.

A useful property of the Frobenius norm that is obvious from the definition is

$$\|A\|_F = \sqrt{\mathrm{tr}(A^T A)}$$
$$= \sqrt{\langle A, A \rangle};$$

that is,

- the Frobenius norm is the norm that arises from the matrix inner product (see page 74).

From the commutativity of an inner product, we have $\|A^T\|_F = \|A\|_F$. We have seen that the L_2 matrix norm also has this property.

Similar to defining the angle between two vectors in terms of the inner product and the norm arising from the inner product, we define the *angle* between two matrices A and B of the same size and shape as

$$\mathrm{angle}(A, B) = \cos^{-1}\left(\frac{\langle A, B \rangle}{\|A\|_F \|B\|_F}\right). \tag{3.233}$$

If Q is an $n \times m$ orthogonal matrix, then

$$\|Q\|_F = \sqrt{m} \tag{3.234}$$

(see equation (3.169)).

If A and B are orthogonally similar (see equation (3.191)), then

$$\|A\|_F = \|B\|_F;$$

that is, the Frobenius norm is an orthogonally invariant norm. To see this, let $A = Q^T B Q$, where Q is an orthogonal matrix. Then

$$\begin{aligned}
\|A\|_F^2 &= \mathrm{tr}(A^T A) \\
&= \mathrm{tr}(Q^T B^T Q Q^T B Q) \\
&= \mathrm{tr}(B^T B Q Q^T) \\
&= \mathrm{tr}(B^T B) \\
&= \|B\|_F^2.
\end{aligned}$$

(The norms are nonnegative, of course, and so equality of the squares is sufficient.)

Parseval's Identity

Several important properties result because the Frobenius norm arises from an inner product. For example, following the Fourier expansion in terms of the singular value decomposition, equation (3.219), we mentioned that the singular values have the general properties of Fourier coefficients; for example, they satisfy Parseval's identity, equation (2.38), on page 29. This identity states that the sum of the squares of the Fourier coefficients is equal to the square of the norm that arises from the inner product used in the Fourier expansion. Hence, we have the important property of the Frobenius norm that the square of the norm is the sum of squares of the singular values of the matrix:

$$\|A\|_{\mathrm{F}}^2 = \sum d_i^2. \tag{3.235}$$

3.9.3 Matrix Norm Inequalities

There is an equivalence among any two matrix norms similar to that of expression (2.17) for vector norms (over finite-dimensional vector spaces). If $\|\cdot\|_a$ and $\|\cdot\|_b$ are matrix norms, then there are positive numbers r and s such that, for any matrix A,

$$r\|A\|_b \leq \|A\|_a \leq s\|A\|_b. \tag{3.236}$$

We will not prove this result in general but, in Exercise 3.28, ask the reader to do so for matrix norms induced by vector norms. These induced norms include the matrix L_p norms of course.

If A is an $n \times m$ real matrix, we have some specific instances of (3.236):

$$\|A\|_\infty \leq \sqrt{m}\,\|A\|_{\mathrm{F}}, \tag{3.237}$$

$$\|A\|_{\mathrm{F}} \leq \sqrt{\min(n, m)}\,\|A\|_2, \tag{3.238}$$

$$\|A\|_2 \leq \sqrt{m}\,\|A\|_1, \tag{3.239}$$

$$\|A\|_1 \leq \sqrt{n}\,\|A\|_2, \tag{3.240}$$

$$\|A\|_2 \leq \|A\|_{\mathrm{F}}, \tag{3.241}$$

$$\|A\|_{\mathrm{F}} \leq \sqrt{n}\,\|A\|_\infty. \tag{3.242}$$

See Exercises 3.29 and 3.30 on page 143. Compare these inequalities with those for L_p vector norms on page 18. Recall specifically that for vector L_p norms we had the useful fact that for a given x and for $p \geq 1$, $\|x\|_p$ is a nonincreasing function of p; and specifically we had inequality (2.12):

$$\|x\|_\infty \leq \|x\|_2 \leq \|x\|_1.$$

3.9.4 The Spectral Radius

The spectral radius is the appropriate measure of the condition of a square matrix for certain iterative algorithms. Except in the case of symmetric matrices, as shown in equation (3.231), the spectral radius is not a norm (see Exercise 3.31a).

We have for any norm $\| \cdot \|$ and any square matrix A that

$$\rho(A) \leq \|A\|. \tag{3.243}$$

To see this, we consider the associated eigenvalue and eigenvector c_i and v_i and form the matrix $V = [v_i|0|\cdots|0]$, so $c_i V = AV$, and by the consistency property of any matrix norm,

$$\begin{aligned}
|c_i|\|V\| &= \|c_i V\| \\
&= \|AV\| \\
&\leq \|A\|\,\|V\|,
\end{aligned}$$

or

$$|c_i| \leq \|A\|,$$

(see also Exercise 3.31b).

The inequality (3.243) and the L_1 and L_∞ norms yield useful bounds on the eigenvalues and the maximum absolute row and column sums of matrices: the modulus of any eigenvalue is no greater than the largest sum of absolute values of the elements in any row or column.

The inequality (3.243) and equation (3.231) also yield a minimum property of the L_2 norm of a symmetric matrix A:

$$\|A\|_2 \leq \|A\|.$$

3.9.5 Convergence of a Matrix Power Series

We define the convergence of a sequence of matrices in terms of the convergence of a sequence of their norms, just as we did for a sequence of vectors (on page 20). We say that a sequence of matrices $A_1, A_2, \ldots$ (of the same shape) converges to the matrix A with respect to the norm $\| \cdot \|$ if the sequence of real numbers $\|A_1 - A\|$, $\|A_2 - A\|$, $\ldots$ converges to 0. Because of the equivalence property of norms, the choice of the norm is irrelevant. Also, because of inequality (3.243), we see that the convergence of the sequence of spectral radii $\rho(A_1 - A), \rho(A_2 - A), \ldots$ to 0 must imply the convergence of $A_1, A_2, \ldots$ to A.

Conditions for Convergence of a Sequence of Powers

For a square matrix A, we have the important fact that

$$A^k \rightarrow 0, \quad \text{if } \|A\| < 1, \tag{3.244}$$

where 0 is the square zero matrix of the same order as A and $\|\cdot\|$ is any matrix norm. (The consistency property is required.) This convergence follows from inequality (3.220) because that yields $\lim_{k\to\infty} \|A^k\| \leq \lim_{k\to\infty} \|A\|^k$, and so if $\|A\| < 1$, then $\lim_{k\to\infty} \|A^k\| = 0$.

Now consider the spectral radius. Because of the spectral decomposition, we would expect the spectral radius to be related to the convergence of a sequence of powers of a matrix. If $A^k \rightarrow 0$, then for any conformable vector x, $A^k x \rightarrow 0$; in particular, for the eigenvector $v_1 \neq 0$ corresponding to the dominant eigenvalue c_1, we have $A^k v_1 = c_1^k v_1 \rightarrow 0$. For $c_1^k v_1$ to converge to zero, we must have $|c_1| < 1$; that is, $\rho(A) < 1$. We can also show the converse:

$$A^k \rightarrow 0 \quad \text{if } \rho(A) < 1. \tag{3.245}$$

We will do this by defining a norm $\|\cdot\|_d$ in terms of the L_1 matrix norm in such a way that $\rho(A) < 1$ implies $\|A\|_d < 1$. Then we can use equation (3.244) to establish the convergence.

Let $A = QTQ^{\mathrm{T}}$ be the Schur factorization of the $n \times n$ matrix A, where Q is orthogonal and T is upper triangular with the same eigenvalues as A, $c_1, \ldots, c_n$. Now for any $d > 0$, form the diagonal matrix $D = \mathrm{diag}(d^1, \ldots, d^n)$. Notice that DTD^{-1} is an upper triangular matrix and its diagonal elements (which are its eigenvalues) are the same as the eigenvalues of T and A. Consider the column sums of the absolute values of the elements of DTD^{-1}:

$$|c_j| + \sum_{i=1}^{j-1} d^{-(j-i)} |t_{ij}|.$$

Now, because $|c_j| \leq \rho(A)$ for given $\epsilon > 0$, by choosing d large enough, we have

$$|c_j| + \sum_{i=1}^{j-1} d^{-(j-i)} |t_{ij}| < \rho(A) + \epsilon,$$

or

$$\|DTD^{-1}\|_1 = \max_j \left(|c_j| + \sum_{i=1}^{j-1} d^{-(j-i)} |t_{ij}| \right) < \rho(A) + \epsilon.$$

Now define $\| \cdot \|_d$ for any $n \times n$ matrix X, where Q is the orthogonal matrix in the Schur factorization and D is as defined above, as

$$\|X\|_d = \|(QD^{-1})^{-1}X(QD^{-1})\|_1. \tag{3.246}$$

Now $\| \cdot \|_d$ is a norm (Exercise 3.32). Furthermore,

$$\begin{aligned}
\|A\|_d &= \|(QD^{-1})^{-1}A(QD^{-1})\|_1 \\
&= \|DTD^{-1}\|_1 \\
&< \rho(A) + \epsilon,
\end{aligned}$$

and so if $\rho(A) < 1$, ϵ and d can be chosen so that $\|A\|_d < 1$, and by equation (3.244) above, we have $A^k \to 0$; hence, we conclude that

$$A^k \to 0 \quad \text{if and only if } \rho(A) < 1. \tag{3.247}$$

From inequality (3.243) and the fact that $\rho(A^k) = \rho(A)^k$, we have $\rho(A) \leq \|A^k\|^{1/k}$. Now, for any $\epsilon > 0$, $\rho\big(A/(\rho(A) + \epsilon)\big) < 1$ and so

$$\lim_{k \to \infty} \big(A/(\rho(A) + \epsilon)\big)^k = 0$$

from expression (3.247); hence,

$$\lim_{k \to \infty} \frac{\|A^k\|}{(\rho(A) + \epsilon)^k} = 0.$$

There is therefore a positive integer M_ϵ such that $\|A^k\|/(\rho(A) + \epsilon)^k < 1$ for all $k > M_\epsilon$, and hence $\|A^k\|^{1/k} < (\rho(A) + \epsilon)$ for $k > M_\epsilon$. We have therefore, for any $\epsilon > 0$,

$$\rho(A) \leq \|A^k\|^{1/k} < \rho(A) + \epsilon \quad \text{for } k > M_\epsilon,$$

and thus

$$\lim_{k \to \infty} \|A^k\|^{1/k} = \rho(A). \tag{3.248}$$

Convergence of a Power Series; Inverse of $I - A$

Consider the power series in an $n \times n$ matrix such as in equation (3.140) on page 94,

$$I + A + A^2 + A^3 + \cdots .$$

In the standard fashion for dealing with series, we form the partial sum

$$S_k = I + A + A^2 + A^3 + \cdots A^k$$

and consider $\lim_{k \to \infty} S_k$. We first note that

$$(I - A)S_k = I - A^{k+1}$$

and observe that if $A^{k+1} \to 0$, then $S_k \to (I - A)^{-1}$, which is equation (3.140). Therefore,

$$(I - A)^{-1} = I + A + A^2 + A^3 + \cdots \quad \text{if } \|A\| < 1. \tag{3.249}$$

Nilpotent Matrices

The condition in equation (3.236) is not necessary; that is, if $A^k \to 0$, it may be the case that, for some norm, $\|A\| > 1$. A simple example is

$$A = \begin{bmatrix} 0 & 2 \\ 0 & 0 \end{bmatrix}.$$

For this matrix, $A^2 = 0$, yet $\|A\|_1 = \|A\|_2 = \|A\|_\infty = \|A\|_F = 2$.

A matrix like A above such that its product with itself is 0 is called *nilpotent*. More generally, for a square matrix A, if $A^k = 0$ for some positive integer k, but $A^{k-1} \neq 0$, A is said to be *nilpotent of index k*. Strictly speaking, a nilpotent matrix is nilpotent of index 2, but often the term "nilpotent" without qualification is used to refer to a matrix that is nilpotent of any index. A simple example of a matrix that is nilpotent of index 3 is

$$A = \begin{bmatrix} 0 & 0 & 0 \\ 1 & 0 & 0 \\ 0 & 1 & 0 \end{bmatrix}.$$

It is easy to see that if $A_{n \times n}$ is nilpotent, then

$$\text{tr}(A) = 0, \tag{3.250}$$

$$\rho(A) = 0, \tag{3.251}$$

(that is, all eigenvalues of A are 0), and

$$\text{rank}(A) = n - 1. \tag{3.252}$$

You are asked to supply the proofs of these statements in Exercise 3.33.

In applications, for example in time series or other stochastic processes, because of expression (3.247), the spectral radius is often the most useful. Stochastic processes may be characterized by whether the absolute value of the dominant eigenvalue (spectral radius) of a certain matrix is less than 1. Interesting special cases occur when the dominant eigenvalue is equal to 1.

3.10 Approximation of Matrices

In Section 2.2.6, we discussed the problem of approximating a given vector in terms of vectors from a lower dimensional space. Likewise, it is often of interest to approximate one matrix by another. In statistical applications, we may wish to find a matrix of smaller rank that contains a large portion of the information content of a matrix of larger rank ("dimension reduction"as on page 345; or variable selection as in Section 9.4.2, for example), or we may want to impose conditions on an estimate that it have properties known to be possessed by the estimand (positive definiteness of the correlation matrix, for example, as in Section 9.4.6). In numerical linear algebra, we may wish to find a matrix that is easier to compute or that has properties that ensure more stable computations.

Metric for the Difference of Two Matrices

A natural way to assess the goodness of the approximation is by a norm of the difference (that is, by a *metric induced by a norm*), as discussed on page 22. If $\widetilde{A}$ is an approximation to A, we measure the quality of the approximation by $\|A - \widetilde{A}\|$ for some norm. In the following, we will measure the goodness of the approximation using the norm that arises from the inner product (the Frobenius norm).

Best Approximation with a Matrix of Given Rank

Suppose we want the best approximation to an $n \times m$ matrix A of rank r by a matrix $\widetilde{A}$ in $\mathbb{R}^{n \times m}$ but with smaller rank, say k; that is, we want to find $\widetilde{A}$ of rank k such that

$$\|A - \widetilde{A}\|_{\mathrm{F}} \tag{3.253}$$

is a minimum for all $\widetilde{A} \in \mathbb{R}^{n \times m}$ of rank k.

We have an orthogonal basis in terms of the singular value decomposition, equation (3.219), for some subspace of $\mathbb{R}^{n \times m}$, and we know that the Fourier coefficients provide the best approximation for any subset of k basis matrices, as in equation (2.43). This Fourier fit would have rank k as required, but it would be the best only within that set of expansions. (This is the limitation imposed in equation (2.43).) Another approach to determine the best fit could be developed by representing the columns of the approximating matrix as linear combinations of the given matrix A and then expanding $\|A - \widetilde{A}\|_{\mathrm{F}}^2$. Neither the Fourier expansion nor the restriction $\mathcal{V}(\widetilde{A}) \subset \mathcal{V}(A)$ permit us to address the question of what is the overall best approximation of rank k within $\mathbb{R}^{n \times m}$. As we see below, however, there is a minimum of expression (3.253) that occurs within $\mathcal{V}(A)$, and a minimum is at the truncated Fourier expansion in the singular values (equation (3.219)).

To state this more precisely, let A be an $n \times m$ matrix of rank r with singular value decomposition

$$A = U \begin{bmatrix} D_r & 0 \\ 0 & 0 \end{bmatrix} V^{\mathrm{T}},$$

where $D_r = \mathrm{diag}(d_1, \ldots, d_r)$, and the singular values are indexed so that $d_1 \geq \cdots \geq d_r > 0$. Then, for all $n \times m$ matrices X with rank $k < r$,

$$\|A - X\|_{\mathrm{F}}^2 \geq \sum_{i=k+1}^{r} d_i^2, \tag{3.254}$$

and this minimum occurs for $X = \widetilde{A}$, where

$$\widetilde{A} = U \begin{bmatrix} D_k & 0 \\ 0 & 0 \end{bmatrix} V^{\mathrm{T}}. \tag{3.255}$$

To see this, for any X, let Q be an $n \times k$ matrix whose columns are an orthonormal basis for $\mathcal{V}(X)$, and let $X = QY$, where Y is a $k \times n$ matrix, also of rank k. The minimization problem now is

$$\min_Y \|A - QY\|_{\mathrm{F}}$$

with the restriction $\mathrm{rank}(Y) = k$.

Now, expanding, completing the Gramian and using its nonnegative definiteness, and permuting the factors within a trace, we have

$$
\begin{aligned}
\|A - QY\|_{\mathrm{F}}^2 &= \mathrm{tr}\left((A - QY)^{\mathrm{T}}(A - QY)\right) \\
&= \mathrm{tr}\left(A^{\mathrm{T}}A\right) + \mathrm{tr}\left(Y^{\mathrm{T}}Y - A^{\mathrm{T}}QY - Y^{\mathrm{T}}Q^{\mathrm{T}}A\right) \\
&= \mathrm{tr}\left(A^{\mathrm{T}}A\right) + \mathrm{tr}\left((Y - Q^{\mathrm{T}}A)^{\mathrm{T}}(Y - Q^{\mathrm{T}}A)\right) - \mathrm{tr}\left(A^{\mathrm{T}}QQ^{\mathrm{T}}A\right) \\
&\geq \mathrm{tr}\left(A^{\mathrm{T}}A\right) - \mathrm{tr}\left(Q^{\mathrm{T}}AA^{\mathrm{T}}Q\right).
\end{aligned}
$$

The squares of the singular values of A are the eigenvalues of $A^{\mathrm{T}}A$, and so $\mathrm{tr}(A^{\mathrm{T}}A) = \sum_{i=1}^r d_i^2$. The eigenvalues of $A^{\mathrm{T}}A$ are also the eigenvalues of AA^{T}, and so, from inequality (3.213), $\mathrm{tr}(Q^{\mathrm{T}}AA^{\mathrm{T}}Q) \leq \sum_{i=1}^k d_i^2$, and so

$$\|A - X\|_{\mathrm{F}}^2 \geq \sum_{i=1}^r d_i^2 - \sum_{i=1}^k d_i^2;$$

hence, we have inequality (3.254). (This technique of "completing the Gramian" when an orthogonal matrix is present in a sum is somewhat similar to the technique of completing the square; it results in the difference of two Gramian matrices, which are defined in Section 3.3.7.)

Direct expansion of $\|A - \widetilde{A}\|_{\mathrm{F}}^2$ yields

$$\mathrm{tr}\left(A^{\mathrm{T}}A\right) - 2\mathrm{tr}\left(A^{\mathrm{T}}\widetilde{A}\right) + \mathrm{tr}\left(\widetilde{A}^{\mathrm{T}}\widetilde{A}\right) = \sum_{i=1}^r d_i^2 - \sum_{i=1}^k d_i^2,$$

and hence $\widetilde{A}$ is the best rank k approximation to A under the Frobenius norm.

Equation (3.255) can be stated another way: the best approximation of A of rank k is

$$\widetilde{A} = \sum_{i=1}^k d_i u_i v_i^{\mathrm{T}}. \tag{3.256}$$

This result for the best approximation of a given matrix by one of lower rank was first shown by Eckart and Young (1936). On page 271, we will discuss a bound on the difference between two symmetric matrices whether of the same or different ranks.

In applications, the rank k may be stated a priori or we examine a sequence $k = r - 1,\ r - 2,\ \ldots$, and determine the norm of the best fit at each rank. If s_k is the norm of the best approximating matrix, the sequence

s_{r-1}, s_{r-2}, ... may suggest a value of k for which the reduction in rank is sufficient for our purposes and the loss in closeness of the approximation is not too great. Principal components analysis is a special case of this process (see Section 9.3).

Exercises

3.1. Vector spaces of matrices.
 a) Exhibit a basis set for $\mathrm{I\!R}^{n \times m}$ for $n \geq m$.
 b) Does the set of $n \times m$ diagonal matrices form a vector space? (The answer is yes.) Exhibit a basis set for this vector space (assuming $n \geq m$).
 c) Exhibit a basis set for the vector space of $n \times n$ symmetric matrices.
 d) Show that the cardinality of any basis set for the vector space of $n \times n$ symmetric matrices is $n(n + 1)/2$.

3.2. By expanding the expression on the left-hand side, derive equation (3.64) on page 70.

3.3. Show that for any quadratic form $x^{\mathrm{T}} A x$ there is a symmetric matrix A_s such that $x^{\mathrm{T}} A_s x = x^{\mathrm{T}} A x$. (The proof is by construction, with $A_s = \frac{1}{2}(A + A^{\mathrm{T}})$, first showing A_s is symmetric and then that $x^{\mathrm{T}} A_s x = x^{\mathrm{T}} A x$.)

3.4. Give conditions on a, b, and c for the matrix below to be positive definite.

$$\begin{bmatrix} a & b \\ b & c \end{bmatrix}.$$

3.5. Show that the Mahalanobis distance defined in equation (3.67) is a metric (that is, show that it satisfies the properties listed on page 22).

3.6. Verify the relationships for Kronecker products shown in equations (3.70) through (3.74) on page 73.
 Make liberal use of equation (3.69) and previously verified equations.

3.7. Cauchy-Schwarz inequalities for matrices.
 a) Prove the Cauchy-Schwarz inequality for the dot product of matrices ((3.80), page 75), which can also be written as

$$(\mathrm{tr}(A^{\mathrm{T}} B))^2 \leq \mathrm{tr}(A^{\mathrm{T}} A)\mathrm{tr}(B^{\mathrm{T}} B).$$

 b) Prove the Cauchy-Schwarz inequality for determinants of matrices A and B of the same shape:

$$|(A^{\mathrm{T}} B)|^2 \leq |A^{\mathrm{T}} A||B^{\mathrm{T}} B|.$$

 Under what conditions is equality achieved?
 c) Let A and B be matrices of the same shape, and define

$$p(A, B) = |A^{\mathrm{T}} B|.$$

 Is $p(\cdot, \cdot)$ an inner product? Why or why not?

3.8. Prove that a square matrix that is either row or column diagonally dominant is nonsingular.

3.9. Prove that a positive definite matrix is nonsingular.

3.10. Let A be an $n \times m$ matrix.
   a) Under what conditions does A have a Hadamard multiplicative inverse?
   b) If A has a Hadamard multiplicative inverse, what is it?

3.11. The affine group $\mathcal{AL}(n)$.
   a) What is the identity in $\mathcal{AL}(n)$?
   b) Let (A, v) be an element of $\mathcal{AL}(n)$. What is the inverse of (A, v)?

3.12. Verify the relationships shown in equations (3.133) through (3.139) on page 93. Do this by multiplying the appropriate matrices. For example, the first equation is verified by the equations

$$(I + A^{-1})A(I + A)^{-1} = (A + I)(I + A)^{-1} = (I + A)(I + A)^{-1} = I.$$

Make liberal use of equation (3.132) and previously verified equations. Of course it is much more interesting to derive relationships such as these rather than merely to verify them. The verification, however, often gives an indication of how the relationship would arise naturally.

3.13. By writing $AA^{-1} = I$, derive the expression for the inverse of a partitioned matrix given in equation (3.145).

3.14. Show that the expression given for the generalized inverse in equation (3.165) on page 101 is correct.

3.15. Show that the expression given in equation (3.167) on page 102 is a Moore-Penrose inverse of A. (Show that properties 1 through 4 hold.)

3.16. Write formal proofs of the properties of eigenvalues/vectors listed on page 107.

3.17. Let A be a square matrix with an eigenvalue c and corresponding eigenvector v. Consider the matrix polynomial in A

$$p(A) = b_0 I + b_1 A + \cdots + b_k A^k.$$

Show that if (c, v) is an eigenpair of A, then $p(c)$, that is,

$$b_0 + b_1 c + \cdots + b_k c^k,$$

is an eigenvalue of $p(A)$ with corresponding eigenvector v. (Technically, the symbol $p(\cdot)$ is overloaded in these two instances.)

3.18. Write formal proofs of the properties of eigenvalues/vectors listed on page 110.

3.19. a) Show that the unit vectors are eigenvectors of a diagonal matrix.
   b) Give an example of two similar matrices whose eigenvectors are not the same.

   Hint: In equation (3.190), let A be a 2×2 diagonal matrix (so you know its eigenvalues and eigenvectors) with unequal values along the diagonal, and let P be a 2×2 upper triangular matrix, so that you can invert it. Form B and check the eigenvectors.

3.20. Let A be a diagonalizable matrix (not necessarily symmetric) with a spectral decomposition of the form of equation (3.205), $A = \sum_i c_i P_i$. Let c_j be a simple eigenvalue with associated left and right eigenvectors y_j and x_j, respectively. (Note that because A is not symmetric, it may have nonreal eigenvalues and eigenvectors.)
   a) Show that $y_j^{\mathrm{H}} x_j \neq 0$.
   b) Show that the projection matrix P_j is $x_j y_j^{\mathrm{H}} / y_j^{\mathrm{H}} x_j$.

3.21. If A is nonsingular, show that for any (conformable) vector x

$$(x^{\mathrm{T}} A x)(x^{\mathrm{T}} A^{-1} x) \geq (x^{\mathrm{T}} x)^2.$$

   Hint: Use the square roots and the Cauchy-Schwarz inequality.

3.22. Prove that the induced norm (page 129) is a matrix norm; that is, prove that it satisfies the consistency property.

3.23. Prove the inequality (3.222) for an induced matrix norm on page 129:

$$\|Ax\| \leq \|A\| \, \|x\|.$$

3.24. Prove that, for the square matrix A,

$$\|A\|_2^2 = \rho(A^{\mathrm{T}} A).$$

   Hint: Show that $\|A\|_2^2 = \max x^{\mathrm{T}} A^{\mathrm{T}} A x$ for any normalized vector x.

3.25. Let Q be an $n \times n$ orthogonal matrix, and let x be an n-vector.
   a) Prove equation (3.228):

$$\|Qx\|_2 = \|x\|_2.$$

   Hint: Write $\|Qx\|_2$ as $\sqrt{(Qx)^{\mathrm{T}} Qx}$.
   b) Give examples to show that this does not hold for other norms.

3.26. The triangle inequality for matrix norms: $\|A + B\| \leq \|A\| + \|B\|$.
   a) Prove the triangle inequality for the matrix L_1 norm.
   b) Prove the triangle inequality for the matrix L_∞ norm.
   c) Prove the triangle inequality for the matrix Frobenius norm.

3.27. Prove that the Frobenius norm satisfies the consistency property.

3.28. If $\|\cdot\|_a$ and $\|\cdot\|_b$ are matrix norms induced respectively by the vector norms $\|\cdot\|_{v_a}$ and $\|\cdot\|_{v_b}$, prove inequality (3.236); that is, show that there are positive numbers r and s such that, for any A,

$$r\|A\|_b \leq \|A\|_a \leq s\|A\|_b.$$

3.29. Use the Cauchy-Schwarz inequality to prove that for any square matrix A with real elements,

$$\|A\|_2 \leq \|A\|_{\mathrm{F}}.$$

3.30. Prove inequalities (3.237) through (3.242), and show that the bounds
are sharp by exhibiting instances of equality.

3.31. The spectral radius, $\rho(A)$.

a) We have seen by an example that $\rho(A) = 0$ does not imply $A = 0$.
What about other properties of a matrix norm? For each, either
show that the property holds for the spectral radius or, by means
of an example, that it does not hold.

b) Use the outer product of an eigenvector and the one vector to show
that for any norm $\| \cdot \|$ and any matrix A, $\rho(A) \leq \|A\|$.

3.32. Show that the function $\| \cdot \|_d$ defined in equation (3.246) is a norm.

Hint: Just verify the properties on page 128 that define a norm.

3.33. Prove equations (3.250) through (3.252).

3.34. Prove equations (3.254) and (3.255) under the restriction that $\mathcal{V}(X) \subset$
$\mathcal{V}(A)$; that is, where $X = BL$ for a matrix B whose columns span $\mathcal{V}(A)$.

4

Vector/Matrix Derivatives and Integrals

The operations of differentiation and integration of vectors and matrices are logical extensions of the corresponding operations on scalars. There are three objects involved in this operation:

- the variable of the operation;
- the operand (the function being differentiated or integrated); and
- the result of the operation.

In the simplest case, all three of these objects are of the same type, and they are scalars. If either the variable or the operand is a vector or a matrix, however, the structure of the result may be more complicated. This statement will become clearer as we proceed to consider specific cases.

In this chapter, we state or show the form that the derivative takes in terms of simpler derivatives. We state high-level rules for the nature of the differentiation in terms of simple partial differentiation of a scalar with respect to a scalar. We do not consider whether or not the derivatives exist. In general, if the simpler derivatives we write that comprise the more complicated object exist, then the derivative of that more complicated object exists. Once a shape of the derivative is determined, definitions or derivations in ϵ-δ terms could be given, but we will refrain from that kind of formal exercise. The purpose of this chapter is not to develop a calculus for vectors and matrices but rather to consider some cases that find wide applications in statistics. For a more careful treatment of differentiation of vectors and matrices, the reader is referred to Rogers (1980) or to Magnus and Neudecker (1999). Anderson (2003), Muirhead (1982), and Nachbin (1965) cover various aspects of integration with respect to vector or matrix differentials.

4.1 Basics of Differentiation

It is useful to recall the heuristic interpretation of a derivative. A derivative of a function is the infinitesimal rate of change of the function with respect to the variable with which the differentiation is taken. If both the function and the variable are scalars, this interpretation is unambiguous. If, however, the operand of the differentiation, Φ, is a more complicated function, say a vector or a matrix, and/or the variable of the differentiation, Ξ, is a more complicated object, the changes are more difficult to measure. Change in the value both of the function,

$$\delta\Phi = \Phi_{\mathrm{new}} - \Phi_{\mathrm{old}},$$

and of the variable,

$$\delta\Xi = \Xi_{\mathrm{new}} - \Xi_{\mathrm{old}},$$

could be measured in various ways; for example, by using various norms, as discussed in Sections 2.1.5 and 3.9. (Note that the subtraction is not necessarily ordinary scalar subtraction.)

Furthermore, we cannot just divide the function values by $\delta\Xi$. We do not have a definition for division by that kind of object. We need a mapping, possibly a norm, that assigns a positive real number to $\delta\Xi$. We can define the change in the function value as just the simple difference of the function evaluated at the two points. This yields

$$\lim_{\|\delta\Xi\|\to 0} \frac{\Phi(\Xi + \delta\Xi) - \Phi(\Xi)}{\|\delta\Xi\|}. \tag{4.1}$$

So long as we remember the complexity of $\delta\Xi$, however, we can adopt a simpler approach. Since for both vectors and matrices, we have definitions of multiplication by a scalar and of addition, we can simplify the limit in the usual definition of a derivative, $\delta\Xi \to 0$. Instead of using $\delta\Xi$ as the element of change, we will use $t\Upsilon$, where t is a scalar and Υ is an element to be added to Ξ. The limit then will be taken in terms of $t \to 0$. This leads to

$$\lim_{t\to 0} \frac{\Phi(\Xi + t\Upsilon) - \Phi(\Xi)}{t} \tag{4.2}$$

as a formula for the derivative of Φ with respect to Ξ.

The expression (4.2) may be a useful formula for evaluating a derivative, but we must remember that it is not the derivative. The type of object of this formula is the same as the type of object of the function, Φ; it does not accommodate the type of object of the argument, Ξ, unless Ξ is a scalar. As we will see below, for example, if Ξ is a vector and Φ is a scalar, the derivative must be a vector, yet in that case the expression (4.2) is a scalar.

The expression (4.1) is rarely directly useful in evaluating a derivative, but it serves to remind us of both the generality and the complexity of the concept. Both Φ and its arguments could be functions, for example. (In functional analysis, various kinds of functional derivatives are defined, such as a Gâteaux derivative. These derivatives find applications in developing robust statistical methods; see Shao, 2003, for example.) In this chapter, we are interested in the combinations of three possibilities for Φ, namely scalar, vector, and matrix, and the same three possibilities for Ξ and Υ.

Continuity

It is clear from the definition of continuity that for the derivative of a function to exist at a point, the function must be continuous at that point. A function of a vector or a matrix is continuous if it is continuous for each element of the vector or matrix. Just as scalar sums and products are continuous, vector/matrix sums and all of the types of vector/matrix products we have discussed are continuous. A continuous function of a continuous function is continuous.

Many of the vector/matrix functions we have discussed are clearly continuous. For example, the L_p vector norms in equation (2.11) are continuous over the nonnegative reals but not over the reals unless p is an even (positive) integer. The determinant of a matrix is continuous, as we see from the definition of the determinant and the fact that sums and scalar products are continuous. The fact that the determinant is a continuous function immediately yields the result that cofactors and hence the adjugate are continuous. From the relationship between an inverse and the adjugate (equation (3.131)), we see that the inverse is a continuous function.

Notation and Properties

We write the differential operator with respect to the dummy variable x as $\partial/\partial x$ or $\partial/\partial x^{\mathrm{T}}$. We usually denote differentiation using the symbol for "partial" differentiation, ∂, whether the operator is written ∂x_i for differentiation with respect to a specific scalar variable or ∂x for differentiation with respect to the array x that contains all of the individual elements. Sometimes, however, if the differentiation is being taken with respect to the whole array (the vector or the matrix), we use the notation d/dx.

The operand of the differential operator $\partial/\partial x$ is a function of x. (If it is not a function of x — that is, if it is a constant function with respect to x — then the operator evaluates to 0.) The result of the operation, written $\partial f/\partial x$, is also a function of x, with the same domain as f, and we sometimes write $\partial f(x)/\partial x$ to emphasize this fact. The value of this function at the fixed point x_0 is written as $\partial f(x_0)/\partial x$. (The derivative of the constant $f(x_0)$ is identically 0, but it is not necessary to write $\partial f(x)/\partial x|_{x_0}$ because $\partial f(x_0)/\partial x$ is interpreted as the value of the function $\partial f(x)/\partial x$ at the fixed point x_0.)

If $\partial/\partial x$ operates on f, and $f : S \rightarrow T$, then $\partial/\partial x : S \rightarrow U$. The nature of S, or more directly the nature of x, whether it is a scalar, a vector, or a matrix, and the nature of T determine the structure of the result U. For example, if x is an n-vector and $f(x) = x^{\mathrm{T}}x$, then

$$f : \mathbb{R}^n \rightarrow \mathbb{R}$$

and

$$\partial f/\partial x : \mathbb{R}^n \rightarrow \mathbb{R}^n,$$

as we will see. The outer product, $h(x) = xx^{\mathrm{T}}$, is a mapping to a higher rank array, but the derivative of the outer product is a mapping to an array of the same rank; that is,

$$h : \mathrm{I\!R}^n \to \mathrm{I\!R}^{n \times n}$$

and

$$\partial h / \partial x : \mathrm{I\!R}^n \to \mathrm{I\!R}^n.$$

(Note that "rank" here means the number of dimensions; see page 5.)

As another example, consider $g(\cdot) = \det(\cdot)$, so

$$g : \mathrm{I\!R}^{n \times n} \mapsto \mathrm{I\!R}.$$

In this case,

$$\partial g / \partial X : \mathrm{I\!R}^{n \times n} \mapsto \mathrm{I\!R}^{n \times n};$$

that is, the derivative of the determinant of a square matrix is a square matrix, as we will see later.

Higher-order differentiation is a composition of the $\partial / \partial x$ operator with itself or of the $\partial / \partial x$ operator and the $\partial / \partial x^{\mathrm{T}}$ operator. For example, consider the familiar function in linear least squares

$$f(b) = (y - Xb)^{\mathrm{T}}(y - Xb).$$

This is a mapping from $\mathrm{I\!R}^m$ to $\mathrm{I\!R}$. The first derivative with respect to the m-vector b is a mapping from $\mathrm{I\!R}^m$ to $\mathrm{I\!R}^m$, namely $2X^{\mathrm{T}}Xb - 2X^{\mathrm{T}}y$. The second derivative with respect to b^{T} is a mapping from $\mathrm{I\!R}^m$ to $\mathrm{I\!R}^{m \times m}$, namely, $2X^{\mathrm{T}}X$. (Many readers will already be familiar with these facts. We will discuss the general case of differentiation with respect to a vector in Section 4.2.2.)

We see from expression (4.1) that differentiation is a linear operator; that is, if $\mathcal{D}(\Phi)$ represents the operation defined in expression (4.1), Ψ is another function in the class of functions over which $\mathcal{D}$ is defined, and a is a scalar that does not depend on the variable Ξ, then $\mathcal{D}(a\Phi + \Psi) = a\mathcal{D}(\Phi) + \mathcal{D}(\Psi)$. This yields the familiar rules of differential calculus for derivatives of sums or constant scalar products. Other usual rules of differential calculus apply, such as for differentiation of products and composition (the chain rule). We can use expression (4.2) to work these out. For example, for the derivative of the product $\Phi\Psi$, after some rewriting of terms, we have the numerator

$$\Phi(\Xi)\big(\Psi(\Xi + t\Upsilon) - \Psi(\Xi)\big)$$
$$+ \Psi(\Xi)\big(\Phi(\Xi + t\Upsilon) - \Phi(\Xi)\big)$$
$$+ \big(\Phi(\Xi + t\Upsilon) - \Phi(\Xi)\big)\big(\Psi(\Xi + t\Upsilon) - \Psi(\Xi)\big).$$

Now, dividing by t and taking the limit, assuming that as

$$t \to 0,$$

$$(\Phi(\Xi + t\Upsilon) - \Phi(\Xi)) \to 0,$$

we have

$$\mathcal{D}(\Phi\Psi) = \mathcal{D}(\Phi)\Psi + \Phi\mathcal{D}(\Psi), \tag{4.3}$$

where again $\mathcal{D}$ represents the differentiation operation.

Differentials

For a differentiable scalar function of a scalar variable, $f(x)$, the *differential of f at c with increment u* is $u\,\mathrm{d}f/\mathrm{d}x|_c$. This is the linear term in a truncated Taylor series expansion:

$$f(c + u) = f(c) + u\frac{\mathrm{d}}{\mathrm{d}x}f(c) + r(c, u). \tag{4.4}$$

Technically, the differential is a function of both x and u, but the notation $\mathrm{d}f$ is used in a generic sense to mean the differential of f. For vector/matrix functions of vector/matrix variables, the differential is defined in a similar way. The structure of the differential is the same as that of the function; that is, for example, the differential of a matrix-valued function is a matrix.

4.2 Types of Differentiation

In the following sections we consider differentiation with respect to different types of objects first, and we consider differentiation of different types of objects.

4.2.1 Differentiation with Respect to a Scalar

Differentiation of a structure (vector or matrix, for example) with respect to a scalar is quite simple; it just yields the ordinary derivative of each element of the structure in the same structure. Thus, the derivative of a vector or a matrix with respect to a scalar variable is a vector or a matrix, respectively, of the derivatives of the individual elements.

Differentiation with respect to a vector or matrix, which we will consider below, is often best approached by considering differentiation with respect to the individual elements of the vector or matrix, that is, with respect to scalars.

Derivatives of Vectors with Respect to Scalars

The derivative of the vector $y(x) = (y_1, \ldots, y_n)$ with respect to the scalar x is the vector

$$\partial y/\partial x = (\partial y_1/\partial x, \ldots, \partial y_n/\partial x). \tag{4.5}$$

The second or higher derivative of a vector with respect to a scalar is likewise a vector of the derivatives of the individual elements; that is, it is an array of higher rank.

Derivatives of Matrices with Respect to Scalars

The derivative of the matrix $Y(x) = (y_{ij})$ with respect to the scalar x is the matrix

$$\partial Y(x)/\partial x = (\partial y_{ij}/\partial x). \tag{4.6}$$

The second or higher derivative of a matrix with respect to a scalar is likewise a matrix of the derivatives of the individual elements.

Derivatives of Functions with Respect to Scalars

Differentiation of a function of a vector or matrix that is linear in the elements of the vector or matrix involves just the differentiation of the elements, followed by application of the function. For example, the derivative of a trace of a matrix is just the trace of the derivative of the matrix. On the other hand, the derivative of the determinant of a matrix is not the determinant of the derivative of the matrix (see below).

Higher-Order Derivatives with Respect to Scalars

Because differentiation with respect to a scalar does not change the rank of the object ("rank" here means rank of an array or "shape"), higher-order derivatives $\partial^k/\partial x^k$ with respect to scalars are merely objects of the same rank whose elements are the higher-order derivatives of the individual elements.

4.2.2 Differentiation with Respect to a Vector

Differentiation of a given object with respect to an n-vector yields a vector for each element of the given object. The basic expression for the derivative, from formula (4.2), is

$$\lim_{t \to 0} \frac{\Phi(x + ty) - \Phi(x)}{t} \tag{4.7}$$

for an arbitrary conformable vector y. The arbitrary y indicates that the derivative is omnidirectional; it is the rate of change of a function of the vector in any direction.

Derivatives of Scalars with Respect to Vectors; The Gradient

The derivative of a scalar-valued function with respect to a vector is a vector of the partial derivatives of the function with respect to the elements of the vector. If $f(x)$ is a scalar function of the vector $x = (x_1, \ldots, x_n)$,

$$\frac{\partial f}{\partial x} = \left(\frac{\partial f}{\partial x_1}, \ldots, \frac{\partial f}{\partial x_n} \right), \tag{4.8}$$

if those derivatives exist. This vector is called the *gradient* of the scalar-valued function, and is sometimes denoted by $g_f(x)$ or $\nabla f(x)$, or sometimes just g_f or ∇f:

$$g_f = \nabla f = \frac{\partial f}{\partial x}. \tag{4.9}$$

The notation g_f or ∇f implies differentiation with respect to "all" arguments of f, hence, if f is a scalar-valued function of a vector argument, they represent a vector.

This derivative is useful in finding the maximum or minimum of a function. Such applications arise throughout statistical and numerical analysis. In Section 6.3.2, we will discuss a method of solving linear systems of equations by formulating the problem as a minimization problem.

Inner products, bilinear forms, norms, and variances are interesting scalar-valued functions of vectors. In these cases, the function Φ in equation (4.7) is scalar-valued and the numerator is merely $\Phi(x + ty) - \Phi(x)$. Consider, for example, the quadratic form $x^T A x$. Using equation (4.7) to evaluate $\partial x^T A x / \partial x$, we have

$$\lim_{t \to 0} \frac{(x + ty)^T A (x + ty) - x^T A x}{t}$$

$$= \lim_{t \to 0} \frac{x^T A x + t y^T A x + t y^T A^T x + t^2 y^T A y - x^T A x}{t} \tag{4.10}$$

$$= y^T (A + A^T) x,$$

for an arbitrary y (that is, "in any direction"), and so $\partial x^T A x / \partial x = (A + A^T) x$.

This immediately yields the derivative of the square of the Euclidean norm of a vector, $\|x\|_2^2$, and the derivative of the Euclidean norm itself by using the chain rule. Other L_p vector norms may not be differentiable everywhere because of the presence of the absolute value in their definitions. The fact that the Euclidean norm is differentiable everywhere is one of its most important properties.

The derivative of the quadratic form also immediately yields the derivative of the variance. The derivative of the correlation, however, is slightly more difficult because it is a ratio (see Exercise 4.2).

The operator $\partial / \partial x^T$ applied to the scalar function f results in g_f^T.

The second derivative of a scalar-valued function with respect to a vector is a derivative of the first derivative, which is a vector. We will now consider derivatives of vectors with respect to vectors.

Derivatives of Vectors with Respect to Vectors; The Jacobian

The derivative of an m-vector-valued function of an n-vector argument consists of nm scalar derivatives. These derivatives could be put into various structures. Two obvious structures are an $n \times m$ matrix and an $m \times n$ matrix. For a function $f : S \subset \mathrm{I\!R}^n \to \mathrm{I\!R}^m$, we define $\partial f^{\mathrm{T}}/\partial x$ to be the $n \times m$ matrix, which is the natural extension of $\partial/\partial x$ applied to a scalar function, and $\partial f/\partial x^{\mathrm{T}}$ to be its transpose, the $m \times n$ matrix. Although the notation $\partial f^{\mathrm{T}}/\partial x$ is more precise because it indicates that the elements of f correspond to the columns of the result, we often drop the transpose in the notation. We have

$$
\frac{\partial f}{\partial x} = \frac{\partial f^{\mathrm{T}}}{\partial x} \quad \text{by convention}
$$

$$
= \left[\frac{\partial f_1}{\partial x} \cdots \frac{\partial f_m}{\partial x} \right]
$$

$$
= \begin{bmatrix}
\frac{\partial f_1}{\partial x_1} & \frac{\partial f_2}{\partial x_1} & \cdots & \frac{\partial f_m}{\partial x_1} \\
\frac{\partial f_1}{\partial x_2} & \frac{\partial f_2}{\partial x_2} & \cdots & \frac{\partial f_m}{\partial x_2} \\
& & \cdots & \\
\frac{\partial f_1}{\partial x_n} & \frac{\partial f_2}{\partial x_n} & \cdots & \frac{\partial f_m}{\partial x_n}
\end{bmatrix} \tag{4.11}
$$

if those derivatives exist. This derivative is called the *matrix gradient* and is denoted by G_f or ∇f for the vector-valued function f. (Note that the ∇ symbol can denote either a vector or a matrix, depending on whether the function being differentiated is scalar-valued or vector-valued.)

The $m \times n$ matrix $\partial f/\partial x^{\mathrm{T}} = (\nabla f)^{\mathrm{T}}$ is called the *Jacobian* of f and is denoted by J_f:

$$
\mathrm{J}_f = \mathrm{G}_f^{\mathrm{T}} = (\nabla f)^{\mathrm{T}}. \tag{4.12}
$$

The absolute value of the determinant of the Jacobian appears in integrals involving a change of variables. (Occasionally, the term "Jacobian" is used to refer to the absolute value of the determinant rather than to the matrix itself.)

To emphasize that the quantities are functions of x, we sometimes write $\partial f(x)/\partial x$, $\mathrm{J}_f(x)$, $\mathrm{G}_f(x)$, or $\nabla f(x)$.

Derivatives of Matrices with Respect to Vectors

The derivative of a matrix with respect to a vector is a three-dimensional object that results from applying equation (4.8) to each of the elements of the matrix. For this reason, it is simpler to consider only the partial derivatives of the matrix Y with respect to the individual elements of the vector x; that is, $\partial Y/\partial x_i$. The expressions involving the partial derivatives can be thought of as defining one two-dimensional layer of a three-dimensional object.

Using the rules for differentiation of powers that result directly from the definitions, we can write the partial derivatives of the inverse of the matrix Y as

$$
\frac{\partial}{\partial x} Y^{-1} = -Y^{-1} \left(\frac{\partial}{\partial x} Y \right) Y^{-1} \tag{4.13}
$$

(see Exercise 4.3).

Beyond the basics of differentiation of constant multiples or powers of a variable, the two most important properties of derivatives of expressions are the linearity of the operation and the chaining of the operation. These yield rules that correspond to the familiar rules of the differential calculus. A simple result of the linearity of the operation is the rule for differentiation of the trace:

$$\frac{\partial}{\partial x}\mathrm{tr}(Y) = \mathrm{tr}\left(\frac{\partial}{\partial x}Y\right).$$

Higher-Order Derivatives with Respect to Vectors; The Hessian

Higher-order derivatives are derivatives of lower-order derivatives. As we have seen, a derivative of a given function with respect to a vector is a more complicated object than the original function. The simplest higher-order derivative with respect to a vector is the second-order derivative of a scalar-valued function. Higher-order derivatives may become uselessly complicated.

In accordance with the meaning of derivatives of vectors with respect to vectors, the second derivative of a scalar-valued function with respect to a vector is a matrix of the partial derivatives of the function with respect to the elements of the vector. This matrix is called the *Hessian*, and is denoted by H_f or sometimes by $\nabla\nabla f$ or $\nabla^2 f$:

$$\mathrm{H}_f = \frac{\partial^2 f}{\partial x \partial x^{\mathrm{T}}} = \begin{bmatrix} \frac{\partial^2 f}{\partial x_1^2} & \frac{\partial^2 f}{\partial x_1 \partial x_2} & \cdots & \frac{\partial^2 f}{\partial x_1 \partial x_m} \\ \frac{\partial^2 f}{\partial x_2 \partial x_1} & \frac{\partial^2 f}{\partial x_2^2} & \cdots & \frac{\partial^2 f}{\partial x_2 \partial x_m} \\ & & \cdots & \\ \frac{\partial^2 f}{\partial x_m \partial x_1} & \frac{\partial^2 f}{\partial x_m \partial x_2} & \cdots & \frac{\partial^2 f}{\partial x_m^2} \end{bmatrix}. \tag{4.14}$$

To emphasize that the Hessian is a function of x, we sometimes write $\mathrm{H}_f(x)$ or $\nabla\nabla f(x)$ or $\nabla^2 f(x)$.

Summary of Derivatives with Respect to Vectors

As we have seen, the derivatives of functions are complicated by the problem of measuring the change in the function, but often the derivatives of functions with respect to a vector can be determined by using familiar scalar differentiation. In general, we see that

- the derivative of a scalar (a quadratic form) with respect to a vector is a vector and
- the derivative of a vector with respect to a vector is a matrix.

Table 4.1 lists formulas for the vector derivatives of some common expressions. The derivative $\partial f/\partial x^{\mathrm{T}}$ is the transpose of $\partial f/\partial x$.

Table 4.1. Formulas for Some Vector Derivatives

$f(x)$	$\partial f/\partial x$
ax	a
$b^{\mathrm{T}}x$	b
$x^{\mathrm{T}}b$	b^{T}
$x^{\mathrm{T}}x$	$2x$
xx^{T}	$2x^{\mathrm{T}}$
$b^{\mathrm{T}}Ax$	$A^{\mathrm{T}}b$
$x^{\mathrm{T}}Ab$	$b^{\mathrm{T}}A$
$x^{\mathrm{T}}Ax$	$(A+A^{\mathrm{T}})x$
	$2Ax$, if A is symmetric
$\exp(-\frac{1}{2}x^{\mathrm{T}}Ax)$	$-\exp(-\frac{1}{2}x^{\mathrm{T}}Ax)Ax$, if A is symmetric
$\|x\|_2^2$	$2x$
$\mathrm{V}(x)$	$2x/(n-1)$

In this table, x is an n-vector, a is a constant scalar, b is a constant conformable vector, and A is a constant conformable matrix.

4.2.3 Differentiation with Respect to a Matrix

The derivative of a function with respect to a matrix is a matrix with the same shape consisting of the partial derivatives of the function with respect to the elements of the matrix. This rule defines what we mean by differentiation with respect to a matrix.

By the definition of differentiation with respect to a matrix X, we see that the derivative $\partial f/\partial X^{\mathrm{T}}$ is the transpose of $\partial f/\partial X$. For scalar-valued functions, this rule is fairly simple. For example, consider the trace. If X is a square matrix and we apply this rule to evaluate $\partial\,\mathrm{tr}(X)/\partial X$, we get the identity matrix, where the nonzero elements arise only when $j = i$ in $\partial(\sum x_{ii})/\partial x_{ij}$. If AX is a square matrix, we have for the (i, j) term in $\partial\,\mathrm{tr}(AX)/\partial X$, $\partial\sum_i\sum_k a_{ik}x_{ki}/\partial x_{ij} = a_{ji}$, and so $\partial\,\mathrm{tr}(AX)/\partial X = A^{\mathrm{T}}$, and likewise, inspecting $\partial\sum_i\sum_k x_{ik}x_{ki}/\partial x_{ij}$, we get $\partial\,\mathrm{tr}(X^{\mathrm{T}}X)/\partial X = 2X^{\mathrm{T}}$. Likewise for the scalar-valued $a^{\mathrm{T}}Xb$, where a and b are conformable constant vectors, for $\partial\sum_m(\sum_k a_k x_{km})b_m/\partial x_{ij} = a_i b_j$, so $\partial a^{\mathrm{T}}Xb/\partial X = ab^{\mathrm{T}}$.

Now consider $\partial|X|/\partial X$. Using an expansion in cofactors (equation (3.21) or (3.22)), the only term in $|X|$ that involves x_{ij} is $x_{ij}(-1)^{i+j}|X_{-(i)(j)}|$, and the cofactor $(x_{(ij)}) = (-1)^{i+j}|X_{-(i)(j)}|$ does not involve x_{ij}. Hence, $\partial|X|/\partial x_{ij} = (x_{(ij)})$, and so $\partial|X|/\partial X = (\mathrm{adj}(X))^{\mathrm{T}}$ from equation (3.24). Using equation (3.131), we can write this as $\partial|X|/\partial X = |X|X^{-\mathrm{T}}$.

The chain rule can be used to evaluate $\partial\log|X|/\partial X$.

Applying the rule stated at the beginning of this section, we see that the derivative of a matrix Y with respect to the matrix X is

$$\frac{\mathrm{d}Y}{\mathrm{d}X} = Y \otimes \frac{\mathrm{d}}{\mathrm{d}X}. \tag{4.15}$$

Table 4.2 lists some formulas for the matrix derivatives of some common expressions. The derivatives shown in Table 4.2 can be obtained by evaluating expression (4.15), possibly also using the chain rule.

Table 4.2. Formulas for Some Matrix Derivatives

General X

$f(X)$	$\partial f / \partial X$
$a^{\mathrm{T}} X b$	ab^{T}
$\mathrm{tr}(AX)$	A^{T}
$\mathrm{tr}(X^{\mathrm{T}} X)$	$2X^{\mathrm{T}}$
BX	$I_n \otimes B$
XC	$C^{\mathrm{T}} \otimes I_m$
BXC	$C^{\mathrm{T}} \otimes B$

Square and Possibly Invertible X

$f(X)$	$\partial f / \partial X$
$\mathrm{tr}(X)$	I_n
$\mathrm{tr}(X^k)$	kX^{k-1}
$\mathrm{tr}(BX^{-1}C)$	$-(X^{-1}CBX^{-1})^{\mathrm{T}}$
$|X|$	$|X|X^{-\mathrm{T}}$
$\log|X|$	$X^{-\mathrm{T}}$
$|X|^k$	$k|X|^k X^{-\mathrm{T}}$
$BX^{-1}C$	$-(X^{-1}C)^{\mathrm{T}} \otimes BX^{-1}$

In this table, X is an $n \times m$ matrix, a is a constant n-vector, b is a constant m-vector, A is a constant $m \times n$ matrix, B is a constant $p \times n$ matrix, and C is a constant $m \times q$ matrix.

There are some interesting applications of differentiation with respect to a matrix in maximum likelihood estimation. Depending on the structure of the parameters in the distribution, derivatives of various types of objects may be required. For example, the determinant of a variance-covariance matrix, in the sense that it is a measure of a volume, often occurs as a normalizing factor in a probability density function; therefore, we often encounter the need to differentiate a determinant with respect to a matrix.

4.3 Optimization of Functions

Because a derivative measures the rate of change of a function, a point at which the derivative is equal to 0 is a stationary point, which may be a maximum or a minimum of the function. Differentiation is therefore a very useful tool for finding the optima of functions, and so, for a given function $f(x)$, the gradient vector function, $g_f(x)$, and the Hessian matrix function, $H_f(x)$, play important roles in optimization methods.

We may seek either a maximum or a minimum of a function. Since maximizing the scalar function $f(x)$ is equivalent to minimizing $-f(x)$, we can always consider optimization of a function to be minimization of a function. Thus, we generally use terminology for the problem of finding a minimum of a function. Because the function may have many ups and downs, we often use the phrase *local minimum* (or local maximum or local optimum).

Except in the very simplest of cases, the optimization method must be iterative, moving through a sequence of points, $x^{(0)}, x^{(1)}, x^{(2)}, \ldots$, that approaches the optimum point arbitrarily closely. At the point $x^{(k)}$, the direction of *steepest descent* is clearly $-g_f(x^{(k)})$, but because this direction may be continuously changing, the steepest descent direction may not be the best direction in which to seek the next point, $x^{(k+1)}$.

4.3.1 Stationary Points of Functions

The first derivative helps only in finding a stationary point. The matrix of second derivatives, the Hessian, provides information about the nature of the stationary point, which may be a local minimum or maximum, a saddlepoint, or only an inflection point.

The so-called second-order optimality conditions are the following (see a general text on optimization for their proofs).

- If (but not only if) the stationary point is a local minimum, then the Hessian is nonnegative definite.
- If the Hessian is positive definite, then the stationary point is a local minimum.
- Likewise, if the stationary point is a local maximum, then the Hessian is nonpositive definite, and if the Hessian is negative definite, then the stationary point is a local maximum.
- If the Hessian has both positive and negative eigenvalues, then the stationary point is a saddlepoint.

4.3.2 Newton's Method

We consider a differentiable scalar-valued function of a vector argument, $f(x)$. By a Taylor series about a stationary point x_*, truncated after the second-order term

$$f(x) \approx f(x_*) + (x - x_*)^{\mathrm{T}} \mathrm{g}_f(x_*) + \frac{1}{2}(x - x_*)^{\mathrm{T}} \mathrm{H}_f(x_*)(x - x_*), \qquad (4.16)$$

because $\mathrm{g}_f(x_*) = 0$, we have a general method of finding a stationary point for the function $f(\cdot)$, called Newton's method. If x is an m-vector, $\mathrm{g}_f(x)$ is an m-vector and $\mathrm{H}_f(x)$ is an $m \times m$ matrix.

Newton's method is to choose a starting point $x^{(0)}$, then, for $k = 0, 1, \ldots$, to solve the linear systems

$$\mathrm{H}_f\big(x^{(k)}\big)p^{(k+1)} = -\mathrm{g}_f\big(x^{(k)}\big) \qquad (4.17)$$

for $p^{(k+1)}$, and then to update the point in the domain of $f(\cdot)$ by

$$x^{(k+1)} = x^{(k)} + p^{(k+1)}. \qquad (4.18)$$

The two steps are repeated until there is essentially no change from one iteration to the next. If $f(\cdot)$ is a quadratic function, the solution is obtained in one iteration because equation (4.16) is exact. These two steps have a very simple form for a function of one variable (see Exercise 4.4a).

Linear Least Squares

In a least squares fit of a linear model

$$y = X\beta + \epsilon, \qquad (4.19)$$

where y is an n-vector, X is an $n \times m$ matrix, and β is an m-vector, we replace β by a variable b, define the residual vector

$$r = y - Xb, \qquad (4.20)$$

and minimize its Euclidean norm,

$$f(b) = r^{\mathrm{T}}r, \qquad (4.21)$$

with respect to the variable b. We can solve this optimization problem by taking the derivative of this sum of squares and equating it to zero. Doing this, we get

$$\frac{\mathrm{d}(y - Xb)^{\mathrm{T}}(y - Xb)}{\mathrm{d}b} = \frac{\mathrm{d}(y^{\mathrm{T}}y - 2b^{\mathrm{T}}X^{\mathrm{T}}y + b^{\mathrm{T}}X^{\mathrm{T}}Xb)}{\mathrm{d}b}$$

$$= -2X^{\mathrm{T}}y + 2X^{\mathrm{T}}Xb$$
$$= 0,$$

which yields the normal equations

$$X^{\mathrm{T}}Xb = X^{\mathrm{T}}y.$$

The solution to the normal equations is a stationary point of the function (4.21). The Hessian of $(y - Xb)^{\mathrm{T}}(y - Xb)$ with respect to b is $2X^{\mathrm{T}}X$ and

$$X^{\mathrm{T}}X \succeq 0.$$

Because the matrix of second derivatives is nonnegative definite, the value of b that solves the system of equations arising from the first derivatives is a local minimum of equation (4.21). We discuss these equations further in Sections 6.7 and 9.2.2.

Quasi-Newton Methods

All gradient-descent methods determine the path $p^{(k)}$ to take in the kth step by a system of equations of the form

$$R^{(k)}p^{(k)} = -\mathrm{g}_f\big(x^{(k-1)}\big).$$

In the steepest-descent method, $R^{(k)}$ is the identity, I, in these equations. For functions with eccentric contours, the steepest-descent method traverses a zigzag path to the minimum. In Newton's method, $R^{(k)}$ is the Hessian evaluated at the previous point, $\mathrm{H}_f\big(x^{(k-1)}\big)$, which results in a more direct path to the minimum. Aside from the issues of consistency of the resulting equation and the general problems of reliability, a major disadvantage of Newton's method is the computational burden of computing the Hessian, which requires $\mathrm{O}(m^2)$ function evaluations, and solving the system, which requires $\mathrm{O}(m^3)$ arithmetic operations, at each iteration.

Instead of using the Hessian at each iteration, we may use an approximation, $B^{(k)}$. We may choose approximations that are simpler to update and/or that allow the equations for the step to be solved more easily. Methods using such approximations are called *quasi-Newton* methods or *variable metric* methods.

Because

$$\mathrm{H}_f\big(x^{(k)}\big)\big(x^{(k)} - x^{(k-1)}\big) \approx \mathrm{g}_f\big(x^{(k)}\big) - \mathrm{g}_f\big(x^{(k-1)}\big),$$

we choose $B^{(k)}$ so that

$$B^{(k)}\big(x^{(k)} - x^{(k-1)}\big) = \mathrm{g}_f\big(x^{(k)}\big) - \mathrm{g}_f\big(x^{(k-1)}\big). \tag{4.22}$$

This is called the *secant condition*.

We express the secant condition as

$$B^{(k)}s^{(k)} = y^{(k)}, \tag{4.23}$$

where

$$s^{(k)} = x^{(k)} - x^{(k-1)}$$

and
$$y^{(k)} = g_f(x^{(k)}) - g_f(x^{(k-1)}),$$
as above.

The system of equations in (4.23) does not fully determine $B^{(k)}$ of course. Because $B^{(k)}$ should approximate the Hessian, we may require that it be symmetric and positive definite.

The most common approach in quasi-Newton methods is first to choose a reasonable starting matrix $B^{(0)}$ and then to choose subsequent matrices by additive updates,
$$B^{(k+1)} = B^{(k)} + B_a^{(k)}, \tag{4.24}$$
subject to preservation of symmetry and positive definiteness. An approximate Hessian $B^{(k)}$ may be used for several iterations before it is updated; that is, $B_a^{(k)}$ may be taken as 0 for several successive iterations.

4.3.3 Optimization of Functions with Restrictions

Instead of the simple least squares problem of determining a value of b that minimizes the sum of squares, we may have some restrictions that b must satisfy; for example, we may have the requirement that the elements of b sum to 1. More generally, consider the least squares problem for the linear model (4.19) with the requirement that b satisfy some set of linear restrictions, $Ab = c$, where A is a full-rank $k \times m$ matrix (with $k \leq m$). (The rank of A must be less than m or else the constraints completely determine the solution to the problem. If the rank of A is less than k, however, some rows of A and some elements of b could be combined into a smaller number of constraints. We can therefore assume A is of full row rank. Furthermore, we assume the linear system is consistent (that is, $\mathrm{rank}([A|c]) = k$) for otherwise there could be no solution.) We call any point b that satisfies $Ab = c$ a *feasible point*.

We write the constrained optimization problem as

$$\begin{aligned} \min_b \; & f(b) = (y - Xb)^{\mathrm{T}}(y - Xb) \\ \text{s.t.} \; & Ab = c. \end{aligned} \tag{4.25}$$

If b_c is any feasible point (that is, $Ab_c = c$), then any other feasible point can be represented as $b_c + p$, where p is any vector in the null space of A, $\mathcal{N}(A)$. From our discussion in Section 3.5.2, we know that the dimension of $\mathcal{N}(A)$ is $m - k$, and its order is m. If N is an $m \times m - k$ matrix whose columns form a basis for $\mathcal{N}(A)$, all feasible points can be generated by $b_c + Nz$, where $z \in \mathrm{I\!R}^{m-k}$. Hence, we need only consider the restricted variables

$$b = b_c + Nz$$

and the "reduced" function

$$h(z) = f(b_c + Nz).$$

The argument of this function is a vector with only $m - k$ elements instead of m elements as in the unconstrained problem. The unconstrained minimum of h, however, is the solution of the original constrained problem.

The Reduced Gradient and Reduced Hessian

If we assume differentiability, the gradient and Hessian of the reduced function can be expressed in terms of the original function:

$$\begin{aligned} g_h(z) &= N^\mathrm{T} g_f(b_c + Nz) \\ &= N^\mathrm{T} g_f(b) \end{aligned} \tag{4.26}$$

and

$$\begin{aligned} \mathrm{H}_h(z) &= N^\mathrm{T} \mathrm{H}_f(b_c + Nz)N \\ &= N^\mathrm{T} \mathrm{H}_f(b)N. \end{aligned} \tag{4.27}$$

In equation (4.26), $N^\mathrm{T} g_f(b)$ is called the *reduced gradient* or *projected gradient*, and $N^\mathrm{T} \mathrm{H}_f(b)N$ in equation (4.27) is called the *reduced Hessian* or *projected Hessian*.

The properties of stationary points are related to the derivatives referred to above are the conditions that determine a minimum of this reduced objective function; that is, b_* is a minimum if and only if

- $N^\mathrm{T} g_f(b_*) = 0$,
- $N^\mathrm{T} \mathrm{H}_f(b_*)N$ is positive definite, and
- $Ab_* = c$.

These relationships then provide the basis for the solution of the optimization problem.

Lagrange Multipliers

Because the $m \times m$ matrix $[N|A^\mathrm{T}]$ spans $\mathbb{R}^m$, we can represent the vector $g_f(b_*)$ as a linear combination of the columns of N and A^T, that is,

$$\begin{aligned} g_f(b_*) &= [N|A^\mathrm{T}] \begin{pmatrix} z_* \\ \lambda_* \end{pmatrix} \\ &= \begin{pmatrix} Nz_* \\ A^\mathrm{T}\lambda_* \end{pmatrix}, \end{aligned}$$

where z_* is an $(m - k)$-vector and λ_* is a k-vector. Because $\nabla h(z_*) = 0$, Nz_* must also vanish (that is, $Nz_* = 0$), and thus, at the optimum, the nonzero elements of the gradient of the objective function are linear combinations of the rows of the constraint matrix, $A^\mathrm{T}\lambda_*$. The k elements of the linear combination vector λ_* are called *Lagrange multipliers*.

The Lagrangian

Let us now consider a simple generalization of the constrained problem above and an abstraction of the results above so as to develop a general method. We consider the problem

$$\min_x f(x)$$
$$\text{s.t. } c(x) = 0, \tag{4.28}$$

where f is a scalar-valued function of an m-vector variable and c is a k-vector-valued function of the variable. There are some issues concerning the equation $c(x) = 0$ that we will not go into here. Obviously, we have the same concerns as before; that is, whether $c(x) = 0$ is consistent and whether the individual equations $c_i(x) = 0$ are independent. Let us just assume they are and proceed. (Again, we refer the interested reader to a more general text on optimization.)

Motivated by the results above, we form a function that incorporates a dot product of Lagrange multipliers and the function $c(x)$:

$$F(x) = f(x) + \lambda^{\mathrm{T}} c(x). \tag{4.29}$$

This function is called the *Lagrangian*. The solution, (x_*, λ_*), of the optimization problem occurs at a stationary point of the Lagrangian,

$$g_f(x_*) = \begin{pmatrix} 0 \\ J_c(x_*)^{\mathrm{T}} \lambda_* \end{pmatrix}. \tag{4.30}$$

Thus, at the optimum, the gradient of the objective function is a linear combination of the columns of the Jacobian of the constraints.

Another Example: The Rayleigh Quotient

The important equation (3.208) on page 122 can also be derived by using differentiation. This equation involves maximization of the Rayleigh quotient (equation (3.209)),

$$x^{\mathrm{T}} A x / x^{\mathrm{T}} x$$

under the constraint that $x \neq 0$. In this function, this constraint is equivalent to the constraint that $x^{\mathrm{T}} x$ equal a fixed nonzero constant, which is canceled in the numerator and denominator. We can arbitrarily require that $x^{\mathrm{T}} x = 1$, and the problem is now to determine the maximum of $x^{\mathrm{T}} A x$ subject to the constraint $x^{\mathrm{T}} x = 1$. We now formulate the Lagrangian

$$x^{\mathrm{T}} A x - \lambda(x^{\mathrm{T}} x - 1), \tag{4.31}$$

differentiate, and set it equal to 0, yielding

$$A x - \lambda x = 0.$$

This implies that a stationary point of the Lagrangian occurs at an eigenvector and that the value of $x^{\mathrm{T}}Ax$ is an eigenvalue. This leads to the conclusion that the maximum of the ratio is the maximum eigenvalue. We also see that the second order necessary condition for a local maximum is satisfied; $A - \lambda I$ is nonpositive definite when λ is the maximum eigenvalue. (We can see this using the spectral decomposition of A and then subtracting λI.) Note that we do not have the sufficient condition that $A - \lambda I$ is negative definite ($A - \lambda I$ is obviously singular), but the fact that it is a maximum is established by inspection of the finite set of stationary points.

Optimization without Differentiation

In the previous example, differentiation led us to a stationary point, but we had to establish by inspection that the stationary point is a maximum. In optimization problems generally, and in constrained optimization problems particularly, it is often easier to use other methods to determine the optimum.

A constrained minimization problem we encounter occasionally is

$$\min_{X} \left(\log |X| + \mathrm{tr}(X^{-1}A) \right) \tag{4.32}$$

for a given positive definite matrix A and subject to X being positive definite. The derivatives given in Table 4.2 could be used. The derivatives set equal to 0 immediately yield $X = A$. This means that $X = A$ is a stationary point, but whether or not it is a minimum would require further analysis. As is often the case with such problems, an alternate approach leaves no such pesky complications. Let A and X be $n \times n$ positive definite matrices, and let $c_1, \ldots, c_n$ be the eigenvalues of $X^{-1}A$. Now, by property 7 on page 107 these are also the eigenvalues of $X^{-1/2}AX^{-1/2}$, which is positive definite (see inequality (3.122) on page 89). Now, consider the expression (4.32) with general X minus the expression with $X = A$:

$$
\begin{aligned}
\log |X| + \mathrm{tr}(X^{-1}A) - \log |A| - \mathrm{tr}(A^{-1}A) &= \log |XA^{-1}| + \mathrm{tr}(X^{-1}A) - \mathrm{tr}(I) \\
&= -\log |X^{-1}A| + \mathrm{tr}(X^{-1}A) - n \\
&= -\log \left(\prod_i c_i \right) + \sum_i c_i - n \\
&= \sum_i (-\log c_i + c_i - 1) \\
&\geq 0
\end{aligned}
$$

because if $c > 0$, then $\log c \leq c - 1$, and the minimun occurs when each $c_i = 1$; that is, when $X^{-1}A = I$. Thus, the minimum of expression (4.32) occurs uniquely at $X = A$.

4.4 Multiparameter Likelihood Functions

For a sample $y = (y_1, \ldots, y_n)$ from a probability distribution with probability density function $p(\cdot;\theta)$, the *likelihood function* is

$$L(\theta; y) = \prod_{i=1}^{n} p(y_i; \theta), \qquad (4.33)$$

and the *log-likelihood function* is $l(\theta; y) = \log(L(\theta; y))$. It is often easier to work with the log-likelihood function.

The log-likelihood is an important quantity in information theory and in unbiased estimation. If Y is a random variable with the given probability density function with the r-vector parameter θ, the *Fisher information* matrix that Y contains about θ is the $r \times r$ matrix

$$I(\theta) = \mathrm{Cov}_\theta \left(\frac{\partial l(t, Y)}{\partial t_i}, \frac{\partial l(t, Y)}{\partial t_j} \right), \qquad (4.34)$$

where Cov_θ represents the variance-covariance matrix of the functions of Y formed by taking expectations for the given θ. (I use different symbols here because the derivatives are taken with respect to a *variable*, but the θ in Cov_θ cannot be the variable of the differentiation. This distinction is somewhat pedantic, and sometimes I follow the more common practice of using the same symbol in an expression that involves both Cov_θ and $\partial l(\theta, Y)/\partial \theta_i$.)

For example, if the distribution is the d-variate normal distribution with mean d-vector μ and $d \times d$ positive definite variance-covariance matrix Σ, the likelihood, equation (4.33), is

$$L(\mu, \Sigma; y) = \frac{1}{\left((2\pi)^{d/2}|\Sigma|^{1/2}\right)^n} \exp\left(-\frac{1}{2} \sum_{i=1}^{n} (y_i - \mu)^{\mathrm{T}} \Sigma^{-1} (y_i - \mu) \right).$$

(Note that $|\Sigma|^{1/2} = |\Sigma^{\frac{1}{2}}|$. The square root matrix $\Sigma^{\frac{1}{2}}$ is often useful in transformations of variables.)

Anytime we have a quadratic form that we need to simplify, we should recall equation (3.63): $x^{\mathrm{T}} A x = \mathrm{tr}(A x x^{\mathrm{T}})$. Using this, and because, as is often the case, the log-likelihood is easier to work with, we write

$$l(\mu, \Sigma; y) = c - \frac{n}{2} \log |\Sigma| - \frac{1}{2}\mathrm{tr}\left(\Sigma^{-1} \sum_{i=1}^{n} (y_i - \mu)(y_i - \mu)^{\mathrm{T}} \right), \qquad (4.35)$$

where we have used c to represent the constant portion. Next, we use the Pythagorean equation (2.47) or equation (3.64) on the outer product to get

$$l(\mu, \Sigma; y) = c - \frac{n}{2} \log |\Sigma| - \frac{1}{2}\mathrm{tr}\left(\Sigma^{-1} \sum_{i=1}^{n} (y_i - \bar{y})(y_i - \bar{y})^{\mathrm{T}} \right)$$

$$- \frac{n}{2}\mathrm{tr}\left(\Sigma^{-1}(\bar{y} - \mu)(\bar{y} - \mu)^{\mathrm{T}} \right). \qquad (4.36)$$

In maximum likelihood estimation, we seek the maximum of the likelihood function (4.33) with respect to θ while we consider y to be fixed. If the maximum occurs within an open set and if the likelihood is differentiable, we might be able to find the maximum likelihood estimates by differentiation. In the log-likelihood for the d-variate normal distribution, we consider the parameters μ and Σ to be variables. To emphasize that perspective, we replace the parameters μ and Σ by the variables $\hat{\mu}$ and $\widehat{\Sigma}$. Now, to determine the maximum, we could take derivatives with respect to $\hat{\mu}$ and $\widehat{\Sigma}$, set them equal to 0, and solve for the maximum likelihood estimates. Some subtle problems arise that depend on the fact that for any constant vector a and scalar b, $\Pr(a^{\mathrm{T}}X = b) = 0$, but we do not interpret the likelihood as a probability. In Exercise 4.5b you are asked to determine the values of $\hat{\mu}$ and $\widehat{\Sigma}$ using properties of traces and positive definite matrices without resorting to differentiation. (This approach does not avoid the subtle problems, however.)

Often in working out maximum likelihood estimates, students immediately think of differentiating, setting to 0, and solving. As noted above, this requires that the likelihood function be differentiable, that it be concave, and that the maximum occur at an interior point of the parameter space. Keeping in mind exactly what the problem is — one of finding a maximum — often leads to the correct solution more quickly.

4.5 Integration and Expectation

Just as we can take derivatives with respect to vectors or matrices, we can also take antiderivatives or definite integrals with respect to vectors or matrices. Our interest is in integration of functions weighted by a multivariate probability density function, and for our purposes we will be interested only in definite integrals.

Again, there are three components:

- the differential (the variable of the operation) and its domain (the range of the integration),
- the integrand (the function), and
- the result of the operation (the integral).

In the simplest case, all three of these objects are of the same type; they are scalars. In the happy cases that we consider, each definite integral within the nested sequence exists, so convergence and order of integration are not issues. (The implication of these remarks is that while there is a much bigger field of mathematics here, we are concerned about the relatively simple cases that suffice for our purposes.)

In some cases of interest involving vector-valued random variables, the differential is the vector representing the values of the random variable and the integrand has a scalar function (the probability density) as a factor. In one type of such an integral, the integrand is only the probability density function, and the integral evaluates to a probability, which of course is a scalar. In another type of such an integral, the integrand is a vector representing the values of the random variable times the probability density function. The integral in this case evaluates to a vector, namely the expectation of the random variable over the domain of the integration. Finally, in an example of a third type of such an integral, the integrand is an outer product with itself of a vector representing the values of the random variable minus its mean times the probability density function. The integral in this case evaluates to a variance-covariance matrix. In each of these cases, the integral is the same type of object as the integrand.

4.5.1 Multidimensional Integrals and Integrals Involving Vectors and Matrices

An integral of the form $\int f(v)\, dv$, where v is a vector, can usually be evaluated as a multiple integral with respect to each differential dv_i. Likewise, an integral of the form $\int f(M)\, dM$, where M is a matrix can usually be evaluated by "unstacking" the columns of dM, evaluating the integral as a multiple integral with respect to each differential dm_{ij}, and then possibly "restacking" the result.

Multivariate integrals (that is, integrals taken with respect to a vector or a matrix) define probabilities and expectations in multivariate probability distributions. As with many well-known univariate integrals, such as $\Gamma(\cdot)$, that relate to univariate probability distributions, there are standard multivariate integrals, such as the multivariate gamma, $\Gamma_d(\cdot)$, that relate to multivariate probability distributions. Using standard integrals often facilitates the computations.

Change of Variables; Jacobians

When evaluating an integral of the form $\int f(x)\, dx$, where x is a vector, for various reasons we may form a one-to-one differentiable transformation of the variables of integration; that is, of x. We write x as a function of the new variables; that is, $x = g(y)$, and so $y = g^{-1}(x)$. A simple fact from elementary multivariable calculus is

$$\int_{R(x)} f(x)\, dx = \int_{R(y)} f(g(y))\, |\det(\mathrm{J}_g(y))|\, dy, \qquad (4.37)$$

where $R(y)$ is the image of $R(x)$ under g^{-1} and $\mathrm{J}_g(y)$ is the Jacobian of g (see equation (4.12)). (This is essentially a chain rule result for $dx = d(g(y)) = \mathrm{J}_g dy$ under the interpretation of dx and dy as positive differential elements and the interpretation of $|\det(\mathrm{J}_g)|$ as a volume element, as discussed on page 57.)

In the simple case of a full rank linear transformation of a vector, the Jacobian is constant, and so for $y = Ax$ with A a fixed matrix, we have

$$\int f(x)\,dx = |\det(A)|^{-1} \int f(A^{-1}y)\,dy.$$

(Note that we write $\det(A)$ instead of $|A|$ for the determinant if we are to take the absolute value of it because otherwise we would have $||A||$, which is a symbol for a norm. However, $|\det(A)|$ is not a norm; it lacks each of the properties listed on page 16.)

In the case of a full rank linear transformation of a matrix variable of integration, the Jacobian is somewhat more complicated, but the Jacobian is constant for a fixed transformation matrix. For a transformation $Y = AX$, we determine the Jacobian as above by considering the columns of X one by one. Hence, if X is an $n \times m$ matrix and A is a constant nonsingular matrix, we have

$$\int f(X)\,dX = |\det(A)|^{-m} \int f(A^{-1}Y)\,dY.$$

For a transformation of the form $Z = XB$, we determine the Jacobian by considering the rows of X one by one.

4.5.2 Integration Combined with Other Operations

Integration and another finite linear operator can generally be performed in any order. For example, because the trace is a finite linear operator, integration and the trace can be performed in either order:

$$\int \operatorname{tr}(A(x))\,dx = \operatorname{tr}\left(\int A(x)\,dx\right).$$

For a scalar function of two vectors x and y, it is often of interest to perform differentiation with respect to one vector and integration with respect to the other vector. In such cases, it is of interest to know when these operations can be interchanged. The answer is given in the following theorem, which is a consequence of the Lebesgue dominated convergence theorem. Its proof can be found in any standard text on real analysis.

Let $\mathcal{X}$ be an open set, and let $f(x, y)$ and $\partial f/\partial x$ be scalar-valued functions that are continuous on $\mathcal{X} \times \mathcal{Y}$ for some set $\mathcal{Y}$. Now suppose there are scalar functions $g_0(y)$ and $g_1(y)$ such that

$$\left.\begin{array}{l} |f(x,y)| \le g_0(y) \\[2ex] \|\tfrac{\partial}{\partial x} f(x,y)\| \le g_1(y) \end{array}\right\} \quad \text{for all } (x,y) \in \mathcal{X} \times \mathcal{Y},$$

$$\int_{\mathcal{Y}} g_0(y)\,dy < \infty,$$

and

$$\int_{\mathcal{Y}} g_1(y)\,dy < \infty.$$

Then

$$\frac{\partial}{\partial x} \int_y f(x,y)\,\mathrm{d}y = \int_y \frac{\partial}{\partial x} f(x,y)\,\mathrm{d}y. \qquad (4.38)$$

An important application of this interchange is in developing the information inequality. (This inequality is not germane to the present discussion; it is only noted here for readers who may already be familiar with it.)

4.5.3 Random Variables

A vector random variable is a function from some sample space into $\mathrm{I\!R}^n$, and a matrix random variable is a function from a sample space into $\mathrm{I\!R}^{n \times m}$. (Technically, in each case, the function is required to be *measurable* with respect to a *measure* defined in the context of the sample space and an appropriate collection of subsets of the sample space.) Associated with each random variable is a distribution function whose derivative with respect to an appropriate measure is nonnegative and integrates to 1 over the full space formed by $\mathrm{I\!R}$.

Vector Random Variables

The simplest kind of vector random variable is one whose elements are independent. Such random vectors are easy to work with because the elements can be dealt with individually, but they have limited applications. More interesting random vectors have a multivariate structure that depends on the relationships of the distributions of the individual elements. The simplest nondegenerate multivariate structure is of second degree; that is, a covariance or correlation structure. The probability density of a random vector with a multivariate structure generally is best represented by using matrices. In the case of the multivariate normal distribution, the variances and covariances together with the means completely characterize the distribution. For example, the fundamental integral that is associated with the d-variate normal distribution, sometimes called *Aitken's integral*,

$$\int_{\mathrm{I\!R}^d} \mathrm{e}^{-(x-\mu)^{\mathrm{T}} \Sigma^{-1}(x-\mu)/2}\,\mathrm{d}x = (2\pi)^{d/2}|\Sigma|^{1/2}, \qquad (4.39)$$

provides that constant. The rank of the integral is the same as the rank of the integrand. ("Rank" is used here in the sense of "number of dimensions".) In this case, the integrand and the integral are scalars.

Equation (4.39) is a simple result that follows from the evaluation of the individual single integrals after making the change of variables $y_i = x_i - \mu_i$. It can also be seen by first noting that because Σ^{-1} is positive definite, as in equation (3.215), it can be written as $P^{\mathrm{T}} \Sigma^{-1} P = I$ for some nonsingular matrix P. Now, after the translation $y = x - \mu$, which leaves the integral unchanged, we make the linear change of variables $z = P^{-1}y$, with the associated Jacobian $|\det(P)|$, as in equation (4.37). From $P^{\mathrm{T}} \Sigma^{-1} P = I$, we have

$|\det(P)| = (\det(\Sigma))^{1/2} = |\Sigma|^{1/2}$ because the determinant is positive. Aitken's integral therefore is

$$\int_{\mathbb{R}^d} e^{-y^{\mathrm{T}}\Sigma^{-1}y/2}\,\mathrm{d}y = \int_{\mathbb{R}^d} e^{-(Pz)^{\mathrm{T}}\Sigma^{-1}Pz/2}\,(\det(\Sigma))^{1/2}\mathrm{d}z$$

$$= \int_{\mathbb{R}^d} e^{-z^{\mathrm{T}}z/2}\,\mathrm{d}z\,(\det(\Sigma))^{1/2}$$

$$= (2\pi)^{d/2}(\det(\Sigma))^{1/2}.$$

The expected value of a function f of the vector-valued random variable X is

$$\mathrm{E}(f(X)) = \int_{D(X)} f(x)p_X(x)\,\mathrm{d}x, \tag{4.40}$$

where $D(X)$ is the support of the distribution, $p_X(x)$ is the probability density function evaluated at x, and x $\mathrm{d}x$ are dummy vectors whose elements correspond to those of X. Interpreting $\int_{D(X)}$ $\mathrm{d}x$ as a nest of univariate integrals, the result of the integration of the vector $f(x)p_X(x)$ is clearly of the same type as $f(x)$. For example, if $f(x) = x$, the expectation is the mean, which is a vector. For the normal distribution, we have

$$\mathrm{E}(X) = (2\pi)^{-d/2}|\Sigma|^{-1/2}\int_{\mathbb{R}^d} xe^{-(x-\mu)^{\mathrm{T}}\Sigma^{-1}(x-\mu)/2}\,\mathrm{d}x$$

$$= \mu.$$

For the variance of the vector-valued random variable X,

$$\mathrm{V}(X),$$

the function f in expression (4.40) above is the matrix $(X - \mathrm{E}(X))(X - \mathrm{E}(X))^{\mathrm{T}}$, and the result is a matrix. An example is the normal variance:

$$\mathrm{V}(X) = \mathrm{E}\left((X - \mathrm{E}(X))(X - \mathrm{E}(X))^{\mathrm{T}}\right)$$

$$= (2\pi)^{-d/2}|\Sigma|^{-1/2}\int_{\mathbb{R}^d}\left((x - \mu)(x - \mu)^{\mathrm{T}}\right)e^{-(x-\mu)^{\mathrm{T}}\Sigma^{-1}(x-\mu)/2}\,\mathrm{d}x$$

$$= \Sigma.$$

Matrix Random Variables

While there are many random variables of interest that are vectors, there are only a few random matrices whose distributions have been studied. One, of course, is the Wishart distribution; see Exercise 4.8. An integral of the Wishart probability density function over a set of nonnegative definite matrices is the probability of the set.

A simple distribution for random matrices is one in which the individual elements have identical and independent normal distributions. This distribution of matrices was named the BMvN distribution by Birkhoff and Gulati (1979) (from the last names of three mathematicians who used such random matrices in numerical studies). Birkhoff and Gulati (1979) showed that if the elements of the $n \times n$ matrix X are i.i.d. $N(0, \sigma^2)$, and if Q is an orthogonal matrix and R is an upper triangular matrix with positive elements on the diagonal such that $QR = X$, then Q has the *Haar distribution*. (The factorization $X = QR$ is called the QR decomposition and is discussed on page 190 If X is a random matrix as described, this factorization exists with probability 1.) The Haar(n) distribution is uniform over the space of $n \times n$ orthogonal matrices.

The measure

$$\mu(D) = \int_D H^{\mathrm{T}} \, \mathrm{d}H, \tag{4.41}$$

where D is a subset of the orthogonal group $\mathcal{O}(n)$ (see page 105), is called the *Haar measure*. This measure is used to define a kind of "uniform" probability distribution for orthogonal factors of random matrices. For any $Q \in \mathcal{O}(n)$, let QD represent the subset of $\mathcal{O}(n)$ consisting of the matrices $\tilde{H} = QH$ for $H \in D$ and DQ represent the subset of matrices formed as HQ. From the integral, we see

$$\mu(QD) = \mu(DQ) = \mu(D),$$

so the Haar measure is invariant to multiplication within the group. The measure is therefore also called the *Haar invariant measure* over the orthogonal group. (See Muirhead, 1982, for more properties of this measure.)

A common matrix integral is the complete d-variate gamma function, denoted by $\Gamma_d(x)$ and defined as

$$\Gamma_d(x) = \int_D \mathrm{e}^{-\mathrm{tr}(A)} |A|^{x-(d+1)/2} \, \mathrm{d}A, \tag{4.42}$$

where D is the set of all $d \times d$ positive definite matrices, $A \in D$, and $x > (d-1)/2$. A multivariate gamma distribution can be defined in terms of the integrand. (There are different definitions of a multivariate gamma distribution.) The multivariate gamma function also appears in the probability density function for a Wishart random variable (see Muirhead, 1982, or Carmeli, 1983, for example).

Exercises

4.1. Use equation (4.6), which defines the derivative of a matrix with respect to a scalar, to show the product rule equation (4.3) directly:

$$\frac{\partial YW}{\partial x} = \frac{\partial Y}{\partial x} W + Y \frac{\partial W}{\partial x}.$$

4.2. For the n-vector x, compute the gradient $g_V(x)$, where $V(x)$ is the variance of x, as given in equation (2.53).

Hint: Use the chain rule.

4.3. For the square, nonsingular matrix Y, show that

$$\frac{\partial Y^{-1}}{\partial x} = -Y^{-1}\frac{\partial Y}{\partial x}Y^{-1}.$$

Hint: Differentiate $YY^{-1} = I$.

4.4. Newton's method.

You should not, of course, just blindly pick a starting point and begin iterating. How can you be sure that your solution is a local optimum? Can you be sure that your solution is a global optimum? It is often a good idea to make some plots of the function. In the case of a function of a single variable, you may want to make plots in different scales. For functions of more than one variable, profile plots may be useful (that is, plots of the function in one variable with all the other variables held constant).

  a) Use Newton's method to determine the maximum of the function
  $f(x) = \sin(4x) - x^4/12$.

  b) Use Newton's method to determine the minimum of

$$f(x_1, x_2) = 2x_1^4 + 3x_1^3 + 2x_1^2 + x_2^2 - 4x_1x_2.$$

  What is the Hessian at the minimum?

4.5. Consider the log-likelihood $l(\mu, \Sigma; y)$ for the d-variate normal distribution, equation (4.35). Be aware of the subtle issue referred to in the text. It has to do with whether $\sum_{i=1}^{n}(y_i - \bar{y})(y_i - \bar{y})^{\mathrm{T}}$ is positive definite.

  a) Replace the parameters μ and Σ by the variables $\hat{\mu}$ and $\widehat{\Sigma}$, take derivatives with respect to $\hat{\mu}$ and $\widehat{\Sigma}$, set them equal to 0, and solve for the maximum likelihood estimates. What assumptions do you have to make about n and d?

  b) Another approach to maximizing the expression in equation (4.35) is to maximize the last term with respect to $\hat{\mu}$ (this is the only term involving μ) and then, with the maximizing value substituted, to maximize

$$-\frac{n}{2}\log|\Sigma| - \frac{1}{2}\mathrm{tr}\left(\Sigma^{-1}\sum_{i=1}^{n}(y_i - \bar{y})(y_i - \bar{y})^{\mathrm{T}}\right).$$

  Use this approach to determine the maximum likelihood estimates $\hat{\mu}$ and $\widehat{\Sigma}$.

4.6. Let

$$D = \left\{\begin{bmatrix} c & -s \\ s & c \end{bmatrix} : -1 \le c \le 1, c^2 + s^2 = 1\right\}.$$

Evaluate the Haar measure $\mu(D)$. (This is the class of 2×2 rotation matrices; see equation (5.3), page 177.)

4.7. Write a Fortran or C program to generate $n \times n$ random orthogonal matrices with the Haar uniform distribution. Use the following method due to Heiberger (1978), which was modified by Stewart (1980). (See also Tanner and Thisted, 1982.)

  a) Generate $n-1$ independent i-vectors, $x_2, x_3, \ldots, x_n$, from $N_i(0, I_i)$. (x_i is of length i.)
  b) Let $r_i = \|x_i\|_2$, and let $\widetilde{H}_i$ be the $i \times i$ reflection matrix that transforms x_i into the i-vector $(r_i, 0, 0, \ldots, 0)$.
  c) Let H_i be the $n \times n$ matrix

$$\begin{bmatrix} I_{n-i} & 0 \\ 0 & \widetilde{H}_i \end{bmatrix},$$

and form the diagonal matrix,

$$J = \mathrm{diag}\big((-1)^{b_1}, (-1)^{b_2}, \ldots, (-1)^{b_n}\big),$$

where the b_i are independent realizations of a Bernoulli random variable.
  d) Deliver the orthogonal matrix $Q = JH_1 H_2 \cdots H_n$.
The matrix Q generated in this way is orthogonal and has a Haar distribution.
Can you think of any way to test the goodness-of-fit of samples from this algorithm? Generate a sample of 1,000 2×2 random orthogonal matrices, and assess how well the sample follows a Haar uniform distribution.

4.8. The probability density for the Wishart distribution is proportional to

$$e^{\mathrm{tr}(\Sigma^{-1}W/2)}|W|^{(n-d-1)/2},$$

where W is a $d \times d$ nonnegative definite matrix, the parameter Σ is a fixed $d \times d$ positive definite matrix, and the parameter n is positive. (Often n is restricted to integer values greater than d.) Determine the constant of proportionality.

5

Matrix Transformations and Factorizations

In most applications of linear algebra, problems are solved by transformations of matrices. A given matrix that represents some transformation of a vector is transformed so as to determine one vector given another vector. The simplest example of this is in working with the linear system $Ax = b$. The matrix A is transformed through a succession of operations until x is determined easily by the transformed A and b. Each operation is a pre- or postmultiplication by some other matrix. Each matrix formed as a product must be *equivalent* to A; therefore each transformation matrix must be of full rank. In eigen-problems, we likewise perform a sequence of pre- or postmultiplications. In this case, each matrix formed as a product must be *similar* to A; therefore each transformation matrix must be orthogonal. We develop transformations of matrices by transformations on the individual rows or columns.

Factorizations

Invertible transformations result in a factorization of the matrix. If B is a $k \times n$ matrix and C is an $n \times k$ matrix such that $CB = I_n$, for a given $n \times m$ matrix A the transformation $BA = D$ results in a factorization: $A = CD$. In applications of linear algebra, we determine C and D such that $A = CD$ and such that C and D have useful properties for the problem being addressed. This is also called a decomposition of the matrix. We will use the terms "matrix factorization" and "matrix decomposition" interchangeably. Most methods for eigenanalysis and for solving linear systems proceed by factoring the matrix, as we see in Chapters 6 and 7.

In Chapter 3, we discussed some factorizations, including

- the full rank factorization (equation (3.112)) of a general matrix,
- the equivalent canonical factorization (equation (3.117)) of a general matrix,
- the similar canonical factorization (equation (3.193)) or "diagonal factorization" of a diagonalizable matrix (which is necessarily square),

- the orthogonally similar canonical factorization (equation (3.197)) of a symmetric matrix (which is necessarily diagonalizable),
- the square root (equation (3.216)) of a nonnegative definite matrix (which is necessarily symmetric), and
- the singular value factorization (equation (3.218)) of a general matrix.

In this chapter, we consider some general matrix transformations and then introduce three additional factorizations:

- the LU (and LR and LDU) factorization of a general matrix,
- the QR factorization of a general matrix, and
- the Cholesky factorization of a nonnegative definite matrix.

These factorizations are useful both in theory and in practice. Another factorization that is very useful in proving various theorems, but that we will not discuss in this book, is the *Jordan decomposition*. For a discussion of this factorization, see Horn and Johnson (1991), for example.

5.1 Transformations by Orthogonal Matrices

In previous chapters, we observed some interesting properties of orthogonal matrices. From equation (3.228), for example, we see that orthogonal transformations preserve lengths of vectors.

If Q is an orthogonal matrix (that is, if $Q^\mathrm{T}Q = I$), then, for vectors x and y, we have

$$\langle Qx, Qy \rangle = (xQ)^\mathrm{T}(Qy) = x^\mathrm{T}Q^\mathrm{T}Qy = x^\mathrm{T}y = \langle x, y \rangle,$$

and hence,

$$\arccos\left(\frac{\langle Qx, Qy \rangle}{\|Qx\|_2 \, \|Qy\|_2}\right) = \arccos\left(\frac{\langle x, y \rangle}{\|x\|_2 \, \|y\|_2}\right). \tag{5.1}$$

Thus we see that orthogonal transformations also preserve angles.

As noted previously, permutation matrices are orthogonal, and we have used them extensively in rearranging the columns and/or rows of matrices. We have noted the fact that if Q is an orthogonal matrix and

$$B = Q^\mathrm{T}AQ,$$

then A and B have the same eigenvalues (and A and B are said to be orthogonally similar). By forming the transpose, we see immediately that the transformation $Q^\mathrm{T}AQ$ preserves symmetry; that is, if A is symmetric, then B is symmetric.

From equation (3.229), we see that $\|Q^{-1}\|_2 = 1$. This has important implications for the accuracy of numerical computations. (Using computations with orthogonal matrices will not make problems more "ill-conditioned".)

We often use orthogonal transformations that preserve lengths and angles while rotating $\mathrm{I\!R}^n$ or reflecting regions of $\mathrm{I\!R}^n$. The transformations are appropriately called rotators and reflectors, respectively.

5.2 Geometric Transformations

In many important applications of linear algebra, a vector represents a point in space, with each element of the vector corresponding to an element of a coordinate system, usually a Cartesian system. A set of vectors describes a geometric object. Algebraic operations are geometric transformations that rotate, deform, or translate the object. While these transformations are often used in the two or three dimensions that correspond to the easily perceived physical space, they have similar applications in higher dimensions. Thinking about operations in linear algebra in terms of the associated geometric operations often provides useful intuition.

Invariance Properties of Transformations

Important characteristics of these transformations are what they leave *unchanged*; that is, their *invariance properties* (see Table 5.1). All of these transformations we will discuss are *linear transformations* because they preserve straight lines.

Table 5.1. Invariance Properties of Transformations

Transformation	Preserves
linear	lines
affine	lines, collinearity
shearing	lines, collinearity
scaling	lines, angles (and, hence, collinearity)
translation	lines, angles, lengths
rotation	lines, angles, lengths
reflection	lines, angles, lengths

We have seen that an orthogonal transformation preserves lengths of vectors (equation (3.228)) and angles between vectors (equation (5.1)). Such a transformation that preserves lengths and angles is called an *isometric transformation*. Such a transformation also preserves areas and volumes.

Another isometric transformation is a *translation*, which for a vector x is just the addition of another vector:

$$\tilde{x} = x + t.$$

A transformation that preserves angles is called an *isotropic transformation*. An example of an isotropic transformation that is not isometric is a uniform scaling or dilation transformation, $\tilde{x} = ax$, where a is a scalar.

The transformation $\tilde{x} = Ax$, where A is a diagonal matrix with not all elements the same, does preserve angles; it is an *anisotropic* scaling. Another anisotropic transformation is a *shearing transformation*, $\tilde{x} = Ax$, where A is the same as an identity matrix, except for a single row or column that has a one on the diagonal but possibly nonzero elements in the other positions; for example,

$$\begin{bmatrix} 1 & 0 & a_1 \\ 0 & 1 & a_1 \\ 0 & 0 & 1 \end{bmatrix}.$$

Although they do not preserve angles, both anisotropic scaling and shearing transformations preserve parallel lines. A transformation that preserves parallel lines is called an *affine transformation*. Preservation of parallel lines is equivalent to preservation of collinearity, and so an alternative characterization of an affine transformation is one that preserves collinearity. More generally, we can combine nontrivial scaling and shearing transformations to see that the transformation Ax for any nonsingular matrix A is affine. It is easy to see that addition of a constant vector to all vectors in a set preserves collinearity within the set, so a more general affine transformation is $\tilde{x} = Ax + t$ for a nonsingular matrix A and a vector t.

A *projective transformation*, which uses the homogeneous coordinate system of the projective plane (see Section 5.2.3), preserves straight lines, but does not preserve parallel lines. Projective transformations are very useful in computer graphics. In those applications we do not always want parallel lines to project onto the display plane as parallel lines.

5.2.1 Rotations

The simplest rotation of a vector can be thought of as the rotation of a plane defined by two coordinates about the other principal axes. Such a rotation changes two elements of all vectors in that plane and leaves all the other elements, representing the other coordinates, unchanged. This rotation can be described in a two-dimensional space defined by the coordinates being changed, without reference to the other coordinates.

Consider the rotation of the vector x through the angle θ into $\tilde{x}$. The length is preserved, so we have $\|\tilde{x}\| = \|x\|$. Referring to Figure 5.1, we can write

$$\tilde{x}_1 = \|x\| \cos(\phi + \theta),$$
$$\tilde{x}_2 = \|x\| \sin(\phi + \theta).$$

Now, from elementary trigonometry, we know

$$\cos(\phi + \theta) = \cos\phi\cos\theta - \sin\phi\sin\theta,$$
$$\sin(\phi + \theta) = \sin\phi\cos\theta + \cos\phi\sin\theta.$$

Because $\cos\phi = x_1/\|x\|$ and $\sin\phi = x_2/\|x\|$, we can combine these equations to get

$$\begin{aligned} \tilde{x}_1 &= x_1\cos\theta - x_2\sin\theta, \\ \tilde{x}_2 &= x_1\sin\theta + x_2\cos\theta. \end{aligned} \tag{5.2}$$

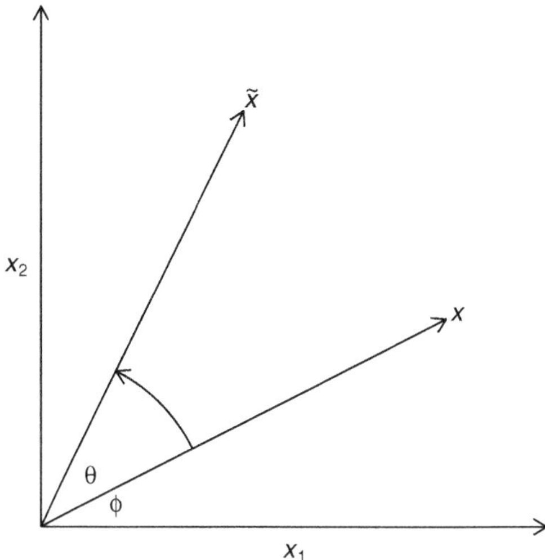

Fig. 5.1. Rotation of x

Hence, multiplying x by the orthogonal matrix

$$\begin{bmatrix} \cos\theta & -\sin\theta \\ \sin\theta & \cos\theta \end{bmatrix} \tag{5.3}$$

performs the rotation of x.

This idea easily extends to the rotation of a plane formed by two coordinates about all of the other (orthogonal) principal axes. By convention, we assume clockwise rotations for axes that increase in the direction from which the system is viewed. For example, if there were an x_3 axis in Figure 5.1, it would point toward the viewer. (This is called a "right-hand" coordinate system.)

The rotation matrix about principal axes is the same as an identity matrix with two diagonal elements changed to $\cos\theta$ and the corresponding off-diagonal elements changed to $\sin\theta$ and $-\sin\theta$.

To rotate a 3-vector, x, about the x_2 axis in a right-hand coordinate system, we would use the rotation matrix

$$\begin{bmatrix} \cos\theta & 0 & \sin\theta \\ 0 & 1 & 0 \\ -\sin\theta & 0 & \cos\theta \end{bmatrix}.$$

A rotation of any hyperplane in n-space can be formed by n successive rotations of hyperplanes formed by two principal axes. (In 3-space, this fact is known as *Euler's rotation theorem*. We can see this to be the case, in 3-space or in general, by construction.)

A rotation of an arbitrary plane can be defined in terms of the direction cosines of a vector in the plane before and after the rotation. In a coordinate geometry, rotation of a plane can be viewed equivalently as a rotation of the coordinate system in the opposite direction. This is accomplished by rotating the unit vectors e_i into $\tilde{e}_i$.

A special type of transformation that rotates a vector to be perpendicular to a principal axis is called a Givens rotation. We discuss the use of this type of transformation in Section 5.4 on page 182.

5.2.2 Reflections

Let u and v be orthonormal vectors, and let x be a vector in the space spanned by u and v, so

$$x = c_1 u + c_2 v$$

for some scalars c_1 and c_2. The vector

$$\tilde{x} = -c_1 u + c_2 v \tag{5.4}$$

is a *reflection* of x through the line defined by the vector v, or $u^{\perp}$.

First consider a reflection that transforms a vector

$$x = (x_1, x_2, \ldots, x_n)$$

into a vector collinear with a unit vector,

$$\tilde{x} = (0, \ldots, 0, \tilde{x}_i, 0, \ldots, 0)$$
$$= \pm \|x\|_2 e_i. \tag{5.5}$$

Geometrically, in two dimensions we have the picture shown in Figure 5.2, where $i = 1$. Which vector x is rotated through (that is, which is u and which is v) depends on the choice of the sign in $\pm\|x\|_2$. The choice that was made yields the $\tilde{x}$ shown in the figure, and from the figure, this can be seen to be correct. If the opposite choice is made, we get the $\tilde{\tilde{x}}$ shown. In the simple two-dimensional case, this is equivalent to reversing our choice of u and v.

5.2.3 Translations; Homogeneous Coordinates

Translations are relatively simple transformations involving the addition of vectors. Rotations, as we have seen, and other geometric transformations such as shearing, as we have indicated, involve multiplication by an appropriate matrix. In applications where several geometric transformations are to be made, it would be convenient if translations could also be performed by matrix multiplication. This can be done by using *homogeneous coordinates*.

Homogeneous coordinates, which form the natural coordinate system for projective geometry, have a very simple relationship to Cartesian coordinates.

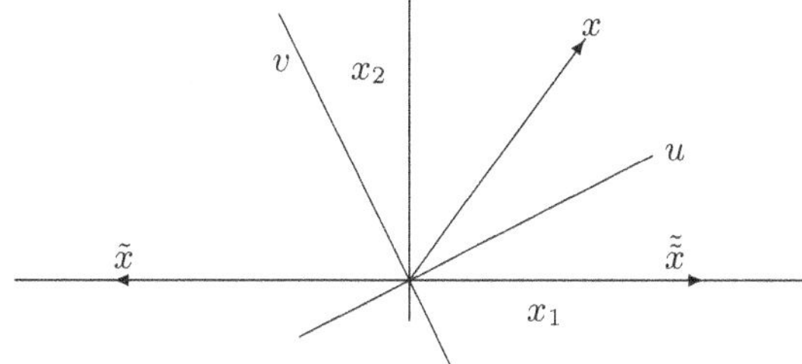

Fig. 5.2. Reflections of x about $u^{\perp}$

The point with Cartesian coordinates $(x_1, x_2, \ldots, x_d)$ is represented in homogeneous coordinates as $(x_0^{\mathrm{h}}, x_1^{\mathrm{h}}, \ldots, x_d^{\mathrm{h}})$, where, for arbitrary x_0^{h} not equal to zero, $x_1^{\mathrm{h}} = x_0^{\mathrm{h}} x_1$, and so on. Because the point is the same, the two different symbols represent the same thing, and we have

$$(x_1, \ldots, x_d) = (x_0^{\mathrm{h}}, x_1^{\mathrm{h}}, \ldots, x_d^{\mathrm{h}}). \tag{5.6a}$$

Alternatively, the hyperplane coordinate may be added at the end, and we have

$$(x_1, \ldots, x_d) = (x_1^{\mathrm{h}}, \ldots, x_d^{\mathrm{h}}, x_0^{\mathrm{h}}). \tag{5.6b}$$

Each value of x_0^{h} corresponds to a hyperplane in the ordinary Cartesian coordinate system. The most common choice is $x_0^{\mathrm{h}} = 1$, and so $x_i^{\mathrm{h}} = x_i$. The special plane $x_0^{\mathrm{h}} = 0$ does not have a meaning in the Cartesian system, but in projective geometry it corresponds to a hyperplane at infinity.

We can easily effect the translation $\tilde{x} = x + t$ by first representing the point x as $(1, x_1, \ldots, x_d)$ and then multiplying by the $(d+1) \times (d+1)$ matrix

$$T = \begin{bmatrix} 1 & 0 & \cdots & 0 \\ t_1 & 1 & \cdots & 0 \\ & \cdots & & \\ t_d & 0 & \cdots & 1 \end{bmatrix}.$$

We will use the symbol x^{h} to represent the vector of corresponding homogeneous coordinates:

$$x^{\mathrm{h}} = (1, x_1, \ldots, x_d).$$

We must be careful to distinguish the point x from the vector that represents the point. In Cartesian coordinates, there is a natural correspondence and the symbol x representing a point may also represent the vector $(x_1, \ldots, x_d)$. The vector of homogeneous coordinates of the result Tx^{h} corresponds to the Cartesian coordinates of $\tilde{x}$, $(x_1 + t_1, \ldots, x_d + t_d)$, which is the desired result.

Homogeneous coordinates are used extensively in computer graphics not only for the ordinary geometric transformations but also for projective transformations, which model visual properties. Riesenfeld (1981) and Mortenson (1997) describe many of these applications. See Exercise 5.2 for a simple example.

5.3 Householder Transformations (Reflections)

We have briefly discussed geometric transformations that reflect a vector through another vector. We now consider some properties and uses of these transformations.

Consider the problem of reflecting x through the vector u. As before, we assume that u and v are orthogonal vectors and that x lies in a space spanned by u and v, and $x = c_1 u + c_2 v$. Form the matrix

$$H = I - 2uu^{\mathrm{T}}, \tag{5.7}$$

and note that

$$\begin{aligned}
Hx &= c_1 u + c_2 v - 2c_1 uu^{\mathrm{T}} u - 2c_2 uu^{\mathrm{T}} v \\
&= c_1 u + c_2 v - 2c_1 u^{\mathrm{T}} uu - 2c_2 u^{\mathrm{T}} vu \\
&= -c_1 u + c_2 v \\
&= \tilde{x},
\end{aligned}$$

as in equation (5.4). The matrix H is a *reflector*; it has transformed x into its reflection $\tilde{x}$ about u.

A reflection is also called a Householder reflection or a Householder transformation, and the matrix H is called a Householder matrix or a Householder reflector. The following properties of H are immediate:

- $Hu = -u$.
- $Hv = v$ for any v orthogonal to u.
- $H = H^{\mathrm{T}}$ (symmetric).
- $H^{\mathrm{T}} = H^{-1}$ (orthogonal).

Because H is orthogonal, if $Hx = \tilde{x}$, then $\|x\|_2 = \|\tilde{x}\|_2$ (see equation (3.228)), so $\tilde{x}_1 = \pm\|x\|_2$.

The matrix uu^{T} is symmetric, idempotent, and of rank 1. (A transformation by a matrix of the form $A - vw^{\mathrm{T}}$ is often called a "rank-one" update, because vw^{T} is of rank 1. Thus, a Householder reflection is a special rank-one update.)

Zeroing Elements in a Vector

The usefulness of Householder reflections results from the fact that it is easy to construct a reflection that will transform a vector x into a vector $\tilde{x}$ that has zeros in all but one position, as in equation (5.5). To construct the reflector of x into $\tilde{x}$, first form the normalized vector $(x - \tilde{x})$:

$$v = x - \tilde{x}/\|\tilde{x}\|_2.$$

We know $\|\tilde{x}\|_2$ to within a sign, and *we choose the sign so as not to add quantities of different signs and possibly similar magnitudes*. (See the discussions of catastrophic cancellation beginning on page 397, in Chapter 10.) Hence, we have

$$q = (x_1, \ \ldots, \ x_{i-1}, \ x_i + \mathrm{sign}(x_i)\|x\|_2, \ x_{i+1}, \ \ldots, \ x_n), \tag{5.8}$$

then

$$u = q/\|q\|_2, \tag{5.9}$$

and finally

$$H = I - 2uu^{\mathrm{T}}. \tag{5.10}$$

Consider, for example, the vector

$$x = (3, 1, 2, 1, 1),$$

which we wish to transform into

$$\tilde{x} = (\widetilde{x_1}, 0, 0, 0, 0).$$

We have

$$\|x\| = 4,$$

so we form the vector

$$u = \frac{1}{\sqrt{56}}(7, 1, 2, 1, 1)$$

and the reflector

$$H = I - 2uu^{\mathrm{T}}$$

$$= \begin{bmatrix} 1 & 0 & 0 & 0 & 0 \\ 0 & 1 & 0 & 0 & 0 \\ 0 & 0 & 1 & 0 & 0 \\ 0 & 0 & 0 & 1 & 0 \\ 0 & 0 & 0 & 0 & 1 \end{bmatrix} - \frac{1}{28} \begin{bmatrix} 49 & 7 & 14 & 7 & 7 \\ 7 & 1 & 2 & 1 & 1 \\ 14 & 2 & 4 & 2 & 2 \\ 7 & 1 & 2 & 1 & 1 \\ 7 & 1 & 2 & 1 & 1 \end{bmatrix}$$

$$= \frac{1}{28} \begin{bmatrix} -21 & -7 & -14 & -7 & -7 \\ -7 & 27 & -2 & -1 & -1 \\ -14 & -2 & 24 & -2 & -2 \\ -7 & -1 & -2 & 27 & -1 \\ -7 & -1 & -2 & -1 & 27 \end{bmatrix}$$

to yield $Hx = (-4, 0, 0, 0, 0)$.

Carrig and Meyer (1997) describe two variants of the Householder transformations that take advantage of computer architectures that have a cache memory or that have a bank of floating-point registers whose contents are immediately available to the computational unit.

5.4 Givens Transformations (Rotations)

We have briefly discussed geometric transformations that rotate a vector in such a way that a specified element becomes 0 and only one other element in the vector is changed. Such a method may be particularly useful if only part of the matrix to be transformed is available. These transformations are called *Givens transformations*, or *Givens rotations*, or sometimes *Jacobi transformations*.

The basic idea of the rotation, which is a special case of the rotations discussed on page 176, can be seen in the case of a vector of length 2. Given the vector $x = (x_1, x_2)$, we wish to rotate it to $\tilde{x} = (\tilde{x}_1, 0)$. As with a reflector, $\tilde{x}_1 = \|x\|$. Geometrically, we have the picture shown in Figure 5.3.

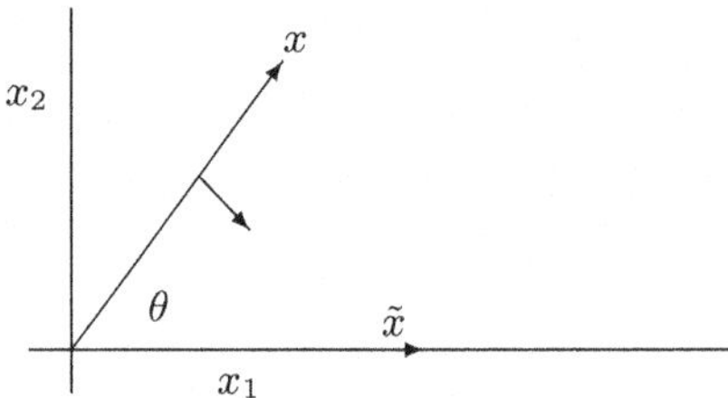

Fig. 5.3. Rotation of x onto a Coordinate Axis

It is easy to see that the orthogonal matrix

$$Q = \begin{bmatrix} \cos\theta & \sin\theta \\ -\sin\theta & \cos\theta \end{bmatrix} \tag{5.11}$$

will perform this rotation of x if $\cos\theta = x_1/r$ and $\sin\theta = x_2/r$, where $r = \|x\| = \sqrt{x_1^2 + x_2^2}$. (This is the same matrix as in equation (5.3), except that the rotation is in the opposite direction.) Notice that θ is not relevant; we only need real numbers c and s such that $c^2 + s^2 = 1$.

We have

$$\tilde{x}_1 = \frac{x_1^2}{r} + \frac{x_2^2}{r}$$
$$= \|x\|,$$

$$\tilde{x}_2 = -\frac{x_2 x_1}{r} + \frac{x_1 x_2}{r}$$
$$= 0;$$

that is,

$$Q \begin{pmatrix} x_1 \\ x_2 \end{pmatrix} = \begin{pmatrix} \|x\| \\ 0 \end{pmatrix}.$$

Zeroing One Element in a Vector

As with the Householder reflection that transforms a vector

$$x = (x_1, x_2, x_3, \ldots, x_n)$$

into a vector

$$\tilde{x}_H = (\tilde{x}_{H1}, 0, 0, \ldots, 0),$$

it is easy to construct a Givens rotation that transforms x into

$$\tilde{x}_G = (\tilde{x}_{G1}, 0, x_3, \ldots, x_n).$$

We can construct an orthogonal matrix G_{pq} similar to that shown in equation (5.11) that will transform the vector

$$x = (x_1, \ldots, x_p, \ldots, x_q, \ldots, x_n)$$

into

$$\tilde{x} = (x_1, \ldots, \tilde{x}_p, \ldots, 0, \ldots, x_n).$$

The orthogonal matrix that will do this is

$$
G_{pq}(\theta) =
\begin{bmatrix}
1 & 0 & \cdots & 0 & 0 & 0 & \cdots & 0 & 0 & 0 & \cdots & 0 \\
0 & 1 & \cdots & 0 & 0 & 0 & \cdots & 0 & 0 & 0 & \cdots & 0 \\
& & \ddots & & & & & & & & & \\
0 & 0 & \cdots & 1 & 0 & 0 & \cdots & 0 & 0 & 0 & \cdots & 0 \\
0 & 0 & \cdots & 0 & c & 0 & \cdots & 0 & s & 0 & \cdots & 0 \\
0 & 0 & \cdots & 0 & 0 & 1 & \cdots & 0 & 0 & 0 & \cdots & 0 \\
& & & & & & \ddots & & & & & \\
0 & 0 & \cdots & 0 & 0 & 0 & \cdots & 1 & 0 & 0 & \cdots & 0 \\
0 & 0 & \cdots & 0 & -s & 0 & \cdots & 0 & c & 0 & \cdots & 0 \\
0 & 0 & \cdots & 0 & 0 & 0 & \cdots & 0 & 0 & 1 & \cdots & 0 \\
& & & & & & & & & & \ddots & \\
0 & 0 & \cdots & 0 & 0 & 0 & \cdots & 0 & 0 & 0 & \cdots & 1
\end{bmatrix},
\tag{5.12}
$$

where the entries in the pth and q^{th} rows and columns are

$$c = \frac{x_p}{r}$$

and

$$s = \frac{x_q}{r},$$

where $r = \sqrt{x_p^2 + x_q^2}$. A rotation matrix is the same as an identity matrix with four elements changed.

Considering x to be the pth column in a matrix X, we can easily see that $G_{pq}X$ results in a matrix with a zero as the qth element of the pth column, and all except the pth and q^{th} rows and columns of $G_{pq}X$ are the same as those of X.

Givens Rotations That Preserve Symmetry

If X is a symmetric matrix, we can preserve the symmetry by a transformation of the form $Q^{\mathrm{T}}XQ$, where Q is any orthogonal matrix. The elements of a Givens rotation matrix that is used in this way and with the objective of forming zeros in two positions in X simultaneously would be determined in the same way as above, but the elements themselves would not be the same. We illustrate that below, while at the same time considering the problem of transforming a value into something other than zero.

Givens Rotations to Transform to Other Values

Consider a symmetric matrix X that we wish to transform to the symmetric matrix $\widetilde{X}$ that has all rows and columns except the pth and q^{th} the same as those in X, and we want a specified value in the $(pp)^{\mathrm{th}}$ position of $\widetilde{X}$, say $\widetilde{x}_{pp} = a$. We seek a rotation matrix G such that $\widetilde{X} = G^{\mathrm{T}}XG$. We have

$$\begin{bmatrix} c & s \\ -s & c \end{bmatrix}^{\mathrm{T}} \begin{bmatrix} x_{pp} & x_{pq} \\ x_{pq} & x_{qq} \end{bmatrix} \begin{bmatrix} c & s \\ -s & c \end{bmatrix} = \begin{bmatrix} a & \widetilde{x}_{pq} \\ \widetilde{x}_{pq} & \widetilde{x}_{qq} \end{bmatrix} \tag{5.13}$$

and

$$c^2 + s^2 = 1.$$

Hence

$$a = c^2 x_{pp} - 2cs x_{pq} + s^2 x_{qq}. \tag{5.14}$$

Writing $t = s/c$ (the tangent), we have the quadratic

$$(x_{qq} - 1)t^2 - 2x_{pq}t + x_{pp} - a = 0 \tag{5.15}$$

with roots

$$t = \frac{x_{pq} \pm 2\sqrt{x_{pq}^2 - (x_{pp} - a)(x_{qq} - 1)}}{(x_{qq} - 1)}. \tag{5.16}$$

The roots are real if and only if

$$x_{pq}^2 \geq (x_{pp} - a)(x_{qq} - 1).$$

If the roots are real, we choose the nonnegative one. (We evaluate equation (5.16); see the discussion of equation (10.3) on page 398.) We then form

$$c = \frac{1}{\sqrt{1 + t^2}} \tag{5.17}$$

and

$$s = ct. \tag{5.18}$$

The rotation matrix G formed from c and s will transform X into $\widetilde{X}$.

Fast Givens Rotations

Often in applications we need to perform a succession of Givens transformations. The overall number of computations can be reduced using a succession of "fast Givens rotations". We write the matrix Q in equation (5.11) as CT,

$$\begin{bmatrix} \cos\theta & \sin\theta \\ -\sin\theta & \cos\theta \end{bmatrix} = \begin{bmatrix} \cos\theta & 0 \\ 0 & \cos\theta \end{bmatrix} \begin{bmatrix} 1 & \tan\theta \\ -\tan\theta & 1 \end{bmatrix}, \qquad (5.19)$$

and instead of working with matrices such as Q, which require four multiplications and two additions, we work with matrices such as T, involving the tangents, which require only two multiplications and two additions. After a number of computations with such matrices, the diagonal matrices of the form of C are accumulated and multiplied together.

The diagonal elements in the accumulated C matrices in the fast Givens rotations can become widely different in absolute values, so to avoid excessive loss of accuracy, it is usually necessary to rescale the elements periodically.

5.5 Factorization of Matrices

It is often useful to represent a matrix A in a factored form,

$$A = BC,$$

where B and C have some specified desirable properties, such as being triangular. Most direct methods of solving linear systems discussed in Chapter 6 are based on factorizations (or, equivalently, "decompositions") of the matrix of coefficients. Matrix factorizations are also performed for reasons other than to solve a linear system, such as in eigenanalysis. Matrix factorizations are generally performed by a sequence of transformations and their inverses. The major important matrix factorizations are:

- full rank factorization (for any matrix);
- diagonal or similar canonical factorization (for diagonalizable matrices);
- orthogonally similar canonical factorization (for symmetric matrices);
- LU factorization and LDU factorization (for nonnegative definite matrices and some others, including nonsquare matrices);
- QR factorization (for any matrix);
- singular value decomposition, SVD, (for any matrix);
- square root factorization (for nonnegative definite matrices); and
- Cholesky factorization (for nonnegative definite matrices).

We have already discussed the full rank, the diagonal canonical, the orthogonally similar canonical, the SVD, and the square root factorizations. In the next few sections we will introduce the LU, LDU, QR, and Cholesky factorizations.

5.6 *LU* and *LDU* Factorizations

For any matrix (whether square or not) that can be expressed as LU, where L is unit lower triangular and U is upper triangular, the product LU is called the LU factorization. If the matrix is not square, or if the matrix is not of full rank, L and/or U will be of trapezoidal form. An LU factorization exists and is unique for nonnegative definite matrices. For more general matrices, the factorization may not exist, and the conditions for the existence are not so easy to state (see Harville, 1997, for example).

Use of Outer Products

An LU factorization is accomplished by a sequence of Gaussian eliminations that are constructed so as to generate 0s below the main diagonal in a given column (see page 66).

Applying these operations to a given matrix A yields a sequence of matrices $A^{(k)}$ with increasing numbers of columns that contain 0s below the main diagonal. Each step in Gaussian elimination is equivalent to multiplication of the current matrix, $A^{(k-1)}$, by some matrix L_k. If we encounter a zero on the diagonal, or possibly for other numerical considerations, we may need to rearrange rows or columns of $A^{(k-1)}$ (see page 209), but if we ignore that for the time being, the L_k matrix has a particularly simple form and is easy to construct. It is the product of axpy elementary matrices similar to those in equation (3.50) on page 66, where the multipliers are determined so as to zero out the column below the main diagonal:

$$L_k = E_{n,k}\left(-a_{n,k}^{(k-1)}/a_{kk}^{(k-1)}\right) \cdots E_{k+1,k}\left(-a_{k+1,k}^{(k-1)}/a_{kk}^{(k-1)}\right); \qquad (5.20)$$

that is,

$$L_k = \begin{bmatrix} 1 \cdots & & 0 & 0 \cdots 0 \\ & \ddots & & \\ 0 \cdots & & 1 & 0 \cdots 0 \\ 0 \cdots & & -\dfrac{a_{k+1,k}^{(k-1)}}{a_{kk}^{(k-1)}} & 1 \cdots 0 \\ & & & \ddots \\ 0 \cdots & & -\dfrac{a_{nk}^{(k-1)}}{a_{kk}^{(k-1)}} & 0 \cdots 1 \end{bmatrix}. \qquad (5.21)$$

Each L_k is nonsingular, with a determinant of 1. The whole process of forward reduction can be expressed as a matrix product,

$$U = L_{n-1}L_{n-2}\ldots L_2 L_1 A, \qquad (5.22)$$

and by the way we have performed the forward reduction, U is an upper triangular matrix. The matrix $L_{n-1}L_{n-2}\ldots L_2 L_1$ is nonsingular and is unit lower triangular (all 1s on the diagonal). Its inverse therefore is also unit lower triangular. Call its inverse L; that is,

$$L = (L_{n-1}L_{n-2}\ldots L_2 L_1)^{-1}. \tag{5.23}$$

The forward reduction is equivalent to expressing A as LU,

$$A = LU; \tag{5.24}$$

hence this process is called an *LU factorization* or an *LU decomposition*.

The diagonal elements of the lower triangular matrix L in the *LU* factorization are all 1s by the method of construction. If an *LU* factorization exists, it is clear that the upper triangular matrix, U, can be made unit upper triangular (all 1s on the diagonal) by putting the diagonal elements of the original U into a diagonal matrix D and then writing the factorization as LDU, where U is now a unit upper triangular matrix.

The computations leading up to equation (5.24) involve a sequence of equivalent matrices, as discussed in Section 3.3.5. Those computations are outer products involving a column of L_k and rows of $A^{(k-1)}$.

Use of Inner Products

The *LU* factorization can also be performed by using inner products. From equation (5.24), we see

$$a_{ij} = \sum_{k=1}^{i-1} l_{ik}u_{kj} + u_{ij},$$

so

$$l_{ij} = \frac{a_{ij} - \sum_{k=1}^{j-1} l_{ik}u_{kj}}{u_{jj}} \quad \text{for } i = j+1, j+2, \ldots, n. \tag{5.25}$$

The use of computations implied by equation (5.25) is called the Doolittle method or the Crout method. (There is a slight difference between the Doolittle method and the Crout method: the Crout method yields a decomposition in which the 1s are on the diagonal of the U matrix rather than the L matrix.) Whichever method is used to form the *LU* decomposition, $n^3/3$ multiplications and additions are required.

Properties

If a nonsingular matrix has an *LU* factorization, L and U are unique. It is neither necessary nor sufficient that a matrix be nonsingular for it to have an *LU* factorization. An example of a singular matrix that has an *LU* factorization is any upper triangular/trapezoidal matrix with all zeros on the diagonal. In this case, U can be chosen as the matrix itself and L chosen as the identity. For example,

$$A = \begin{bmatrix} 0 & 1 & 1 \\ 0 & 0 & 0 \end{bmatrix}$$

$$= \begin{bmatrix} 1 & 0 \\ 0 & 1 \end{bmatrix} \begin{bmatrix} 0 & 1 & 1 \\ 0 & 0 & 0 \end{bmatrix}$$

$$= LU. \tag{5.26}$$

In this case, A is an upper trapezoidal matrix and so is U.

An example of a nonsingular matrix that does not have an LU factorization is an identity matrix with permuted rows or columns:

$$\begin{bmatrix} 0 & 1 \\ 1 & 0 \end{bmatrix}.$$

A sufficient condition for an $n \times m$ matrix A to have an LU factorization is that for $k = 1, 2, \ldots, \min(n-1, m)$, each $k \times k$ principal submatrix of A, A_k, be nonsingular. Note that this fact also provides a way of constructing a singular matrix that has an LU factorization. Furthermore, for $k = 1, 2, \ldots, \min(n, m)$,

$$\det(A_k) = u_{11} u_{22} \cdots u_{kk}.$$

5.7 QR Factorization

A very useful factorization is

$$A = QR, \tag{5.27}$$

where Q is orthogonal and R is upper triangular or trapezoidal. This is called the QR factorization.

Forms of the Factors

If A is square and of full rank, R has the form

$$\begin{bmatrix} X & X & X \\ 0 & X & X \\ 0 & 0 & X \end{bmatrix}. \tag{5.28}$$

If A is nonsquare, R is nonsquare, with an upper triangular submatrix. If A has more columns than rows, R is trapezoidal and can be written as $[R_1 \mid R_2]$, where R_1 is upper triangular.

If A is $n \times m$ with more rows than columns, which is the case in common applications of QR factorization, then

$$R = \begin{bmatrix} R_1 \\ 0 \end{bmatrix}, \tag{5.29}$$

where R_1 is $m \times m$ upper triangular.

When A has more rows than columns, we can likewise partition Q as $[Q_1 \,|\, Q_2]$, and we can use a version of Q that contains only relevant rows or columns,

$$A = Q_1 R_1, \tag{5.30}$$

where Q_1 is an $n \times m$ matrix whose columns are orthonormal. This form is called a "skinny" QR. It is more commonly used than one with a square Q.

Relation to the Moore-Penrose Inverse

It is interesting to note that the Moore-Penrose inverse of A with full column rank is immediately available from the QR factorization:

$$A^{+} = \begin{bmatrix} R_1^{-1} & 0 \end{bmatrix} Q^{\mathrm{T}}. \tag{5.31}$$

Nonfull Rank Matrices

If A is square but not of full rank, R has the form

$$\begin{bmatrix} \mathrm{X} & \mathrm{X} & \mathrm{X} \\ 0 & \mathrm{X} & \mathrm{X} \\ 0 & 0 & 0 \end{bmatrix}. \tag{5.32}$$

In the common case in which A has more rows than columns, if A is not of full (column) rank, R_1 in equation (5.29) will have the form shown in matrix (5.32).

If A is not of full rank, we apply permutations to the columns of A by multiplying on the right by a permutation matrix. The permutations can be taken out by a second multiplication on the right. If A is of rank r ($\leq m$), the resulting decomposition consists of three matrices: an orthogonal Q, a T with an $r \times r$ upper triangular submatrix, and a permutation matrix E_π^{T},

$$A = QTE_\pi^{\mathrm{T}}. \tag{5.33}$$

The matrix T has the form

$$T = \begin{bmatrix} T_1 & T_2 \\ 0 & 0 \end{bmatrix}, \tag{5.34}$$

where T_1 is upper triangular and is $r \times r$. The decomposition in equation (5.33) is not unique because of the permutation matrix. The choice of the permutation matrix is the same as the pivoting that we discussed in connection with Gaussian elimination. A generalized inverse of A is immediately available from equation (5.33):

$$A^{-} = P \begin{bmatrix} T_1^{-1} & 0 \\ 0 & 0 \end{bmatrix} Q^{\mathrm{T}}. \tag{5.35}$$

Additional orthogonal transformations can be applied from the right-hand side of the $n \times m$ matrix A in the form of equation (5.33) to yield

$$A = QRU^{\mathrm{T}}, \tag{5.36}$$

where R has the form

$$R = \begin{bmatrix} R_1 & 0 \\ 0 & 0 \end{bmatrix}, \tag{5.37}$$

where R_1 is $r \times r$ upper triangular, Q is $n \times n$ and as in equation (5.33), and U^{T} is $n \times m$ and orthogonal. (The permutation matrix in equation (5.33) is also orthogonal, of course.) The decomposition (5.36) is unique, and it provides the unique Moore-Penrose generalized inverse of A:

$$A^+ = U \begin{bmatrix} R_1^{-1} & 0 \\ 0 & 0 \end{bmatrix} Q^{\mathrm{T}}. \tag{5.38}$$

It is often of interest to know the rank of a matrix. Given a decomposition of the form of equation (5.33), the rank is obvious, and in practice, this QR decomposition with pivoting is a good way to determine the rank of a matrix. The QR decomposition is said to be "rank-revealing". The computations are quite sensitive to rounding, however, and the pivoting must be done with some care (see Hong and Pan, 1992; Section 2.7.3 of Björck, 1996; and Bischof and Quintana-Ortí, 1998a,b).

The QR factorization is particularly useful in computations for overdetermined systems, as we will see in Section 6.7 on page 222, and in other computations involving nonsquare matrices.

Formation of the QR Factorization

There are three good methods for obtaining the QR factorization: Householder transformations or reflections; Givens transformations or rotations; and the (modified) Gram-Schmidt procedure. Different situations may make one of these procedures better than the two others. The Householder transformations described in the next section are probably the most commonly used. If the data are available only one row at a time, the Givens transformations discussed in Section 5.7.2 are very convenient. Whichever method is used to compute the QR decomposition, at least $2n^3/3$ multiplications and additions are required. The operation count is therefore about twice as great as that for an LU decomposition.

5.7.1 Householder Reflections to Form the QR Factorization

To use reflectors to compute a QR factorization, we form in sequence the reflector for the ith column that will produce 0s below the (i, i) element.

For a convenient example, consider the matrix

$$A = \begin{bmatrix} 3 & -\frac{98}{28} & \text{X X X} \\ 1 & \frac{122}{28} & \text{X X X} \\ 2 & -\frac{8}{28} & \text{X X X} \\ 1 & \frac{66}{28} & \text{X X X} \\ 1 & \frac{10}{28} & \text{X X X} \end{bmatrix}.$$

The first transformation would be determined so as to transform $(3,1,2,1,1)$ to $(\text{X},0,0,0,0)$. We use equations (5.8) through (5.10) to do this. Call this first Householder matrix P_1. We have

$$P_1 A = \begin{bmatrix} -4 & 1 & \text{X X X} \\ 0 & 5 & \text{X X X} \\ 0 & 1 & \text{X X X} \\ 0 & 3 & \text{X X X} \\ 0 & 1 & \text{X X X} \end{bmatrix}.$$

We now choose a reflector to transform $(5,1,3,1)$ to $(-6,0,0,0)$. We do not want to disturb the first column in $P_1 A$ shown above, so we form P_2 as

$$P_2 = \begin{bmatrix} 1 & 0 & \dots & 0 \\ 0 & & & \\ \vdots & & H_2 & \\ 0 & & & \end{bmatrix}.$$

Forming the vector $(11,1,3,1)/\sqrt{132}$ and proceeding as before, we get the reflector

$$\begin{aligned} H_2 &= I - \frac{1}{66}(11,1,3,1)(11,1,3,1)^{\text{T}} \\ &= \frac{1}{66} \begin{bmatrix} -55 & -11 & -33 & -11 \\ -11 & 65 & -3 & -1 \\ -33 & -3 & 57 & -3 \\ -11 & -1 & -3 & 65 \end{bmatrix}. \end{aligned}$$

Now we have

$$P_2 P_1 A = \begin{bmatrix} -4 & \text{X X X X} \\ 0 & -6 & \text{X X X} \\ 0 & 0 & \text{X X X} \\ 0 & 0 & \text{X X X} \\ 0 & 0 & \text{X X X} \end{bmatrix}.$$

Continuing in this way for three more steps, we would have the QR decomposition of A with $Q^{\text{T}} = P_5 P_4 P_3 P_2 P_1$.

The number of computations for the QR factorization of an $n \times n$ matrix using Householder reflectors is $2n^3/3$ multiplications and $2n^3/3$ additions.

5.7.2 Givens Rotations to Form the QR Factorization

Just as we built the QR factorization by applying a succession of Householder reflections, we can also apply a succession of Givens rotations to achieve the factorization. If the Givens rotations are applied directly, the number of computations is about twice as many as for the Householder reflections, but if fast Givens rotations are used and accumulated cleverly, the number of computations for Givens rotations is not much greater than that for Householder reflections. As mentioned on page 185, it is necessary to monitor the differences in the magnitudes of the elements in the C matrix and often necessary to rescale the elements. This additional computational burden is excessive unless done carefully (see Bindel et al., 2002, for a description of an efficient method).

5.7.3 Gram-Schmidt Transformations to Form the QR Factorization

Gram-Schmidt transformations yield a set of orthonormal vectors that span the same space as a given set of linearly independent vectors, $\{x_1, x_2, \ldots, x_m\}$. Application of these transformations is called Gram-Schmidt orthogonalization. If the given linearly independent vectors are the columns of a matrix A, the Gram-Schmidt transformations ultimately yield the QR factorization of A. The basic Gram-Schmidt transformation is shown in equation (2.34) on page 27.

The Gram-Schmidt algorithm for forming the QR factorization is just a simple extension of equation (2.34); see Exercise 5.9 on page 200.

5.8 Singular Value Factorization

Another factorization useful in solving linear systems is the singular value decomposition, or SVD, shown in equation (3.218) on page 127. For the $n \times m$ matrix A, this is

$$A = UDV^{\mathrm{T}},$$

where U is an $n \times n$ orthogonal matrix, V is an $m \times m$ orthogonal matrix, and D is a diagonal matrix of the singular values. The SVD is "rank-revealing": the number of nonzero singular values is the rank of the matrix.

Golub and Kahan (1965) showed how to use a QR-type factorization to compute a singular value decomposition. This method, with refinements as presented in Golub and Reinsch (1970), is the best algorithm for singular value decomposition. We discuss this method in Section 7.7 on page 253.

5.9 Factorizations of Nonnegative Definite Matrices

There are factorizations that may not exist except for nonnegative definite matrices, or may exist only for such matrices. The LU decomposition, for example, exists and is unique for a nonnegative definite matrix; but may not exist for general matrices. In this section we discuss two important factorizations for nonnegative definite matrices, the square root and the Cholesky factorization.

5.9.1 Square Roots

On page 125, we defined the square root of a nonnegative definite matrix in the natural way and introduced the notation $A^{\frac{1}{2}}$ as the square root of the nonnegative definite $n \times n$ matrix A:

$$A = \left(A^{\frac{1}{2}} \right)^2. \tag{5.39}$$

Because A is symmetric, it has a diagonal factorization, and because it is nonnegative definite, the elements of the diagonal matrix are nonnegative. In terms of the orthogonal diagonalization of A, as on page 125 we write $A^{\frac{1}{2}} = VC^{\frac{1}{2}}V^{\mathrm{T}}$.

We now show that this square root of a nonnegative definite matrix is unique among nonnegative definite matrices. Let A be a (symmetric) nonnegative definite matrix and $A = VCV^{\mathrm{T}}$, and let B be a symmetric nonnegative definite matrix such that $B^2 = A$. We want to show that $B = VC^{\frac{1}{2}}V^{\mathrm{T}}$ or that $B - VC^{\frac{1}{2}}V^{\mathrm{T}} = 0$. Form

$$\left(B - VC^{\frac{1}{2}}V^{\mathrm{T}} \right)\left(B - VC^{\frac{1}{2}}V^{\mathrm{T}} \right) = B^2 - VC^{\frac{1}{2}}V^{\mathrm{T}}B - BVC^{\frac{1}{2}}V^{\mathrm{T}} + \left(VC^{\frac{1}{2}}V^{\mathrm{T}} \right)^2$$

$$= 2A - VC^{\frac{1}{2}}V^{\mathrm{T}}B - \left(VC^{\frac{1}{2}}V^{\mathrm{T}}B \right)^{\mathrm{T}}. \tag{5.40}$$

Now, we want to show that $VC^{\frac{1}{2}}V^{\mathrm{T}}B = A$. The argument below follows Harville (1997). Because B is nonnegative definite, we can write $B = UDU^{\mathrm{T}}$ for an orthogonal $n \times n$ matrix U and a diagonal matrix D with nonnegative elements, $d_1, \ldots d_n$. We first want to show that $V^{\mathrm{T}}UD = C^{\frac{1}{2}}V^{\mathrm{T}}U$. We have

$$V^{\mathrm{T}}UD^2 = V^{\mathrm{T}}UDU^{\mathrm{T}}UDU^{\mathrm{T}}U$$
$$= V^{\mathrm{T}}B^2U$$
$$= V^{\mathrm{T}}AU$$
$$= V^{\mathrm{T}}(VC^{\frac{1}{2}}V^{\mathrm{T}})^2U$$
$$= V^{\mathrm{T}}VC^{\frac{1}{2}}V^{\mathrm{T}}VC^{\frac{1}{2}}V^{\mathrm{T}}U$$
$$= CV^{\mathrm{T}}U.$$

Now consider the individual elements in these matrices. Let z_{ij} be the $(ij)^{\text{th}}$ element of $V^{\text{T}}U$, and since D^2 and C are diagonal matrices, the $(ij)^{\text{th}}$ element of $V^{\text{T}}UD^2$ is $d_j^2 z_{ij}$ and the corresponding element of $CV^{\text{T}}U$ is $c_i z_{ij}$, and these two elements are equal, so $d_j z_{ij} = \sqrt{c_i} z_{ij}$. These, however, are the $(ij)^{\text{th}}$ elements of $V^{\text{T}}UD$ and $C^{\frac{1}{2}}V^{\text{T}}U$, respectively; hence $V^{\text{T}}UD = C^{\frac{1}{2}}V^{\text{T}}U$. We therefore have

$$VC^{\frac{1}{2}}V^{\text{T}}B = VC^{\frac{1}{2}}V^{\text{T}}UDU^{\text{T}} = VC^{\frac{1}{2}}C^{\frac{1}{2}}V^{\text{T}}UU^{\text{T}} = VCV^{\text{T}} = A.$$

We conclude that $VC^{\frac{1}{2}}V^{\text{T}}$ is the unique square root of A.

If A is positive definite, it has an inverse, and the unique square root of the inverse is denoted as $A^{-\frac{1}{2}}$.

5.9.2 Cholesky Factorization

If the matrix A is symmetric and *positive definite* (that is, if $x^{\text{T}}Ax > 0$ for all $x \neq 0$), another important factorization is the *Cholesky decomposition*. In this factorization,

$$A = T^{\text{T}}T, \tag{5.41}$$

where T is an upper triangular matrix with positive diagonal elements. We occasionally denote the Cholesky factor of A (that is, T in the expression above) as A_{C}. (Notice on page 34 and later on page 293 that we use a lowercase c subscript to represent a centered vector or matrix.)

The factor T in the Cholesky decomposition is sometimes called the *square root*, but we have defined a different matrix as the square root, $A^{\frac{1}{2}}$ (page 125 and Section 5.9.1). The Cholesky factor is more useful in practice, but the square root has more applications in the development of the theory.

A factor of the form of T in equation (5.41) is unique up to the sign, just as a square root is. To make the Cholesky factor unique, we require that the diagonal elements be positive. The elements along the diagonal of T will be square roots. Notice, for example, that t_{11} is $\sqrt{a_{11}}$.

Algorithm 5.1 is a method for constructing the Cholesky factorization. The algorithm serves as the basis for a constructive proof of the existence and uniqueness of the Cholesky factorization (see Exercise 5.5 on page 199). The uniqueness is seen by factoring the principal square submatrices.

Algorithm 5.1 Cholesky Factorization

1. Let $t_{11} = \sqrt{a_{11}}$.
2. For $j = 2, \ldots, n$, let $t_{1j} = a_{1j}/t_{11}$.
3. For $i = 2, \ldots, n$,
   {

       let $t_{ii} = \sqrt{a_{ii} - \sum_{k=1}^{i-1} t_{ki}^2}$, and
       for $j = i+1, \ldots, n$,
         {

$$\text{let } t_{ij} = (a_{ij} - \textstyle\sum_{k=1}^{i-1} t_{ki}t_{kj})/t_{ii}.$$
$$\}$$
$$\}$$

There are other algorithms for computing the Cholesky decomposition. The method given in Algorithm 5.1 is sometimes called the inner product formulation because the sums in step 3 are inner products. The algorithms for computing the Cholesky decomposition are numerically stable. Although the order of the number of computations is the same, there are only about half as many computations in the Cholesky factorization as in the LU factorization. Another advantage of the Cholesky factorization is that there are only $n(n+1)/2$ unique elements as opposed to $n^2 + n$ in the LU decomposition.

The Cholesky decomposition can also be formed as $\widetilde{T}^{\mathrm{T}} D \widetilde{T}$, where D is a diagonal matrix that allows the diagonal elements of $\widetilde{T}$ to be computed without taking square roots. This modification is sometimes called a *Banachiewicz factorization* or *root-free Cholesky*. The Banachiewicz factorization can be formed in essentially the same way as the Cholesky factorization shown in Algorithm 5.1: just put 1s along the diagonal of T and store the squared quantities in a vector d.

Cholesky Decomposition of Singular Nonnegative Definite Matrices

Any symmetric nonnegative definite matrix has a decomposition similar to the Cholesky decomposition for a positive definite matrix. If A is $n \times n$ with rank r, there exists a unique matrix T such that $A = T^{\mathrm{T}}T$, where T is an upper triangular matrix with r positive diagonal elements and $n - r$ rows containing all zeros. The algorithm is the same as Algorithm 5.1, except that in step 3 if $t_{ii} = 0$, the entire row is set to zero. The algorithm serves as a constructive proof of the existence and uniqueness.

Relations to Other Factorizations

For a symmetric matrix, the LDU factorization is $U^{\mathrm{T}} DU$; hence, we have for the Cholesky factor

$$T = D^{\frac{1}{2}} U,$$

where $D^{\frac{1}{2}}$ is the matrix whose elements are the square roots of the corresponding elements of D. (This is consistent with our notation above for Cholesky factors; $D^{\frac{1}{2}}$ is the Cholesky factor of D, and it is symmetric.)

The LU and Cholesky decompositions generally are applied to square matrices. However, many of the linear systems that occur in scientific applications are *overdetermined*; that is, there are more equations than there are variables, resulting in a nonsquare coefficient matrix.

For the $n \times m$ matrix A with $n \geq m$, we can write

$$A^{\mathrm{T}}A = R^{\mathrm{T}}Q^{\mathrm{T}}QR$$
$$= R^{\mathrm{T}}R, \tag{5.42}$$

so we see that the matrix R in the QR factorization is (or at least can be) the same as the matrix T in the Cholesky factorization of $A^{\mathrm{T}}A$. There is some ambiguity in the Q and R matrices, but if the diagonal entries of R are required to be nonnegative, the ambiguity disappears and the matrices in the QR decomposition are unique.

An overdetermined system may be written as

$$Ax \approx b,$$

where A is $n \times m$ $(n \geq m)$, or it may be written as

$$Ax = b + e,$$

where e is an n-vector of possibly arbitrary "errors". Because not all equations can be satisfied simultaneously, we must define a meaningful "solution". A useful solution is an x such that e has a small norm. The most common definition is an x such that e has the least Euclidean norm; that is, such that the sum of squares of the e_is is minimized.

It is easy to show that such an x satisfies the square system $A^{\mathrm{T}}Ax = A^{\mathrm{T}}b$, the "normal equations". This expression is important and allows us to analyze the overdetermined system (not just to solve for the x but to gain some better understanding of the system). It is easy to show that if A is of full rank (i.e., of rank m, all of its columns are linearly independent, or, redundantly, "full column rank"), then $A^{\mathrm{T}}A$ is positive definite. Therefore, we could apply either Gaussian elimination or the Cholesky decomposition to obtain the solution.

As we have emphasized many times before, however, *useful conceptual expressions are not necessarily useful as computational formulations*. That is sometimes true in this case also. In Section 6.1, we will discuss issues relating to the expected accuracy in the solutions of linear systems. There we will define a "condition number". Larger values of the condition number indicate that the expected accuracy is less. We will see that the condition number of $A^{\mathrm{T}}A$ is the square of the condition number of A. Given these facts, we conclude that it may be better to work directly on A rather than on $A^{\mathrm{T}}A$, which appears in the normal equations. We discuss solutions of overdetermined systems in Section 6.7, beginning on page 222, and in Section 6.8, beginning on page 229. Overdetermined systems are also a main focus of the statistical applications in Chapter 9.

5.9.3 Factorizations of a Gramian Matrix

The sums of squares and cross products matrix, the Gramian matrix $X^{\mathrm{T}}X$, formed from a given matrix X, arises often in linear algebra. We discuss properties of the sums of squares and cross products matrix beginning on page 287. Now we consider some additional properties relating to various factorizations.

First we observe that $X^\mathrm{T}X$ is symmetric and hence has an orthogonally similar canonical factorization,

$$X^\mathrm{T}X = VCV^\mathrm{T}.$$

We have already observed that $X^\mathrm{T}X$ is nonnegative definite, and so it has the LU factorization

$$X^\mathrm{T}X = LU,$$

with L lower triangular and U upper triangular, and it has the Cholesky factorization

$$X^\mathrm{T}X = T^\mathrm{T}T$$

with T upper triangular. With $L = T^\mathrm{T}$ and $U = T$, both factorizations are the same. In the LU factorization, the diagonal elements of either L or U are often constrained to be 1, and hence the two factorizations are usually different.

It is instructive to relate the factors of the $m \times m$ matrix $X^\mathrm{T}X$ to the factors of the $n \times m$ matrix X. Consider the QR factorization

$$X = QR,$$

where R is upper triangular. Then $X^\mathrm{T}X = (QR)^\mathrm{T}QR = R^\mathrm{T}R$, so R is the Cholesky factor T because the factorizations are unique (again, subject to the restrictions that the diagonal elements be nonnegative).

Consider the SVD factorization

$$X = UDV^\mathrm{T}.$$

We have $X^\mathrm{T}X = (UDV^\mathrm{T})^\mathrm{T}UDV^\mathrm{T} = VD^2V^\mathrm{T}$, which is the orthogonally similar canonical factorization of $X^\mathrm{T}X$. The eigenvalues of $X^\mathrm{T}X$ are the squares of the singular values of X, and the condition number of $X^\mathrm{T}X$ (which we define in Section 6.1) is the square of the condition number of X.

5.10 Incomplete Factorizations

Often instead of an exact factorization, an approximate or "incomplete" factorization may be more useful because of its computational efficiency. This may be the case in the context of an iterative algorithm in which a matrix is being successively transformed, and, although a factorization is used in each step, the factors from a previous iteration are adequate approximations. Another common situation is in working with sparse matrices. Many exact operations on a sparse matrix yield a dense matrix; however, we may want to preserve the sparsity, even at the expense of losing exact equalities. When a zero position in a sparse matrix becomes nonzero, this is called "fill-in".

For example, instead of an LU factorization of a sparse matrix A, we may seek lower and upper triangular factors $\widetilde{L}$ and $\widetilde{U}$, such that

$$A \approx \widetilde{L}\widetilde{U}, \tag{5.43}$$

and if $a_{ij} = 0$, then $\tilde{l}_{ij} = \tilde{u}_{ij} = 0$. This approximate factorization is easily accomplished by modifying the Gaussian elimination step that leads to the outer product algorithm of equations (5.22) and (5.23).

More generally, we may choose a set of indices $S = \{(p,q)\}$ and modify the elimination step to be

$$a_{ij}^{(k+1)} \leftarrow \begin{cases} a_{ij}^{(k)} - a_{ij}^{(k)} a_{ij}^{(k)} a_{ij}^{(k)} & \text{if } (i,j) \in S \\ a_{ij} & \text{otherwise.} \end{cases} \tag{5.44}$$

Note that a_{ij} does not change unless (i,j) is in S. This allows us to preserve 0s in L and U corresponding to given positions in A.

Exercises

5.1. Consider the transformation of the 3-vector x that first rotates the vector $30°$ about the x_1 axis, then rotates the vector $45°$ about the x_2 axis, and then translates the vector by adding the 3-vector y. Find the matrix A that effects these transformations by a single multiplication. Use the vector x^{h} of homogeneous coordinates that corresponds to the vector x. (Thus, A is 4×4.)

5.2. Homogeneous coordinates are often used in mapping three-dimensional graphics to two dimensions. The perspective plot function `persp` in R, for example, produces a 4×4 matrix for projecting three-dimensional points represented in homogeneous coordinates onto two-dimensional points in the displayed graphic. R uses homogeneous coordinates in the form of equation (5.6b) rather than equation (5.6a). If the matrix produced is T and if a^{h} is the representation of a point (x_a, y_a, z_a) in homogeneous coordinates, in the form of equation (5.6b), then $a^{\text{h}}T$ yields transformed homogeneous coordinates that correspond to the projection onto the two-dimensional coordinate system of the graphical display. Consider the two graphs in Figure 5.4. The graph on the left in the unit cube was produced by the simple R statements

```
x<-c(0,1)
y<-c(0,1)
z<-matrix(c(0,0,1,1),nrow=2)
persp(x, y, z, theta = 45, phi = 30)
```

(The angles `theta` and `phi` are the azimuthal and latitudinal viewing angles, respectively, in degrees.) The graph on the right is the same with a heavy line going down the middle of the surface; that is, from the point $(0.5, 0, 0)$ to $(0.5, 1, 1)$. Obtain the transformation matrix necessary to identify the rotated points and produce the graph on the right.

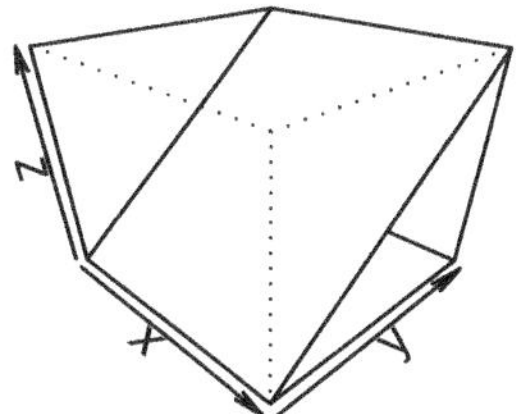 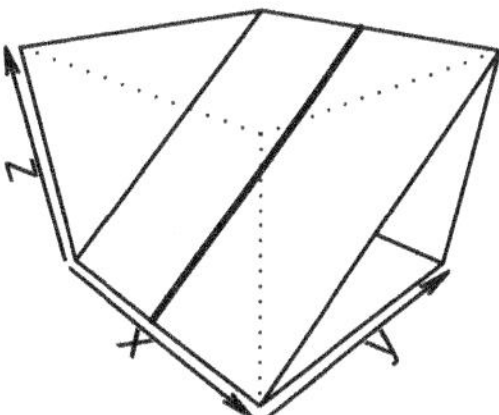

Fig. 5.4. Illustration of the Use of Homogeneous Coordinates to Locate Three-Dimensional Points on a Two-Dimensional Graph

5.3. Determine the rotation matrix that rotates 3-vectors through an angle of $30°$ in the plane $x_1 + x_2 + x_3 = 0$.

5.4. Let $A = LU$ be the LU decomposition of the $n \times n$ matrix A.

    a) Suppose we multiply the jth column of A by c_j, $j = 1, 2, \ldots n$, to form the matrix A_c. What is the LU decomposition of A_c? Try to express your answer in a compact form.

    b) Suppose we multiply the ith row of A by c_i, $i = 1, 2, \ldots n$, to form the matrix A_r. What is the LU decomposition of A_r? Try to express your answer in a compact form.

    c) What application might these relationships have?

5.5. Show that if A is positive definite, there exists a unique upper triangular matrix T with positive diagonal elements such that

$$A = T^{\text{T}}T.$$

Hint: Show that $a_{ii} > 0$. Show that if A is partitioned into square submatrices A_{11} and A_{22},

$$A = \begin{bmatrix} A_{11} & A_{12} \\ A_{21} & A_{22} \end{bmatrix},$$

that A_{11} and A_{22} are positive definite. Use Algorithm 5.1 (page 194) to show the existence of a T, and finally show that T is unique.

5.6. Let X_1, X_2, and X_3 be independent random variables identically distributed as standard normals.

    a) Determine a matrix A such that the random vector

$$A \begin{bmatrix} X_1 \\ X_2 \\ X_3 \end{bmatrix}$$

has a multivariate normal distribution with variance-covariance matrix

$$\begin{bmatrix} 4 & 2 & 8 \\ 2 & 10 & 7 \\ 8 & 7 & 21 \end{bmatrix}.$$

b) Is your solution unique? (The answer is no.) Determine a different solution.

5.7. Generalized inverses.

a) Prove equation (5.35) on page 189 (generalized inverse of a nonfull rank matrix).

b) Prove equation (5.38) on page 190, (Moore-Penrose inverse of a non-full rank matrix).

5.8. Determine the Givens transformation matrix that will rotate the matrix

$$A = \begin{bmatrix} 3 & 5 & 6 \\ 6 & 1 & 2 \\ 8 & 6 & 7 \\ 2 & 3 & 1 \end{bmatrix}$$

so that the second column becomes $(5, \tilde{a}_{22}, 6, 0)$ (see also Exercise 12.3).

5.9. Gram-Schmidt transformations.

a) Use Gram-Schmidt transformations to determine an orthonormal basis for the space spanned by the vectors

$$\begin{aligned} v_1 &= (3, 6, 8, 2), \\ v_2 &= (5, 1, 6, 3), \\ v_3 &= (6, 2, 7, 1). \end{aligned}$$

b) Write out a formal algorithm for computing the QR factorization of the $n \times m$ full rank matrix A. Assume $n \geq m$.

c) Write a Fortran or C subprogram to implement the algorithm you described.

Solution of Linear Systems

One of the most common problems in numerical computing is to solve the linear system

$$Ax = b;$$

that is, for given A and b, to find x such that the equation holds. The system is said to be *consistent* if there exists such an x, and in that case a solution x may be written as $A^- b$, where A^- is some inverse of A. If A is square and of full rank, we can write the solution as $A^{-1}b$.

It is important to distinguish the expression $A^{-1}b$ or A^+b, which represents the solution, from the method of computing the solution. We would never compute A^{-1} just so we could multiply it by b to form the solution $A^{-1}b$.

There are two general methods of solving a system of linear equations: direct methods and iterative methods. A direct method uses a fixed number of computations that would in exact arithmetic lead to the solution; an iterative method generates a sequence of approximations to the solution. Iterative methods often work well for very large sparse matrices. We first consider a characteristic of the problem that affects how easy it is to solve the system accurately.

6.1 Condition of Matrices

Data are said to be "ill-conditioned" for a particular problem or computation if the data are likely to cause difficulties in the computations, such as severe loss of precision. More generally, the term "ill-conditioned" is applied to a problem in which small changes to the input result in large changes in the output. In the case of a linear system

$$Ax = b,$$

the problem of solving the system is ill-conditioned if small changes to some elements of A or b will cause large changes in the solution x.

Consider, for example, the system of equations

$$1.000x_1 + 0.500x_2 = 1.500,$$
$$0.667x_1 + 0.333x_2 = 1.000. \tag{6.1}$$

The solution is easily seen to be $x_1 = 1.000$ and $x_2 = 1.000$.

Now consider a small change in the right-hand side:

$$1.000x_1 + 0.500x_2 = 1.500,$$
$$0.667x_1 + 0.333x_2 = 0.999. \tag{6.2}$$

This system has solution $x_1 = 0.000$ and $x_2 = 3.000$.

Alternatively, consider a small change in one of the elements of the coefficient matrix:

$$1.000x_1 + 0.500x_2 = 1.500,$$
$$0.667x_1 + 0.334x_2 = 1.000. \tag{6.3}$$

The solution now is $x_1 = 2.000$ and $x_2 = -1.000$.

In both cases, small changes of the order of 10^{-3} in the input (the elements of the coefficient matrix or the right-hand side) result in relatively large changes (of the order of 1) in the output (the solution). Solving the system (either one of them) is an ill-conditioned problem.

The nature of the data that cause ill-conditioning depends on the type of problem. In this case, the problem is that the lines represented by the equations are almost parallel, as seen in Figure 6.1, and so their point of intersection is very sensitive to slight changes in the coefficients defining the lines.

The problem can also be described in terms of the angle between the lines. When the angle is small, but not necessarily 0, we refer to the condition as "collinearity". (This term is somewhat misleading because, strictly speaking, it should indicate that the angle is exactly 0.) In this example, the cosine of the angle between the lines, from equation (2.32), is $1 - 2 \times 10^{-7}$. In general, collinearity (or "multicollinearity") exists whenever the angle between any line (that is, vector) and the subspace spanned by any other set of vectors is small.

For a specific problem such as solving a system of equations, we may quantify the condition of the matrix by a *condition number*. To develop this quantification for the problem of solving linear equations, consider a linear system $Ax = b$, with A nonsingular and $b \neq 0$, as above. Now perturb the system slightly by adding a small amount, δb, to b, and let $\tilde{b} = b + \delta b$. The system

$$A\tilde{x} = \tilde{b}$$

has a solution $\tilde{x} = \delta x + x = A^{-1}\tilde{b}$. (Notice that δb and δx do not necessarily represent scalar multiples of the respective vectors.) If the system is well-conditioned, for any reasonable norm, if $\|\delta b\|/\|b\|$ is small, then $\|\delta x\|/\|x\|$ is likewise small.

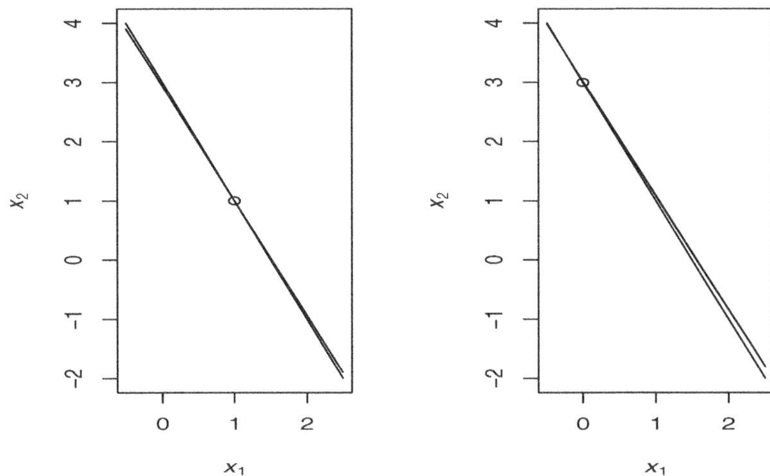

Fig. 6.1. Almost Parallel Lines: Ill-Conditioned Coefficient Matrices, Equations (6.1) and (6.2)

From $\delta x = A^{-1}\delta b$ and the inequality (3.222) (page 129), for an induced norm on A, we have

$$\|\delta x\| \leq \|A^{-1}\|\,\|\delta b\|.\tag{6.4}$$

Likewise, because $b = Ax$, we have

$$\frac{1}{\|x\|} \leq \|A\|\frac{1}{\|b\|},\tag{6.5}$$

and equations (6.4) and (6.5) together imply

$$\frac{\|\delta x\|}{\|x\|} \leq \|A\|\,\|A^{-1}\|\frac{\|\delta b\|}{\|b\|}.\tag{6.6}$$

This provides a bound on the change in the solution $\|\delta x\|/\|x\|$ in terms of the perturbation $\|\delta b\|/\|b\|$.

The bound in equation (6.6) motivates us to define the *condition number with respect to inversion* denoted by $\kappa(\cdot)$ as

$$\kappa(A) = \|A\|\,\|A^{-1}\|\tag{6.7}$$

for nonsingular A. In the context of linear algebra, the condition number with respect to inversion is so dominant in importance that we generally just refer to it as the "condition number". A condition number is a useful measure of the condition of A for the problem of solving a linear system of equations. There are other condition numbers useful in numerical analysis, however, such as the condition number for computing the sample variance (see equation (10.8) on page 411) or a condition number for a root of a function.

We can write equation (6.6) as

$$\frac{\|\delta x\|}{\|x\|} \leq \kappa(A)\frac{\|\delta b\|}{\|b\|}, \tag{6.8}$$

and, following a development similar to that above, write

$$\frac{\|\delta b\|}{\|b\|} \leq \kappa(A)\frac{\|\delta x\|}{\|x\|}. \tag{6.9}$$

These inequalities, as well as the other ones we write in this section, are sharp, as we can see by letting $A = I$.

Because the condition number is an upper bound on a quantity that we would not want to be large, a large condition number is "bad".

Notice that our definition of the condition number does not specify the norm; it only requires that the norm be an induced norm. (An equivalent definition does not rely on the norm being an induced norm.) We sometimes specify a condition number with regard to a particular norm, and just as we sometimes denote a specific norm by a special symbol, we may use a special symbol to denote a specific condition number. For example, $\kappa_p(A)$ may denote the condition number of A in terms of an L_p norm. Most of the properties of condition numbers (but not their actual values) are independent of the norm used.

The coefficient matrix in equations (6.1) and (6.2) is

$$A = \begin{bmatrix} 1.000 & 0.500 \\ 0.667 & 0.333 \end{bmatrix},$$

and its inverse is

$$A^{-1} = \begin{bmatrix} -666 & 1000 \\ 1344 & -2000 \end{bmatrix}.$$

It is easy to see that

$$\|A\|_1 = 1.667$$

and

$$\|A^{-1}\|_1 = 3000;$$

hence,

$$\kappa_1(A) = 5001.$$

Likewise,

$$\|A\|_\infty = 1.500$$

and

$$\|A^{-1}\|_\infty = 3344;$$

hence,

$$\kappa_\infty(A) = 5016.$$

Notice that the condition numbers are not exactly the same, but they are close. Although we used this matrix in an example of ill-conditioning, these condition numbers, although large, are not so large as to cause undue concern for numerical computations. Indeed, solving the systems of equations (6.1), (6.2), and (6.3) would not cause problems for a computer program. Notice also that the condition numbers are of the order of magnitude of the ratio of the output perturbation to the input perturbation in those equations.

An interesting relationship for the L_2 condition number is

$$\kappa_2(A) = \frac{\max_{x \neq 0} \frac{\|Ax\|}{\|x\|}}{\min_{x \neq 0} \frac{\|Ax\|}{\|x\|}} \tag{6.10}$$

(see Exercise 6.1, page 238). The numerator and denominator in equation (6.10) look somewhat like the maximum and minimum eigenvalues, as we have suggested. Indeed, the L_2 condition number is just the ratio of the largest eigenvalue in absolute value to the smallest (see page 131). The L_2 condition number is also called the *spectral condition number*.

The eigenvalues of the coefficient matrix in equations (6.1) and (6.2) are 1.333375 and -0.0003750, and so

$$\kappa_2(A) = 3555.67,$$

which is the same order of magnitude as $\kappa_\infty(A)$ and $\kappa_1(A)$ computed above.

Some useful facts about condition numbers are:

- $\kappa(A) = \kappa(A^{-1}),$
- $\kappa(cA) = \kappa(A), \quad$ for $c \neq 0,$
- $\kappa(A) \geq 1,$
- $\kappa_1(A) = \kappa_\infty(A^{\mathrm{T}}),$
- $\kappa_2(A^{\mathrm{T}}) = \kappa_2(A),$
- $\kappa_2(A^{\mathrm{T}}A) = \kappa_2^2(A)$
  $\geq \kappa_2(A),$ and
- if A and B are orthogonally similar (equation (3.191)), then

$$\|A\|_2 = \|B\|_2$$

and

$$\kappa_2(A) = \kappa_2(B)$$

(see equation (3.228)).

Even though the condition number provides a very useful indication of the condition of the problem of solving a linear system of equations, it can be misleading at times. Consider, for example, the coefficient matrix

$$A = \begin{bmatrix} 1 & 0 \\ 0 & \epsilon \end{bmatrix},$$

where $\epsilon < 1$. The condition numbers are

$$\kappa_1(A) = \kappa_2(A) = \kappa_\infty(A) = \frac{1}{\epsilon},$$

and so if ϵ is small, the condition number is large. It is easy to see, however, that small changes to the elements of A or b in the system $Ax = b$ do not cause undue changes in the solution (our heuristic definition of ill-conditioning). In fact, the simple expedient of multiplying the second row of A by $1/\epsilon$ (that is, multiplying the second equation, $a_{21}x_1 + a_{22}x_2 = b_2$, by $1/\epsilon$) yields a linear system that is very well-conditioned.

This kind of apparent ill-conditioning is called *artificial ill-conditioning*. It is due to the different rows (or columns) of the matrix having a very different *scale*; the condition number can be changed just by scaling the rows or columns. This usually does not make a linear system any better or any worse conditioned.

In Section 6.4 we relate the condition number to bounds on the numerical accuracy of the solution of a linear system of equations.

The relationship between the size of the matrix and its condition number is interesting. In general, we would expect the condition number to increase as the size increases. This is the case, but the nature of the increase depends on the type of elements in the matrix. If the elements are randomly and independently distributed as normal or uniform with a mean of zero and variance of one, the increase in the condition number is approximately linear in the size of the matrix (see Exercise 10.22, page 427).

Our definition of condition number given above is for nonsingular matrices. We can formulate a useful alternate definition that extends to singular matrices and to nonsquare matrices: the *condition number* of a matrix is the ratio of the largest singular value to the smallest nonzero singular value, and of course this is the same as the definition for square nonsingular matrices. This is also called the *spectral condition number*.

The condition number, like the determinant, is not easy to compute (see page 440 in Section 11.4).

In the ridge regression discussed on page 291, when X is of full rank, we can see that the condition number of the matrix $X^{\mathrm{T}}X + \lambda I$ is smaller than that of $X^{\mathrm{T}}X$:

$$\frac{\max(d_i + \lambda)}{\min(d_i + \lambda)} < \frac{\max(d_i)}{\min(d_i)}$$

for $\lambda > 0$.

6.2 Direct Methods for Consistent Systems

There are two general approaches to solving the linear system $Ax = b$. One class of methods is *direct* in the sense that the solution is obtained in a preset number of steps. The number of steps generally depends only on the size of the matrix. Other methods, called *iterative*, arrive at the solution through a sequence of steps whose number depends on some criterion that indicates the solution has been obtained. (Note, of course, that a purported solution can be checked out very quickly by simple multiplication.)

6.2.1 Gaussian Elimination and Matrix Factorizations

The most common direct method for the solution of linear systems is Gaussian elimination. The basic idea in this method is to form equivalent sets of equations, beginning with the system to be solved, $Ax = b$, or

$$a_{1*}^{\mathrm{T}}x = b_1$$
$$a_{2*}^{\mathrm{T}}x = b_2$$
$$\ldots = \ldots$$
$$a_{n*}^{\mathrm{T}}x = b_n,$$

where a_{j*} is the jth row of A. An equivalent set of equations can be formed by a sequence of *elementary operations* on the equations in the given set.

These elementary operations on equations are essentially the same as the elementary operations on the rows of matrices discussed in Section 3.2.3 and in Section 5.6. There are three kinds of elementary operations: an interchange of two equations,

$$a_{j*}^{\mathrm{T}}x = b_j \leftarrow a_{k*}^{\mathrm{T}}x = b_k,$$
$$a_{k*}^{\mathrm{T}}x = b_k \leftarrow a_{j*}^{\mathrm{T}}x = b_j,$$

which affects two equations simultaneously, a scalar multiplication of a given equation,

$$a_{j*}^{\mathrm{T}}x = b_j \quad \leftarrow \quad ca_{j*}^{\mathrm{T}}x = cb_j,$$

and a replacement of a single equation with a sum of it and a scalar multiple of another equation,

$$a_{j*}^{\mathrm{T}}x = b_j \quad \leftarrow \quad a_{j*}^{\mathrm{T}}x + ca_{k*}^{\mathrm{T}}x = b_j + cb_k.$$

The interchange operation can be accomplished by premultiplication by an elementary permutation matrix (see page 62):

$$E_{jk}Ax = E_{jk}b.$$

The scalar multiplication can be performed by premultiplication by an elementary transformation matrix $E_j(c)$, and the axpy operation can be effected by premultiplication by an $E_{jk}(c)$ elementary transformation matrix.

The elementary operation on the equation

$$a_{2*}^{\mathrm{T}}x = b_2$$

in which the first equation is combined with it using $c_1 = -a_{21}/a_{11}$ and $c_2 = 1$ will yield an equation with a zero coefficient for x_1. Generalizing this, we perform elementary operations on the second through the nth equations to yield a set of equivalent equations in which all but the first have zero coefficients for x_1.

Next, we perform elementary operations using the second equation with the third through the nth equations, so that the new third through the nth equations have zero coefficients for x_2. This is the kind of sequence of multiplications by elementary operator matrices shown in equation (3.50) on page 66 and grouped together as L_k in equation (5.20) on page 186.

The sequence of equivalent equations, beginning with $Ax = b$, is

$$
(0) \quad
\begin{array}{rcl}
a_{11}x_1 + a_{12}x_2 + \quad \cdots \quad + \quad a_{1n}x_n &=& b_1 \\
a_{21}x_1 + a_{22}x_2 + \quad \cdots \quad + \quad a_{2n}x_n &=& b_2 \\
\vdots \quad + \quad \vdots \qquad\qquad\quad \vdots \quad\quad \vdots & & \\
a_{n1}x_1 + a_{n2}x_2 + \quad \cdots \quad + \quad a_{nn}x_n &=& b_n
\end{array}
\tag{6.11}
$$

then $A^{(1)}x = b^{(1)}$, or $L_1 Ax = L_1 b$,

$$
(1) \quad
\begin{array}{rcl}
a_{11}x_1 + a_{12}x_2 + \quad \cdots \quad + \quad a_{1n}x_n &=& b_1 \\
a_{22}^{(1)}x_2 + \quad \cdots \quad + \quad a_{2n}^{(1)}x_n &=& b_2^{(1)} \\
\vdots \quad + \quad \cdots \quad + \quad \vdots \quad\quad \vdots & & \\
a_{n2}^{(1)}x_2 + \quad \cdots \quad + \quad a_{nn}^{(1)}x_n &=& b_n^{(1)}
\end{array}
\tag{6.12}
$$

$$\vdots \quad \vdots$$

and finally $A^{(n)}x = b^{(n)}$, or $L_{n-1} \cdots L_1 Ax = L_{n-1} \cdots L_1 b$, or $Ux = L_{n-1} \cdots L_1 b$,

$$
(n-1) \quad
\begin{array}{rcl}
a_{11}x_1 + a_{12}x_2 + \quad \cdots \quad + \quad a_{1n}x_n &=& b_1 \\
a_{22}^{(1)}x_2 + \quad \cdots \quad + \quad a_{2n}^{(1)}x_n &=& b_2^{(1)} \\
\vdots \quad \vdots \quad\quad \vdots & & \\
a_{n-1,n-1}^{(n-2)}x_{n-1} + a_{n-1,n}^{(n-2)}x_n &=& b_{n-1}^{(n-2)} \\
a_{nn}^{(n-1)}x_n &=& b_n^{(n-1)}
\end{array}
\tag{6.13}
$$

Recalling equation (5.23), we see that the last system is $Ux = L^{-1}b$. This system is easy to solve because the coefficient matrix is upper triangular. The last equation in the system yields

$$
x_n = \frac{b_n^{(n-1)}}{a_{nn}^{(n-1)}}.
$$

By back substitution, we get

$$x_{n-1} = \frac{(b_{n-1}^{(n-2)} - a_{n-1,n}^{(n-2)} x_n)}{a_{n-1,n-1}^{(n-2)}},$$

and we obtain the rest of the xs in a similar manner. This back substitution is equivalent to forming

$$x = U^{-1}L^{-1}b, \tag{6.14}$$

or $x = A^{-1}b$ with $A = LU$.

Gaussian elimination consists of two steps: the forward reduction, which is of order $O(n^3)$, and the back substitution, which is of order $O(n^2)$.

Pivoting

The only obvious problem with this method arises if some of the $a_{kk}^{(k-1)}$s used as divisors are zero (or very small in magnitude). These divisors are called "pivot elements".

Suppose, for example, we have the equations

$$0.0001x_1 + x_2 = 1,$$
$$x_1 + x_2 = 2.$$

The solution is $x_1 = 1.0001$ and $x_2 = 0.9999$. Suppose we are working with three digits of precision (so our solution is $x_1 = 1.00$ and $x_2 = 1.00$). After the first step in Gaussian elimination, we have

$$0.0001x_1 + \qquad x_2 = \qquad 1,$$
$$-10,000x_2 = -10,000,$$

and so the solution by back substitution is $x_2 = 1.00$ and $x_1 = 0.000$. The L_2 condition number of the coefficient matrix is 2.618, so even though the coefficients vary greatly in magnitude, we certainly would not expect any difficulty in solving these equations.

A simple solution to this potential problem is to interchange the equation having the small leading coefficient with an equation below it. Thus, in our example, we first form

$$x_1 + x_2 = 2,$$
$$0.0001x_1 + x_2 = 1,$$

so that after the first step we have

$$x_1 + x_2 = 2,$$
$$x_2 = 1,$$

and the solution is $x_2 = 1.00$ and $x_1 = 1.00$, which is correct to three digits.

Another strategy would be to interchange the column having the small leading coefficient with a column to its right. Both the row interchange and the column interchange strategies could be used simultaneously, of course. These processes, which obviously do not change the solution, are called *pivoting*. The equation or column to move into the active position may be chosen in such a way that the magnitude of the new diagonal element is the largest possible.

Performing only row interchanges, so that at the kth stage the equation with

$$\max_{i=k}^{n} |a_{ik}^{(k-1)}|$$

is moved into the kth row, is called *partial pivoting*. Performing both row interchanges and column interchanges, so that

$$\max_{i=k;j=k}^{n;n} |a_{ij}^{(k-1)}|$$

is moved into the kth diagonal position, is called *complete pivoting*. See Exercises 6.2a and 6.2b.

It is always important to distinguish descriptions of effects of actions from the actions that are actually carried out in the computer. Pivoting is "interchanging" rows or columns. We would usually do something like that in the computer only when we are finished and want to produce some output. In the computer, a row or a column is determined by the index identifying the row or column. All we do for pivoting is to keep track of the indices that we have permuted.

There are many more computations required in order to perform complete pivoting than are required to perform partial pivoting. Gaussian elimination with complete pivoting can be shown to be stable; that is, the algorithm yields an exact solution to a slightly perturbed system, $(A + \delta A)x = b$. (We discuss stability on page 409.) For Gaussian elimination with partial pivoting, there are examples that show that it is not stable. These examples are somewhat contrived, however, and experience over many years has indicated that Gaussian elimination with partial pivoting is stable for most problems occurring in practice. For this reason, together with the computational savings, Gaussian elimination with partial pivoting is one of the most commonly used methods for solving linear systems. See Golub and Van Loan (1996) for a further discussion of these issues.

There are two modifications of partial pivoting that result in stable algorithms. One is to add one step of iterative refinement (see Section 6.5, page 219) following each pivot. It can be shown that Gaussian elimination with partial pivoting together with one step of iterative refinement is unconditionally stable (Skeel, 1980). Another modification is to consider two columns for possible interchange in addition to the rows to be interchanged. This does not require nearly as many computations as complete pivoting does. Higham (1997) shows that this method, suggested by Bunch and Kaufman (1977) and used in LINPACK and LAPACK, is stable.

Nonfull Rank and Nonsquare Systems

The existence of an x that solves the linear system $Ax = b$ depends on that system being consistent; it does not depend on A being square or of full rank. The methods discussed above apply in this case. (See the discussion of LU and QR factorizations for nonfull rank and nonsquare matrices on pages 188 and 189.) In applications, it is often annoying that many software developers do not provide capabilities for handling such systems. Many of the standard programs for solving systems provide solutions only if A is square and of full rank. This is a poor design decision.

6.2.2 Choice of Direct Method

Direct methods of solving linear systems all use some form of matrix factorization, as discussed in Chapter 5. The LU factorization is the most commonly used method to solve a linear system.

For certain patterned matrices, other direct methods may be more efficient. If a given matrix initially has a large number of zeros, it is important to preserve the zeros in the same positions (or in other known positions) in the matrices that result from operations on the given matrix. This helps to avoid unnecessary computations. The iterative methods discussed in the next section are often more useful for sparse matrices.

Another important consideration is how easily an algorithm lends itself to implementation on advanced computer architectures. Many of the algorithms for linear algebra can be vectorized easily. It is now becoming more important to be able to parallelize the algorithms. The iterative methods discussed in the next section can often be parallelized more easily.

6.3 Iterative Methods for Consistent Systems

In iterative methods for solving the linear system $Ax = b$, we begin with starting point $x^{(0)}$, which we consider to be an approximate solution, and then move through a sequence of successive approximations $x^{(1)}, x^{(2)}, \ldots$, that ultimately (it is hoped!) converge to a solution. The user must specify a convergence criterion to determine when the approximation is close enough to the solution. The criterion may be based on successive changes in the solution $x^{(k)} - x^{(k-1)}$ or on the difference $\|Ax^{(k)} - b\|$.

Iterative methods may be particularly useful for very large systems because it may not be necessary to have the entire A matrix available for computations in each step. These methods are also useful for sparse systems. Also, as mentioned above, the iterative algorithms can often be parallelized (see Heath, Ng, and Peyton, 1991).

6.3.1 The Gauss-Seidel Method with Successive Overrelaxation

One of the simplest iterative procedures is the *Gauss-Seidel method*. In this method, we begin with an initial approximation to the solution, $x^{(0)}$. We then compute an update for the first element of x:

$$x_1^{(1)} = \frac{1}{a_{11}} \left(b_1 - \sum_{j=2}^{n} a_{1j} x_j^{(0)} \right).$$

Continuing in this way for the other elements of x, we have for $i = 1, \ldots, n$

$$x_i^{(1)} = \frac{1}{a_{ii}} \left(b_i - \sum_{j=1}^{i-1} a_{ij} x_j^{(1)} - \sum_{j=i+1}^{n} a_{ij} x_j^{(0)} \right),$$

where no sums are performed if the upper limit is smaller than the lower limit. After getting the approximation $x^{(1)}$, we then continue this same kind of iteration for $x^{(2)}, x^{(3)}, \ldots$.

We continue the iterations until a convergence criterion is satisfied. As we discuss in Section 10.3.3, this criterion may be of the form

$$\Delta \left(x^{(k)}, x^{(k-1)} \right) \leq \epsilon,$$

where $\Delta \left(x^{(k)}, x^{(k-1)} \right)$ is a measure of the difference of $x^{(k)}$ and $x^{(k-1)}$, such as $\|x^{(k)} - x^{(k-1)}\|$. We may also base the convergence criterion on $\|r^{(k)} - r^{(k-1)}\|$, where $r^{(k)} = b - Ax^{(k)}$.

The Gauss-Seidel iterations can be thought of as beginning with a rearrangement of the original system of equations as

$$
\begin{array}{llll}
a_{11}x_1 & & = b_1 - a_{12}x_2 \cdots - a_{1n}x_n \\
a_{21}x_1 & + \quad a_{22}x_2 & = b_2 & \cdots - a_{2n}x_n \\
\quad \vdots & + \quad \vdots & \quad \vdots & \quad \vdots \\
a_{(n-1)1}x_1 & + a_{(n-1)2}x_2 + \cdots & = b_{n-1} & - a_{nn}x_n \\
a_{n1}x_1 & + \quad a_{n2}x_2 \quad + \cdots + a_{nn}x_n = b_n
\end{array}
$$

In this form, we identify three matrices: a diagonal matrix D, a lower triangular L with 0s on the diagonal, and an upper triangular U with 0s on the diagonal:

$$(D + L)x = b - Ux.$$

We can write this entire sequence of Gauss-Seidel iterations in terms of these three fixed matrices:

$$x^{(k+1)} = (D + L)^{-1}\left(-Ux^{(k)} + b\right). \tag{6.15}$$

This method will converge for any arbitrary starting value $x^{(0)}$ if and only if the spectral radius of $(D + L)^{-1}U$ is less than 1. (See Golub and Van Loan, 1996, for a proof of this.) Moreover, the rate of convergence increases with decreasing spectral radius.

Successive Overrelaxation

The Gauss-Seidel method may be unacceptably slow, so it may be modified so that the update is a weighted average of the regular Gauss-Seidel update and the previous value. This kind of modification is called *successive overrelaxation*, or *SOR*. Instead of equation (6.15), the update is given by

$$\frac{1}{\omega}(D + L)\,x^{(k+1)} = \frac{1}{\omega}\big((1 - \omega)D - \omega U\big)x^{(k)} + b, \tag{6.16}$$

where the relaxation parameter ω is usually chosen to be between 0 and 1. For $\omega = 1$ the method is the ordinary Gauss-Seidel method; see Exercises 6.2c, 6.2d, and 6.2e.

6.3.2 Conjugate Gradient Methods for Symmetric Positive Definite Systems

In the Gauss-Seidel methods the convergence criterion is based on successive differences in the solutions $x^{(k)}$ and $x^{(k-1)}$ or in the residuals $r^{(k)}$ and $r^{(k-1)}$. Other iterative methods focus directly on the magnitude of the residual

$$r^{(k)} = b - Ax^{(k)}. \tag{6.17}$$

We seek a value $x^{(k)}$ such that the residual is small (in some sense). Methods that minimize $\|r^{(k)}\|_2$ are called minimal residual (MINRES) methods or generalized minimal residual (GMRES) methods.

For a system with a symmetric positive definite coefficient matrix A, it turns out that the best iterative method is based on minimizing the conjugate L_2 norm (see equation (3.66))

$$\|r^{(k)\mathrm{T}}A^{-1}r^{(k)}\|_2.$$

A method based on this minimization problem is called a conjugate gradient method.

The Conjugate Gradient Method

The problem of solving the linear system $Ax = b$ is equivalent to finding the minimum of the function

$$f(x) = \frac{1}{2}x^{\mathrm{T}}Ax - x^{\mathrm{T}}b. \tag{6.18}$$

By setting the derivative of f to 0, we see that a stationary point of f occurs at the point x where $Ax = b$ (see Section 4.3).

If A is positive definite, the (unique) minimum of f is at $x = A^{-1}b$, and the value of f at the minimum is $-\frac{1}{2}b^{\mathrm{T}}Ab$. The minimum point can be approached iteratively by starting at a point $x^{(0)}$, moving to a point $x^{(1)}$ that yields a smaller value of the function, and continuing to move to points yielding smaller values of the function. The kth point is $x^{(k-1)} + \alpha^{(k-1)} p^{(k-1)}$, where $\alpha^{(k-1)}$ is a scalar and $p^{(k-1)}$ is a vector giving the direction of the movement. Hence, for the kth point, we have the linear combination

$$x^{(k)} = x^{(0)} + \alpha^{(1)} p^{(1)} + \cdots + \alpha^{(k-1)} p^{(k-1)}.$$

At the point $x^{(k)}$, the function f decreases most rapidly in the direction of the negative gradient, $-\nabla f(x^{(k)})$, which is just the residual,

$$-\nabla f(x^{(k)}) = r^{(k)}.$$

If this residual is 0, no movement is indicated because we are at the solution.

Moving in the direction of steepest descent may cause a very slow convergence to the minimum. (The curve that leads to the minimum on the quadratic surface is obviously not a straight line. The direction of steepest descent changes as we move to a new point $x^{(k+1)}$.) A good choice for the sequence of directions $p^{(1)}, p^{(2)}, \ldots$ is such that

$$(p^{(k)})^{\mathrm{T}} A p^{(i)} = 0, \quad \text{for } i = 1, \ldots, k - 1. \tag{6.19}$$

Such a vector $p^{(k)}$ is A-conjugate to $p^{(1)}, p^{(2)}, \ldots p^{(k-1)}$ (see page 71). Given a current point $x^{(k)}$ and a direction to move $p^{(k)}$ to the next point, we must also choose a distance $\alpha^{(k)} \|p^{(k)}\|$ to move in that direction. We then have the next point,

$$x^{(k+1)} = x^{(k)} + \alpha^{(k)} p^{(k)}. \tag{6.20}$$

(Notice that here, as often in describing algorithms in linear algebra, we use Greek letters, such as α, to denote scalar quantities.)

We choose the directions as in Newton steps, so the first direction is $Ar^{(0)}$ (see Section 4.3.2). The paths defined by the directions $p^{(1)}, p^{(2)}, \ldots$ in equation (6.19) are called the conjugate gradients. A conjugate gradient method for solving the linear system is shown in Algorithm 6.1.

Algorithm 6.1 The Conjugate Gradient Method for Solving the Symmetric Positive Definite System $Ax = b$, Starting with $x^{(0)}$

0. Input stopping criteria, ϵ and $k_{\max}$.
   Set $k = 0$; $r^{(k)} = b - Ax^{(k)}$; $s^{(k)} = Ar^{(k)}$; $p^{(k)} = s^{(k)}$; and $\gamma^{(k)} = \|s^{(k)}\|^2$.
1. If $\gamma^{(k)} \leq \epsilon$, set $x = x^{(k)}$ and terminate.
2. Set $q^{(k)} = Ap^{(k)}$.
3. Set $\alpha^{(k)} = \dfrac{\gamma^{(k)}}{\|q^{(k)}\|^2}$.
4. Set $x^{(k+1)} = x^{(k)} + \alpha^{(k)} p^{(k)}$.
5. Set $r^{(k+1)} = r^{(k)} - \alpha^{(k)} q^{(k)}$.
6. Set $s^{(k+1)} = Ar^{(k+1)}$.
7. Set $\gamma^{(k+1)} = \|s^{(k+1)}\|^2$.

8. Set $p^{(k+1)} = s^{(k+1)} + \frac{\gamma^{(k+1)}}{\gamma^{(k)}} p^{(k)}$.

9. If $k < k_{\max}$,

  set $k = k + 1$ and go to step 1;

 otherwise

  issue message that

  "algorithm did not converge in $k_{\max}$ iterations". ∎

There are various ways in which the computations in Algorithm 6.1 could be arranged. Although any vector norm could be used in Algorithm 6.1, the L_2 norm is the most common one.

This method, like other iterative methods, is more appropriate for large systems. ("Large" in this context means bigger than 1000×1000.)

In exact arithmetic, the conjugate gradient method should converge in n steps for an $n \times n$ system. In practice, however, its convergence rate varies widely, even for systems of the same size. Its convergence rate generally decreases with increasing L_2 condition number (which is a function of the maximum and minimum nonzero eigenvalues), but that is not at all the complete story. The rate depends in a complicated way on all of the eigenvalues. The more spread out the eigenvalues are, the slower the rate. For different systems with roughly the same condition number, the convergence is faster if all eigenvalues are in two clusters around the maximum and minimum values. See Greenbaum and Strakoš (1992) for an analysis of the convergence rates.

Krylov Methods

Notice that the steps in the conjugate gradient algorithm involve the matrix A only through linear combinations of its rows or columns; that is, in any iteration, only a vector of the form Av or $A^{\mathrm{T}}w$ is used. The conjugate gradient method and related procedures, called *Lanczos methods*, move through a *Krylov space* in the progression to the solution. A Krylov space is the k-dimensional vector space of order n generated by the $n \times n$ matrix A and the vector v by forming the basis $\{v, Av, A^2v, \ldots, A^{k-1}v\}$. We often denote this space as $\mathcal{K}_k(A, v)$ or just as $\mathcal{K}_k$:

$$\mathcal{K}_k = \mathcal{V}(\{v, Av, A^2v, \ldots, A^{k-1}v\}). \tag{6.21}$$

Methods for computing eigenvalues are often based on Krylov spaces.

GMRES Methods

The conjugate gradient method seeks to minimize the residual vector in equation (6.17), $r^{(k)} = b - Ax^{(k)}$, and the convergence criterion is based on the linear combinations of the columns of the coefficient matrix formed by that vector, $\|Ar^{(k)}\|$.

The generalized minimal residual (GMRES) method of Saad and Schultz (1986) for solving $Ax = b$ begins with an approximate solution $x^{(0)}$ and takes $x^{(k)}$ as $x^{(k-1)} + z^{(k)}$, where $z^{(k)}$ is the solution to the minimization problem,

$$\min_{z \in \mathcal{K}_k(A, r^{(k-1)})} \|r^{(k-1)} - Az\|,$$

where, as before, $r^{(k)} = b - Ax^{(k)}$. This minimization problem is a constrained least squares problem. In the original implementations, the convergence of GMRES could be very slow, but modifications have speeded it up considerably. See Walker (1988) and Walker and Zhou (1994) for details of the methods. Brown and Walker (1997) consider the behavior of GMRES when the coefficient matrix is singular and give conditions for GMRES to converge to a solution of minimum length (the solution corresponding to the Moore-Penrose inverse; see Section 6.7.3, page 227).

Preconditioning

As we mentioned above, the convergence rate of the conjugate gradient method depends on the distribution of the eigenvalues in rather complicated ways. The ratio of the largest to the smallest (that is, the L_2 condition number is important) and the convergence rate for the conjugate gradient method is slower for larger L_2 condition numbers. The rate also is slower if the eigenvalues are spread out, especially if there are several eigenvalues near the largest or smallest. This phenomenon is characteristic of other Krylov space methods.

One way of addressing the problem of slow convergence of iterative methods is by *preconditioning*; that is, by replacing the system $Ax = b$ with another system,

$$M^{-1}Ax = M^{-1}b, \tag{6.22}$$

where M is very similar (by some measure) to A, but the system $M^{-1}Ax = M^{-1}b$ has a better condition for the problem at hand. We choose M to be symmetric and positive definite, and such that $Mx = b$ is easy to solve. If M is an approximation of A, then $M^{-1}A$ should be well-conditioned; its eigenvalues should all be close to each other.

A problem with applying the conjugate gradient method to the preconditioned system $M^{-1}Ax = M^{-1}b$ is that $M^{-1}A$ may not be symmetric. We can form an equivalent symmetric system, however, by decomposing the symmetric positive definite M as $M = VCV^{\mathrm{T}}$ and then

$$M^{-1/2} = V\mathrm{diag}(1/\sqrt{c_{11}}, \ldots, 1/\sqrt{c_{nn}})V^{\mathrm{T}},$$

as in equation (3.216), after inverting the positive square roots of C. Multiplying both sides of $M^{-1}Ax = M^{-1}b$ by $M^{1/2}$, inserting the factor $M^{-1/2}M^{1/2}$, and arranging terms yields

$$(M^{-1/2}AM^{-1/2})M^{1/2}x = M^{-1/2}b.$$

This can all be done and Algorithm 6.1 can be modified without explicit formation of and multiplication by $M^{1/2}$. The preconditioned conjugate gradient method is shown in Algorithm 6.2.

Algorithm 6.2 The Preconditioned Conjugate Gradient Method for Solving the Symmetric Positive Definite System $Ax = b$, Starting with $x^{(0)}$

0. Input stopping criteria, ϵ and $k_{\max}$.
   Set $k = 0$; $r^{(k)} = b - Ax^{(k)}$; $s^{(k)} = Ar^{(k)}$; $p^{(k)} = M^{-1}s^{(k)}$; $y^{(k)} = M^{-1}r^{(k)}$; and $\gamma^{(k)} = y^{(k)\mathrm{T}}s^{(k)}$.
1. If $\gamma^{(k)} \leq \epsilon$, set $x = x^{(k)}$ and terminate.
2. Set $q^{(k)} = Ap^{(k)}$.
3. Set $\alpha^{(k)} = \frac{\gamma^{(k)}}{\|q^{(k)}\|^2}$.
4. Set $x^{(k+1)} = x^{(k)} + \alpha^{(k)}p^{(k)}$.
5. Set $r^{(k+1)} = r^{(k)} - \alpha^{(k)}q^{(k)}$.
6. Set $s^{(k+1)} = Ar^{(k+1)}$.
7. Set $y^{(k+1)} = M^{-1}r^{(k+1)}$.
8. Set $\gamma^{(k+1)} = y^{(k+1)\mathrm{T}}s^{(k+1)}$.
9. Set $p^{(k+1)} = M^{-1}s^{(k+1)} + \frac{\gamma^{(k+1)}}{\gamma^{(k)}}p^{(k)}$.
10. If $k < k_{\max}$,
        set $k = k + 1$ and go to step 1;
    otherwise
        issue message that
        "algorithm did not converge in $k_{\max}$ iterations".

The choice of an appropriate matrix M is not an easy problem, and we will not consider the results here. Benzi (2002) provides a survey of preconditioning methods. We will also mention the preconditioned conjugate gradient method in Section 7.1.3, but there, again, we will refer the reader to other sources for details.

6.3.3 Multigrid Methods

Iterative methods have important applications in solving differential equations. The solution of differential equations by a finite difference discretization involves the formation of a grid. The solution process may begin with a fairly coarse grid on which a solution is obtained. Then a finer grid is formed, and the solution is interpolated from the coarser grid to the finer grid to be used as a starting point for a solution over the finer grid. The process is then continued through finer and finer grids. If all of the coarser grids are used throughout the process, the technique is a *multigrid* method. There are many variations of exactly how to do this. Multigrid methods are useful solution techniques for differential equations.

6.4 Numerical Accuracy

The condition numbers we defined in Section 6.1 are useful indicators of the accuracy we may expect when solving a linear system $Ax = b$. Suppose the entries of the matrix A and the vector b are accurate to approximately p decimal digits, so we have the system

$$(A + \delta A)(x + \delta x) = b + \delta b$$

with

$$\frac{\|\delta A\|}{\|A\|} \approx 10^{-p}$$

and

$$\frac{\|\delta b\|}{\|b\|} \approx 10^{-p}.$$

Assume A is nonsingular, and suppose that the condition number with respect to inversion, $\kappa(A)$, is approximately 10^t, so

$$\kappa(A)\frac{\|\delta A\|}{\|A\|} \approx 10^{t-p}.$$

Ignoring the approximation of b (that is, assuming $\delta b = 0$), we can write

$$\delta x = -A^{-1}\delta A(x + \delta x),$$

which, together with the triangular inequality and inequality (3.222) on page 129, yields the bound

$$\|\delta x\| \leq \|A^{-1}\|\,\|\delta A\|\left(\|x\| + \|\delta x\|\right).$$

Using equation (6.7) with this, we have

$$\|\delta x\| \leq \kappa(A)\frac{\|\delta A\|}{\|A\|}\left(\|x\| + \|\delta x\|\right)$$

or

$$\left(1 - \kappa(A)\frac{\|\delta A\|}{\|A\|}\right)\|\delta x\| \leq \kappa(A)\frac{\|\delta A\|}{\|A\|}\|x\|.$$

If the condition number is not too large relative to the precision (that is, if $10^{t-p} \ll 1$), then we have

$$\frac{\|\delta x\|}{\|x\|} \approx \kappa(A)\frac{\|\delta A\|}{\|A\|}$$
$$\approx 10^{t-p}. \tag{6.23}$$

Expression (6.23) provides a rough bound on the accuracy of the solution in terms of the precision of the data and the condition number of the coefficient matrix. This result must be used with some care, however. Rust (1994), among others, points out failures of the condition number for setting bounds on the accuracy of the solution.

Another consideration in the practical use of expression (6.23) is the fact that the condition number is usually not known, and methods for computing it suffer from the same rounding problems as the solution of the linear system itself. In Section 11.4, we describe ways of estimating the condition number, but as the discussion there indicates, these estimates are often not very reliable.

We would expect the norms in the expression (6.23) to be larger for larger size problems. The approach taken above addresses a type of "total" error. It may be appropriate to scale the norms to take into account the number of elements. Chaitin-Chatelin and Frayssé (1996) discuss error bounds for individual elements of the solution vector and condition measures for elementwise error.

Another approach to determining the accuracy of a solution is to use random perturbations of A and/or b and then to estimate the effects of the perturbations on x. Stewart (1990) discusses ways of doing this. Stewart's method estimates error measured by a norm, as in expression (6.23). Kenney and Laub (1994) and Kenney, Laub, and Reese (1998) describe an estimation method to address elementwise error.

Higher accuracy in computations for solving linear systems can be achieved in various ways: multiple precision, interval arithmetic, and residue arithmetic. Stallings and Boullion (1972) and Keller-McNulty and Kennedy (1986) describe ways of using residue arithmetic in some linear computations for statistical applications.

Another way of improving the accuracy is by using iterative refinement, which we now discuss.

6.5 Iterative Refinement

Once an approximate solution $x^{(0)}$ to the linear system $Ax = b$ is available, iterative refinement can yield a solution that is closer to the true solution. The residual

$$r = b - Ax^{(0)}$$

is used for iterative refinement. Clearly, if $h = A^+ r$, then $x^{(0)} + h$ is a solution to the original system.

The problem considered here is not just an iterative solution to the linear system, as we discussed in Section 6.3. Here, we assume $x^{(0)}$ was computed accurately given the finite precision of the computer. In this case, it is likely that r cannot be computed accurately enough to be of any help. If, however, r can be computed using a higher precision, then a useful value of h can be computed. This process can then be iterated as shown in Algorithm 6.3.

Algorithm 6.3 Iterative Refinement of the Solution to $Ax = b$, Starting with $x^{(0)}$

0. Input stopping criteria, ϵ and $k_{\max}$.
   Set $k = 0$.
1. Compute $r^{(k)} = b - Ax^{(k)}$ in higher precision.
2. Compute $h^{(k)} = A^+ r^{(k)}$.
3. Set $x^{(k+1)} = x^{(k)} + h^{(k)}$.
4. If $\|h^{(k)}\| \leq \epsilon \|x^{(k+1)}\|$, then
       set $x = x^{(k+1)}$ and terminate; otherwise,
       if $k < k_{\max}$,
           set $k = k + 1$ and go to step 1;
       otherwise,
           issue message that
           "algorithm did not converge in $k_{\max}$ iterations". ∎

In step 2, if A is of full rank then A^+ is A^{-1}. Also, as we have emphasized already, the fact that we write *an expression such as A^+r does not mean that we compute A^+*. The norm in step 4 is usually chosen to be the ∞ norm. The algorithm may not converge, so it is necessary to have an alternative exit criterion, such as a maximum number of iterations.

The use of iterative refinement as a general-purpose method is severely limited by the need for higher precision in step 1. On the other hand, if computations in higher precision can be performed, they can be applied to step 2 — or just in the original computations for $x^{(0)}$. In terms of both accuracy and computational efficiency, using higher precision throughout is usually better.

6.6 Updating a Solution to a Consistent System

In applications of linear systems, it is often the case that after the system $Ax = b$ has been solved, the right-hand side is changed and the system $Ax = c$ must be solved. If the linear system $Ax = b$ has been solved by a direct method using one of the factorizations discussed in Chapter 5, the factors of A can be used to solve the new system $Ax = c$. If the right-hand side is a small perturbation of b, say $c = b + \delta b$, an iterative method can be used to solve the new system quickly, starting from the solution to the original problem.

If the coefficient matrix in a linear system $Ax = b$ is perturbed to result in the system $(A + \delta A)x = b$, it may be possible to use the solution x_0 to the original system efficiently to arrive at the solution to the perturbed system. One way, of course, is to use x_0 as the starting point in an iterative procedure. Often, in applications, the perturbations are of a special type, such as

$$\widetilde{A} = A - uv^{\mathrm{T}},$$

where u and v are vectors. (This is a "rank-one" perturbation of A, and when the perturbed matrix is used as a transformation, it is called a "rank-one" update. As we have seen, a Householder reflection is a special rank-one update.) Assuming A is an $n \times n$ matrix of full rank, it is easy to write $\widetilde{A}^{-1}$ in terms of A^{-1}:

$$\widetilde{A}^{-1} = A^{-1} + \alpha(A^{-1}u)(v^{\mathrm{T}}A^{-1}) \tag{6.24}$$

with

$$\alpha = \frac{1}{1 - v^{\mathrm{T}}A^{-1}u}.$$

These are called the Sherman-Morrison formulas (from Sherman and Morrison, 1950). $\widetilde{A}^{-1}$ exists so long as $v^{\mathrm{T}}A^{-1}u \neq 1$. Because $x_0 = A^{-1}b$, the solution to the perturbed system is

$$\tilde{x}_0 = x_0 + \frac{(A^{-1}u)(v^{\mathrm{T}}x_0)}{(1 - v^{\mathrm{T}}A^{-1}u)}.$$

If the perturbation is more than rank one (that is, if the perturbation is

$$\widetilde{A} = A - UV^{\mathrm{T}}, \tag{6.25}$$

where U and V are $n \times m$ matrices with $n \geq m$), a generalization of the Sherman-Morrison formula, sometimes called the Woodbury formula, is

$$\widetilde{A}^{-1} = A^{-1} + A^{-1}U(I_m - V^{\mathrm{T}}A^{-1}U)^{-1}V^{\mathrm{T}}A^{-1} \tag{6.26}$$

(from Woodbury, 1950). The solution to the perturbed system is easily seen to be

$$\tilde{x}_0 = x_0 + A^{-1}U(I_m - V^{\mathrm{T}}A^{-1}U)^{-1}V^{\mathrm{T}}x_0.$$

As we have emphasized many times, we rarely compute the inverse of a matrix, and so the Sherman-Morrison-Woodbury formulas are not used directly. Having already solved $Ax = b$, it should be easy to solve another system, say $Ay = u_i$, where u_i is a column of U. If m is relatively small, as it is in most applications of this kind of update, there are not many systems $Ay = u_i$ to solve. Solving these systems, of course, yields $A^{-1}U$, the most formidable component of the Sherman-Morrison-Woodbury formula. The system to solve is of order m also.

Occasionally the updating matrices in equation (6.25) may be used with a weighting matrix, so we have $\widetilde{A} = A - UWV^{\mathrm{T}}$. An extension of the Sherman-Morrison-Woodbury formula is

$$(A - UWV^{\mathrm{T}})^{-1} = A^{-1} + A^{-1}U(W^{-1} - V^{\mathrm{T}}A^{-1}U)^{-1}V^{\mathrm{T}}A^{-1}. \tag{6.27}$$

This is sometimes called the Hemes formula. (The attributions of discovery are somewhat murky, and statements made by historians of science of the form "___ was the first to ___" must be taken with a grain of salt; not every discovery has resulted in an available publication. This is particularly true in numerical analysis, where scientific programmers often just develop a method in the process of writing code and have neither the time nor the interest in getting a publication out of it.)

Another situation that requires an update of a solution occurs when the system is augmented with additional equations and more variables:

$$\begin{bmatrix} A & A_{12} \\ A_{21} & A_{22} \end{bmatrix} \begin{bmatrix} x \\ x_+ \end{bmatrix} = \begin{bmatrix} b \\ b_+ \end{bmatrix}.$$

A simple way of obtaining the solution to the augmented system is to use the solution x_0 to the original system in an iterative method. The starting point for a method based on Gauss-Seidel or a conjugate gradient method can be taken as $(x_0, 0)$, or as $(x_0, x_+^{(0)})$ if a better value of $x_+^{(0)}$ is known.

In many statistical applications, the systems are overdetermined, with A being $n \times m$ and $n > m$. In the next section, we consider the general problem of solving overdetermined systems by using least squares, and then in Section 6.7.4 we discuss updating a least squares solution to an overdetermined system.

6.7 Overdetermined Systems; Least Squares

In applications, linear systems are often used as models of relationships between one observable variable, a "response", and another group of observable variables, "predictor variables". The model is unlikely to fit exactly any set of observed values of responses and predictor variables. This may be due to effects of other predictor variables that are not included in the model, measurement error, the relationship among the variables being nonlinear, or some inherent randomness in the system. In such applications, we generally take a larger number of observations than there are variables in the system; thus, with each set of observations on the response and associated predictors making up one equation, we have a system with more equations than variables.

An overdetermined system may be written as

$$Xb \approx y, \tag{6.28}$$

where X is $n \times m$ and $\operatorname{rank}(X|y) > m$; that is, the system is not consistent. We have changed the notation slightly from the consistent systems $Ax = b$ that we have been using because now we have in mind statistical applications and in those the notation $y \approx X\beta$ is more common. The problem is to determine a value of b that makes the approximation close in some sense. In applications of linear systems, we refer to this as "fitting" the system, which is referred to as a "model".

Overdetermined systems abound in fitting equations to data. The usual linear regression model is an overdetermined system and we discuss regression problems further in Section 9.2.2. We should not confuse *statistical inference* with *fitting equations to data*, although the latter task is a component of the former activity. In this section, we consider some of the more mechanical and computational aspects of the problem.

Accounting for an Intercept

Given a set of observations, the ith row of the system $Xb \approx y$ represents the linear relationship between y_i and the corresponding xs in the vector x_i:

$$y_i \approx b_1 x_{1i} + \cdots + b_m x_{mi}.$$

A different formulation of the relationship between y_i and the corresponding xs might include an intercept term:

$$y_i \approx \tilde{b}_0 + \tilde{b}_1 x_{1i} + \cdots + \tilde{b}_m x_{mi}.$$

There are two ways to incorporate this intercept term. One way is just to include a column of 1s in the X matrix. This approach makes the matrix X in equation (6.28) $n \times (m + 1)$, or else it means that we merely redefine x_{1i} to be the constant 1. Another way is to assume that the model is an exact fit for some set of values of y and the xs. If we assume that the model fits $y = 0$ and $x = 0$ exactly, we have a model without an intercept (that is, with a zero intercept).

Often, a reasonable assumption is that the model may have a nonzero intercept, but it fits the means of the set of observations; that is, the equation is exact for $y = \bar{y}$ and $x = \bar{x}$, where the jth element of $\bar{x}$ is the mean of the jth column vector of X. (Students with some familiarity with the subject may think this is a natural consequence of fitting the model. It is not unless the model fitting is by ordinary least squares.) If we require that the fitted equation be exact for the means (or if this happens naturally, as in the case of ordinary least squares), we may center each column by subtracting its mean from each element in the same manner as we centered vectors on page 34. In place of y, we have the vector $y - \bar{y}$. The matrix formed by centering all of the columns of a given matrix is called a centered matrix, and if the original matrix is X, we represent the centered matrix as X_{c} in a notation analogous to what we introduced for centered vectors. If we represent the matrix whose i^{th} column is the constant mean of the ith column of X as $\overline{X}$,

$$X_{\text{c}} = X - \overline{X}.$$

Using the centered data provides two linear systems: a set of approximate equations in which the intercept is ignored and an equation that fits the point that is assumed to be satisfied exactly:

$$\bar{y} = \overline{X}b.$$

In the rest of this section, we will generally ignore the question of an intercept. Except in a method discussed on page 237, the X can be considered to include a column of 1s, to be centered, or to be adjusted by any other point. We will return to this idea of centering the data in Section 8.6.3.

6.7.1 Least Squares Solution of an Overdetermined System

Although there may be no b that will make the system in (6.28) an equation, the system can be written as the equation

$$Xb = y - r, \tag{6.29}$$

where r is an n-vector of possibly arbitrary residuals or "errors".

A *least squares* solution $\widehat{b}$ to the system in (6.28) is one such that the Euclidean norm of the vector of residuals is minimized; that is, the solution to the problem

$$\min_b \|y - Xb\|_2. \tag{6.30}$$

The least squares solution is also called the "ordinary least squares" (OLS) fit.

By rewriting the square of this norm as

$$(y - Xb)^{\mathrm{T}}(y - Xb), \tag{6.31}$$

differentiating, and setting it equal to 0, we see that the minimum (of both the norm and its square) occurs at the $\widehat{b}$ that satisfies the square system

$$X^{\mathrm{T}}X\widehat{b} = X^{\mathrm{T}}y. \tag{6.32}$$

The system (6.32) is called the *normal equations*. The matrix $X^{\mathrm{T}}X$ is called the Gram matrix or the Gramian (see Section 8.6.1). Its condition determines the expected accuracy of a solution to the least squares problem. As we mentioned in Section 6.1, however, because the condition number of $X^{\mathrm{T}}X$ is the square of the condition number of X, it may be better to work directly on X in (6.28) rather than to use the normal equations. The normal equations are useful expressions, however, whether or not they are used in the computations. This is another case where *a formula does not define an algorithm,* as with other cases we have encountered many times. We should note, of course, that any information about the stability of the problem that the Gramian may provide can be obtained from X directly.

Special Properties of Least Squares Solutions

The least squares fit to the overdetermined system has a very useful property with two important consequences. The least squares fit partitions the space into two interpretable orthogonal spaces, as we see from equation (6.32). It is clear that residual vector $y - X\widehat{b}$ is orthogonal to each column in X:

$$X^{\mathrm{T}}(y - X\widehat{b}) = 0. \tag{6.33}$$

A consequence of this fact for models that include an intercept is that the sum of the residuals is 0. (The residual vector is orthogonal to the 1 vector.) Another consequence for models that include an intercept is that the least squares solution provides an exact fit to the mean.

These properties are so familiar to statisticians that some think that they are essential characteristics of any regression modeling; they are not. We will see in later sections that they do not hold for other approaches to fitting the basic model $y \approx Xb$. The least squares solution, however, has some desirable statistical properties under fairly common distributional assumptions, as we discuss in Chapter 9.

Weighted Least Squares

One of the simplest variations on fitting the linear model $Xb \approx y$ is to allow different weights on the observations; that is, instead of each row of X and corresponding element of y contributing equally to the fit, the elements of X and y are possibly weighted differently. The relative weights can be put into an n-vector w and the squared norm in equation (6.31) replaced by a quadratic form in $\mathrm{diag}(w)$. More generally, we form the quadratic form as

$$(y - Xb)^{\mathrm{T}}W(y - Xb), \tag{6.34}$$

where W is a positive definite matrix. Because the weights apply to both y and Xb, there is no essential difference in the weighted or unweighted versions of the problem.

The use of the QR factorization for the overdetermined system in which the weighted norm (6.34) is to be minimized is similar to the development above. It is exactly what we get if we replace $y - Xb$ in equation (6.36) by $W_{\mathrm{C}}(y - Xb)$, where W_{C} is the Cholesky factor of W.

Numerical Accuracy in Overdetermined Systems

In Section 6.4, we discussed numerical accuracy in computations for solving a consistent (square) system of equations and showed how bounds on the numerical error could be expressed in terms of the condition number of the coefficient matrix, which we had defined (on page 203) as the ratio of norms of the coefficient matrix and its inverse. One of the most useful versions of this condition number is the one using the L_2 matrix norm, which is called the spectral condition number. This is the most commonly used condition number, and we generally just denote it by $\kappa(\cdot)$. The spectral condition number is the ratio of the largest eigenvalue in absolute value to the smallest in absolute value, and this extends easily to a definition of the spectral condition number that applies both to nonsquare matrices and to singular matrices: the *condition number* of a matrix is the ratio of the largest singular value to the smallest nonzero singular value. As we saw on page 290, the nonzero singular values of X are the square roots of the nonzero eigenvalues of $X^{\mathrm{T}}X$; hence

$$\kappa(X^{\mathrm{T}}X) = (\kappa(X))^2. \tag{6.35}$$

The condition number of $X^{\mathrm{T}}X$ is a measure of the numerical accuracy we can expect in solving the normal equations (6.32). Because the condition number of X is smaller, we have an indication that it might be better not to form the normal equations unless we must. It might be better to work just with X. That is the approach we will take in the next sections.

6.7.2 Least Squares with a Full Rank Coefficient Matrix

If the $n \times m$ matrix X is of full column rank, the least squares solution, from equation (6.32), is $\widehat{b} = (X^{\mathrm{T}}X)^{-1}X^{\mathrm{T}}y$ and is obviously unique. A good way to compute this is to form the QR factorization of X.

First we write $X = QR$, as in equation (5.27) on page 188, where R is as in equation (5.29),

$$R = \begin{bmatrix} R_1 \\ 0 \end{bmatrix},$$

with R_1 an $m \times m$ upper triangular matrix. The residual norm (6.31) can be written as

$$\begin{aligned}
(y - Xb)^{\mathrm{T}}(y - Xb) &= (y - QRb)^{\mathrm{T}}(y - QRb) \\
&= (Q^{\mathrm{T}}y - Rb)^{\mathrm{T}}(Q^{\mathrm{T}}y - Rb) \\
&= (c_1 - R_1 b)^{\mathrm{T}}(c_1 - R_1 b) + c_2^{\mathrm{T}} c_2, \tag{6.36}
\end{aligned}$$

where c_1 is a vector with m elements and c_2 is a vector with $n - m$ elements, such that

$$Q^{\mathrm{T}}y = \begin{pmatrix} c_1 \\ c_2 \end{pmatrix}. \tag{6.37}$$

Because quadratic forms are nonnegative, the minimum of the residual norm in equation (6.36) occurs when $(c_1 - R_1 b)^{\mathrm{T}}(c_1 - R_1 b) = 0$; that is, when $(c_1 - R_1 b) = 0$, or

$$R_1 b = c_1. \tag{6.38}$$

We could also use the same technique of differentiation to find the minimum of equation (6.36) that we did to find the minimum of equation (6.31).

Because R_1 is triangular, the system is easy to solve: $\widehat{b} = R_1^{-1}c_1$. From equation (5.31), we have

$$X^+ = \begin{bmatrix} R_1^{-1} & 0 \end{bmatrix},$$

and so we have

$$\widehat{b} = X^+ y. \tag{6.39}$$

We also see from equation (6.36) that the minimum of the residual norm is $c_2^{\mathrm{T}} c_2$. This is called the *residual sum of squares* in the least squares fit.

6.7.3 Least Squares with a Coefficient Matrix Not of Full Rank

If X is not of full rank (that is, if X has rank $r < m$), the least squares solution is not unique, and in fact a solution is any vector $\widehat{b} = (X^{\mathrm{T}}X)^- X^{\mathrm{T}}y$, where $(X^{\mathrm{T}}X)^-$ is any generalized inverse. This is a solution to the normal equations (6.32). The residual corresponding to this solution is

$$y - X(X^{\mathrm{T}}X)^- X^{\mathrm{T}}y = (I - X(X^{\mathrm{T}}X)^- X^{\mathrm{T}})y.$$

The residual vector is invariant to the choice of generalized inverse, as we see from equation (8.51) on page 289.

An Optimal Property of the Solution Using the Moore-Penrose Inverse

The solution corresponding to the Moore-Penrose inverse is unique because, as we have seen, that generalized inverse is unique. That solution is interesting for another reason, however: the b from the Moore-Penrose inverse has the minimum L_2-norm of all solutions.

To see that this solution has minimum norm, first factor X, as in equation (5.36) on page 190,

$$X = QRU^{\mathrm{T}},$$

and form the Moore-Penrose inverse as in equation (5.38):

$$X^+ = U \begin{bmatrix} R_1^{-1} & 0 \\ 0 & 0 \end{bmatrix} Q^{\mathrm{T}}.$$

Then

$$\widehat{b} = X^+ y \tag{6.40}$$

is a least squares solution, just as in the full rank case. Now, let

$$Q^{\mathrm{T}}y = \begin{pmatrix} c_1 \\ c_2 \end{pmatrix},$$

as in equation (6.37), except ensure that c_1 has exactly r elements and c_2 has $n - r$ elements, and let

$$U^{\mathrm{T}}b = \begin{pmatrix} z_1 \\ z_2 \end{pmatrix},$$

where z_1 has r elements. We proceed as in the equations (6.36). We seek to minimize $\|y - Xb\|_2$ (which is the square root of the expression in equations (6.36)); and because multiplication by an orthogonal matrix does not change the norm, we have

$$\|y - Xb\|_2 = \|Q^{\mathrm{T}}(y - XUU^{\mathrm{T}}b)\|_2$$

$$= \left\| \begin{pmatrix} c_1 \\ c_2 \end{pmatrix} - \begin{bmatrix} R_1 & 0 \\ 0 & 0 \end{bmatrix} \begin{pmatrix} z_1 \\ z_2 \end{pmatrix} \right\|_2$$

$$= \left\| \begin{pmatrix} c_1 - R_1 z_1 \\ c_2 \end{pmatrix} \right\|_2. \tag{6.41}$$

The residual norm is minimized for $z_1 = R_1^{-1}c_1$ and z_2 arbitrary. However, if $z_2 = 0$, then $\|z\|_2$ is also minimized. Because $U^{\mathrm{T}}b = z$ and U is orthogonal, $\|\widehat{b}\|_2 = \|z\|_2$, and so $\|\widehat{b}\|_2$ is the minimum among all least squares solutions.

6.7.4 Updating a Least Squares Solution of an Overdetermined System

In the last section, we considered the problem of updating a given solution to be a solution to a perturbed consistent system. An overdetermined system is often perturbed by adding either some rows or some columns to the coefficient matrix X. This corresponds to including additional equations in the system,

$$\begin{bmatrix} X \\ X_+ \end{bmatrix} b \approx \begin{bmatrix} y \\ y_+ \end{bmatrix},$$

or to adding variables,

$$\begin{bmatrix} X & X_+ \end{bmatrix} \begin{bmatrix} b \\ b_+ \end{bmatrix} \approx y.$$

In either case, if the QR decomposition of X is available, the decomposition of the augmented system can be computed readily. Consider, for example, the addition of k equations to the original system $Xb \approx y$, which has n approximate equations. With the QR decomposition, for the original full rank system, putting $Q^{\mathrm{T}}X$ and $Q^{\mathrm{T}}y$ as partitions in a matrix, we have

$$\begin{bmatrix} R_1 & c_1 \\ 0 & c_2 \end{bmatrix} = Q^{\mathrm{T}} \begin{bmatrix} X & y \end{bmatrix}.$$

Augmenting this with the additional rows yields

$$\begin{bmatrix} R & c_1 \\ 0 & c_2 \\ X_+ & y_+ \end{bmatrix} = \begin{bmatrix} Q^{\mathrm{T}} & 0 \\ 0 & I \end{bmatrix} \begin{bmatrix} X & y \\ X_+ & y_+ \end{bmatrix}. \tag{6.42}$$

All that is required now is to apply orthogonal transformations, such as Givens rotations, to the system (6.42) to produce

$$\begin{bmatrix} R_* & c_{1*} \\ 0 & c_{2*} \end{bmatrix},$$

where R_* is an $m \times m$ upper triangular matrix and c_{1*} is an m-vector as before but c_{2*} is an $(n - m + k)$-vector.

The updating is accomplished by applying m rotations to system (6.42) so as to zero out the $(n+q)^{\text{th}}$ row for $q = 1, 2, \ldots, k$. These operations go through an outer loop with $p = 1, 2, \ldots, n$ and an inner loop with $q = 1, 2, \ldots, k$. The operations rotate R through a sequence $R^{(p,q)}$ into R_*, and they rotate X_+ through a sequence $X_+^{(p,q)}$ into 0. At the p, q step, the rotation matrix Q_{pq} corresponding to equation (5.12) on page 183 has

$$\cos\theta = \frac{R_{pp}^{(p,q)}}{r}$$

and

$$\sin\theta = \frac{\left(X_+^{(p,q)}\right)_{qp}}{r},$$

where

$$r = \sqrt{\left(R_{pp}^{(p,q)}\right)^2 + \left(\left(X_+^{(p,q)}\right)_{qp}\right)^2}.$$

Gentleman (1974) and Miller (1992) give Fortran programs that implement this kind of updating. The software, which was published in *Applied Statistics*, is available in `statlib` (see page 505).

6.8 Other Solutions of Overdetermined Systems

The basic form of an overdetermined linear system may be written as in equation (6.28) as

$$Xb \approx y,$$

where X is $n \times m$ and $\text{rank}(X|y) > m$.

As in equation (6.29) in Section 6.7.1, we can write this as an equation,

$$Xb = y - r,$$

where r is a vector of residuals. Fitting the equation $y = Xb$ means minimizing r; that is, minimizing some norm of r.

There are various norms that may provide a reasonable fit. In Section 6.7, we considered use of the L_2 norm; that is, an ordinary least squares (OLS) fit. There are various other ways of approaching the problem, and we will briefly consider a few of them in this section.

As we have stated before, *we should not confuse statistical inference with fitting equations to data, although the latter task is a component of the former activity.* Applications in statistical data analysis are discussed in Chapter 9. In those applications, we need to make statements (that is, assumptions) about relevant probability distributions. These probability distributions, together with the methods used to collect the data, may indicate specific methods for fitting the equations to the given data. In this section, we continue to address the more mechanical aspects of the problem of fitting equations to data.

6.8.1 Solutions that Minimize Other Norms of the Residuals

A solution to an inconsistent, overdetermined system

$$Xb \approx y,$$

where X is $n \times m$ and $\mathrm{rank}(X|y) > m$, is some value b that makes $y - Xb$ close to zero. We define "close to zero" in terms of a norm on $y - Xb$. The most common norm, of course, is the L_2 norm as in expression (6.30), and the minimization of this norm is straightforward, as we have seen. In addition to the simple analytic properties of the L_2 norm, the least squares solution has some desirable statistical properties under fairly common distributional assumptions, as we have seen.

Minimum L_1 Norm Fitting; Least Absolute Values

A common alternative norm is the L_1 norm. The minimum L_1 norm solution is called the *least absolute values* fit or the *LAV* fit. It is not as affected by outlying observations as the least squares fit is.

Consider a simple example. Assume we have observations on a response, $y = (0, 3, 4, 0, 8)$, and on a single predictor variable, $x = (1, 3, 4, 6, 7)$. We have $\bar{y} = 3$ and $\bar{x} = 4.2$. We write the model equation as

$$y \approx b_0 + b_1 x. \tag{6.43}$$

The model with the data is

$$\begin{bmatrix} 0 \\ 3 \\ 4 \\ 0 \\ 8 \end{bmatrix} = \begin{bmatrix} 1 & 1 \\ 1 & 3 \\ 1 & 4 \\ 1 & 6 \\ 1 & 7 \end{bmatrix} b + r. \tag{6.44}$$

A least squares solution yields $\hat{b}_0 = -0.3158$ and $\hat{b}_1 = 0.7895$. With these values, equation (6.43) goes through the mean, 3, and $(1, 4.2)$. The residual vector from this fit is orthogonal to 1 (that is, the sum of the residuals is 0) and to x.

A solution that minimizes the L_1 norm is $\tilde{b}_0 = -1.333$ and $\tilde{b}_1 = 1.333$. The LAV fit may not be unique (although it is in this case). We immediately note that the least absolute deviations fit does not go through the mean of the data, nor is the residual vector orthogonal to the 1 vector and to x. The LAV fit does go through two points in the dataset, however. (This is a special case of one of several interesting properties of LAV fits, which we will not discuss here. The interested reader is referred to Kennedy and Gentle, 1980, Chapter 11, for discussion of some of these properties, as well as assumptions about probability distributions that result in desirable statistical properties for LAV estimators.) A plot of the data, the two fitted lines, and the residuals is shown in Figure 6.2.

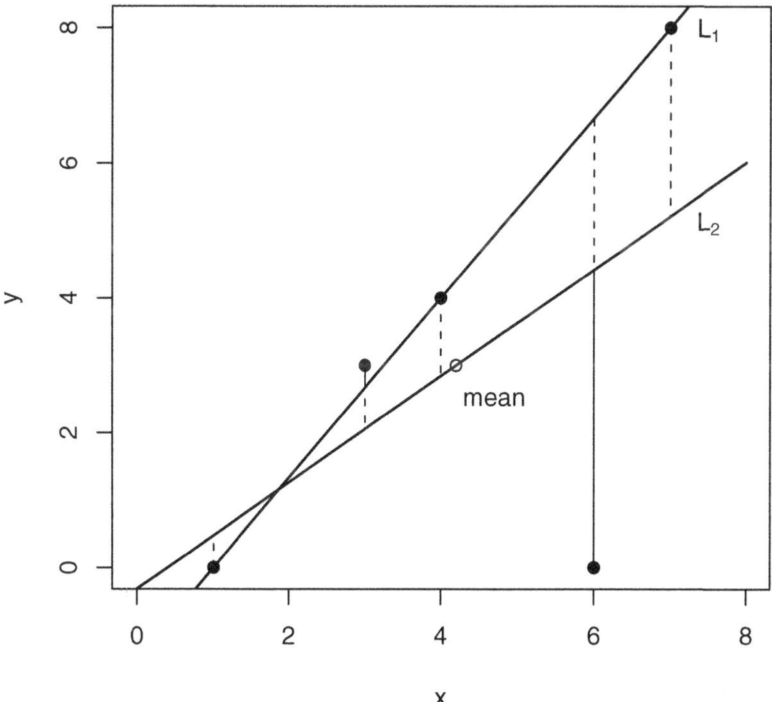

Fig. 6.2. OLS and Minimum L_1 Norm Fits

The problem of minimizing the L_1 norm can be formulated as the linear programming problem

$$\min_{b} \quad 1^{\mathrm{T}}(e^{+} + e^{-})$$

$$\text{s.t. } Xb + Ie^{+} - Ie^{-} = y$$

$$e^{+}, e^{-} \geq 0 \tag{6.45}$$

$$b \quad \text{unrestricted,}$$

where e^{+} and e^{-} are nonnegative n-vectors. There are special algorithms that take advantage of the special structure of the problem to speed up the basic linear programming simplex algorithm (see Kennedy and Gentle, 1980, Chapter 11).

Minimum L_{∞} Norm Fitting; Minimax

Another norm that may be useful in some applications is the L_{∞} norm. A solution minimizing that norm is called the *least maximum deviation* fit. A least maximum deviation fit is greatly affected by outlying observations. As with the LAV fit, the least maximum deviation fit does not necessarily go through the mean of the data. The least maximum deviation fit also may not be unique.

This problem can also be formulated as a linear programming problem, and as with the least absolute deviations problem, there are special algorithms that take advantage of the special structure of the problem to speed up the basic linear programming simplex algorithm. Again, the interested reader is referred to Kennedy and Gentle (1980, Chapter 11) for a discussion of some of the properties of minimum L_{∞} norm fits, as well as assumptions about probability distributions that result in desirable statistical properties for minimum L_{∞} norm estimators.

L_p Norms and Iteratively Reweighted Least Squares

More general L_p norms may also be of interest. For $1 < p < 2$, if no element of $y - Xb$ is zero, we can formulate the pth power of the norm as

$$\|y - Xb\|_p^p = (y - Xb)^{\mathrm{T}} W (y - Xb), \tag{6.46}$$

where

$$W = \mathrm{diag}(|\, y_i - x_i^{\mathrm{T}} b|^{2-p}) \tag{6.47}$$

and x_i^{T} is the ith row of X. The formulation (6.46) leads to the iteratively reweighted least squares (IRLS) algorithm, in which a sequence of weights $W^{(k)}$ and weighted least squares solutions $b^{(k)}$ are formed as in Algorithm 6.4.

Algorithm 6.4 Iteratively Reweighted Least Squares

0. Input a threshold for a zero residual, ϵ_1 (which may be data dependent), and a large value for weighting zero residuals, w_{big}. Input stopping criteria, ϵ_2 and $k_{\max}$.
   Set $k = 0$. Choose an initial value $b^{(k)}$, perhaps as the OLS solution.
1. Compute the diagonal matrix $W^{(k)}$: $w_i^{(k)} = |y_i - x_i^{\mathrm{T}} b^{(k)}|^{2-p}$, except, if $|y_i - x_i^{\mathrm{T}} b^{(k)}| < \epsilon_1$, set $w_i^{(k)} = w_{\text{big}}$.
2. Compute $b^{(k+1)}$ by solving the weighted least squares problem: minimize $(y - Xb)^{\mathrm{T}} W^{(k)} (y - Xb)$.
3. If $\|b^{(k+1)} - b^{(k)}\| \le \epsilon_2$, set $b = b^{(k+1)}$ and terminate.
4. If $k < k_{\max}$,
   set $k = k + 1$ and go to step 1;
   otherwise,
   issue message that
   "algorithm did not converge in $k_{\max}$ iterations".

Compute $b^{(1)}$ by minimizing equation (6.46) with $W = W^{(0)}$; then compute $W^{(1)}$ from $b^{(1)}$, and iterate in this fashion. This method is easy to implement and will generally work fairly well, except for the problem of zero (or small) residuals. The most effective way of dealing with zero residuals is to set them to some large value.

Algorithm 6.4 will work for LAV fitting, although the algorithms based on linear programming alluded to above are better for this task. As mentioned above, LAV fits generally go through some observations; that is, they fit them exactly, yielding zero residuals. This means that in using Algorithm 6.4, the manner of dealing with zero residuals may become an important aspect of the efficiency of the algorithm.

6.8.2 Regularized Solutions

Overdetermined systems often arise because of a belief that some response y is linearly related to some other set of variables. This relation is expressed in the system

$$y \approx Xb.$$

The fact that $y \ne Xb$ for any b results because the relationship is not exact. There is perhaps some error in the measurements. It is also possible that there is some other variable not included in the columns of X. In addition, there may be some underlying randomness that could never be accounted for.

In any application in which we fit an overdetermined system, it is likely that the given values of X and y are only a sample (not necessarily a random sample) from some universe of interest. Whatever value of b provides the best fit (in terms of the criterion chosen) may not provide the best fit if some other equally valid values of X and y were used. The given dataset is fit optimally, but the underlying phenomenon of interest may not be modeled very well.

The given dataset may suggest relationships among the variables that are not present in the larger universe of interest. Some element of the "true" b may be zero, but in the best fit for a given dataset, the value of that element may be significantly different from zero. Deciding on the evidence provided by a given dataset that there is a relationship among certain variables when indeed there is no relationship in the broader universe is an example of *overfitting*.

There are various approaches we may take to avoid overfitting, but there is no panacea. The problem is inherent in the process.

One approach to overfitting is *regularization*. In this technique, we restrain the values of b in some way. Minimizing $\|y - Xb\|$ may yield a b with large elements, or values that are likely to vary widely from one dataset to another. One way of "regularizing" the solution is to minimize also some norm of b. The general formulation of the problem then is

$$\min_b(\|y - Xb\|_r + \lambda\|b\|_b), \tag{6.48}$$

where λ is some appropriately chosen nonnegative number. The norm on the residuals, $\|\cdot\|_r$, and that on the solution vector b, $\|\cdot\|_b$, are often chosen to be the same, and, of course, most often, they are chosen as the L_2 norm. If both norms are the L_2 norm, the fitting is called Tikhonov regularization. In statistical applications, this leads to "ridge regression". If $\|\cdot\|_r$ is the L_2 norm and $\|\cdot\|_b$ is the L_1 norm, the statistical method is called the "lasso". We discuss these formulations briefly in Section 9.4.4.

As an example, let us consider the data in equation (6.44) for the equation

$$y = b_0 + b_1 x.$$

We found the least squares solution to be $\hat{b}_0 = -0.3158$ and $\hat{b}_1 = 0.7895$, which fits the means and has a residual vector that is orthogonal to 1 and to x. Now let us regularize the least squares fit with an L_2 norm on b and with $\lambda = 5$. (The choice of λ depends on the scaling of the data and a number of other things we will not consider here. Typically, in an application, various values of λ are considered.) Again, we face the question of treating b_0 and b_1 differently. The regularization, which is a shrinkage, can be applied to both or just to b_1. Furthermore, we have the question of whether we want to force the equation to fit some data point exactly. In statistical applications, it is common not to apply the shrinkage to the intercept term and to force the fitted equation to fit the means exactly. Doing that, we get $\hat{b}_{1_\lambda} = 0.6857$, which is shrunken from the value of $\hat{b}_1$, and $\hat{b}_{0_\lambda} = 0.1200$, which is chosen so as to fit the mean. A plot of the data and the two fitted lines is shown in Figure 6.3.

6.8.3 Minimizing Orthogonal Distances

In writing the equation $Xb = y + r$ in place of the overdetermined linear system $Xb \approx y$, we are allowing adjustments to y so as to get an equation.

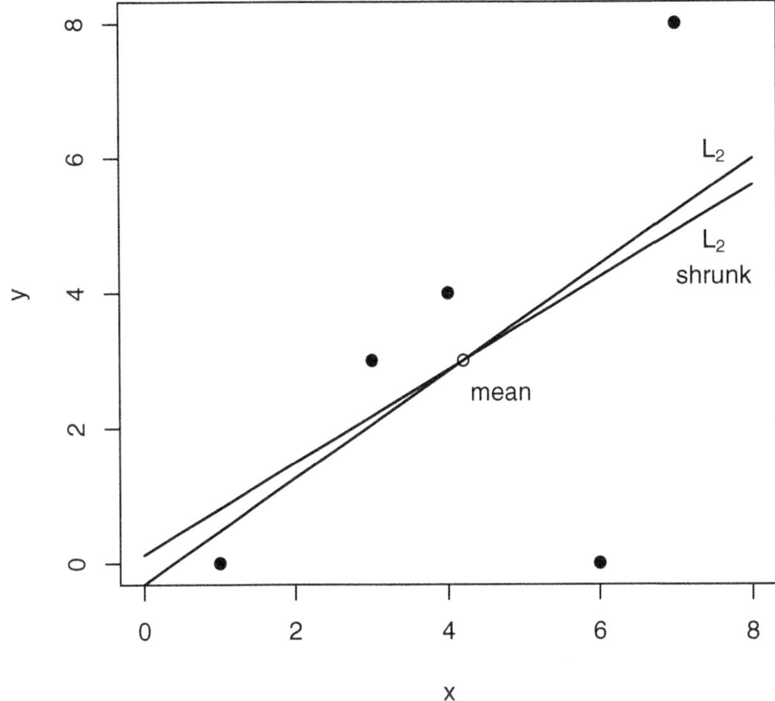

Fig. 6.3. OLS and L_2 Norm Regularized Minimum L_2 Norm Fits

Another way of making an equation out of the overdetermined linear system $Xb \approx y$ is to write it as

$$(X + E)b = y + r; \tag{6.49}$$

that is, to allow adjustments to both X and y. Both X and E are in $\mathbb{R}^{n \times m}$ (and we assume $n > m$).

In fitting the linear model only with adjustments to y, we determine b so as to minimize some norm of r. Likewise, with adjustments to both X and y, we seek b so as to minimize some norm of the matrix E and the vector r. There are obviously several ways to approach this. We could take norms of E and r separately and consider some weighted combination of the norms. Another way is to adjoin r to E and minimize some norm of the $n \times (m+1)$ matrix $[E|r]$.

A common approach is to minimize $\|[E|r]\|_{\mathrm{F}}$. This, of course, is the sum of squares of all elements in $[E|r]$. The method is therefore sometimes called "total least squares".

If it exists, the minimum of $\|[E|r]\|_{\mathrm{F}}$ is achieved at

$$b = -v_{2*}/v_{22}, \tag{6.50}$$

where

$$[X|y] = UDV^{\mathrm{T}} \tag{6.51}$$

is the singular value decomposition (see equation (3.218) on page 127), and V is partitioned as

$$V = \begin{bmatrix} V_{11} & v_{*2} \\ v_{2*} & v_{22} \end{bmatrix}.$$

If E has some special structure, the problem of minimizing the orthogonal residuals may not have a solution. Golub and Van Loan (1980) show that a sufficient condition for a solution to exist is that $d_m \geq d_{m+1}$. (Recall that the ds in the SVD are nonnegative and they are indexed so as to be nonincreasing.) Golub and Van Loan (1980) also show that the solution is unique if $d_m > d_{m+1}$.

Again, as an example, let us consider the data in equation (6.44) for the equation

$$y = b_0 + b_1 x.$$

We found the least squares solution to be $\hat{b}_0 = -0.3158$ and $\hat{b}_1 = 0.7895$, which fits the mean and has a residual vector that is orthogonal to 1 and to x. Now we determine a fit so that the L_2 norm of the orthogonal residuals is minimized. Again, we will force the equation to fit the mean exactly. We get $\hat{b}_{0_{\mathrm{orth}}} = -4.347$ and $\hat{b}_{0_{\mathrm{orth}}} = 1.749$. A plot of the data, the two fitted lines, and the residuals is shown in Figure 6.4.

The orthogonal residuals can be weighted in the usual way by premultiplication by a Cholesky factor of a weight matrix, as discussed on page 225.

If some norm other than the L_2 norm is to be minimized, an iterative approach must be used. Ammann and Van Ness (1988, 1989) describe an iterative method that is applicable to any norm, so long as a method is available to compute a value of b that minimizes the norm of the usual vertical distances in a model such as equation (9.9). The method is simple. We first fit $y = Xb$, minimizing the vertical distances in the usual way; we then rotate y into $\tilde{y}$ and X into $\widetilde{X}$, so that the fitted plane is horizontal. Next, we fit $\tilde{y} = \widetilde{X}b$ and repeat. After continuing this way until the fits in the rotated spaces do not change from step to step, we adjust the fitted b back to the original unrotated space. Because of these rotations, if we assume that the model fits some point exactly, we must adjust y and X accordingly (see the discussion on page 223). In the following, we assume that the model fits the means exactly, so we center the data. We let m be the number of columns in the centered data matrix. (The centered matrix does not contain a column of 1s. If the formulation of the model $y = Xb$ includes an intercept term, then X is $n \times (m+1)$.)

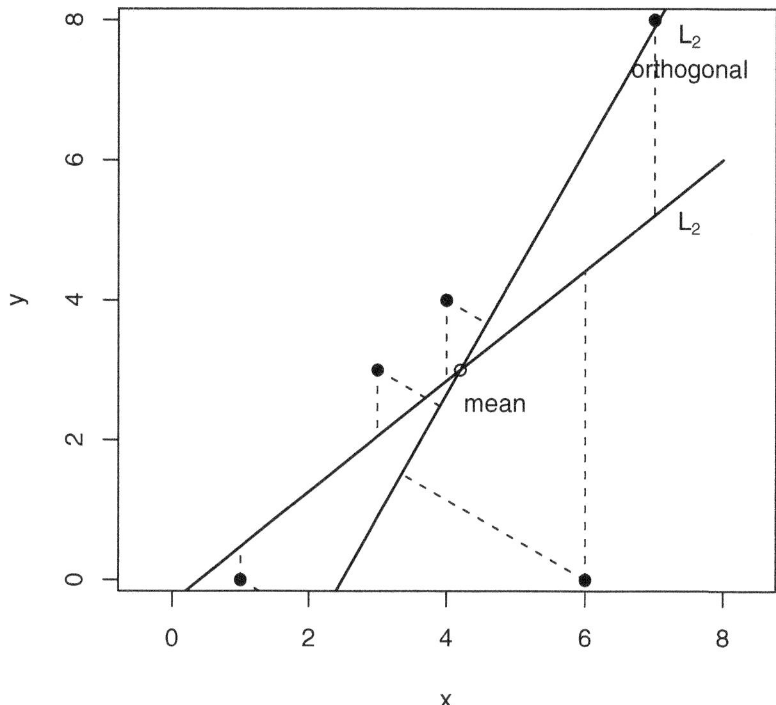

Fig. 6.4. OLS and Minimum Orthogonal L_2-Norm Fits

Algorithm 6.5 Iterative Orthogonal Residual Fitting through the Means

0. Input stopping criteria, ϵ and $k_{\max}$.
   Set $k = 1$, $y_{\mathrm{c}}^{(0)} = y_{\mathrm{c}}$, $X_{\mathrm{c}}^{(0)} = X_{\mathrm{c}}$, and $D^{(0)} = I_{m+1}$.
1. Determine a value $b_{\mathrm{c}}^{(k)}$ that minimizes the norm of $\left(y_{\mathrm{c}}^{(k-1)} - X_{\mathrm{c}}^{(k-1)} b_{\mathrm{c}}^{(k)} \right)$.
2. If $b_{\mathrm{c}}^{(k)} \leq \epsilon$, go to step 7.
3. Determine a rotation matrix $Q^{(k)}$ that makes the kth fit horizontal.
4. Transform the matrix $\left[y_{\mathrm{c}}^{(k-1)} | X_{\mathrm{c}}^{(k-1)} \right]$ $\left[y_{\mathrm{c}}^{(k)} | X_{\mathrm{c}}^{(k)} \right]$ by a rotation matrix:

$$\left[y_{\mathrm{c}}^{(k)} | X_{\mathrm{c}}^{(k)} \right] = \left[y_{\mathrm{c}}^{(k-1)} | X_{\mathrm{c}}^{(k-1)} \right] Q^{(k)}.$$

5. Transform $D^{(k-1)}$ by the same rotation: $D^{(k)} = D^{(k-1)} Q^{(k)}$.
6. If $k < k_{\max}$,
   set $k = k + 1$ and go to step 1;
   otherwise,
   issue message that
   "algorithm did not converge in $k_{\max}$ iterations".

7. For $j = 2, \ldots, m$, choose $b_j = d_{j,m+1}/d_{m+1,m+1}$ (So long as the rotations have not produced a vertical plane in the unrotated space, $d_{m+1,m+1}$ will not be zero.)

8. Compute $b_1 = \bar{y} - \sum_{j=2}^{k} b_j * \bar{x}_j$ (where $\bar{x}_j$ is the mean of the jth column of the original uncentered X).   ∎

An appropriate rotation matrix for Algorithm 6.5 is Q in the QR decomposition of

$$\begin{bmatrix} I_m & 0 \\ (b^{(k)})^{\text{T}} & 1 \end{bmatrix}.$$

Note that forcing the fit to go through the means, as we do in Algorithm 6.5, is not usually done for norms other than the L_2 norm (see Figure 6.2).

Exercises

6.1. Let A be nonsingular, and let $\kappa(A) = \|A\|\,\|A^{-1}\|$.
    a) Prove equation (6.10):

$$\kappa_2(A) = \frac{\max_{x \neq 0} \frac{\|Ax\|}{\|x\|}}{\min_{x \neq 0} \frac{\|Ax\|}{\|x\|}}.$$

    b) Using the relationship above, explain heuristically why $\kappa(A)$ is called the "condition number" of A.

6.2. Consider the system of linear equations

$$\begin{aligned}
x_1 + 4x_2 + x_3 &= 12, \\
2x_1 + 5x_2 + 3x_3 &= 19, \\
x_1 + 2x_2 + 2x_3 &= 9.
\end{aligned}$$

    a) Solve the system using Gaussian elimination with partial pivoting.
    b) Solve the system using Gaussian elimination with complete pivoting.
    c) Determine the D, L, and U matrices of the Gauss-Seidel method (equation (6.15), page 212) and determine the spectral radius of

$$(D + L)^{-1}U.$$

    d) Do two steps of the Gauss-Seidel method starting with $x^{(0)} = (1, 1, 1)$, and evaluate the L_2 norm of the difference of two successive approximate solutions.
    e) Do two steps of the Gauss-Seidel method with successive overrelaxation using $\omega = 0.1$, starting with $x^{(0)} = (1, 1, 1)$, and evaluate the L_2 norm of the difference of two successive approximate solutions.

f) Do two steps of the conjugate gradient method starting with $x^{(0)} = (1, 1, 1)$, and evaluate the L_2 norm of the difference of two successive approximate solutions.

6.3. The normal equations.

a) For any matrix X with real elements, show that $X^\mathrm{T}X$ is nonnegative definite.

b) For any $n \times m$ matrix X with real elements and with $n < m$, show that $X^\mathrm{T}X$ is not positive definite.

c) Let X be an $n \times m$ matrix of full column rank. Show that $X^\mathrm{T}X$ is positive definite.

6.4. Solving an overdetermined system $Xb = y$, where X is $n \times m$.

a) Count how many multiplications and additions are required to form $X^\mathrm{T}X$. (A multiplication or addition such as this is performed in floating point on a computer, so the operation is called a "flop". Sometimes a flop is considered a combined operation of multiplication and addition; at other times, each is considered a separate flop. See page 415. The distinction is not important here; just count the total number.)

b) Count how many flops are required to form $X^\mathrm{T}y$.

c) Count how many flops are required to solve $X^\mathrm{T}Xb = X^\mathrm{T}y$ using a Cholesky decomposition.

d) Count how many flops are required to form a QR decomposition of X using reflectors.

e) Count how many flops are required to form a $Q^\mathrm{T}y$.

f) Count how many flops are required to solve $R_1 b = c_1$ (equation (6.38), page 226).

g) If n is large relative to m, what is the ratio of the total number of flops required to form and solve the normal equations using the Cholesky method to the total number required to solve the system using a QR decomposition? Why is the QR method generally preferred?

6.5. Verify equation (6.46).

7

Evaluation of Eigenvalues and Eigenvectors

Before we discuss methods for computing eigenvalues, we mention an interesting observation. A given n^{th}-degree polynomial $p(c)$ is the characteristic polynomial of some matrix. The companion matrix of equation (3.177) is one such matrix. Thus, given a general polynomial p, we can form a matrix A whose eigenvalues are the roots of the polynomial; and likewise, given a square matrix, we can write a polynomial in its eigenvalues. It is a well-known fact in the theory of equations that there is no general formula for the roots of a polynomial of degree greater than 4. This means that we cannot expect to have a direct method for calculating eigenvalues of any given matrix. The eigenvalues of some matrices can be evaluated directly of course. The eigenvalues of a diagonal matrix are merely the diagonal elements. In that case, however, the characteristic polynomial is of the factored form $\prod(a_{ii} - c)$, whose roots are immediately obtainable. For general eigenvalue computations, however, we must use an iterative method.

In statistical applications, the matrices whose eigenvalues are of interest are often symmetric. Symmetric matrices are diagonalizable and have only real eigenvalues. (As usual, we will assume the matrices themselves are real.) The problem of determining the eigenvalues of a symmetric matrix therefore is simpler than the corresponding problem for a general matrix. In many statistical applications, the symmetric matrices of interest are nonnegative definite. Nonsymmetric matrices of interest in statistical applications are often irreducible nonnegative matrices. Either of these properties can also allow use of simpler methods.

In this chapter, we describe various methods for computing eigenvalues. A given method may have some desirable property for particular applications, and in some cases, the methods may be used in combination. Some of the methods rely on sequences that converge to a particular eigenvalue or eigenvector. The power method, discussed in Section 7.2, is of this type; one eigenpair at a time is computed. Other methods are based on sequences of similar matrices that converge to a diagonal matrix. An example of such a method is called the *LR method*. This method, which we will not consider in detail, is based on a factorization of A into left and right factors, F_L and F_R, and the fact that if c is an eigenvalue of $F_L F_R$, then it is also an eigenvalue of $F_R F_L$ (property 7, page 107). If $A = L^{(0)} U^{(0)}$ is an LU decomposition of A with 1s on the diagonal of either $L^{(0)}$ or $U^{(0)}$, iterations of LU decompositions of the similar matrices

$$L^{(k+1)} U^{(k+1)} = U^{(k)} L^{(k)},$$

under some conditions, will converge to a similar diagonal matrix. The sufficient conditions for convergence include nonnegative definiteness.

Most methods for extracting eigenvalues and eigenvectors use a sequence of orthogonally similar transformations that eventually yield a product of orthogonal matrices and a diagonal matrix.

7.1 General Computational Methods

For whatever approach is taken for finding eigenpairs, there are some general methods that may speed up the process or that may help in achieving higher numerical accuracy.

7.1.1 Eigenvalues from Eigenvectors and Vice Versa

Some methods for eigenanalysis yield the eigenvalues, and other methods yield the eigenvectors. Given one member of an eigenpair, we usually want to find the other member.

If we are given an eigenvector v of the matrix A, there must be some element v_j that is not zero. For any nonzero element of the eigenvector, the eigenvalue corresponding to v is

$$(Av)_j / v_j. \tag{7.1}$$

Likewise, if the eigenvalue c is known, a corresponding eigenvector is any solution to the singular system

$$(A - cI)v = 0. \tag{7.2}$$

(It is relevant to note that the system is singular because most standard software packages will refuse to solve singular systems whether or not they are consistent!)

An eigenvector associated with the eigenvalue c can be found using equation (7.2) if we know the position of any nonzero element in the vector. Suppose, for example, it is known that $v_1 \neq 0$. We can set $v_1 = 1$ and form another system to solve for the remaining elements of v by writing

$$\begin{bmatrix} a_{11} - 1 & a_1^T \\ a_2 & A_{22} - cI_{n-1} \end{bmatrix} \begin{bmatrix} 1 \\ v_2 \end{bmatrix} = \begin{bmatrix} 0 \\ 0 \end{bmatrix}, \tag{7.3}$$

where v_2 is an $(n-1)$-vector and a_1^{T} and a_2 are the remaining elements in the first row and first column, respectively, of A. Rearranging this, we get the $(n-1) \times (n-1)$ system

$$(A_{22} - cI_{n-1})v_2 = -a_2. \tag{7.4}$$

The locations of any zero elements in the eigenvector are critical for using this method. To form a system as in equation (7.3), the position of some nonzero element must be known. Another problem in using this method arises when the geometric multiplicity of the eigenvalue is greater than 1. In that case, the system in equation (7.4) is also singular, and the process must be repeated to form an $(n-2) \times (n-2)$ system. If the multiplicity of the eigenvalue is k, the first full rank system encountered while continuing in this way is the one that is $(n-k) \times (n-k)$.

7.1.2 Deflation

Whenever an eigenpair and an associated left eigenvalue of a real matrix A are available, another matrix can be formed for which all the other nonzero eigenvalues and corresponding eigenvectors are the same as for A.

Suppose c_i is an eigenvalue of A with associated right and left eigenvectors v_i and w_i, respectively. Now, suppose that c_j is a nonzero eigenvalue of A such that $c_j \neq c_i$. Let v_j and w_j be, respectively, right and left eigenvectors associated with c_j. Now,

$$\langle Av_i, w_j \rangle = \langle c_i v_i, w_j \rangle = c_i \langle v_i, w_j \rangle,$$

but also

$$\langle Av_i, w_j \rangle = \langle v_i, A^{\mathrm{T}} w_j \rangle = \langle v_i, c_j w_j \rangle = c_j \langle v_i, w_j \rangle.$$

But if

$$c_i \langle v_i, w_j \rangle = c_j \langle v_i, w_j \rangle$$

and $c_j \neq c_i$, then $\langle v_i, w_j \rangle = 0$. Consider the matrix

$$B = A - c_i v_i w_i^{\mathrm{H}}. \tag{7.5}$$

We see that

$$\begin{aligned} Bw_j &= Aw_j - c_i v_i w_i^{\mathrm{H}} w_j \\ &= Aw_j \\ &= c_j w_j, \end{aligned}$$

so c_j and w_j are, respectively, an eigenvalue and an eigenvector of B.

The matrix B has some of the flavor of the sum of some terms in a spectral decomposition of A. (Recall that the spectral decomposition is guaranteed to exist only for matrices with certain properties. In Chapter 3, we stated the existence for diagonalizable matrices but derived it only for symmetric matrices.)

The ideas above lead to a useful method for finding eigenpairs of a diagonalizable matrix. (The method also works if we begin with a simple eigenvalue.) We will show the details only for a real symmetric matrix.

Deflation of Symmetric Matrices

Let A be an $n \times n$ symmetric matrix. A therefore is diagonalizable, its eigenvalues and eigenvectors are real, and the left and right eigenvalues are the same.

Let (c, v), with $v^{\mathrm{T}} v = 1$, be an eigenpair of A. Now let X be an $n \times n - 1$ matrix whose columns form an orthogonal basis for $\mathcal{V}(A - v v^{\mathrm{T}})$. One easy way of doing this is to choose $n - 1$ of the n unit vectors of order n such that none are equal to v and then, beginning with v, use Gram-Schmidt transformations to orthogonalize the vectors, using Algorithm 2.1 on page 28. (Assuming v is not a unit vector, we merely choose $e_1, \ldots, e_{n-1}$ together with v as the starting set of linearly independent vectors.) Now let $P = [v|X]$. We have

$$P^{-1} = \left[\begin{array}{c} v^{\mathrm{T}} \\ X^{\mathrm{T}}(I - v v^{\mathrm{T}}) \end{array} \right],$$

as we see by direct multiplication, and

$$P^{-1} A P = \left[\begin{array}{cc} c & 0 \\ 0 & B \end{array} \right], \tag{7.6}$$

where B is the $(n - 1) \times (n - 1)$ matrix $X^{\mathrm{T}} A X$.

Clearly, B is symmetric and the eigenvalues of B are the same as the other $n - 1$ eigenvalues of A. The important point is that B is $(n - 1) \times (n - 1)$.

7.1.3 Preconditioning

The convergence of iterative methods applied to a linear system $Ax = b$ can often be speeded up by replacing the system by an equivalent system $M^{-1} A x = M^{-1} b$. The iterations then depend on the properties, such as the relative magnitudes of the eigenvalues, of $M^{-1} A$ rather than A. The replacement of the system $Ax = b$ by $M^{-1} A x = M^{-1} b$ is called *preconditioning*. (It is also sometimes called *left preconditioning*, and the use of the system $A M^{-1} y = b$ with $y = M x$ is called *right preconditioning*. Either or both kinds of preconditioning may be used in a given iterative algorithm.) The matrix M is called a *preconditioner*.

Determining an effective preconditioner matrix M is not straightforward. Obviously, if $M = A$, the preconditioned system is simpler than the original system, but determining M^{-1} is as difficult as dealing with the original system.

In general, the objective would be to determine $M^{-1}A$ so that it is "close" to I. The salient properties of I are that it is normal (see page 274) and its eigenvalues are clustered.

There are various kinds of preconditioning; some work better as an adjunct to one algorithm, and others work better in conjunction with some other algorithm. In the case of a sparse matrix A, for example an incomplete factorization $A \approx \widetilde{L}\widetilde{U}$ where both $\widetilde{L}$ and $\widetilde{U}$ are sparse, $M = \widetilde{L}\widetilde{U}$ may be a good preconditioner. We will not consider any of the details here. Later (page 216) we will consider preconditioning in the context of an iterative algorithm for solving linear systems. Benzi (2002) provides a good survey of techniques, but the effort to identify general methods remains incomplete and is an area of active research.

7.2 Power Method

The power method is a straightforward method that can be used for a real diagonalizable matrix with a simple dominant eigenvalue. An important type of matrix that satisfies this condition is an irreducible nonnegative square matrix (see Section 8.7.2).

Let A be a real $n \times n$ diagonalizable matrix with a simple dominant eigenvalue. Index the eigenvalues c_i so that $|c_1| > |c_2| \geq \cdots |c_n|$, with corresponding unit eigenvectors v_i. Note that the requirement for the dominant eigenvalue that $c_1 > c_2$ implies that c_1 and the dominant eigenvector v_1 are unique and that c_1 is real (because otherwise $\bar{c}_1$ would also be an eigenvalue, and that would violate the requirement).

Now let x be an n-vector that is not orthogonal to v_1. Because A is assumed to be diagonalizable, the eigenvectors are linearly independent and so x can be represented as a linear combination of the eigenvectors,

$$x = b_1 v_1 + \cdots + b_n v_n. \tag{7.7}$$

Because x is not orthogonal to v_1, $b_1 \neq 0$. The power method is based on a sequence

$$x, Ax, A^2 x, \ldots.$$

(This sequence is a finite Krylov space generating set; see equation (6.21).) From the relationships above and the definition of eigenvalues and eigenvectors, we have

$$
\begin{aligned}
Ax &= b_1 A v_1 + \cdots + b_n A v_n \\
&= b_1 c_1 v_1 + \cdots + b_n c_n v_n \\
A^2 x &= b_1 c_1^2 v_1 + \cdots + b_n c_n^2 v_n \\
\cdots &= \cdots \\
A^j x &= b_1 c_1^j v_1 + \cdots + b_n c_n^j v_n \\
&= c_1^j \left(b_1 v_1 + \cdots + b_n \left(\frac{c_n}{c_1}\right)^j v_n \right).
\end{aligned}
\tag{7.8}
$$

To simplify the notation, let

$$
u^{(j)} = A^j x / c_1^j
\tag{7.9}
$$

(or, equivalently, $u^{(j)} = A u^{(j-1)} / c_1$). From equations (7.8) and the fact that $|c_1| > |c_i|$ for $i > 1$, we see that $u^{(j)} \to b_1 v_1$, which is the nonnormalized dominant eigenvector.

We have the bound

$$
\begin{aligned}
\left\| u^{(j)} - b_1 v_1 \right\| &= \left\| b_2 \left(\frac{c_2}{c_1}\right)^j v_2 + \cdots \right. \\
&\qquad \left. \cdots + b_n \left(\frac{c_n}{c_1}\right)^j v_n \right\| \\
&\leq |b_2| \left|\frac{c_2}{c_1}\right|^j \|v_2\| + \cdots \\
&\qquad \cdots + |b_n| \left|\frac{c_n}{c_1}\right|^j \|v_n\| \\
&\leq \left(|b_2| + \cdots + |b_n| \right) \left|\frac{c_2}{c_1}\right|^j.
\end{aligned}
\tag{7.10}
$$

The last expression results from the fact that $|c_2| \geq |c_i|$ for $i > 2$ and that the v_i are unit vectors.

From equation (7.10), we see that the norm of the difference of $u^{(j)}$ and $b_1 v_1$ decreases by a factor of approximately $|c_2/c_1|$ with each iteration; hence, this ratio is an important indicator of the rate of convergence of $u^{(j)}$ to the dominant eigenvector.

If $|c_1| > |c_2| > |c_3|$, $b_2 \neq 0$, and $b_1 \neq 0$, the power method converges linearly (see page 418); that is,

$$
0 < \lim_{j \to \infty} \frac{\|x^{(j+1)} - b_1 v_1\|}{\|x^{(j)} - b_1 v_1\|} < 1
\tag{7.11}
$$

(see Exercise 7.1c, page 256).

If an approximate value of the eigenvector v_1 is available and x is taken to be that approximate value, the convergence will be faster. If an approximate value of the dominant eigenvalue, $\widehat{c}_1$, is available, starting with any $y^{(0)}$, a few iterations on

$$(A - \widehat{c}_1 I)y^{(k)} = y^{(k-1)}$$

may yield a better starting value for x. Once the eigenvector associated with the dominant eigenvalue is determined, the eigenvalue c_1 can easily be determined, as described above.

In some applications, only the dominant eigenvalue is of interest. If other eigenvalues are needed, however, we find them one at a time by deflation.

If A is nonsingular, we can also use the power method on A^{-1} to determine the smallest eigenvalue of A.

7.3 Jacobi Method

The Jacobi method for determining the eigenvalues of a simple symmetric matrix A uses a sequence of orthogonal similarity transformations that eventually results in the transformation

$$A = PCP^{-1}$$

(see equation (3.193) on page 116) or

$$C = P^{-1}AP,$$

where C is diagonal. Recall that similar matrices have the same eigenvalues.

The matrices for the similarity transforms are the Givens rotation or Jacobi rotation matrices discussed on page 182. The general form of one of these orthogonal matrices, $G_{pq}(\theta)$, given in equation (5.12) on page 183, is the identity matrix with $\cos\theta$ in the $(p,p)^{\text{th}}$ and $(q,q)^{\text{th}}$ positions, $\sin\theta$ in the $(p,q)^{\text{th}}$ position, and $-\sin\theta$ in the $(q,p)^{\text{th}}$ position:

$$G_{pq}(\theta) = \begin{array}{c} \\ \\ p \\ \\ q \\ \\ \end{array}\begin{array}{cc} \quad\quad p \quad\quad\quad q \quad\quad \\ \left[\begin{array}{ccccc} I & 0 & 0 & 0 & 0 \\ 0 & \cos\theta & 0 & \sin\theta & 0 \\ 0 & 0 & I & 0 & 0 \\ 0 & -\sin\theta & 0 & \cos\theta & 0 \\ 0 & 0 & 0 & 0 & I \end{array}\right]. \end{array}$$

The Jacobi iteration is

$$A^{(k)} = G^{\text{T}}_{p_k q_k}(\theta_k)A^{(k-1)}G_{p_k q_k}(\theta_k),$$

where p_k, q_k, and θ_k are chosen so that the $A^{(k)}$ is "more diagonal" than $A^{(k-1)}$. Specifically, the iterations will be chosen so as to reduce the sum of the squares of the off-diagonal elements, which for any square matrix A is

$$\|A\|_{\mathrm{F}}^2 - \sum_i a_{ii}^2.$$

The orthogonal similarity transformations preserve the Frobenius norm

$$\left\|A^{(k)}\right\|_{\mathrm{F}} = \left\|A^{(k-1)}\right\|_{\mathrm{F}}.$$

Because the rotation matrices change only the elements in the $(p,p)^{\mathrm{th}}$, $(q,q)^{\mathrm{th}}$, and $(p,q)^{\mathrm{th}}$ positions (and also the $(q,p)^{\mathrm{th}}$ position since both matrices are symmetric), we have

$$\left(a_{pp}^{(k)}\right)^2 + \left(a_{qq}^{(k)}\right)^2 + 2\left(a_{pq}^{(k)}\right)^2 = \left(a_{pp}^{(k-1)}\right)^2 + \left(a_{qq}^{(k-1)}\right)^2 + 2\left(a_{pq}^{(k-1)}\right)^2.$$

The off-diagonal sum of squares at the kth stage in terms of that at the $(k-1)^{\mathrm{th}}$ stage is

$$\left\|A^{(k)}\right\|_{\mathrm{F}}^2 - \sum_i \left(a_{ii}^{(k)}\right)^2 = \left\|A^{(k)}\right\|_{\mathrm{F}}^2 - \sum_{i\neq p,q} \left(a_{ii}^{(k)}\right)^2 - \left(\left(a_{pp}^{(k)}\right)^2 + \left(a_{qq}^{(k)}\right)^2\right)$$

$$= \left\|A^{(k-1)}\right\|_{\mathrm{F}}^2 - \sum_i \left(a_{ii}^{(k-1)}\right)^2 - 2\left(a_{pq}^{(k-1)}\right)^2 + 2\left(a_{pq}^{(k)}\right)^2.$$

$$(7.12)$$

Hence, for a given index pair, (p,q), at the kth iteration, the sum of the squares of the off-diagonal elements is minimized by choosing the rotation matrix so that

$$a_{pq}^{(k)} = 0. \tag{7.13}$$

As we saw on page 183, it is easy to determine the angle θ so as to introduce a zero in a single Givens rotation. Here, we are using the rotations in a similarity transformation, so it is a little more complicated.

The requirement that $a_{pq}^{(k)} = 0$ implies

$$a_{pq}^{(k-1)}\left(\cos^2\theta - \sin^2\theta\right) + \left(a_{pp}^{(k-1)} - a_{qq}^{(k-1)}\right)\cos\theta\sin\theta = 0. \tag{7.14}$$

Using the trigonometric identities

$$\cos(2\theta) = \cos^2\theta - \sin^2\theta$$
$$\sin(2\theta) = 2\cos\theta\sin\theta,$$

in equation (7.14), we have

$$\tan(2\theta) = \frac{2a_{pq}^{(k-1)}}{a_{pp}^{(k-1)} - a_{qq}^{(k-1)}},$$

which yields a unique angle in $[-\pi/4,\ \pi/4]$. Of course, the quantities we need are $\cos\theta$ and $\sin\theta$, not the angle itself. First, using the identity

$$\tan\theta = \frac{\tan(2\theta)}{1 + \sqrt{1 + \tan^2(2\theta)}},$$

we get $\tan\theta$ from $\tan(2\theta)$; and then from $\tan\theta$ we can compute the quantities required for the rotation matrix $G_{pq}(\theta)$:

$$\cos\theta = \frac{1}{\sqrt{1 + \tan^2\theta}},$$
$$\sin\theta = \cos\theta\tan\theta.$$

Convergence occurs when the off-diagonal elements are sufficiently small. The quantity (7.12) using the Frobenius norm is the usual value to compare with a convergence criterion, ϵ.

From equation (7.13), we see that the best index pair, (p, q), is such that

$$\left|a_{pq}^{(k-1)}\right| = \max_{i<j}\left|a_{ij}^{(k-1)}\right|.$$

If this choice is made, the Jacobi method can be shown to converge (see Watkins, 2002). The method with this choice is called the *classical Jacobi method*.

For an $n \times n$ matrix, the number of operations to identify the maximum off-diagonal is $O(n^2)$. The computations for the similarity transform itself are only $O(n)$ because of the sparsity of the rotators. Of course, the computations for the similarity transformations are more involved than those to identify the maximum off-diagonal, so, for small n, the classical Jacobi method should be used. If n is large, however, it may be better not to spend time looking for the maximum off-diagonal. Various *cyclic Jacobi* methods have been proposed in which the pairs (p, q) are chosen systematically without regard to the magnitude of the off-diagonal being zeroed. Depending on the nature of the cyclic Jacobi method, it may or may not be guaranteed to converge. For certain schemes, quadratic convergence has been proven; for at least one other scheme, an example showing failure of convergence has been given. See Watkins (2002) for a discussion of the convergence issues.

The Jacobi method is one of the oldest algorithms for computing eigenvalues, and has recently become important again because it lends itself to easy implementation on parallel processors (see Zhou and Brent, 2003).

Notice that at the kth iteration, only two rows and two columns of $A^{(k)}$ are modified. This is what allows the Jacobi method to be performed in parallel. We can form $\lfloor n/2 \rfloor$ pairs and do $\lfloor n/2 \rfloor$ rotations simultaneously. Thus, each parallel iteration consists of a choice of a set of index pairs and then a batch of rotations. Although, as we have indicated, the convergence may depend on which rows are chosen for the rotations, if we are to achieve much efficiency by performing the operations in parallel, we cannot spend much time in deciding how to form the pairs for the rotations. Various schemes have been suggested for forming the pairs for a parallel iteration. A simple scheme, called "mobile Jacobi" (see Watkins, 2002), is:

1. Perform $\lfloor n/2 \rfloor$ rotations using the pairs

$$(1,2), \ (3,4), \ (5,6), \ldots .$$

2. Interchange all rows and columns that were rotated.
3. Perform $\lfloor (n-1)/2 \rfloor$ rotations using the pairs

$$(2,3), \ (4,5), \ (6,7), \ldots .$$

4. Interchange all rows and columns that were rotated.
5. If convergence has not been achieved, go to 1.

The notation above that specifies the pairs refers to the rows and columns at the current state; that is, after the interchanges up to that point. The interchange operation is a similarity transformation using an elementary permutation matrix (see page 63), and hence the eigenvalues are left unchanged by this operation. The method described above is a good one, but there are other ways of forming pairs. Some of the issues to consider are discussed by Luk and Park (1989), who analyzed and compared some proposed schemes.

7.4 QR Method

The most common algorithm for extracting eigenvalues is the QR method. While the power method and the Jacobi method require diagonalizable matrices, which restricts their practical use to symmetric matrices, the QR method can be used for nonsymmetric matrices. It is simpler for symmetric matrices, of course, because the eigenvalues are real. Also, for symmetric matrices the computer storage is less, the computations are fewer, and some transformations are particularly simple.

The QR method requires that the matrix first be transformed into upper Hessenberg form (see page 44). A matrix can be reduced to Hessenberg form in a finite number of similarity transformations using either Householder reflections or Givens rotations.

The Hessenberg form for a symmetric matrix is tridiagonal. The Hessenberg form allows a large savings in the subsequent computations, even for nonsymmetric matrices.

The QR method for determining the eigenvalues is iterative and produces a sequence of Hessenberg matrices $A^{(0)}, A^{(1)}, \ldots, A^{(n)}$, where $A^{(n)}$ is a triangular matrix. An upper Hessenberg matrix is formed and its eigenvalues are extracted by a process called "chasing", which consists of steps that alternate between creating nonzero entries in positions $(i+2, i)$, $(i+3, i)$, and $(i+3, i+1)$ and restoring these entries to zero, as the nonzero entries are moved farther down the matrix. For example,

$$
\begin{bmatrix}
X & X & X & X & X & X & X \\
X & X & X & X & X & X & X \\
0 & X & X & X & X & X & X \\
0 & Y & X & X & X & X & X \\
0 & Y & Y & X & X & X & X \\
0 & 0 & 0 & 0 & X & X & X \\
0 & 0 & 0 & 0 & 0 & X & X
\end{bmatrix}
\rightarrow
\begin{bmatrix}
X & X & X & X & X & X & X \\
X & X & X & X & X & X & X \\
0 & X & X & X & X & X & X \\
0 & 0 & X & X & X & X & X \\
0 & 0 & Y & X & X & X & X \\
0 & 0 & Y & Y & X & X & X \\
0 & 0 & 0 & 0 & 0 & X & X
\end{bmatrix}.
$$

In the jth step of the QR method, a bulge is created and is chased down the matrix by similarity transformations, usually Givens transformations,

$$
G_k^{-1} A^{(j-1,k)} G_k.
$$

The transformations are based on the eigenvalues of 2×2 matrices in the lower right-hand part of the matrix.

There are some variations on the way the chasing occurs. Haag and Watkins (1993) describe an efficient modified QR algorithm that uses both Givens transformations and Gaussian elimination transformations, with or without pivoting. For the $n \times n$ Hessenberg matrix $A^{(0,0)}$, the first step of the Haag-Watkins procedure begins with a 3×3 Householder reflection matrix, $\widetilde{G}_0$, whose first column is

$$
(A^{(0,0)} - \sigma_1 I)(A^{(0,0)} - \sigma_2 I)e_1,
$$

where σ_1 and σ_2 are the eigenvalues of the 2×2 matrix

$$
\begin{bmatrix}
a_{n-1,n-1} & a_{n-1,n} \\
a_{n-1,n} & a_{n,n}
\end{bmatrix},
$$

and e_1 is the first unit vector of length n. The $n \times n$ matrix G_0 is $\mathrm{diag}(\widetilde{G}_0, I)$. The initial transformation $G_0^{-1} A^{(0,0)} G_0$ creates a bulge with nonzero elements $a_{31}^{(0,1)}$, $a_{41}^{(0,1)}$, and $a_{42}^{(0,1)}$.

After the initial transformation, the Haag-Watkins procedure makes $n-3$ transformations

$$
A^{(0,k+1)} = G_k^{-1} A^{(0,k)} G_k,
$$

for $k = 1, 2, \ldots, n-3$, that chase the bulge diagonally down the matrix, so that $A^{(0,k+1)}$ differs from Hessenberg form only by the nonzero elements $a_{k+3,k+1}^{(0,k+1)}$, $a_{k+4,k+1}^{(0,k+1)}$, and $a_{k+4,k+2}^{(0,k+1)}$. To accomplish this, the matrix G_k differs from the identity only in rows and columns $k+1$, $k+2$, and $k+3$. The transformation

$$
G_k^{-1} A^{(0,k)}
$$

annihilates the entries $a_{k+2,k}^{(0,k)}$ and $a_{k+3,k}^{(0,k)}$, and the transformation

$$
(G_k^{-1} A^{(0,k)}) G_k
$$

produces $A^{(0,k+1)}$ with two new nonzero elements, $a^{(0,k+1)}_{k+4,k+1}$ and $a^{(0,k+1)}_{k+4,k+2}$. The final transformation in the first step, for $k = n - 2$, annihilates $a^{(0,k)}_{n,n-2}$. The transformation matrix G_{n-2} differs from the identity only in rows and columns $n - 1$ and n. These steps are iterated until the matrix becomes triangular. As the subdiagonal elements converge to zero, the shifts for use in the first transformation of a step (corresponding to σ_1 and σ_2) are determined by 2×2 submatrices higher on the diagonal. Special consideration must be given to situations in which these submatrices contain zero elements. For this, the reader is referred to Watkins (2002) or Golub and Van Loan (1996).

This description has just indicated the general flavor of the QR method. There are different variations on the overall procedure and then many computational details that must be observed. In the Haag-Watkins procedure, for example, the G_ks are not unique, and their form can affect the efficiency and the stability of the algorithm. Haag and Watkins (1993) describe criteria for the selection of the G_ks. They also discuss some of the details of programming the algorithm.

7.5 Krylov Methods

In the power method, we encountered the sequence

$$x, Ax, A^2x, \ldots.$$

This sequence is a finite Krylov space generating set. As we mentioned on page 215, several methods for computing eigenvalues are often based on a Krylov space,

$$\mathcal{K}_k = \mathcal{V}(\{v, Av, A^2v, \ldots, A^{k-1}v\}).$$

(Aleksei Krylov used these vectors to construct the characteristic polynomial.)

The two most important Krylov methods are the Lanczos tridiagonalization algorithm and the Arnoldi orthogonalization algorithm. We will not discuss these methods here but rather refer the interested reader to Golub and Van Loan (1996).

7.6 Generalized Eigenvalues

In Section 3.8.9, we defined the generalized eigenvalues and eigenvectors by replacing the identity in the definition of ordinary eigenvalues and eigenvectors by a general (square) matrix B:

$$|A - cB| = 0. \tag{7.15}$$

If there exists a finite c such that this determinant is zero, then there is some nonzero, finite vector v such that

$$Av = cBv. \tag{7.16}$$

As we have seen in the case of ordinary eigenvalues, symmetry of the matrix, because of diagonalizability, allows for simpler methods to evaluate the eigenvalues. In the case of generalized eigenvalues, symmetry together with positive definiteness allows us to reformulate the problem to be much simpler. If A and B are symmetric and B is positive definite, we refer to the pair (A, B) as *symmetric*.

If A and B are a symmetric pair, B has a Cholesky decomposition, $B = T^{\mathrm{T}}T$, where T is an upper triangular matrix with positive diagonal elements. We can therefore rewrite equation (7.16) as

$$T^{-\mathrm{T}}AT^{-1}u = cu, \tag{7.17}$$

where $u = Tv$. Note that because A is symmetric, $T^{-\mathrm{T}}AT^{-1}$ is symmetric, and since c is an eigenvalue of this matrix, it is real. Its associated eigenvector (with respect to $T^{-\mathrm{T}}AT^{-1}$) is likewise real, and therefore so is the generalized eigenvector v. Because $T^{-\mathrm{T}}AT^{-1}$ is symmetric, the ordinary eigenvectors can be chosen to be orthogonal. (Recall from page 119 that eigenvectors corresponding to distinct eigenvalues *are* orthogonal, and those corresponding to a multiple eigenvalue can be chosen to be orthogonal.) This implies that the generalized eigenvectors of the symmetric pair (A, B) can be chosen to be B-conjugate.

Because of the equivalence of a generalized eigenproblem for a symmetric pair to an ordinary eigenproblem for a symmetric matrix, any of the methods discussed in this chapter can be used to evaluate the generalized eigenpairs of a symmetric pair. The matrices in statistical applications for which the generalized eigenvalues are required are often symmetric pairs. For example, Roy's maximum root statistic, which is used in multivariate analysis, is a generalized eigenvalue of two Wishart matrices.

The generalized eigenvalues of a pair that is not symmetric are more difficult to evaluate. The approach of forming upper Hessenberg matrices, as in the QR method, is also used for generalized eigenvalues. We will not discuss this method here but instead refer the reader to Watkins (2002) for a description of the method, which is called the QZ algorithm.

7.7 Singular Value Decomposition

The standard algorithm for computing the singular value decomposition

$$A = UDV^{\mathrm{T}}$$

is due to Golub and Reinsch (1970) and is built on ideas of Golub and Kahan (1965). The first step in the Golub-Reinsch algorithm for the singular value decomposition of the $n \times m$ matrix A is to reduce A to upper bidiagonal form:

$$A^{(0)} = \begin{bmatrix} X & X & 0 & \cdots & 0 & 0 \\ 0 & X & X & \cdots & 0 & 0 \\ 0 & 0 & X & \cdots & 0 & 0 \\ & & & \ddots & \ddots & \\ 0 & 0 & 0 & \cdots & X & X \\ 0 & 0 & 0 & \cdots & 0 & X \\ 0 & 0 & 0 & \cdots & 0 & 0 \\ \vdots & \vdots & \vdots & \cdots & \vdots & \vdots \\ 0 & 0 & 0 & \cdots & 0 & 0 \end{bmatrix}.$$

We assume $n \geq m$. (If this is not the case, we merely use A^{T}.) This algorithm is basically a factored form of the QR algorithm for the eigenvalues of $A^{(0)\mathrm{T}}A^{(0)}$, which would be symmetric and tridiagonal.

The Golub-Reinsch method produces a sequence of upper bidiagonal matrices, $A^{(0)}, A^{(1)}, A^{(2)}, \ldots$, which converges to the diagonal matrix D. (Each of these has a zero submatrix below the square submatrix.) Similar to the QR method for eigenvalues, the transformation from $A^{(j)}$ to $A^{(j+1)}$ is effected by a sequence of orthogonal transformations,

$$\begin{aligned} A^{(j+1)} &= R_{m-2}^{\mathrm{T}} R_{m-3}^{\mathrm{T}} \cdots R_0^{\mathrm{T}} A^{(j)} T_0 T_1 \cdots T_{m-2} \\ &= R^{\mathrm{T}} A^{(j)} T, \end{aligned}$$

which first introduces a nonzero entry below the diagonal (T_0 does this) and then chases it down the diagonal. After T_0 introduces a nonzero entry in the $(2,1)$ position, R_0^{T} annihilates it and produces a nonzero entry in the $(1,3)$ position; T_1 annihilates the $(1,3)$ entry and produces a nonzero entry in the $(3,2)$ position, which R_1^{T} annihilates, and so on. Each of the R_ks and T_ks are Givens transformations, and, except for T_0, it should be clear how to form them.

If none of the elements along the main diagonal or the diagonal above the main diagonal is zero, then T_0 is chosen as the Givens transformation such that T_0^{T} will annihilate the second element in the vector

$$(a_{11}^2 - \sigma_1, \ a_{11}a_{12}, \ 0, \ \cdots, \ 0),$$

where σ_1 is the eigenvalue of the lower right-hand 2×2 submatrix of $A^{(0)\mathrm{T}}A^{(0)}$ that is closest in value to the (m, m) element of $A^{(0)\mathrm{T}}A^{(0)}$. This is easy to compute (see Exercise 7.6).

If an element along the main diagonal or the diagonal above the main diagonal is zero, we must proceed slightly differently. (Remember that for purposes of computations "zero" generally means "near zero"; that is, to within some set tolerance.)

If an element above the main diagonal is zero, the bidiagonal matrix is separated at that value into a block diagonal matrix, and each block (which is bidiagonal) is treated separately.

If an element on the main diagonal, say a_{kk}, is zero, then a singular value is zero. In this case, we apply a set of Givens transformations from the left. We first use G_1, which differs from the identity only in rows and columns k and $k+1$, to annihilate the $(k, k+1)$ entry and introduce a nonzero in the $(k, k+2)$ position. We then use G_2, which differs from the identity only in rows and columns k and $k+2$, to annihilate the $(k, k+2)$ entry and introduce a nonzero in the $(k, k+3)$ position. Continuing this process, we form a matrix of the form

$$\begin{bmatrix} X & X & 0 & 0 & 0 & 0 & 0 & 0 \\ 0 & X & X & 0 & 0 & 0 & 0 & 0 \\ 0 & 0 & X & Y & 0 & 0 & 0 & 0 \\ 0 & 0 & 0 & 0 & 0 & 0 & 0 & 0 \\ 0 & 0 & 0 & 0 & X & X & 0 & 0 \\ 0 & 0 & 0 & 0 & 0 & X & X & 0 \\ 0 & 0 & 0 & 0 & 0 & 0 & X & X \\ 0 & 0 & 0 & 0 & 0 & 0 & 0 & X \\ \vdots & \vdots & \vdots & \cdots & & \vdots & \vdots & \vdots & \vdots \\ 0 & 0 & 0 & 0 & 0 & 0 & 0 & 0 \end{bmatrix}.$$

The Y in this matrix (in position $(k-1,\ k)$) is then chased up the upper block consisting of the first k rows and columns of the original matrix by using Givens transformations applied from the right. This then yields two block bidiagonal matrices (and a 1×1 0 matrix). We operate on the individual blocks as before.

After the steps have converged to yield a diagonal matrix, $\widetilde{D}$, all of the Givens matrices applied from the left are accumulated into a single matrix and all from the right are accumulated into a single matrix to yield a decomposition

$$A = \widetilde{U}\widetilde{D}\widetilde{V}^{\mathrm{T}}.$$

There is one last thing to do. The elements of $\widetilde{D}$ may not be nonnegative. This is easily remedied by postmultiplying by a diagonal matrix G that is the same as the identity except for having a -1 in any position corresponding to a negative value in $\widetilde{D}$. In addition, we generally form the singular value decomposition is such a way that the elements in D are nonincreasing. The entries in $\widetilde{D}$ can be rearranged by a permutation matrix E_π so they are in nonincreasing order. So we have

$$D = E_\pi^{\mathrm{T}}\widetilde{D}GE_\pi,$$

and the final decomposition is

$$\begin{aligned} A &= \widetilde{U}E_\pi GDE_\pi^{\mathrm{T}}\widetilde{V}^{\mathrm{T}} \\ &= UDV^{\mathrm{T}}. \end{aligned}$$

If $n \geq \frac{5}{3}m$, a modification of this algorithm by Chan (1982a, b) is more efficient than the standard Golub-Reinsch method.

Exercises

7.1. Simple matrices and the power method.
   a) Let A be an $n \times n$ matrix whose elements are generated independently (but not necessarily identically) from real-valued continuous distributions. What is the probability that A is simple?
   b) Under the same conditions as in Exercise 7.1a, and with $n \geq 3$, what is the probability that $|c_{n-2}| < |c_{n-1}| < |c_n|$, where c_{n-2}, c_{n-1}, and c_n are the three eigenvalues with the largest absolute values?
   c) Prove that the power method converges linearly if $|c_{n-2}| < |c_{n-1}| < |c_n|$, $b_{n-1} \neq 0$, and $b_n \neq 0$. (The bs are the coefficients in the expansion of $x^{(0)}$.)

   Hint: Substitute the expansion in equation (7.10) on page 246 into the expression for the convergence ratio in equation (7.11).
   d) Suppose A is simple and the elements of $x^{(0)}$ are generated independently (but not necessarily identically) from continuous distributions. What is the probability that the power method will converge linearly?

7.2. Consider the matrix
$$\begin{bmatrix} 4 & 1 & 2 & 3 \\ 1 & 5 & 3 & 2 \\ 2 & 3 & 6 & 1 \\ 3 & 2 & 1 & 7 \end{bmatrix}.$$

   a) Use the power method to determine the largest eigenvalue and an associated eigenvector of this matrix.
   b) Find a 3×3 matrix, as in equation (7.6), that has the same eigenvalues as the remaining eigenvalues of the matrix above.
   c) Using Givens transformations, reduce the matrix to upper Hessenberg form.

7.3. In the matrix
$$\begin{bmatrix} 2 & 1 & 0 & 0 \\ 1 & 5 & 2 & 0 \\ 3 & 2 & 6 & 1 \\ 0 & 0 & 1 & 8 \end{bmatrix},$$

   determine the Givens transformations to chase the 3 in the $(3, 1)$ position out of the matrix.

7.4. In the matrix

$$\begin{bmatrix} 2 & 1 & 0 & 0 \\ 3 & 5 & 2 & 0 \\ 0 & 0 & 6 & 1 \\ 0 & 0 & 0 & 8 \\ 0 & 0 & 0 & 0 \end{bmatrix},$$

determine the Givens transformations to chase the 3 in the $(2,1)$ position out of the matrix.

7.5. In the QR methods for eigenvectors and singular values, why can we not just use additional orthogonal transformations to triangularize the given matrix (instead of just forming a similar Hessenberg matrix, as in Section 7.4) or to diagonalize the given matrix (instead of just forming the bidiagonal matrix, as in Section 7.7)?

7.6. Determine the eigenvalue σ_1 (on page 254) used in forming the matrix T_0 for initiating the chase in the algorithm for the singular value decomposition. Express it in terms of $a_{m,m}$, $a_{m-1,m-1}$, $a_{m-1,m}$, and $a_{m-1,m-2}$.

## Part II

Applications in Data Analysis

8

Special Matrices and Operations Useful in Modeling and Data Analysis

In previous chapters, we defined a number of special matrices, such as symmetric matrices, banded matrices, elementary operator matrices, and so on. In this chapter, we will discuss some of these matrices in more detail and also introduce some other special matrices and data structures that are useful in statistics.

There are a number of special kinds of matrices that are useful in statistical applications. In statistical applications in which data analysis is the objective, the initial step is the representation of observational data in some convenient form, which often is a matrix. We discuss the representation of observations using matrices in Section 8.1. The matrices for operating on observational data or summarizing the data often have special structures and properties.

One of the most important properties of many matrices occurring in statistical data analysis is nonnegative or positive definiteness; this is the subject of Sections 8.3 and 8.4. Fitted values of a response variable that are associated with given values of covariates in linear models are often projections of the observations onto a subspace determined by the covariates. Projection matrices and Gramian matrices useful in linear models are considered in Sections 8.5 and 8.6. One of the most important properties of many matrices occurring in statistical modeling over time is irreducible nonnegativeness or positiveness; this is the subject of Section 8.7.

8.1 Data Matrices and Association Matrices

There are several ways that data can be organized for representation in the computer. We distinguish logical structures from computer-storage structures. Data structure in computers is an important concern and can greatly affect the efficiency of computer processing. We discuss some simple aspects of the organization for computer storage in Section 11.1, beginning on page 429. In the present section, we consider some general issues of logical organization and structure.

There are two important aspects of data in applications that we will not address here. One is metadata; that is, data about the data. Metadata includes names or labels associated with data, information about how and when the data were collected, information about how the data are stored in the computer, and so on. Another important concern in applications is missing data. In real-world applications it is common to have incomplete data. If the data are stored in some structure that naturally contains a cell or a region for the missing data, the computer representation of the dataset must contain some indication that the cell is empty. For numeric data, the convenient way of doing this is by using "not-a-number", or NaN (see page 386). We briefly discuss issues in handling missing data in Part III. We consider some effects of missing data on the estimation of matrices in Section 9.4.6.

8.1.1 Flat Files

If several features or attributes are observed on each of several entities, a convenient way of organizing the data is as a two-dimensional array with each column corresponding to a specific feature and each row corresponding to a specific observational entity. In the field of statistics, data for the features are stored in "variables", the entities are called "observational units", and a row of the array is called an "observation" (see Figure 8.1).

	Var 1	Var 2	...	Var m
Obs 1	x	x	...	x
Obs 2	x	x	...	x
$\vdots$	$\vdots$	$\vdots$	...	$\vdots$
Obs n	x	x	...	x

Fig. 8.1. Data Appropriate for Representation in a Flat File

The data may be various types of objects, such as names, real numbers, numbers with associated measurement units, sets, vectors, and so on. If the data are represented as real numbers, the data array is a matrix. (Note again our use of the word "matrix"; not just any rectangular array is a matrix in the sense used in this book.) Other types of data can often be made equivalent to a matrix in an intuitive manner.

The flat file arrangement emphasizes the relationships of the data both *within* an observational unit or row and *within* a variable or column. Simple operations on the data matrix may reveal relationships *among* observational units or *among* variables.

8.1.2 Graphs and Other Data Structures

If the numbers of measurements on the observational units varies or if the interest is primarily in simple relationships among observational units or among variables, the flat file structure may not be very useful. Sometimes a graph structure can be used advantageously.

A *graph* is a nonempty set V of points, called *vertices*, together with a set E of unordered pairs of elements of V, called *edges*. (Other definitions of "graph" allow the null set to be a graph.) If we let $\mathcal{G}$ be a graph, we represent it as (V, E). We often represent the set of vertices as $V(\mathcal{G})$ and the set of edges as $E(\mathcal{G})$. An edge is said to be *incident* on each vertex in the edge. The number of vertices (that is, the cardinality of V) is the *order of the graph*, and the number of edges, the cardinality of E, is the *size of the graph*.

An edge in which the two vertices are the same is called a *loop*. (A *simple graph* is sometimes defined as a graph with no loops; that is, one in which each edge contains only pairs of distinct elements.)

A *path* or *walk* is a sequence of edges, $e_1, \ldots, e_n$, such that for $i \geq 2$ one vertex in e_i is a vertex in edge e_{i-1}. Alternatively, a path or walk is defined as a sequence of vertices with common edges.

A graph such that there is a path that includes any pair of vertices is said to be *connected*.

A graph with more than one vertex such that all possible pairs of vertices occur as edges is a *complete graph*.

A *closed path* or *closed walk* is a path such that a vertex in the first edge (or the first vertex in the alternate definition) is in the last edge (or the last vertex).

A *cycle* is a closed path in which all vertices occur exactly twice (or in the alternate definition, in which all vertices except the first and the last are distinct). A graph with no cycles is said to be *acyclic*. An acyclic graph is also called a *tree*. Trees are used extensively in statistics to represent clusters.

The number of edges that contain a given vertex (that is, the number of edges incident on the vertex v) denoted by $d(v)$ is the *degree of the vertex*.

A vertex with degree 0 is said to be *isolated*.

We see immediately that the sum of the degrees of all vertices equals twice the number of edges, that is,

$$\sum d(v_i) = 2\#(E).$$

The sum of the degrees hence must be an even number.

A *regular graph* is one for which $d(v_i)$ is constant for all vertices v_i; more specifically, a graph is *k-regular* if $d(v_i) = k$ for all vertices v_i.

The natural data structure for a graph is a pair of lists, but a graph is often represented graphically (no pun!) as in Figure 8.2, which shows a graph with five vertices seven edges. While a matrix is usually not an appropriate structure for representing raw data from a graph, there are various types of matrices that are useful for studying the data represented by the graph, which we will discuss in Section 8.8.7.

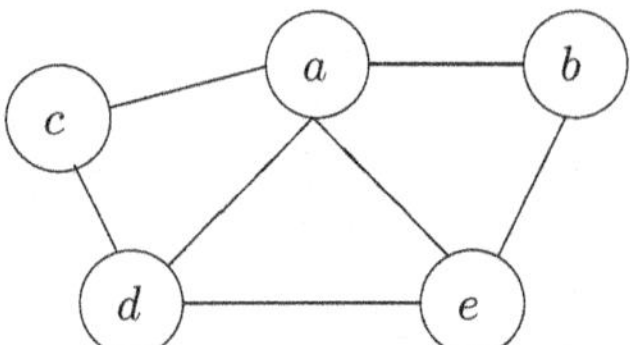

Fig. 8.2. A Simple Graph

If $\mathcal{G}$ is the graph represented in Figure 8.2, the vertices are $V(\mathcal{G}) = \{a, b, c, d, e\}$ and the edges are $E(\mathcal{G}) = \{(a,b), (a,c), (a,d), (a,e), (b,e), (c,d), (d,e)\}$.

The presence of an edge between two vertices can indicate the existence of a relationship between the objects represented by the vertices. The graph represented in Figure 8.2 may represent five observational units for which our primary interest is in their relationships with one another. For example, the observations may be authors of scientific papers, and an edge between two authors may represent the fact that the two have been coauthors on some paper.

The same information represented in the 5-order graph of Figure 8.2 may be represented in a 5×5 rectangular array, as in Figure 8.3.

	a	b	c	d	e
a		Y	Y	Y	Y
b	Y				Y
c	Y			Y	
d	Y		Y		Y
e	Y	Y		Y	

Fig. 8.3. An Alternate Representation

In the graph represented in Figure 8.2, there are no isolated vertices and the graph is connected. (Note that a graph with no isolated vertices is not necessarily connected.) The graph represented in Figure 8.2 is not complete because, for example, there is no edge that contains vertices c and e. The graph is cyclic because of the closed path (defined by vertices) (c, d, e, b, a, c). Note that the closed path (c, d, a, e, b, a, c) is not a cycle.

This use of a graph immediately suggests various extensions of a basic graph. For example, E may be a multiset, with multiple instances of edges containing the same two vertices, perhaps, in the example above, representing multiple papers in which the two authors are coauthors. A graph in which E is a multiset is called a *multigraph*. Instead of just the presence or absence of edges between vertices, a *weighted graph* may be more useful; that is, one in which a real number is associated with a pair of vertices to represent the strength of the relationship, not just presence or absence, between the two vertices. A degenerate weighted graph (that is, an unweighted graph as discussed above) has weights of 0 or 1 between all vertices. A multigraph is a weighted graph in which the weights are restricted to nonnegative integers. Although the data in a weighted graph carry much more information than a graph with only its edges, or even a multigraph that allows strength to be represented by multiple edges, the simplicity of a graph sometimes recommends its use even when there are varying degrees of strength of relationships. A standard approach in applications is to set a threshold for the strength of relationship and to define an edge only when the threshold is exceeded.

Adjacency Matrix; Connectivity Matrix

The connections between vertices in the graphs shown in Figure 8.2 or in Figure 8.4 can be represented in an association matrix called an *adjacency matrix*, a *connectivity matrix*, or an *incidence matrix* to represent edges between vertices, as shown in equation (8.1). (The terms "adjacency", "connectivity", and "incidence" are synonymous. "Adjacency" is perhaps the most commonly used term, but I will naturally use both that term and "connectivity" because of the connotative value of the latter term.) The graph, $\mathcal{G}$, represented in Figure 8.2 has the symmetric adjacency matrix

$$C(\mathcal{G}) = \begin{bmatrix} 0 & 1 & 1 & 1 & 1 \\ 1 & 0 & 0 & 0 & 1 \\ 1 & 0 & 0 & 1 & 0 \\ 1 & 0 & 1 & 0 & 1 \\ 1 & 1 & 0 & 1 & 0 \end{bmatrix}. \tag{8.1}$$

We often use this type of notation; a symbol represents a particular graph, and other objects that relate to the graph make use of that symbol. There is no difference in the connectivity matrix and a table such as in Figure 8.3 except for the metadata.

The diagonal elements of the adjacency matrix for a simple graph (one with no loops) are all 0s.

The relationship can obviously be defined in the other direction; that is, given an $n \times n$ symmetric matrix A, we define the *graph of the matrix* as the graph with n vertices and edges between vertices i and j if $a_{ij} \neq 0$. We often denote the graph of the matrix A by $\mathcal{G}(A)$.

Generally we restrict the elements of the connectivity matrix to be 1 or 0 to indicate only presence or absence of a connection, but not to indicate strength of the connection. In this case, a connectivity matrix is a nonnegative matrix; that is, all of its elements are nonnegative. We indicate that a matrix A is nonnegative by

$$A \geq 0.$$

We discuss the notation and properties of nonnegative (and positive) matrices in Section 8.7.

Digraphs

Another extension of a basic graph is one in which the relationship may not be the same in both directions. This yields a *digraph*, or "directed graph", in which the edges are ordered pairs called directed edges. The vertices in a digraph have two kinds of degree, an *indegree* and an *outdegree*, with the obvious meanings.

Although in this book we are more interested in statistical relationships, the simplest applications of digraphs are for representing networks. Consider, for example, the digraph represented by the network in Figure 8.4. This is a network with five vertices, perhaps representing cities, and directed edges between some of the vertices. The edges could represent airline connections between the cities; for example, there are flights from x to u and from u to x, and from y to z, but not from z to y.

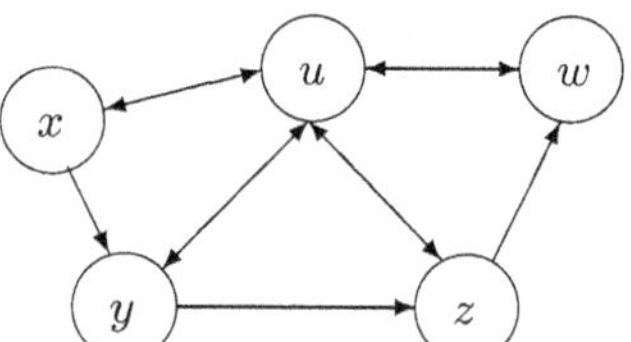

Fig. 8.4. A Simple Digraph

Figure 8.4 represents a digraph with order 5 (there are five vertices) and size 11 (eleven directed edges). A sequence of edges, $e_1, \ldots, e_n$, constituting a path in a digraph must be such that for $i \geq 2$ the first vertex in e_i is the second vertex in edge e_{i-1}. For example, the sequence x, y, z, w, u, x in the graph of Figure 8.4 is a path (in fact, a cycle) but the sequence x, u, w, z, y, x is not a path.

The connectivity matrix for the digraph in Figure 8.4 with nodes ordered as u, w, x, y, z is

$$C = \begin{bmatrix} 0 & 1 & 1 & 1 & 1 \\ 1 & 0 & 0 & 0 & 0 \\ 1 & 0 & 0 & 1 & 0 \\ 1 & 0 & 0 & 0 & 1 \\ 1 & 1 & 0 & 0 & 0 \end{bmatrix}. \tag{8.2}$$

A connectivity matrix for a (nondirected) graph is symmetric, but for a digraph it is not necessarily symmetric. Given an $n \times n$ matrix A, we define the *digraph of the matrix* as the digraph with n vertices and edges from vertex i to j if $a_{ij} \neq 0$. We use the same notation for a digraph as we used above for a graph, $\mathcal{G}(A)$.

In statistical applications, graphs are used for representing symmetric associations. Digraphs are used for representing asymmetric associations or one-way processes such as a stochastic process.

In a simple digraph, the edges only indicate the presence or absence of a relationship, but just as in the case of a simple graph, we can define a weighted digraph by associating nonnegative numbers with each directed edge.

Graphical modeling is useful type for analyzing relationships between elements of a collection of sets. For example, in an analysis of internet traffic, profiles of users may be constructed based on the set of web sites each user visits in relation to the sets visited by other users. For this kind of application, an intersection graph may be useful. An *intersection graph*, for a given collection of sets $\mathcal{S}$, is a graph whose vertices correspond to the sets in $\mathcal{S}$ and whose edges between any two sets have a common element.

The word "graph" is often used without qualification to mean any of these types.

Connectivity of Digraphs

There are two kinds of connected digraphs. A digraph such that there is a (directed) path that includes any pair of vertices is said to be *strongly connected*. A digraph such that there is a path without regard to the direction of any edge that includes any pair of vertices is said to be *weakly connected*. The digraph shown in Figure 8.4 is strongly connected. The digraph shown in Figure 8.5 is weakly connected but not strongly connected.

A digraph that is not weakly connected must have two sets of nodes with no edges between any nodes in one set and any nodes in the other set.

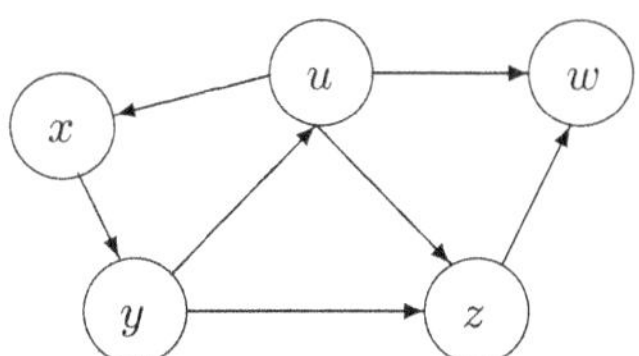

Fig. 8.5. A Digraph that Is Not Strongly Connected

The connectivity matrix of the digraph in Figure 8.5 is

$$C = \begin{bmatrix} 0 & 1 & 1 & 0 & 1 \\ 0 & 0 & 0 & 0 & 0 \\ 0 & 0 & 0 & 1 & 0 \\ 1 & 0 & 0 & 0 & 1 \\ 0 & 1 & 0 & 0 & 0 \end{bmatrix}. \tag{8.3}$$

The matrix of a digraph that is not strongly connected can always be reduced to a special block upper triangular form by row and column permutations; that is, if the digraph $\mathcal{G}$ is not strongly connected, then there exists a permutation matrix E_π such that

$$E_\pi^{\mathrm{T}} A(\mathcal{G}) E_\pi = \begin{bmatrix} B_{11} & B_{12} \\ 0 & B_{22} \end{bmatrix}, \tag{8.4}$$

where B_{11} and B_{22} are square. Such a transformation is called a *symmetric permutation*.

Later we will formally prove this relationship between strong connectivity and this reduced form of the matrix, but first we consider the matrix in equation (8.3). If we interchange the second and fourth columns and rows, we get the reduced form

$$E_{24}^{\mathrm{T}} C E_{24} = \left[\begin{array}{ccc|cc} 0 & 0 & 1 & 1 & 1 \\ 1 & 0 & 0 & 0 & 1 \\ 0 & 1 & 0 & 0 & 0 \\ \hline 0 & 0 & 0 & 0 & 0 \\ 0 & 0 & 0 & 1 & 0 \end{array}\right].$$

Irreducible Matrices

Any nonnegative square matrix that can be permuted into the form in equation (8.4) with square diagonal submatrices is said to be *reducible*; a matrix that cannot be put into that form is *irreducible*. We also use the terms reducible and irreducible to refer to the graph itself.

Strong Connectivity of Digraphs and Irreducibility of Matrices

A nonnegative matrix is irreducible if and only if its digraph is strongly connected. Stated another way, a digraph is not strongly connected if and only if its matrix is reducible.

To see this, first consider a reducible matrix. In its reduced form of equation (8.4), none of the nodes corresponding to the last rows have directed edges leading to any of the nodes corresponding to the first rows; hence, the digraph is not strongly connected.

Now, assume that a given digraph $\mathcal{G}$ is not strongly connected. In that case, there is some node, say the ith node, from which there is no directed path to some other node. Assume that there are $m - 1$ nodes that *can* be reached from node i. If $m = 1$, then we have a trivial partitioning of the $n \times n$ connectivity in which B_{11} of equation (8.4) is $(n - 1) \times (n - 1)$ and B_{22} is a 1×1 0 matrix (that is, 0_1). If $m \geq 1$, perform symmetric permutations so that the row corresponding to node i and all other $m - 1$ nodes are the last m rows of the permuted connectivity matrix. In this case, the first $n - k$ elements in each of those rows must be 0. To see that this must be the case, let $k > n - m$ and $j \leq n - m$ and assume that the element in the $(k, j)^{\mathrm{th}}$

position is nonzero. In that case, there is a path from node i to node k to node j, which is in the set of nodes not reachable from node i; hence the $(k, j)^{\text{th}}$ element (in the permuted matrix) must be 0. The submatrix corresponding to B_{11} is $n - m \times n - m$, and that corresponding to B_{22} is $m \times m$. These properties also hold for connectivity matrices with simple loops (with 1s on the diagonal) and for an augmented connectivity matrix (see page 314).

Reducibility plays an important role in the analysis of Markov chains (see Section 9.7.1).

8.1.3 Probability Distribution Models

Many phenomena are best characterized in terms of a probability distribution. Data in rows of flat files are often assumed to be realizations of vector random variables, some elements of which may have a degenerate distribution (that is, the elements in some columns of the data matrix may be considered to be fixed rather than random). The data in one row are often considered independent of the data in another row. Statistical data analysis is generally concerned with studying various models of relationships among the elements of the vector random variables. For example, the familiar linear regression model relates one variable (one column) to a linear combination of other variables plus a translation and random noise.

A *random graph* of fixed order is a discrete probability space over all possible graphs of that order. For a graph of order n, there are $2^{\binom{n}{2}}$ possible graphs. Asymptotic properties of the probability distribution refer to the increase of the order without limit. Occasionally it is useful to consider the order of the graph to be random also. If the order is unrestricted, the sample space for a random graph of random order is infinite but countable. The number of digraphs of order n is $4^{\binom{n}{2}}$.

Random graphs have many uses in the analysis of large systems of interacting objects; for example, a random intersection graph may be used to make inferences about the clustering of internet users based on the web sites they visit.

8.1.4 Association Matrices

In data analysis, the interesting questions usually involve the relationships among the variables or among the observational units. Matrices formed from the original data matrix for the purpose of measuring these relationships are called *association matrices*. There are basically two types: similarity and dissimilarity matrices. The variance-covariance matrix, which we discuss in Section 8.6.3, is an example of an association matrix that measures similarity. We discuss dissimilarity matrices in Section 8.6.6 and in Section 8.8.7 discuss a type of similarity matrix for data represented in graphs.

In addition to the distinction between similarity and dissimilarity association matrices, we may identify two types of association matrices based on whether the relationships of interest are among the rows (observations) or among the columns (variables or features). In applications, dissimilarity relationships among rows tend to be of more interest, and similarity relationships among columns are usually of more interest. (The applied statistician may think of clustering, multidimensional scaling, or Q factor analysis for the former and correlation analysis, principal components analysis, or factor analysis for the latter.)

8.2 Symmetric Matrices

Most association matrices encountered in applications are real and symmetric. Because real symmetric matrices occur so frequently in statistical applications and because such matrices have so many interesting properties, it is useful to review some of those properties that we have already encountered and to state some additional properties.

First, perhaps, we should iterate a trivial but important fact: the product of symmetric matrices is not, in general, symmetric. A power of a symmetric matrix, however, is symmetric.

We should also emphasize that some of the special matrices we have discussed are assumed to be symmetric because, if they were not, we could define equivalent symmetric matrices. This includes positive definite matrices and more generally the matrices in quadratic forms.

For convenience, here we list some of the important properties of symmetric matrices, many of which concern their eigenvalues. In the following, let A be a symmetric matrix:

- If A is nonsingular, then A^{-1} is also symmetric because $(A^{-1})^{\mathrm{T}} = (A^{\mathrm{T}})^{-1} = A^{-1}$.
- If k is any positive integer, A^k is symmetric.
- If A is nonsingular (so that A^k is defined for nonpositive integers), A^k is symmetric for any integer k.
- All eigenvalues of a (real) symmetric matrix are real (see page 110).
- A is diagonalizable (or simple), and in fact A is orthogonally diagonalizable; that is, it has an orthogonally similar canonical factorization, $A = VCV^{\mathrm{T}}$ (see page 120).
- A has the spectral decomposition $A = \sum_i c_i v_i v_i^{\mathrm{T}}$, where the c_i are the eigenvalues and v_i are the corresponding eigenvectors (see page 121).
- A power of A has the spectral decomposition $A^k = \sum_i c_i^k v_i v_i^{\mathrm{T}}$.
- Any quadratic form $x^{\mathrm{T}}Ax$ can be expressed as $\sum_i b_i^2 c_i$, where the b_i are elements in the vector $V^{-1}x$.
- We have

$$\max_{x \neq 0} \frac{x^{\mathrm{T}} A x}{x^{\mathrm{T}} x} = \max\{c_i\}$$

(see page 122). If A is nonnegative definite, this is the spectral radius $\rho(A)$.

- For the L_2 norm of the symmetric matrix A, we have

$$\|A\|_2 = \rho(A).$$

- For the Frobenius norm of the symmetric matrix A, we have

$$\|A\|_F = \sqrt{\sum c_i^2}.$$

This follows immediately from the fact that A is diagonalizable, as do the facts that

$$\text{tr}(A) = \sum c_i$$

and

$$|A| = \prod c_i$$

(see equations (3.179) and (3.180) on page 110).

Approximation of Symmetric Matrices and an Important Inequality

In Section 3.10, we considered the problem of approximating a given matrix by another matrix of lower rank. There are other situations in statistics in which we need to approximate one matrix by another one. In data analysis, this may be because our given matrix arises from poor observations and we know the "true" matrix has some special properties not possessed by the given matrix computed from the data. A familiar example is a sample variance-covariance matrix computed from incomplete data (see Section 9.4.6). Other examples in statistical applications occur in the simulation of random matrices (see Gentle, 2003, Section 5.3.3). In most cases of interest, the matrix to be approximated is a symmetric matrix.

Consider the difference of two symmetric $n \times n$ matrices, A and $\widetilde{A}$; that is,

$$E = A - \widetilde{A}. \tag{8.5}$$

The matrix of the differences, E, is also symmetric. We measure the "closeness" of A and $\widetilde{A}$ by some norm of E.

The Hoffman-Wielandt theorem gives a lower bound on the Frobenius norm of E in terms of the differences of the eigenvalues of A and $\widetilde{A}$: if the eigenvalues of A are $c_1, \ldots c_n$ and the eigenvalues of $\widetilde{A}$ are $\tilde{c}_1, \ldots \tilde{c}_n$, each set being arranged in nonincreasing order, we have

$$\sum_{i=1}^{n} (c_i - \tilde{c}_i)^2 \leq \|E\|_F^2. \tag{8.6}$$

This fact was proved by Hoffman and Wielandt (1953) using techniques from linear programming. Wilkinson (1965) gives a simpler proof (which he attributes to Wallace Givens) along the following lines.

Because A, $\widetilde{A}$, and E are symmetric, they are all orthogonally diagonalizable. Let the diagonal factorizations of A and E, respectively, be VCV^{T} and $U\mathrm{diag}((e_1,\ldots,e_n))U^{\mathrm{T}}$, where $e_1,\ldots e_n$ are the eigenvalues of E in nonincreasing order. Hence, we have

$$
\begin{aligned}
U\mathrm{diag}((e_1,\ldots,e_n))U^{\mathrm{T}} &= U(A - \widetilde{A})U^{\mathrm{T}} \\
&= U(VCV^{\mathrm{T}} - \widetilde{A})U^{\mathrm{T}} \\
&= UV(C - V^{\mathrm{T}}\widetilde{A}V)V^{\mathrm{T}}U^{\mathrm{T}}.
\end{aligned}
$$

Taking norms of both sides, we have

$$
\sum_{i=1}^{n} e_i^2 = \|C - V^{\mathrm{T}}\widetilde{A}V\|^2. \tag{8.7}
$$

(All norms in the remainder of this section will be the Frobenius norm.) Now, let

$$
f(Q) = \|C - Q^{\mathrm{T}}\widetilde{A}Q\|^2 \tag{8.8}
$$

be a function of any $n \times n$ orthogonal matrix, Q. (Equation (8.7) yields $f(V) = \sum e_i^2$.) To arrive at inequality (8.6), we show that this function is bounded below by the sum of the differences in the squares of the elements of C (which are the eigenvalues of A) and the eigenvalues of $Q^{\mathrm{T}}\widetilde{A}Q$ (which are the eigenvalues of the matrix approximating A).

Because the elements of Q are bounded, $f(\cdot)$ is bounded, and because the set of orthogonal matrices is compact (see page 105) and $f(\cdot)$ is continuous, $f(\cdot)$ must attain its lower bound, say l. To simplify the notation, let

$$
X = Q^{\mathrm{T}}\widetilde{A}Q.
$$

Now suppose that there are r distinct eigenvalues of A (that is, the diagonal element in C):

$$
d_1 > \cdots > d_r.
$$

We can write C as $\mathrm{diag}(d_i I_{m_i})$, where m_i is the multiplicity of d_i. We now partition $Q^{\mathrm{T}}\widetilde{A}Q$ to correspond to the partitioning of C represented by $\mathrm{diag}(d_i I_{m_i})$:

$$
X = \begin{bmatrix} X_{11} & \cdots & X_{1r} \\ \vdots & \vdots & \vdots \\ X_{r1} & \cdots & X_{rr} \end{bmatrix}. \tag{8.9}
$$

In this partitioning, the diagonal blocks, X_{ii}, are $m_i \times m_i$ symmetric matrices. The submatrix X_{ij}, is an $m_i \times m_j$ matrix.

We now proceed in two steps to show that in order for $f(Q)$ to attain its lower bound l, X must be diagonal. First we will show that when $f(Q) = l$, the submatrix X_{ij} in equation (8.9) must be null if $i \neq j$. To this end, let Q_∇ be such that $f(Q_\nabla) = l$, and assume the contrary regarding the corresponding

$X_\triangledown = Q_\triangledown^T \widetilde{A} Q_\triangledown$; that is, assume that in some submatrix $X_{ij_\triangledown}$ where $i \neq j$, there is a nonzero element, say $x_\triangledown$. We arrive at a contradiction by showing that in this case there is another X_0 of the form $Q_0^T \widetilde{A} Q_0$, where Q_0 is orthogonal and such that $f(Q_0) < f(Q_\triangledown)$.

To establish some useful notation, let p and q be the row and column, respectively, of $X_\triangledown$ where this nonzero element $x_\triangledown$ occurs; that is, $x_{pq} = x_\triangledown \neq 0$ and $p \neq q$ because x_{pq} is in $X_{ij_\triangledown}$. (Note the distinction between uppercase letters, which represent submatrices, and lowercase letters, which represent elements of matrices.) Also, because $X_\triangledown$ is symmetric, $x_{qp} = x_\triangledown$. Now let $a_\triangledown = x_{pp}$ and $b_\triangledown = x_{qq}$. We form Q_0 as $Q_\triangledown R$, where R is an orthogonal rotation matrix of the form G_{pq} in equation (5.12). We have, therefore, $\|Q_0^T \widetilde{A} Q_0\|^2 = \|R^T Q_\triangledown^T \widetilde{A} Q_\triangledown R\|^2 = \|Q_\triangledown^T \widetilde{A} Q_\triangledown\|^2$. Let a_0, b_0, and x_0 represent the elements of $Q_0^T \widetilde{A} Q_0$ that correspond to $a_\triangledown$, $b_\triangledown$, and $x_\triangledown$ in $Q_\triangledown^T \widetilde{A} Q_\triangledown$.

From the definition of the Frobenius norm, we have

$$f(Q_0) - f(Q_\triangledown) = 2(a_\triangledown - a_0)d_i + 2(b_\triangledown - b_0)d_j$$

because all other terms cancel. If the angle of rotation is θ, then

$$a_0 = a_\triangledown \cos^2 \theta - 2x_\triangledown \cos \theta \sin \theta + b_\triangledown \sin^2 \theta,$$
$$b_0 = a_\triangledown \sin^2 \theta - 2x_\triangledown \cos \theta \sin \theta + b_\triangledown \cos^2 \theta,$$

and so for a function h of θ we can write

$$\begin{aligned}
h(\theta) &= f(Q_0) - f(Q_\triangledown) \\
&= 2d_i((a_\triangledown - b_\triangledown) \sin^2 \theta + x_\triangledown \sin 2\theta) + 2d_j((b_\triangledown - b_0) \sin^2 \theta - x_\triangledown \sin 2\theta) \\
&= 2d_i((a_\triangledown - b_\triangledown) + 2d_j(b_\triangledown - b_0)) \sin^2 \theta + 2x_\triangledown (d_i - d_j) \sin 2\theta,
\end{aligned}$$

and so

$$\frac{\mathrm{d}}{\mathrm{d}\theta} h(\theta) = 2d_i((a_\triangledown - b_\triangledown) + 2d_j(b_\triangledown - b_0)) \sin 2\theta + 4x_\triangledown (d_i - d_j) \cos 2\theta.$$

The coefficient of $\cos 2\theta$, $4x_\triangledown (d_i - d_j)$, is nonzero because d_i and d_j are distinct, and $x_\triangledown$ is nonzero by the second assumption to be contradicted, and so the derivative at $\theta = 0$ is nonzero. Hence, by the proper choice of a direction of rotation (which effectively interchanges the roles of d_i and d_j), we can make $f(Q_0) - f(Q_\triangledown)$ positive or negative, showing that $f(Q_\triangledown)$ cannot be a minimum if some X_{ij} in equation (8.9) with $i \neq j$ is nonnull; that is, if $Q_\triangledown$ is a matrix such that $f(Q_\triangledown)$ is the minimum of $f(Q)$, then in the partition of $Q_\triangledown^T \widetilde{A} Q_\triangledown$ only the diagonal submatrices $X_{ii_\triangledown}$ can be nonnull:

$$Q_\triangledown^T \widetilde{A} Q_\triangledown = \mathrm{diag}(X_{11_\triangledown}, \ldots, X_{rr_\triangledown}).$$

The next step is to show that each $X_{ii_\triangledown}$ must be diagonal. Because it is symmetric, we can diagonalize it with an orthogonal matrix P_i as

$$P_i^{\mathrm{T}} X_{ii_{\nabla}} P_i = G_i.$$

Now let P be the direct sum of the P_i and form

$$P^{\mathrm{T}} C P - P^{\mathrm{T}} Q_{\nabla}^{\mathrm{T}} \widetilde{A} Q_{\nabla} P = \mathrm{diag}(d_1 I, \ldots, d_r I) - \mathrm{diag}(G_1, \ldots, G_r)$$
$$= C - P^{\mathrm{T}} Q_{\nabla}^{\mathrm{T}} \widetilde{A} Q_{\nabla} P.$$

Hence,

$$f(Q_{\nabla} P) = f(Q_{\nabla}),$$

and so the minimum occurs for a matrix $Q_{\nabla} P$ that reduces $\widetilde{A}$ to a diagonal form. The elements of the G_i must be the $\tilde{c}_i$ in some order, so the minimum of $f(Q)$, which we have denoted by $f(Q_{\nabla})$, is $\sum(c_i - \tilde{c}_{p_i})^2$, where the p_i are a permutation of $1, \ldots, n$. As the final step, we show $p_i = i$. We begin with p_1. Suppose $p_1 \neq 1$ but $p_s = 1$; that is, $\tilde{c}_1 \geq \tilde{c}_{p_1}$. Interchange p_1 and p_s in the permutation. The change in the sum $\sum(c_i - \tilde{c}_{p_i})^2$ is

$$(c_1 - \tilde{c}_1)^2 + (c_s - \tilde{c}_{p_s})^2 - (c_1 - \tilde{c}_{p_s})^2 - (c_s - \tilde{c}_1)^2 = -2(c_s - c_1)(\tilde{c}_{p_1} - \tilde{c}_1)$$
$$\leq 0;$$

that is, the interchange reduces the value of the sum. Similarly, we proceed through the p_i to p_n, getting $p_i = i$.

We have shown, therefore, that the minimum of $f(Q)$ is $\sum_{i=1}^{n}(c_i - \tilde{c}_i)^2$, where both sets of eigenvalues are ordered in nonincreasing value. From equation (8.7), which is $f(V)$, we have the inequality (8.6).

While an upper bound may be of more interest in the approximation problem, the lower bound in the Hoffman-Wielandt theorem gives us a measure of the goodness of the approximation of one matrix by another matrix. Chu (1991) describes various extensions and applications of the Hoffman-Wielandt theorem.

Normal Matrices

A real square matrix A is said to be *normal* if $A^{\mathrm{T}} A = A A^{\mathrm{T}}$. (In general, a square matrix is normal if $A^{\mathrm{H}} A = A A^{\mathrm{H}}$.) Normal matrices include symmetric (and Hermitian), skew symmetric (and Hermitian), and square orthogonal (and unitary) matrices.

There are a number of interesting properties possessed by normal matrices. One property, for example, is that eigenvalues of normal matrices are real. (This follows from properties 12 and 13 on page 110.) Another property of a normal matrix is its characterization in terms of orthogonal similarity to a diagonal matrix formed from its eigenvalues; a square matrix is normal if and only if it can be expressed in the form of equation (3.197), $A = VCV^{\mathrm{T}}$, which we derived for symmetric matrices.

The normal matrices of most interest to us are symmetric matrices, and so when we discuss properties of normal matrices, we will generally consider those properties only as they apply to symmetric matrices.

8.3 Nonnegative Definite Matrices; Cholesky Factorization

We defined nonnegative definite and positive definite matrices on page 70, and discussed some of their properties, particularly in Section 3.8.8. We have seen that these matrices have useful factorizations, in particular, the square root and the Cholesky factorization. In this section, we recall those definitions, properties, and factorizations.

A symmetric matrix A such that any quadratic form involving the matrix is nonnegative is called a *nonnegative definite matrix*. That is, a symmetric matrix A is a nonnegative definite matrix if, for any (conformable) vector x,

$$x^{\mathrm{T}} A x \geq 0. \tag{8.10}$$

(There is a related term, *positive semidefinite matrix*, that is not used consistently in the literature. We will generally avoid the term "semidefinite".)

We denote the fact that A is nonnegative definite by

$$A \succeq 0. \tag{8.11}$$

(Some people use the notation $A \geq 0$ to denote a nonnegative definite matrix, but we have decided to use this notation to indicate that each element of A is nonnegative; see page 48.)

There are several properties that follow immediately from the definition.

- The sum of two (conformable) nonnegative matrices is nonnegative definite.
- All diagonal elements of a nonnegative definite matrix are nonnegative. Hence, if A is nonnegative definite, $\mathrm{tr}(A) \geq 0$.
- Any square submatrix whose principal diagonal is a subset of the principal diagonal of a nonnegative definite matrix is nonnegative definite. In particular, any square principal submatrix of a nonnegative definite matrix is nonnegative definite.

It is easy to show that the latter two facts follow from the definition by considering a vector x with zeros in all positions except those corresponding to the submatrix in question. For example, to see that all diagonal elements of a nonnegative definite matrix are nonnegative, assume the (i, i) element is negative, and then consider the vector x to consist of all zeros except for a 1 in the ith position. It is easy to see that the quadratic form is negative, so the assumption that the (i, i) element is negative leads to a contradiction.

- A diagonal matrix is nonnegative definite if and only if all of the diagonal elements are nonnegative.

This must be true because a quadratic form in a diagonal matrix is the sum of the diagonal elements times the squares of the elements of the vector.

- If A is nonnegative definite, then $A_{-(i_1,\ldots,i_k)(i_1,\ldots,i_k)}$ is nonnegative definite.

Again, we can see this by selecting an x in the defining inequality (8.10) consisting of 1s in the positions corresponding to the rows and columns of A that are retained and 0s elsewhere.

By considering $x^{\mathrm{T}}C^{\mathrm{T}}ACx$ and $y = Cx$, we see that

- if A is nonnegative definite, and C is conformable for the multiplication, then $C^{\mathrm{T}}AC$ is nonnegative definite.

From equation (3.197) and the fact that the determinant of a product is the product of the determinants, we have that

- the determinant of a nonnegative definite matrix is nonnegative.

Finally, for the nonnegative definite matrix A, we have

$$a_{ij}^2 \leq a_{ii}a_{jj}, \tag{8.12}$$

as we see from the definition $x^{\mathrm{T}}Ax \geq 0$ and choosing the vector x to have a variable y in position i, a 1 in position j, and 0s in all other positions. For a symmetric matrix A, this yields the quadratic $a_{ii}y^2 + 2a_{ij}y + a_{jj}$. If this quadratic is to be nonnegative for all y, then the discriminant $4a_{ij}^2 - 4a_{ii}a_{jj}$ must be nonpositive; that is, inequality (8.12) must be true.

Eigenvalues of Nonnegative Definite Matrices

We have seen on page 124 that a real symmetric matrix is nonnegative (positive) definite if and only if all of its eigenvalues are nonnegative (positive).

This fact allows a generalization of the statement above: a triangular matrix is nonnegative (positive) definite if and only if all of the diagonal elements are nonnegative (positive).

The Square Root and the Cholesky Factorization

Two important factorizations of nonnegative definite matrices are the square root,

$$A = (A^{\frac{1}{2}})^2, \tag{8.13}$$

discussed in Section 5.9.1, and the Cholesky factorization,

$$A = T^{\mathrm{T}}T, \tag{8.14}$$

discussed in Section 5.9.2. If T is as in equation (8.14), the symmetric matrix $T + T^{\mathrm{T}}$ is also nonnegative definite, or positive definite if A is. The square root matrix is used often in theoretical developments, such as Exercise 4.5b for example, but the Cholesky factor is more useful in practice.

8.4 Positive Definite Matrices

An important class of nonnegative definite matrices are those that satisfy strict inequalities in the definition involving $x^\mathrm{T}Ax$. These matrices are called positive definite matrices and they have all of the properties discussed above for nonnegative definite matrices as well as some additional useful properties.

A symmetric matrix A is called a *positive definite matrix* if, for any (conformable) vector $x \neq 0$, the quadratic form is positive; that is,

$$x^\mathrm{T}Ax > 0. \tag{8.15}$$

We denote the fact that A is positive definite by

$$A \succ 0. \tag{8.16}$$

(Some people use the notation $A > 0$ to denote a positive definite matrix, but we have decided to use this notation to indicate that each element of A is positive.)

- A positive definite matrix is necessarily nonsingular. (We see this from the fact that no nonzero combination of the columns, or rows, can be 0.) Furthermore, if A is positive definite, then A^{-1} is positive definite. (We showed this is Section 3.8.8, but we can see it in another way: because for any $y \neq 0$ and $x = A^{-1}y$, we have $y^\mathrm{T}A^{-1}y = x^\mathrm{T}y = x^\mathrm{T}Ax > 0$.)

- A diagonally dominant symmetric matrix with positive diagonals is positive definite. The proof of this is Exercise 8.2.

The properties of nonnegative definite matrices noted above hold also for positive definite matrices, generally with strict inequalities. It is obvious that all diagonal elements of a positive definite matrix are positive. Hence, if A is positive definite, $\mathrm{tr}(A) > 0$. Furthermore, as above and for the same reasons, if A is positive definite, then $A_{-(i_1,\ldots,i_k)(i_1,\ldots,i_k)}$ is positive definite. In particular, any square submatrix whose principal diagonal is a subset of the principal diagonal of a positive definite matrix is positive definite, and furthermore, any square principal submatrix of a positive definite matrix is positive definite. Because a quadratic form in a diagonal matrix is the sum of the diagonal elements times the squares of the elements of the vector, a diagonal matrix is positive definite if and only if all of the diagonal elements are positive.

The definition yields a slightly stronger statement regarding the sums involving positive definite matrices than what we could conclude about nonnegative definite matrices:

- The sum of a positive definite matrix and a (conformable) nonnegative definite matrix is positive definite.

That is,

$$x^\mathrm{T}Ax > 0 \; \forall x \neq 0 \quad \text{and} \quad y^\mathrm{T}By \geq 0 \; \forall y \implies z^\mathrm{T}(A + B)z > 0 \; \forall z \neq 0. \tag{8.17}$$

We cannot conclude that the product of two positive definite matrices is positive definite, but we do have the useful fact that

- if A is positive definite, and C is of full rank and conformable for the multiplication AC, then $C^{\mathrm{T}}AC$ is positive definite (see page 89).

From equation (3.197) and the fact that the determinant of a product is the product of the determinants, we have that

- the determinant of a positive definite matrix is positive.

For the positive definite matrix A, we have, analogous to inequality (8.12),

$$a_{ij}^2 < a_{ii}a_{jj}, \qquad (8.18)$$

which we see using the same argument as for that inequality.

We have seen from the definition of positive definiteness and the distribution of multiplication over addition that the sum of a positive definite matrix and a nonnegative definite matrix is positive definite. We can define an ordinal relationship between positive definite and nonnegative definite matrices of the same size. If A is positive definite and B is nonnegative definite of the same size, we say A is *strictly greater than* B and write

$$A \succ B \qquad (8.19)$$

if $A - B$ is positive definite; that is, if $A - B \succ 0$.

We can form a *partial ordering* of nonnegative definite matrices of the same order based on this additive property. We say A is *greater than* B and write

$$A \succeq B \qquad (8.20)$$

if $A - B$ is either the 0 matrix or is nonnegative definite; that is, if $A - B \succeq 0$ (see Exercise 8.1a). The "strictly greater than" relation implies the "greater than" relation. These relations are *partial* in the sense that they do not apply to all pairs of nonnegative matrices; that is, there are pairs of matrices A and B for which neither $A \succeq B$ nor $B \succeq A$.

If $A \succ B$, we also write $B \prec A$; and if $A \succeq B$, we may write $B \preceq A$.

Principal Submatrices of Positive Definite Matrices

A sufficient condition for a symmetric matrix to be positive definite is that the determinant of each of the leading principal submatrices be positive. To see this, first let the $n \times n$ symmetric matrix A be partitioned as

$$A = \begin{bmatrix} A_{n-1} & a \\ a^{\mathrm{T}} & a_{nn} \end{bmatrix},$$

and assume that A_{n-1} is positive definite and that $|A| > 0$. (This is not the same notation that we have used for these submatrices, but the notation is convenient in this context.) From equation (3.147),

$$|A| = |A_{n-1}|(a_{nn} - a^{\mathrm{T}} A_{n-1}^{-1} a).$$

Because A_{n-1} is positive definite, $|A_{n-1}| > 0$, and so $(a_{nn} - a^{\mathrm{T}} A_{n-1}^{-1} a) > 0$; hence, the 1×1 matrix $(a_{nn} - a^{\mathrm{T}} A_{n-1}^{-1} a)$ is positive definite. That any matrix whose leading principal submatrices have positive determinants follows from this by induction, beginning with a 2×2 matrix.

The Convex Cone of Positive Definite Matrices

The class of all $n \times n$ positive definite matrices is a *convex cone* in $\mathrm{I\!R}^{n \times n}$ in the same sense as the definition of a convex cone of vectors (see page 14). If X_1 and X_2 are $n \times n$ positive definite matrices and $a, b \geq 0$, then $aX_1 + bX_2$ is positive definite so long as either $a \neq 0$ or $b \neq 0$.

This class is not closed under Cayley multiplication (that is, in particular, it is not a group with respect to that operation). The product of two positive definite matrices might not even be symmetric.

Inequalities Involving Positive Definite Matrices

Quadratic forms of positive definite matrices and nonnegative matrices occur often in data analysis. There are several useful inequalities involving such quadratic forms.

On page 122, we showed that if $x \neq 0$, for any symmetric matrix A with eigenvalues c_i,

$$\frac{x^{\mathrm{T}} A x}{x^{\mathrm{T}} x} \leq \max\{c_i\}. \tag{8.21}$$

If A is nonnegative definite, by our convention of labeling the eigenvalues, we have $\max\{c_i\} = c_1$. If the rank of A is r, the minimum nonzero eigenvalue is denoted c_r. Letting the eigenvectors associated with $c_1, \ldots, c_r$ be $v_1, \ldots, v_r$ (and recalling that these choices may be arbitrary in the case where some eigenvalues are not simple), by an argument similar to that used on page 122, we have that if A is nonnegative definite of rank r,

$$\frac{v_i^{\mathrm{T}} A v_i}{v_i^{\mathrm{T}} v_i} \geq c_r, \tag{8.22}$$

for $1 \leq i \leq r$.

If A is positive definite and x and y are conformable nonzero vectors, we see that

$$x^{\mathrm{T}} A^{-1} x \geq \frac{(y^{\mathrm{T}} x)^2}{y^{\mathrm{T}} A y} \tag{8.23}$$

by using the same argument as used in establishing the Cauchy-Schwarz inequality (2.10). We first obtain the Cholesky factor T of A (which is, of course, of full rank) and then observe that for every real number t

$$\left(tTy + T^{-\mathrm{T}}x\right)^{\mathrm{T}} \left(tTy + T^{-\mathrm{T}}x\right) \geq 0,$$

and hence the discriminant of the quadratic equation in t must be nonnegative:

$$4\left((Ty)^{\mathrm{T}}T^{-\mathrm{T}}x\right)^2 - 4\left(T^{-\mathrm{T}}x\right)^{\mathrm{T}}\left(T^{-\mathrm{T}} - x\right)(Ty)^{\mathrm{T}}Ty \leq 0.$$

The inequality (8.23) is used in constructing Scheffé simultaneous confidence intervals in linear models.

The Kantorovich inequality for positive numbers has an immediate extension to an inequality that involves positive definite matrices. The Kantorovich inequality, which finds many uses in optimization problems, states, for positive numbers $c_1 \geq c_2 \geq \cdots \geq c_n$ and nonnegative numbers $y_1, \ldots, y_n$ such that $\sum y_i = 1$, that

$$\left(\sum_{i=1}^{n} y_i c_i\right)\left(\sum_{i=1}^{n} y_i c_i^{-1}\right) \leq \frac{(c_1 + c_2)^2}{4c_1 c_2}.$$

Now let A be an $n \times n$ positive definite matrix with eigenvalues $c_1 \geq c_2 \geq \cdots \geq c_n > 0$. We substitute x^2 for y, thus removing the nonnegativity restriction, and incorporate the restriction on the sum directly into the inequality. Then, using the similar canonical factorization of A and A^{-1}, we have

$$\frac{\left(x^{\mathrm{T}}Ax\right)\left(x^{\mathrm{T}}A^{-1}x\right)}{(x^{\mathrm{T}}x)^2} \leq \frac{(c_1 + c_n)^2}{4c_1 c_n}. \tag{8.24}$$

This Kantorovich matrix inequality likewise has applications in optimization; in particular, for assessing convergence of iterative algorithms.

The left-hand side of the Kantorovich matrix inequality also has a lower bound,

$$\frac{\left(x^{\mathrm{T}}Ax\right)\left(x^{\mathrm{T}}A^{-1}x\right)}{(x^{\mathrm{T}}x)^2} \geq 1, \tag{8.25}$$

which can be seen in a variety of ways, perhaps most easily by using the inequality (8.23). (You were asked to prove this directly in Exercise 3.21.)

All of the inequalities (8.21) through (8.25) are sharp. We know that (8.21) and (8.22) are sharp by using the appropriate eigenvectors. We can see the others are sharp by using $A = I$.

There are several variations on these inequalities and other similar inequalities that are reviewed by Marshall and Olkin (1990) and Liu and Neudecker (1996).

8.5 Idempotent and Projection Matrices

An important class of matrices are those that, like the identity, have the property that raising them to a power leaves them unchanged. A matrix A such that

$$AA = A \tag{8.26}$$

is called an *idempotent matrix*. An idempotent matrix is square, and it is either singular or the identity matrix. (It must be square in order to be conformable for the indicated multiplication. If it is not singular, we have $A = (A^{-1}A)A = A^{-1}(AA) = A^{-1}A = I$; hence, an idempotent matrix is either singular or the identity matrix.) An idempotent matrix that is symmetric is called a *projection matrix*.

8.5.1 Idempotent Matrices

Many matrices encountered in the statistical analysis of linear models are idempotent. One such matrix is X^-X (see page 98 and Section 9.2.2). This matrix exists for any $n \times m$ matrix X, and it is square. (It is $m \times m$.)

Because the eigenvalues of A^2 are the squares of the eigenvalues of A, all eigenvalues of an idempotent matrix must be either 0 or 1.

Any vector in the column space of an idempotent matrix A is an eigenvector of A. (This follows immediately from $AA = A$.) More generally, if x and y are vectors in span(A) and a is a scalar, then

$$A(ax + y) = ax + y. \tag{8.27}$$

(To see this, we merely represent x and y as linear combinations of columns (or rows) of A and substitute in the equation.)

The number of eigenvalues that are 1 is the rank of an idempotent matrix. (Exercise 8.3 asks why this is the case.) We therefore have, for an idempotent matrix A,

$$\mathrm{tr}(A) = \mathrm{rank}(A). \tag{8.28}$$

Because the eigenvalues of an idempotent matrix are either 0 or 1, a symmetric idempotent matrix is nonnegative definite.

If A is idempotent and $n \times n$, then

$$\mathrm{rank}(I - A) = n - \mathrm{rank}(A). \tag{8.29}$$

We showed this in equation (3.155) on page 98. (Although there we were considering the special matrix A^-A, the only properties used were the idempotency of A^-A and the fact that $\mathrm{rank}(A^-A) = \mathrm{rank}(A)$.)

Equation (8.29) together with the diagonalizability theorem (equation (3.194)) implies that an idempotent matrix is diagonalizable.

If A is idempotent and V is an orthogonal matrix of the same size, then $V^{\mathrm{T}}AV$ is idempotent (whether or not V is a matrix that diagonalizes A) because

$$(V^{\mathrm{T}}AV)(V^{\mathrm{T}}AV) = V^{\mathrm{T}}AAV = V^{\mathrm{T}}AV. \tag{8.30}$$

If A is idempotent, then $(I - A)$ is also idempotent, as we see by multiplication. This fact and equation (8.29) have generalizations for sums of idempotent matrices that are parts of Cochran's theorem, which we consider below.

Although if A is idempotent so $(I - A)$ is also idempotent and hence is not of full rank (unless $A = 0$), for any scalar $a \neq -1$, $(I + aA)$ is of full rank, and

$$(I + aA)^{-1} = I - \frac{a}{a + 1}A, \tag{8.31}$$

as we see by multiplication.

On page 114, we saw that similar matrices are equivalent (have the same rank). For idempotent matrices, we have the converse: idempotent matrices of the same rank (and size) are similar (see Exercise 8.4).

If A_1 and A_2 are matrices conformable for addition, then $A_1 + A_2$ is idempotent if and only if $A_1 A_2 = A_2 A_1 = 0$. It is easy to see that this condition is sufficient by multiplication:

$$(A_1 + A_2)(A_1 + A_2) = A_1 A_1 + A_1 A_2 + A_2 A_1 + A_2 A_2 = A_1 + A_2.$$

To see that it is necessary, we first observe from the expansion above that $A_1 + A_2$ is idempotent only if $A_1 A_2 + A_2 A_1 = 0$. Multiplying this necessary condition on the left by A_1 yields

$$A_1 A_1 A_2 + A_1 A_2 A_1 = A_1 A_2 + A_1 A_2 A_1 = 0,$$

and multiplying on the right by A_1 yields

$$A_1 A_2 A_1 + A_2 A_1 A_1 = A_1 A_2 A_1 + A_2 A_1 = 0.$$

Subtracting these two equations yields

$$A_1 A_2 = A_2 A_1,$$

and since $A_1 A_2 + A_2 A_1 = 0$, we must have $A_1 A_2 = A_2 A_1 = 0$.

Symmetric Idempotent Matrices

Many of the idempotent matrices in statistical applications are symmetric, and such matrices have some useful properties.

Because the eigenvalues of an idempotent matrix are either 0 or 1, the spectral decomposition of a symmetric idempotent matrix A can be written as

$$V^{\mathrm{T}} A V = \mathrm{diag}(I_r, 0), \tag{8.32}$$

where V is a square orthogonal matrix and $r = \mathrm{rank}(A)$. (This is from equation (3.198) on page 120.)

For symmetric matrices, there is a converse to the fact that all eigenvalues of an idempotent matrix are either 0 or 1. If A is a symmetric matrix all of whose eigenvalues are either 0 or 1, then A is idempotent. We see this from the spectral decomposition of A, $A = V\mathrm{diag}(I_r, 0)V^\mathrm{T}$, and, with $C = \mathrm{diag}(I_r, 0)$, by observing

$$AA = VCV^\mathrm{T}VCV^\mathrm{T} = VCCV^\mathrm{T} = VCV^\mathrm{T} = A,$$

because the diagonal matrix of eigenvalues C contains only 0s and 1s.

If A is symmetric and p is any positive integer,

$$A^{p+1} = A^p \implies A \text{ is idempotent.} \tag{8.33}$$

This follows by considering the eigenvalues of A, $c_1, \ldots, c_n$. The eigenvalues of A^{p+1} are $c_1^{p+1}, \ldots, c_n^{p+1}$ and the eigenvalues of A^p are $c_1^p, \ldots, c_n^p$, but since $A^{p+1} = A^p$, it must be the case that $c_i^{p+1} = c_i^p$ for each $i = 1, \ldots, n$. The only way this is possible is for each eigenvalue to be 0 or 1, and in this case the symmetric matrix must be idempotent.

There are bounds on the elements of a symmetric idempotent matrix. Because A is symmetric and $A^\mathrm{T}A = A$,

$$a_{ii} = \sum_{j=1}^{n} a_{ij}^2; \tag{8.34}$$

hence, $0 \le a_{ii}$. Rearranging equation (8.34), we have

$$a_{ii} = a_{ii}^2 + \sum_{j \ne i} a_{ij}^2, \tag{8.35}$$

so $a_{ii}^2 \le a_{ii}$ or $0 \le a_{ii}(1 - a_{ii})$; that is, $a_{ii} \le 1$. Now, if $a_{ii} = 0$ or $a_{ii} = 1$, then equation (8.35) implies

$$\sum_{j \ne i} a_{ij}^2 = 0,$$

and the only way this can happen is if $a_{ij} = 0$ for all $j \ne i$. So, in summary, if A is an $n \times n$ symmetric idempotent matrix, then

$$0 \le a_{ii} \le 1 \text{ for } i = 1, \ldots, m, \tag{8.36}$$

and

$$\text{if } a_{ii} = 0 \text{ or } a_{ii} = 1, \text{ then } a_{ij} = a_{ji} = 0 \text{ for all } j \ne i. \tag{8.37}$$

Cochran's Theorem

There are various facts that are sometimes called *Cochran's theorem*. The simplest one concerns k symmetric idempotent $n \times n$ matrices, $A_1, \ldots, A_k$, such that

$$I_n = A_1 + \cdots + A_k. \tag{8.38}$$

Under these conditions, we have

$$A_i A_j = 0 \text{ for all } i \neq j. \tag{8.39}$$

We see this by the following argument. For an arbitrary j, as in equation (8.32), for some matrix V, we have

$$V^{\mathrm{T}} A_j V = \mathrm{diag}(I_r, 0),$$

where $r = \mathrm{rank}(A_j)$. Now

$$
\begin{aligned}
I_n &= V^{\mathrm{T}} I_n V \\
&= \sum_{i=1}^{k} V^{\mathrm{T}} A_i V \\
&= \mathrm{diag}(I_r, 0) + \sum_{i \neq j} V^{\mathrm{T}} A_i V,
\end{aligned}
$$

which implies

$$\sum_{i \neq j} V^{\mathrm{T}} A_i V = \mathrm{diag}(0, I_{n-r}). \tag{8.40}$$

Now, from equation (8.30), for each i, $V^{\mathrm{T}} A_i V$ is idempotent, and so from equation (8.36) the diagonal elements are all nonnegative, and hence equation (8.40) implies that for each $i \neq j$, the first r diagonal elements are 0. Furthermore, since these diagonal elements are 0, equation (8.37) implies that all elements in the first r rows and columns are 0. We have, therefore, for each $i \neq j$,

$$V^{\mathrm{T}} A_i V = \mathrm{diag}(0, B_i)$$

for some $(n - r) \times (n - r)$ symmetric idempotent matrix B_i. Now, for any $i \neq j$, consider $A_i A_j$ and form $V^{\mathrm{T}} A_i A_j V$. We have

$$
\begin{aligned}
V^{\mathrm{T}} A_i A_j V &= (V^{\mathrm{T}} A_i V)(V^{\mathrm{T}} A_j V) \\
&= \mathrm{diag}(0, B_i)\mathrm{diag}(I_r, 0) \\
&= 0.
\end{aligned}
$$

Because V is nonsingular, this implies the desired conclusion; that is, that $A_i A_j = 0$ for any $i \neq j$.

We can now extend this result to an idempotent matrix in place of I; that is, for an idempotent matrix A with $A = A_1 + \cdots + A_k$. Rather than stating it simply as in equation (8.39), however, we will state the implications differently.

Let $A_1, \ldots, A_k$ be $n \times n$ symmetric matrices and let

$$A = A_1 + \cdots + A_k. \tag{8.41}$$

Then any two of the following conditions imply the third one:

(a). A is idempotent.
(b). A_i is idempotent for $i = 1, \ldots, k$.
(c). $A_i A_j = 0$ for all $i \neq j$.

This is also called *Cochran's theorem*. (The theorem also applies to non-symmetric matrices if condition (c) is augmented with the requirement that $\text{rank}(A_i^2) = \text{rank}(A_i)$ for all i. We will restrict our attention to symmetric matrices, however, because in most applications of these results, the matrices are symmetric.)

First, if we assume properties (a) and (b), we can show that property (c) follows using an argument similar to that used to establish equation (8.39) for the special case $A = I$. The formal steps are left as an exercise.

Now, let us assume properties (b) and (c) and show that property (a) holds. With properties (b) and (c), we have

$$AA = (A_1 + \cdots + A_k)(A_1 + \cdots + A_k)$$
$$= \sum_{i=1}^{k} A_i A_i + \sum_{i \neq j} \sum_{j=1}^{k} A_i A_j$$
$$= \sum_{i=1}^{k} A_i$$
$$= A.$$

Hence, we have property (a); that is, A is idempotent.

Finally, let us assume properties (a) and (c). Property (b) follows immediately from

$$A_i^2 = A_i A_i = A_i A = A_i AA = A_i^2 A = A_i^3$$

and the implication (8.33).

Any two of the properties (a) through (c) also imply a fourth property for $A = A_1 + \cdots + A_k$ when the A_i are symmetric:

(d). $\text{rank}(A) = \text{rank}(A_1) + \cdots + \text{rank}(A_k)$.

We first note that any two of properties (a) through (c) imply the third one, so we will just use properties (a) and (b). Property (a) gives

$$\text{rank}(A) = \text{tr}(A) = \text{tr}(A_1 + \cdots + A_k) = \text{tr}(A_1) + \cdots + \text{tr}(A_k),$$

and property (b) states that the latter expression is $\text{rank}(A_1) + \cdots + \text{rank}(A_k)$, thus yielding property (d).

There is also a partial converse: properties (a) and (d) imply the other properties.

One of the most important special cases of Cochran's theorem is when $A = I$ in the sum (8.41):

$$I_n = A_1 + \cdots + A_k.$$

The identity matrix is idempotent, so if $\mathrm{rank}(A_1) + \cdots + \mathrm{rank}(A_k) = n$, all the properties above hold.

The most important statistical application of Cochran's theorem is for the distribution of quadratic forms of normally distributed random vectors. These distribution results are also called Cochran's theorem. We briefly discuss it in Section 9.1.3.

Drazin Inverses

A *Drazin inverse* of an operator T is an operator S such that $TS = ST$, $STS = S$, and $T^{k+1}S = T^k$ for any positive integer k.

It is clear that, as an operator, an idempotent matrix is its own Drazin inverse. Interestingly, if A is any square matrix, its Drazin inverse is the matrix $A^k(A^{2k+1})^+A^k$, which is unique for any positive integer k. See Campbell and Meyer (1991) for discussions of properties and applications of Drazin inverses and more on their relationship to the Moore-Penrose inverse.

8.5.2 Projection Matrices: Symmetric Idempotent Matrices

For a given vector space $\mathcal{V}$, a symmetric idempotent matrix A whose columns span $\mathcal{V}$ is said to be a *projection matrix* onto $\mathcal{V}$; in other words, a matrix A is a projection matrix onto $\mathrm{span}(A)$ if and only if A is symmetric and idempotent. (Some authors do not require a projection matrix to be symmetric. In that case, the terms "idempotent" and "projection" are synonymous.)

It is easy to see that, for any vector x, if A is a projection matrix onto $\mathcal{V}$, the vector Ax is in $\mathcal{V}$, and the vector $x - Ax$ is in $\mathcal{V}^\perp$ (the vectors Ax and $x - Ax$ are orthogonal). For this reason, a projection matrix is sometimes called an "orthogonal projection matrix". Note that an orthogonal projection matrix is not an orthogonal matrix, however, unless it is the identity matrix. Stating this in alternative notation, if A is a projection matrix and $A \in \mathbb{R}^{n \times n}$, then A maps $\mathbb{R}^n$ onto $\mathcal{V}(A)$ and $I - A$ is also a projection matrix (called the *complementary projection matrix* of A), and it maps $\mathbb{R}^n$ onto the orthogonal complement, $\mathcal{N}(A)$. These spaces are such that $\mathcal{V}(A) \oplus \mathcal{N}(A) = \mathbb{R}^n$.

In this text, we use the term "projection" to mean "orthogonal projection", but we should note that in some literature "projection" can include "oblique projection". In the less restrictive definition, for vector spaces $\mathcal{V}$, $\mathcal{X}$, and $\mathcal{Y}$, if $\mathcal{V} = \mathcal{X} \oplus \mathcal{Y}$ and $v = x + y$ with $x \in \mathcal{X}$ and $y \in \mathcal{Y}$, then the vector x is called the projection of v onto $\mathcal{X}$ along $\mathcal{Y}$. In this text, to use the unqualified term "projection", we require that $\mathcal{X}$ and $\mathcal{Y}$ be orthogonal; if they are not, then we call x the *oblique projection* of v onto $\mathcal{X}$ along $\mathcal{Y}$. The choice of the more restrictive definition is because of the overwhelming importance of orthogonal projections in statistical applications. The restriction is also consistent with the definition in equation (2.29) of the projection of a vector onto another vector (as opposed to the projection onto a vector space).

Because a projection matrix is idempotent, the matrix projects any of its columns onto itself, and of course it projects the full matrix onto itself: $AA = A$ (see equation (8.27)).

If x is a general vector in $\mathrm{I\!R}^n$, that is, if x has order n and belongs to an n-dimensional space, and A is a projection matrix of rank $r \leq n$, then Ax has order n and belongs to span(A), which is an r-dimensional space. Thus, we say projections are *dimension reductions*.

Useful projection matrices often encountered in statistical linear models are X^+X and XX^+. (Recall that X^-X is an idempotent matrix.) The matrix X^+ exists for any $n \times m$ matrix X, and X^+X is square $(m \times m)$ and symmetric.

Projections onto Linear Combinations of Vectors

On page 25, we gave the projection of a vector y onto a vector x as

$$\frac{x^{\mathrm{T}}y}{x^{\mathrm{T}}x}x.$$

The projection matrix to accomplish this is the "outer/inner products matrix",

$$\frac{1}{x^{\mathrm{T}}x}xx^{\mathrm{T}}. \tag{8.42}$$

The outer/inner products matrix has rank 1. It is useful in a variety of matrix transformations. If x is normalized, the projection matrix for projecting a vector on x is just xx^{T}. The projection matrix for projecting a vector onto a unit vector e_i is $e_ie_i^{\mathrm{T}}$, and $e_ie_i^{\mathrm{T}}y = (0, \ldots, y_i, \ldots, 0)$.

This idea can be used to project y onto the plane formed by two vectors, x_1 and x_2, by forming a projection matrix in a similar manner and replacing x in equation (8.42) with the matrix $X = [x_1|x_2]$. On page 331, we will view linear regression fitting as a projection onto the space spanned by the independent variables.

The angle between vectors we defined on page 26 can be generalized to the angle between a vector and a plane or any linear subspace by defining it as the angle between the vector and the projection of the vector onto the subspace. By applying the definition (2.32) to the projection, we see that the *angle* θ between the vector y and the subspace spanned by the columns of a projection matrix A is determined by the cosine

$$\cos(\theta) = \frac{y^{\mathrm{T}}Ay}{y^{\mathrm{T}}y}. \tag{8.43}$$

8.6 Special Matrices Occurring in Data Analysis

Some of the most useful applications of matrices are in the representation of observational data, as in Figure 8.1 on page 262. If the data are represented as real numbers, the array is a matrix, say X. The rows of the $n \times m$ data matrix X are "observations" and correspond to a vector of measurements on a single *observational unit*, and the columns of X correspond to n measurements of a single variable or feature. In data analysis we may form various association matrices that measure relationships among the variables or the observations that correspond to the columns or the rows of X. Many summary statistics arise from a matrix of the form $X^{\mathrm{T}}X$. (If the data in X are incomplete — that is, if some elements are missing — problems may arise in the analysis. We discuss some of these issues in Section 9.4.6.)

8.6.1 Gramian Matrices

A (real) matrix A such that for some (real) matrix B, $A = B^{\mathrm{T}}B$, is called a *Gramian matrix*. Any nonnegative definite matrix is Gramian (from equation (8.14) and Section 5.9.2 on page 194).

Sums of Squares and Cross Products

Although the properties of Gramian matrices are of interest, our starting point is usually the data matrix X, which we may analyze by forming a Gramian matrix $X^{\mathrm{T}}X$ or XX^{T} (or a related matrix). These Gramian matrices are also called *sums of squares and cross products matrices*. (The term "cross product" does not refer to the cross product of vectors defined on page 33, but rather to the presence of sums over i of the products $x_{ij}x_{ik}$ along with sums of squares x_{ij}^2.) These matrices and other similar ones are useful association matrices in statistical applications.

Some Immediate Properties of Gramian Matrices

Some interesting properties of a Gramian matrix $X^{\mathrm{T}}X$ are:

- $X^{\mathrm{T}}X$ is symmetric.
- $X^{\mathrm{T}}X$ is of full rank if and only if X is of full column rank or, more generally,

$$\operatorname{rank}(X^{\mathrm{T}}X) = \operatorname{rank}(X). \tag{8.44}$$

- $X^{\mathrm{T}}X$ is nonnegative definite and is positive definite if and only if X is of full column rank.
- $X^{\mathrm{T}}X = 0 \implies X = 0$.

These properties (except the first one, which is Exercise 8.7) were shown in the discussion in Section 3.3.7 on page 90.

Each element of a Gramian matrix is the dot products of the columns of the constituent matrix. If x_{*i} and x_{*j} are the ith and j^{th} columns of the matrix X, then

$$(X^{\mathrm{T}}X)_{ij} = x_{*i}^{\mathrm{T}}x_{*j}. \tag{8.45}$$

A Gramian matrix is also the sum of the outer products of the rows of the constituent matrix. If x_{i*} is the ith row of the $n \times m$ matrix X, then

$$X^{\mathrm{T}}X = \sum_{i=1}^{n} x_{i*}x_{i*}^{\mathrm{T}}. \tag{8.46}$$

This is generally the way a Gramian matrix is computed.

By equation (8.14), we see that any Gramian matrix formed from a general matrix X is the same as a Gramian matrix formed from a square upper triangular matrix T:

$$X^{\mathrm{T}}X = T^{\mathrm{T}}T.$$

Another interesting property of a Gramian matrix is that, for any matrices B and C (that are conformable for the operations indicated),

$$BX^{\mathrm{T}}X = CX^{\mathrm{T}}X \quad \Longleftrightarrow \quad BX^{\mathrm{T}} = CX^{\mathrm{T}}. \tag{8.47}$$

The implication from right to left is obvious, and we can see the left to right implication by writing

$$(BX^{\mathrm{T}}X - CX^{\mathrm{T}}X)(B^{\mathrm{T}} - C^{\mathrm{T}}) = (BX^{\mathrm{T}} - CX^{\mathrm{T}})(BX^{\mathrm{T}} - CX^{\mathrm{T}})^{\mathrm{T}},$$

and then observing that if the left-hand side is null, then so is the right-hand side, and if the right-hand side is null, then $BX^{\mathrm{T}} - CX^{\mathrm{T}} = 0$ because $X^{\mathrm{T}}X = 0 \implies X = 0$, as above. Similarly, we have

$$X^{\mathrm{T}}XB = X^{\mathrm{T}}XC \quad \Longleftrightarrow \quad X^{\mathrm{T}}B = X^{\mathrm{T}}C. \tag{8.48}$$

Generalized Inverses of Gramian Matrices

The generalized inverses of $X^{\mathrm{T}}X$ have useful properties. First, we see from the definition, for any generalized inverse $(X^{\mathrm{T}}X)^-$, that $((X^{\mathrm{T}}X)^-)^{\mathrm{T}}$ is also a generalized inverse of $X^{\mathrm{T}}X$. (Note that $(X^{\mathrm{T}}X)^-$ is not necessarily symmetric.) Also, we have, from equation (8.47),

$$X(X^{\mathrm{T}}X)^- X^{\mathrm{T}}X = X. \tag{8.49}$$

This means that $(X^{\mathrm{T}}X)^- X^{\mathrm{T}}$ is a generalized inverse of X.

The Moore-Penrose inverse of X has an interesting relationship with a generalized inverse of $X^{\mathrm{T}}X$:

$$XX^+ = X(X^{\mathrm{T}}X)^- X^{\mathrm{T}}. \tag{8.50}$$

This can be established directly from the definition of the Moore-Penrose inverse.

An important property of $X(X^{\mathrm{T}}X)^- X^{\mathrm{T}}$ is its invariance to the choice of the generalized inverse of $X^{\mathrm{T}}X$. Suppose G is any generalized inverse of $X^{\mathrm{T}}X$. Then, from equation (8.49), we have $X(X^{\mathrm{T}}X)^- X^{\mathrm{T}}X = XGX^{\mathrm{T}}X$, and from the implication (8.47), we have

$$XGX^{\mathrm{T}} = X(X^{\mathrm{T}}X)^- X^{\mathrm{T}}; \tag{8.51}$$

that is, $X(X^{\mathrm{T}}X)^- X^{\mathrm{T}}$ is invariant to the choice of generalized inverse.

Eigenvalues of Gramian Matrices

If the singular value decomposition of X is UDV^{T} (page 127), then the similar canonical factorization of $X^{\mathrm{T}}X$ (equation (3.197)) is $VD^{\mathrm{T}}DV^{\mathrm{T}}$. Hence, we see that the nonzero singular values of X are the square roots of the nonzero eigenvalues of the symmetric matrix $X^{\mathrm{T}}X$. By using DD^{T} similarly, we see that they are also the square roots of the nonzero eigenvalues of XX^{T}.

8.6.2 Projection and Smoothing Matrices

It is often of interest to approximate an arbitrary n-vector in a given m-dimensional vector space, where $m < n$. An $n \times n$ projection matrix of rank m clearly does this.

A Projection Matrix Formed from a Gramian Matrix

An important matrix that arises in analysis of a linear model of the form

$$y = X\beta + \epsilon \tag{8.52}$$

is

$$H = X(X^{\mathrm{T}}X)^{-}X^{\mathrm{T}}, \tag{8.53}$$

where $(X^{\mathrm{T}}X)^{-}$ is any generalized inverse. From equation (8.51), H is invariant to the choice of generalized inverse. By equation (8.50), this matrix can be obtained from the pseudoinverse and so

$$H = XX^{+}. \tag{8.54}$$

In the full rank case, this is uniquely

$$H = X(X^{\mathrm{T}}X)^{-1}X^{\mathrm{T}}. \tag{8.55}$$

Whether or not X is of full rank, H is a projection matrix onto $\mathrm{span}(X)$. It is called the "hat matrix" because it projects the observed response vector, often denoted by y, onto a *predicted* response vector, often denoted by $\widehat{y}$ in $\mathrm{span}(X)$:

$$\widehat{y} = Hy. \tag{8.56}$$

Because H is invariant, this projection is invariant to the choice of generalized inverse. (In the nonfull rank case, however, we generally refrain from referring to the vector Hy as the "predicted response"; rather, we may call it the "fitted response".)

The rank of H is the same as the rank of X, and its trace is the same as its rank (because it is idempotent). When X is of full column rank, we have

$$\mathrm{tr}(H) = \text{number of columns of } X. \tag{8.57}$$

(This can also be seen by using the invariance of the trace to permutations of the factors in a product as in equation (3.55).)

In linear models, $\text{tr}(H)$ is the model degrees of freedom, and the sum of squares due to the model is just $y^{\text{T}}Hy$.

The complementary projection matrix,

$$I - H, \tag{8.58}$$

also has interesting properties that relate to linear regression analysis. In geometrical terms, this matrix projects a vector onto $\mathcal{N}(X^{\text{T}})$, the orthogonal complement of $\text{span}(X)$. We have

$$\begin{aligned} y &= Hy + (I - H)y \\ &= \widehat{y} + r, \end{aligned} \tag{8.59}$$

where $r = (I - H)y \in \mathcal{N}(X^{\text{T}})$. The orthogonal complement is called the residual vector space, and r is called the residual vector. Both the rank and the trace of the orthogonal complement are the number of rows in X (that is, the number of observations) minus the regression degrees of freedom. This quantity is the "residual degrees of freedom" (unadjusted).

These two projection matrices (8.53) or (8.55) and (8.58) partition the total sums of squares:

$$y^{\text{T}}y = y^{\text{T}}Hy + y^{\text{T}}(I - H)y. \tag{8.60}$$

Note that the second term in this partitioning is the Schur complement of $X^{\text{T}}X$ in $[X\,y]^{\text{T}}[X\,y]$ (see equation (3.146) on page 96).

Smoothing Matrices

The hat matrix, either from a full rank X as in equation (8.55) or formed by a generalized inverse as in equation (8.53), *smoothes* the vector y onto the hyperplane defined by the column space of X. It is therefore a *smoothing matrix*. (Note that the rank of the column space of X is the same as the rank of $X^{\text{T}}X$.)

A useful variation of the cross products matrix $X^{\text{T}}X$ is the matrix formed by adding a nonnegative (positive) definite matrix A to it. Because $X^{\text{T}}X$ is nonnegative (positive) definite, $X^{\text{T}}X + A$ is nonnegative definite, as we have seen (page 277), and hence $X^{\text{T}}X + A$ is a Gramian matrix.

Because the square root of the nonnegative definite A exists, we can express the sum of the matrices as

$$X^{\text{T}}X + A = \begin{bmatrix} X \\ A^{\frac{1}{2}} \end{bmatrix}^{\text{T}} \begin{bmatrix} X \\ A^{\frac{1}{2}} \end{bmatrix}. \tag{8.61}$$

In a common application, a positive definite matrix λI, with $\lambda > 0$, is added to $X^{\text{T}}X$, and this new matrix is used as a smoothing matrix. The analogue of the hat matrix (8.55) is

$$H_\lambda = X(X^\mathrm{T}X + \lambda I)^{-1}X^\mathrm{T}, \tag{8.62}$$

and the analogue of the fitted response is

$$\widehat{y}_\lambda = H_\lambda y. \tag{8.63}$$

This has the effect of shrinking the $\widehat{y}$ of equation (8.56) toward 0. (In regression analysis, this is called "ridge regression".)

Any matrix such as H_λ that is used to transform the observed vector y onto a given subspace is called a smoothing matrix.

Effective Degrees of Freedom

Because of the shrinkage in ridge regression (that is, because the fitted model is less dependent just on the data in X) we say the "effective" degrees of freedom of a ridge regression model decreases with increasing λ. We can formally define the *effective model degrees of freedom* of any linear fit $\widehat{y} = H_\lambda y$ as

$$\mathrm{tr}(H_\lambda), \tag{8.64}$$

analogous to the model degrees of freedom in linear regression above. This definition of effective degrees of freedom applies generally in data smoothing. In fact, many smoothing matrices used in applications depend on a single smoothing parameter such as the λ in ridge regression, and so the same notation H_λ is often used for a general smoothing matrix.

To evaluate the effective degrees of freedom in the ridge regression model for a given λ and X, for example, using singular value decomposition of X, $X = UDV^\mathrm{T}$, we have

$$
\begin{aligned}
\mathrm{tr}(X(X^\mathrm{T}X + \lambda I)^{-1}X^\mathrm{T}) \\
&= \mathrm{tr}\left(UDV^\mathrm{T}(VD^2V^\mathrm{T} + \lambda VV^\mathrm{T})^{-1}VDU^\mathrm{T}\right) \\
&= \mathrm{tr}\left(UDV^\mathrm{T}(V(D^2 + \lambda I)V^\mathrm{T})^{-1}VDU^\mathrm{T}\right) \\
&= \mathrm{tr}\left(UD(D^2 + \lambda I)^{-1}DU^\mathrm{T}\right) \\
&= \mathrm{tr}\left(D^2(D^2 + \lambda I)^{-1}\right) \\
&= \sum \frac{d_i^2}{d_i^2 + \lambda}.
\end{aligned}
\tag{8.65}
$$

When $\lambda = 0$, this is the same as the ordinary model degrees of freedom, and when λ is positive, this quantity is smaller, as we would want it to be by the argument above. The $d_i^2/(d_i^2 + \lambda)$ are called shrinkage factors.

If $X^\mathrm{T}X$ is not of full rank, the addition of λI to it also has the effect of yielding a full rank matrix, if $\lambda > 0$, and so the inverse of $X^\mathrm{T}X + \lambda I$ exists even when that of $X^\mathrm{T}X$ does not. In any event, the addition of λI to $X^\mathrm{T}X$ yields a matrix with a better "condition number", which we define in Section 6.1. (On page 206, we return to this model and show that the condition number of $X^\mathrm{T}X + \lambda I$ is better than that of $X^\mathrm{T}X$.)

Residuals from Smoothed Data

Just as in equation (8.59), we can write

$$y = \widehat{y}_\lambda + r_\lambda. \tag{8.66}$$

Notice, however, that H_λ is not in general a projection matrix. Unless H_λ is a projection matrix, however, $\widehat{y}_\lambda$ and r_λ are not orthogonal as are $\widehat{y}$ and r, and we do not have the additive partitioning of the sum of squares as in equation (8.60).

The rank of H_λ is the same as the number of columns of X, but the trace, and hence the model degrees of freedom, is less than this number.

8.6.3 Centered Matrices and Variance-Covariance Matrices

In Section 2.3, we defined the variance of a vector and the covariance of two vectors. These are the same as the "sample variance" and "sample covariance" in statistical data analysis and are related to the variance and covariance of random variables in probability theory. We now consider the variance-covariance matrix associated with a data matrix. We occasionally refer to the variance-covariance matrix simply as the "variance matrix" or just as the "variance".

First, we consider centering and scaling data matrices.

Centering and Scaling of Data Matrices

When the elements in a vector represent similar measurements or observational data on a given phenomenon, summing or averaging the elements in the vector may yield meaningful statistics. In statistical applications, the columns in a matrix often represent measurements on the same feature or on the same variable over different observational units as in Figure 8.1, and so the mean of a column may be of interest.

We may center the column by subtracting its mean from each element in the same manner as we centered vectors on page 34. The matrix formed by centering all of the columns of a given matrix is called a centered matrix, and if the original matrix is X, we represent the centered matrix as X_c in a notation analogous to what we introduced for centered vectors. If we represent the matrix whose i^{th} column is the constant mean of the ith column of X as $\overline{X}$,

$$X_c = X - \overline{X}. \tag{8.67}$$

Here is an R statement to compute this:

```
Xc <- X-rep(1,n)%*%t(apply(X,2,mean))
```

If the unit of a measurement is changed, all elements in a column of the data matrix in which the measurement is used will change. The amount of variation of elements within a column or the relative variation among different columns ideally should not be measured in terms of the basic units of measurement, which can differ irreconcilably from one column to another. (One column could represent scores on an exam and another column could represent weight, for example.)

In analyzing data, it is usually important to scale the variables so that their variations are comparable. We do this by using the standard deviation of the column. If we have also centered the columns, the column vectors are the centered and scaled vectors of the form of those in equation (2.51),

$$x_{\mathrm{cs}} = \frac{x_{\mathrm{c}}}{s_x},$$

where s_x is the standard deviation of x,

$$s_x = \frac{\|x_{\mathrm{c}}\|}{\sqrt{n-1}}.$$

If all columns of the data matrix X are centered and scaled, we denote the resulting matrix as X_{cs}. If s_i represents the standard deviation of the ith column, this matrix is formed as

$$X_{\mathrm{cs}} = X_{\mathrm{c}}\mathrm{diag}(1/s_i). \tag{8.68}$$

Here is an R statement to compute this:

```
Xcs <- Xc%*%diag(1/apply(X,2,sd))
```

If the rows of X are taken as representative of a population of similar vectors, it is often useful to center and scale any vector from that population in the manner of equation (8.68):

$$\tilde{x} = \mathrm{diag}(1/s_i)x_{\mathrm{c}}. \tag{8.69}$$

(Note that x_{c} is a vector of the same order as a row of X_{c}.)

Gramian Matrices Formed from Centered Matrices; Covariance Matrices

An important Gramian matrix is formed as the sums of squares and cross products matrix from a centered matrix and scaled by $(n-1)$, where n is the number of rows of the original matrix:

$$
\begin{aligned}
S_X &= \frac{1}{n-1}X_{\mathrm{c}}^{\mathrm{T}}X_{\mathrm{c}} \\
&= (s_{ij}).
\end{aligned}
\tag{8.70}
$$

This matrix is called the *variance-covariance matrix* associated with the given matrix X, and we denote it by S_X or just S. If x_{*i} and x_{*j} are the vectors corresponding to the ith and j^{th} columns of X, then $s_{ij} = \text{Cov}(x_{*i}, x_{*j})$; that is, the off-diagonal elements are the covariances between the column vectors, as in equation (2.55), and the diagonal elements are *variances* of the column vectors.

This matrix and others formed from it, such as R_X in equation (8.72) below, are called *association matrices* because they are based on measures of association (covariance or correlation) among the columns of X. We could likewise define a Gramian association matrix based on measures of association among the rows of X.

A transformation using the Cholesky factor of S_X or the square root of S_X (assuming S_X is full rank) results in a matrix whose associated variance-covariance is the identity. We call this a sphered matrix:

$$X_{\text{sphered}} = X_{\text{c}} S_X^{-\frac{1}{2}}. \tag{8.71}$$

The matrix S_X is a measure of the anisometry of the space of vectors represented by the rows of X as mentioned in Section 3.2.8. The inverse, S_X^{-1}, in some sense evens out the anisometry. Properties of vectors in the space represented by the rows of X are best assessed following a transformation as in equation (8.69). For example, rather than orthogonality of two vectors u and v, a more interesting relationship would be S_X^{-1}-conjugacy (see equation (3.65)):

$$u^{\text{T}} S_X^{-1} v = 0.$$

Also, the Mahalanobis distance, $\sqrt{(u - v)^{\text{T}} S_X^{-1} (u - v)}$, may be more relevant for measuring the difference in two vectors than the standard Euclidean distance.

Gramian Matrices Formed from Scaled Centered Matrices; Correlation Matrices

If the columns of a centered matrix are standardized (that is, divided by their standard deviations, assuming that each is nonconstant, so that the standard deviation is positive), the scaled cross products matrix is the *correlation matrix*, often denoted by R_X or just R,

$$\begin{aligned} R_X &= \frac{1}{n-1} X_{\text{cs}}^{\text{T}} X_{\text{cs}} \\ &= (r_{ij}), \end{aligned} \tag{8.72}$$

where if x_{*i} and x_{*j} are the vectors corresponding to the ith and j^{th} columns of X, $r_{ij} = \text{Corr}(x_{*i}, x_{*j})$. The correlation matrix can also be expressed as $R_X =$

$X_\mathrm{c}^\mathrm{T} D X_\mathrm{c}$, where D is the diagonal matrix whose k^th diagonal is $1/\sqrt{\mathrm{V}(x_{*k})}$, where $\mathrm{V}(x_{*k})$ is the sample variance of the kth column; that is, $\mathrm{V}(x_{*k}) = \sum_i (x_{ik} - \bar{x}_{*k})^2/(n-1)$. This Gramian matrix R_X is based on measures of association among the columns of X.

The elements along the diagonal of the correlation matrix are all 1, and the off-diagonals are between -1 and 1, each being the correlation between a pair of column vectors, as in equation (2.57). The correlation matrix is nonnegative definite because it is a Gramian matrix.

The trace of an $n \times n$ correlation matrix is n, and therefore the eigenvalues, which are all nonnegative, sum to n.

Without reference to a data matrix, any nonnegative definite matrix with 1s on the diagonal and with all elements less than or equal to 1 in absolute value is called a *correlation matrix*.

8.6.4 The Generalized Variance

The diagonal elements of the variance-covariance matrix S associated with the $n \times m$ data matrix X are the second moments of the centered columns of X, and the off-diagonal elements are pairwise second central moments of the columns. Each element of the matrix provides some information about the spread of a single column or of two columns of X. The determinant of S provides a single overall measure of the spread of the columns of X. This measure, $|S|$, is called the *generalized variance*, or generalized *sample* variance, to distinguish it from an analogous measure of a distributional model.

On page 57, we discussed the equivalence of a determinant and the volume of a parallelotope. The generalized variance captures this, and when the columns or rows of S are more orthogonal to each other, the volume of the parallelotope determined by the columns or rows of S is greater, as shown in Figure 8.6 for $m = 3$.

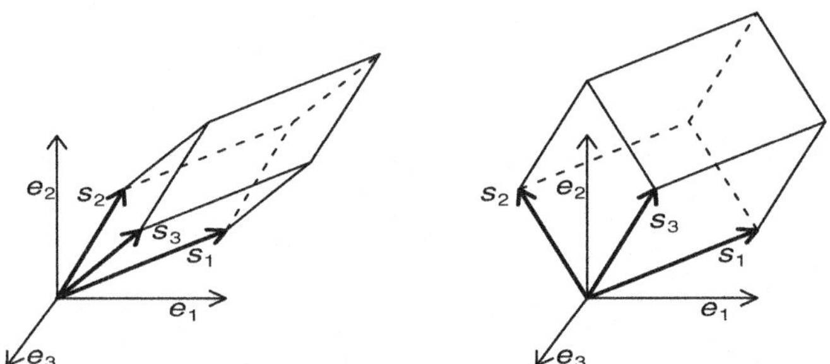

Fig. 8.6. Generalized Variances in Terms of the Columns of S

The columns or rows of S are generally not of much interest in themselves. Our interest is in the relationship of the centered columns of the $n \times m$ data matrix X. Let us consider the case of $m = 2$. Let z_{*1} and z_{*2} represent the centered column vectors of X; that is, for z_{*1}, we have $z_{*1_i} = x_{*1_i} - \bar{x}_1$. Now, as in equation (3.33), consider the parallelogram formed by z_{*1} and z_{*2}. For computing the area, consider z_{*1} as forming the base. The length of the base is

$$\|z_{*1}\| = \sqrt{(n-1)s_{11}},$$

and the height is

$$\|z_{*2}\|\,|\sin(\theta)| = \sqrt{(n-1)s_{22}(1 - r_{12}^2)}.$$

(Recall the relationship between the angle and the correlation from equation (2.58).)

The area of the parallelogram therefore is

$$\text{area} = (n-1)\sqrt{s_{11}s_{22}(1 - r_{12}^2)}.$$

Now, consider S:

$$S = \begin{bmatrix} s_{11} & s_{21} \\ s_{12} & s_{22} \end{bmatrix}$$

$$= \begin{bmatrix} s_{11} & \sqrt{s_{11}s_{22}}\,r_{12} \\ \sqrt{s_{11}s_{22}}\,r_{12} & s_{22} \end{bmatrix}.$$

The determinant of S is therefore

$$s_{11}s_{22}(1 - r_{12}^2),$$

that is,

$$|S| = \frac{1}{(n-1)^m}\text{volume}^2. \tag{8.73}$$

Although we considered only the case $m = 2$, equation (8.73) holds generally, as can be seen by induction on m (see Anderson, 2003).

Comparing Variance-Covariance Matrices

Many standard statistical procedures for comparing groups of data rely on the assumption that the population variance-covariance matrices of the groups are all the same. (The simplest example of this is the two-sample t-test, in which the concern is just that the population variances of the two groups be equal.) Occasionally, the data analyst wishes to test this assumption of homogeneity of variances.

On page 138, we considered the problem of comparing two matrices of the same size. There we defined a metric based on a matrix norm. For the problem of comparing variance-covariance matrices, a measure based on the generalized variances is more commonly used.

In the typical situation, we have an $n \times m$ data matrix X in which the first n_1 rows represent observations from one group, the next n_2 rows represent observations from another group, and the last n_g rows represent observations from the gth group. For each group, we form a sample variance-covariance matrix S_i as in equation (8.70) using the ith submatrix of X. Whenever we have individual variance-covariance matrices in situations similar to this, we define the *pooled variance-covariance matrix*:

$$S_{\mathrm{p}} = \frac{1}{(n-g)} \sum_{i=1}^{g} (n_i - 1)S_i. \tag{8.74}$$

Bartlett suggested a test based on the determinants of $(n_i - 1)S_i$. (From equation (3.27), $|(n_i - 1)S_i| = (n_i - 1)^m |S_i|$.) A similar test suggested by Box also uses the generalized variances. One form of the Box M statistic for testing for homogeneity of variances is

$$M = (n - g) \log(|S_{\mathrm{p}}|) - \sum_{i=1}^{g} (n_i - 1)S_i. \tag{8.75}$$

8.6.5 Similarity Matrices

Covariance and correlation matrices are examples of *similarity association matrices*: they measure the similarity or closeness of the columns of the matrices from which they are formed.

The cosine of the angle between two vectors is related to the correlation between the vectors, so a matrix of the cosine of the angle between the columns of a given matrix would also be a similarity matrix. (The angle is exactly the same as the correlation if the vectors are centered; see equation (2.57).)

Similarity matrices can be formed in many ways, and some are more useful in particular applications than in others. They may not even arise from a standard dataset in the familiar form in statistical applications. For example, we may be interested in comparing text documents. We might form a vector-like object whose elements are the words in the document. The similarity between two such ordered tuples, generally of different lengths, may be a count of the number of words, word pairs, or more general phrases in common between the two documents.

It is generally reasonable that similarity between two objects be symmetric; that is, the first object is as close to the second as the second is to the first. We reserve the term similarity matrix for matrices formed from such measures and, hence, that themselves are symmetric. Occasionally, for example in psychometric applications, the similarities are measured relative to rank order closeness within a set. In such a case, the measure of closeness may not be symmetric. A matrix formed from such measures is called a *directed similarity matrix*.

8.6.6 Dissimilarity Matrices

The elements of similarity generally increase with increasing "closeness". We may also be interested in the dissimilarity. Clearly, any decreasing function of similarity could be taken as a reasonable measure of dissimilarity. There are, however, other measures of dissimilarity that often seem more appropriate. In particular, the properties of a metric (see page 22) may suggest that a dissimilarity be defined in terms of a metric.

In considering either similarity or dissimilarity for a data matrix, we could work with either rows or columns, but the common applications make one or the other more natural for the specific application. Because of the types of the common applications, we will discuss dissimilarities and distances in terms of rows instead of columns; thus, in this section, we will consider the dissimilarity of x_{i*} and x_{j*}, the vectors corresponding to the ith and j^{th} rows of X.

Dissimilarity matrices are also *association matrices* in the general sense of Section 8.1.4.

Dissimilarity or distance can be measured in various ways. A metric is the most obvious measure, although in certain applications other measures are appropriate. The measures may be based on some kind of ranking, for example. If the dissimilarity is based on a metric, the association matrix is often called a *distance matrix*. In most applications, the Euclidean distance, $\|x_{i*} - x_{j*}\|_2$, is the most commonly used metric, but others, especially ones based on L_1 or L_∞ norms, are often useful.

Any hollow matrix with nonnegative elements is a *directed dissimilarity matrix*. A directed dissimilarity matrix is also called a *cost matrix*. If the matrix is symmetric, it is a *dissimilarity matrix*.

An $n \times n$ matrix $D = (d_{ij})$ is an m-dimensional *distance matrix* if there exists m-vectors $x_1 \ldots, x_n$ such that, for some metric Δ, $d_{ij} = \Delta(x_i, x_j)$. A distance matrix is necessarily a dissimilarity matrix. If the metric is the Euclidean distance, the matrix D is called a *Euclidean distance matrix*.

The matrix whose rows are the vectors $x_1^{\text{T}}, \ldots, x_n^{\text{T}}$ is called an associated *configuration matrix* of the given distance matrix. In metric multidimensional scaling, we are given a dissimilarity matrix and seek to find a configuration matrix whose associated distance matrix is closest to the dissimilarity matrix, usually in terms of the Frobenius norm of the difference of the matrices (see Trosset, 2002, for basic definitions and extensions).

8.7 Nonnegative and Positive Matrices

A *nonnegative matrix*, as the name suggests, is a real matrix all of whose elements are nonnegative, and a *positive matrix* is a real matrix all of whose elements are positive. In some other literature, the latter type of matrix is called strictly positive, and a nonnegative matrix with a positive element is called positive.

Many useful matrices are nonnegative. We have already considered various kinds of nonnegative matrices. The adjacency or connectivity of a graph is nonnegative. Dissimilarity matrices, including distance matrices, are nonnegative. Matrices used in modeling stochastic processes are nonnegative.

If A is nonnegative, we write

$$A \geq 0, \tag{8.76}$$

and if it is positive, we write

$$A > 0. \tag{8.77}$$

Notice that $A \geq 0$ and $A \neq 0$ together do not imply $A > 0$.

We write

$$A \geq B$$

to mean $(A - B) \geq 0$ and

$$A > B$$

to mean $(A - B) > 0$. (Recall the definitions of nonnegative definite and positive definite matrices, and, from equations (8.11) and (8.16), the notation used to indicate those properties, $A \succeq 0$ and $A \succ 0$. Furthermore, notice that these definitions and this notation for nonnegative and positive matrices are consistent with analogous definitions and notation involving vectors on page 13. Some authors, however, use the notation of equations (8.76) and (8.77) to mean "nonnegative definite" and "positive definite". We should also note that some authors use somewhat different terms for these and related properties. "Positive" for these authors means nonnegative with at least one positive element, and "strictly positive" means positive as we have defined it.)

Notice that positiveness (nonnegativeness) has nothing to do with positive (nonnegative) definiteness. A positive or nonnegative matrix need not be symmetric or even square, although most such matrices useful in applications are square. A square positive matrix, unlike a positive definite matrix, need not be of full rank.

The following properties are easily verified.

1. If $A \geq 0$ and $u \geq v \geq 0$, then $Au \geq Av$.
2. If $A \geq 0$, $A \neq 0$, and $u > v > 0$, then $Au > Av$.
3. If $A > 0$ and $v \geq 0$, then $Av \geq 0$.
4. If $A > 0$ and A is square, then $\rho(A) > 0$.

Whereas most of the important matrices arising in the analysis of linear models are symmetric, and thus have the properties listed on page 270, many important nonnegative matrices, such as those used in studying stochastic processes, are not necessarily symmetric. The eigenvalues of real symmetric matrices are real, but *the eigenvalues of real nonsymmetric matrices may have an imaginary component.* In the following discussion, we must be careful to remember the meaning of the spectral radius. The definition in equation (3.185) for the spectral radius of the matrix A with eigenvalues c_i,

$$\rho(A) = \max |c_i|,$$

is still correct, but the operator "$|\cdot|$" must be interpreted as the modulus of a complex number.

8.7.1 Properties of Square Positive Matrices

We have the following important properties for square positive matrices. These properties collectively are the conclusions of the *Perron theorem.*

Let A be a square positive matrix and let $r = \rho(A)$. Then:

1. r is an eigenvalue of A. The eigenvalue r is called the *Perron root.* Note that the Perron root is real (although other eigenvalues of A may not be).
2. There is an eigenvector v associated with r such that $v > 0$.
3. The Perron root is simple. (That is, the algebraic multiplicity of the Perron root is 1.)
4. The dimension of the eigenspace of the Perron root is 1. (That is, the geometric multiplicity of $\rho(A)$ is 1.) Hence, if v is an eigenvector associated with r, it is unique except for scaling. This associated eigenvector is called the *Perron vector.* Note that the Perron vector is real (although other eigenvectors of A may not be). The elements of the Perron vector all have the same sign, which we usually take to be positive; that is, $v > 0$.
5. If c_i is any other eigenvalue of A, then $|c_i| < r$. (That is, r is the only eigenvalue on the spectral circle of A.)

We will give proofs only of properties 1 and 2 as examples. Proofs of all of these facts are available in Horn and Johnson (1991).

To see properties 1 and 2, first observe that a positive matrix must have at least one nonzero eigenvalue because the coefficients and the constant in the characteristic equation must all be positive. Now scale the matrix so that its spectral radius is 1 (see page 111). So without loss of generality, let A be a scaled positive matrix with $\rho(A) = 1$. Now let (c, x) be some eigenpair of A such that $|c| = 1$. First, we want to show, for some such c, that $c = \rho(A)$.

Because all elements of A are positive,

$$|x| = |Ax| \leq A|x|,$$

and so

$$A|x| - |x| \geq 0. \tag{8.78}$$

An eigenvector must be nonzero, so we also have

$$A|x| > 0.$$

Now we want to show that $A|x| - |x| = 0$. To that end, suppose the contrary; that is, suppose $A|x| - |x| \neq 0$. In that case, $A(A|x| - |x|) > 0$ from equation (8.78), and so there must be a positive number ϵ such that

$$\frac{A}{1 + \epsilon} A|x| > A|x|$$

or

$$By > y,$$

where $B = A/(1 + \epsilon)$ and $y = A|x|$. Now successively multiplying both sides of this inequality by the positive matrix B, we have

$$B^k y > y \quad \text{for all } k = 1, 2, \ldots.$$

Because $\rho(B) = \rho(A)/(1+\epsilon) < 1$, from equation (3.247) on page 136, we have $\lim_{k \to \infty} B^k = 0$; that is, $\lim_{k \to \infty} B^k y = 0 > y$. This contradicts the fact that $y > 0$. Because the supposition $A|x| - |x| \neq 0$ led to this contradiction, we must have $A|x| - |x| = 0$. Therefore $1 = \rho(A)$ must be an eigenvalue of A, and $|x|$ must be an associated eigenvector; hence, with $v = |x|$, $(\rho(A), v)$ is an eigenpair of A and $v > 0$, and this is the statement made in properties 1 and 2.

The Perron-Frobenius theorem, which we consider below, extends these results to a special class of square nonnegative matrices. (This class includes all positive matrices, so the Perron-Frobenius theorem is an extension of the Perron theorem.)

8.7.2 Irreducible Square Nonnegative Matrices

Nonnegativity of a matrix is not a very strong property. First of all, note that it includes the zero matrix; hence, clearly none of the properties of the Perron theorem can hold. Even a nondegenerate, full rank nonnegative matrix does not necessarily possess those properties. A small full rank nonnegative matrix provides a counterexample for properties 2, 3, and 5:

$$A = \begin{bmatrix} 1 & 1 \\ 0 & 1 \end{bmatrix}.$$

The eigenvalues are 1 and 1; that is, 1 with an algebraic multiplicity of 2 (so property 3 does not hold). There is only one nonnull eigenvector, $(1, -1)$, (so property 2 does not hold, but property 4 holds), but the eigenvector is not positive (or even nonnegative). Of course property 5 cannot hold if property 3 does not hold.

We now consider irreducible square nonnegative matrices. This class includes positive matrices. On page 268, we defined reducibility of a nonnegative square matrix and we saw that a matrix is irreducible if and only if its digraph is strongly connected.

To recall the definition, a nonnegative matrix is said to be *reducible* if by symmetric permutations it can be put into a block upper triangular matrix with square blocks along the diagonal; that is, the nonnegative matrix A is reducible if and only if there is a permutation matrix E_π such that

$$E_\pi^{\mathrm{T}} A E_\pi = \begin{bmatrix} B_{11} & B_{12} \\ 0 & B_{22} \end{bmatrix}, \tag{8.79}$$

where B_{11} and B_{22} are square. A matrix that cannot be put into that form is *irreducible*. An alternate term for reducible is *decomposable*, with the associated term *indecomposable*. (There is an alternate meaning for the term "reducible" applied to a matrix. This alternate use of the term means that the matrix is capable of being expressed by a similarity transformation as the sum of two matrices whose columns are mutually orthogonal.)

We see from the definition in equation (8.79) that a positive matrix is irreducible.

Irreducible matrices have several interesting properties. An $n \times n$ nonnegative matrix A is irreducible if and only if $(I + A)^{n-1}$ is a positive matrix; that is,

$$A \text{ is irreducible} \iff (I + A)^{n-1} > 0. \tag{8.80}$$

To see this, first assume $(I+A)^{n-1} > 0$; thus, $(I+A)^{n-1}$ clearly is irreducible. If A is reducible, then there exists a permutation matrix E_π such that

$$E_\pi^{\mathrm{T}} A E_\pi = \begin{bmatrix} B_{11} & B_{12} \\ 0 & B_{22} \end{bmatrix},$$

and so

$$\begin{aligned}
E_\pi^{\mathrm{T}} (I + A)^{n-1} E_\pi &= \left(E_\pi^{\mathrm{T}} (I + A) E_\pi \right)^{n-1} \\
&= \left(I + E_\pi^{\mathrm{T}} A E_\pi \right)^{n-1} \\
&= \begin{bmatrix} I_{n_1} + B_{11} & B_{12} \\ 0 & I_{n_2} + B_{22} \end{bmatrix}.
\end{aligned}$$

This decomposition of $(I + A)^{n-1}$ cannot exist because it is irreducible; hence we conclude A is irreducible if $(I + A)^{n-1} > 0$.

Now, if A is irreducible, we can see that $(I + A)^{n-1}$ must be a positive matrix either by a strictly linear-algebraic approach or by couching the argument in terms of the digraph $\mathcal{G}(A)$ formed by the matrix, as in the discussion on page 268 that showed that a digraph is strongly connected if (and only if) it is irreducible. We will use the latter approach in the spirit of applications of irreducibility in stochastic processes.

For either approach, we first observe that the $(i, j)^{\mathrm{th}}$ element of $(I+A)^{n-1}$ can be expressed as

$$\left((I + A)^{n-1}\right)_{ij} = \left(\sum_{k=0}^{n-1} \binom{n-1}{k} A^k\right)_{ij}. \tag{8.81}$$

Hence, for $k = 1, \ldots, n-1$, we consider the $(i,j)^{\text{th}}$ entry of A^k. Let $a_{ij}^{(k)}$ represent this quantity.

Given any pair (i,j), for some $l_1, l_2, \ldots, l_{k-1}$, we have

$$a_{ij}^{(k)} = \sum_{l_1, l_2, \ldots, l_{k-1}} a_{1l_1} a_{l_1 l_2} \cdots a_{l_{k-1}j}.$$

Now $a_{ij}^{(k)} > 0$ if and only if $a_{1l_1}, a_{l_1 l_2}, \ldots, a_{l_{k-1}j}$ are all positive; that is, if there is a path $v_1, v_{l_1}, \ldots, v_{l_{k-1}}, v_j$ in $\mathcal{G}(A)$. If A is irreducible, then $\mathcal{G}(A)$ is strongly connected, and hence the path exists. So, for any pair (i,j), we have from equation (8.81) $\left((I + A)^{n-1}\right)_{ij} > 0$; that is, $(I + A)^{n-1} > 0$.

The positivity of $(I + A)^{n-1}$ for an irreducible nonnegative matrix A is a very useful property because it allows us to extend some conclusions of the Perron theorem to irreducible nonnegative matrices.

Properties of Square Irreducible Nonnegative Matrices; the Perron-Frobenius Theorem

If A is a square irreducible nonnegative matrix, then we have the following properties, which are similar to properties 1 through 4 on page 301 for positive matrices. These following properties are the conclusions of the *Perron-Frobenius theorem*.

1. $\rho(A)$ is an eigenvalue of A. This eigenvalue is called the *Perron root*, as before.
2. The Perron root $\rho(A)$ is simple. (That is, the algebraic multiplicity of the Perron root is 1.)
3. The dimension of the eigenspace of the Perron root is 1. (That is, the geometric multiplicity of $\rho(A)$ is 1.)
4. The eigenvector associated with $\rho(A)$ is positive. This eigenvector is called the *Perron vector*, as before.

The relationship (8.80) allows us to prove properties 1 and 4 in a method similar to the proofs of properties 1 and 2 for positive matrices. (This is Exercise 8.9.) Complete proofs of all of these facts are available in Horn and Johnson (1991). See also the solution to Exercise 8.10b on page 498 for a special case.

The one property of square positive matrices that does not carry over to square irreducible nonnegative matrices is property 5: $r = \rho(A)$ is the only eigenvalue on the spectral circle of A. For example, the small irreducible nonnegative matrix

$$A = \begin{bmatrix} 0 & 1 \\ 1 & 0 \end{bmatrix}$$

has eigenvalues 1 and -1, and so both are on the spectral circle.

It turns out, however, that square irreducible nonnegative matrices that have only one eigenvalue on the spectral circle also have other interesting properties that are important, for example, in Markov chains. We therefore give a name to the property:

> A square irreducible nonnegative matrix is said to be *primitive* if it has only one eigenvalue on the spectral circle.

In modeling with Markov chains and other applications, the limiting behavior of A^k is an important property.

On page 135, we saw that $\lim_{k\to\infty} A^k = 0$ if $\rho(A) < 1$. For a primitive matrix, we also have a limit for A^k if $\rho(A) = 1$. (As we have done above, we can scale any matrix with a nonzero eigenvalue to a matrix with a spectral radius of 1.)

If A is a primitive matrix, then we have the useful result

$$\lim_{k\to\infty} \left(\frac{A}{\rho(A)}\right)^k = vw^{\mathrm{T}}, \tag{8.82}$$

where v is an eigenvector of A associated with $\rho(A)$ and w is an eigenvector of A^{T} associated with $\rho(A)$, and w and v are scaled so that $w^{\mathrm{T}}v = 1$. (As we mentioned on page 123, such eigenvectors exist because $\rho(A)$ is a simple eigenvalue. They also exist because of property 4; they are both positive. Note that A is not necessarily symmetric, and so its eigenvectors may include imaginary components; however, the eigenvectors associated with $\rho(A)$ are real, and so we can write w^{T} instead of w^{H}.)

To see equation (8.82), we consider $\left(A - \rho(A)vw^{\mathrm{T}}\right)$. First, if (c_i, v_i) is an eigenpair of $\left(A - \rho(A)vw^{\mathrm{T}}\right)$ and $c_i \neq 0$, then (c_i, v_i) is an eigenpair of A. We can see this by multiplying both sides of the eigen-equation by vw^{T}:

$$\begin{aligned} c_i vw^{\mathrm{T}} v_i &= vw^{\mathrm{T}} \left(A - \rho(A)vw^{\mathrm{T}}\right) v_i \\ &= \left(vw^{\mathrm{T}}A - \rho(A)vw^{\mathrm{T}}vw^{\mathrm{T}}\right) v_i \\ &= \left(\rho(A)vw^{\mathrm{T}} - \rho(A)vw^{\mathrm{T}}\right) v_i \\ &= 0; \end{aligned}$$

hence,

$$\begin{aligned} Av_i &= \left(A - \rho(A)vw^{\mathrm{T}}\right) v_i \\ &= c_i v_i. \end{aligned}$$

Next, we show that

$$\rho\left(A - \rho(A)vw^{\mathrm{T}}\right) < \rho(A). \tag{8.83}$$

If $\rho(A)$ were an eigenvalue of $\left(A - \rho(A)vw^{\mathrm{T}}\right)$, then its associated eigenvector, say w, would also have to be an eigenvector of A, as we saw above. But since as an eigenvalue of A the geometric multiplicity of $\rho(A)$ is 1, for some scalar s, $w = sv$. But this is impossible because that would yield

$$\begin{aligned}
\rho(A)sv &= \left(A - \rho(A)vw^{\mathrm{T}}\right)sv \\
&= sAv - s\rho(A)v \\
&= 0,
\end{aligned}$$

and neither $\rho(A)$ nor sv is zero. But as we saw above, any eigenvalue of $\left(A - \rho(A)vw^{\mathrm{T}}\right)$ is an eigenvalue of A and no eigenvalue of $\left(A - \rho(A)vw^{\mathrm{T}}\right)$ can be as large as $\rho(A)$ in modulus; therefore we have inequality (8.83).

Finally, we recall equation (3.212), with w and v as defined above, and with the eigenvalue $\rho(A)$,

$$\left(A - \rho(A)vw^{\mathrm{T}}\right)^{k} = A^{k} - (\rho(A))^{k}vw^{\mathrm{T}}, \tag{8.84}$$

for $k = 1, 2, \ldots$.

Dividing both sides of equation (8.84) by $(\rho(A))^{k}$ and rearranging terms, we have

$$\left(\frac{A}{\rho(A)}\right)^{k} = vw^{\mathrm{T}} + \frac{\left(A - \rho(A)vw^{\mathrm{T}}\right)}{\rho(A)}. \tag{8.85}$$

Now

$$\rho\left(\frac{\left(A - \rho(A)vw^{\mathrm{T}}\right)}{\rho(A)}\right) = \frac{\rho\left(A - \rho(A)vw^{\mathrm{T}}\right)}{\rho(A)},$$

which is less than 1; hence, from equation (3.245) on page 135, we have

$$\lim_{k \to \infty} \left(\frac{\left(A - \rho(A)vw^{\mathrm{T}}\right)}{\rho(A)}\right)^{k} = 0;$$

so, taking the limit in equation (8.85), we have equation (8.82).

Applications of the Perron-Frobenius theorem are far-ranging. It has implications for the convergence of some iterative algorithms, such as the power method discussed in Section 7.2. The most important applications in statistics are in the analysis of Markov chains, which we discuss in Section 9.7.1.

8.7.3 Stochastic Matrices

A nonnegative matrix A such that

$$P1 = 1 \tag{8.86}$$

is called a *stochastic matrix*. The definition means that $(1, 1)$ is an eigenpair of any stochastic matrix. It is also clear that if P is a stochastic matrix, then $\|P\|_{\infty} = 1$ (see page 130), and because $\rho(P) \le \|P\|$ for any norm (see page 134) and 1 is an eigenvalue of P, we have $\rho(P) = 1$.

A stochastic matrix may not be positive, and it may be reducible or irreducible. (Hence, $(1,1)$ may not be the Perron root and Perron eigenvector.)

If P is a stochastic matrix such that

$$1^{\mathrm{T}}P = 1^{\mathrm{T}}, \tag{8.87}$$

it is called a *doubly stochastic matrix*. If P is a doubly stochastic matrix, $\|P\|_1 = 1$, and, of course, $\|P\|_\infty = 1$ and $\rho(P) = 1$.

A permutation matrix is a doubly stochastic matrix; in fact, it is the simplest and one of the most commonly encountered doubly stochastic matrices. A permutation matrix is clearly reducible.

Stochastic matrices are particularly interesting because of their use in defining a discrete homogeneous Markov chain. In that application, a stochastic matrix and *distribution vectors* play key roles. A distribution vector is a nonnegative matrix whose elements sum to 1; that is, a vector v such that $1^{\mathrm{T}}v = 1$. In Markov chain models, the stochastic matrix is a probability transition matrix from a distribution at time t, π_t, to the distribution at time $t+1$,

$$\pi_{t+1} = P\pi_t.$$

In Section 9.7.1, we define some basic properties of Markov chains. Those properties depend in large measure on whether the transition matrix is reducible or not.

8.7.4 Leslie Matrices

Another type of nonnegative transition matrix, often used in population studies, is a *Leslie matrix*, after P. H. Leslie, who used it in models in demography. A Leslie matrix is a matrix of the form

$$\begin{bmatrix} \alpha_1 & \alpha_2 & \cdots & \alpha_{m-1} & \alpha_m \\ \sigma_1 & 0 & \cdots & 0 & 0 \\ 0 & \sigma_2 & \cdots & 0 & 0 \\ \vdots & \vdots & \vdots & \ddots & \vdots \\ 0 & 0 & \cdots & \sigma_{m-1} & 0 \end{bmatrix}, \tag{8.88}$$

where all elements are nonnegative, and additionally $\sigma_i \leq 1$.

A Leslie matrix is clearly reducible. Furthermore, a Leslie matrix has a single unique positive eigenvalue (see Exercise 8.10), which leads to some interesting properties (see Section 9.7.2).

8.8 Other Matrices with Special Structures

Matrices of a variety of special forms arise in statistical analyses and other applications. For some matrices with special structure, specialized algorithms can increase the speed of performing a given task considerably. Many tasks involving matrices require a number of computations of the order of n^3, where n is the number of rows or columns of the matrix. For some of the matrices discussed in this section, because of their special structure, the order of computations may be n^2. The improvement from $O(n^3)$ to $O(n^2)$ is enough to make some tasks feasible that would otherwise be infeasible because of the time required to complete them. The collection of papers in Olshevsky (2003) describe various specialized algorithms for the kinds of matrices discussed in this section.

8.8.1 Helmert Matrices

A *Helmert matrix* is a square orthogonal matrix that partitions sums of squares. Its main use in statistics is in defining contrasts in general linear models to compare the second level of a factor with the first level, the third level with the average of the first two, and so on. (There is another meaning of "Helmert matrix" that arises from so-called Helmert transformations used in geodesy.)

For example, a partition of the sum $\sum_{i=1}^{n} y_i^2$ into orthogonal sums each involving $\bar{y}_k^2$ and $\sum_{i=1}^{k}(y_i - \bar{y}_k)^2$ is

$$\tilde{y}_i = (i(i+1))^{-1/2}\left(\sum_{j=1}^{i+1} y_j - (i+1)y_{i+1}\right) \quad \text{for } i = 1, \ldots n-1,$$

$$\tag{8.89}$$

$$\tilde{y}_n = n^{-1/2}\sum_{j=1}^{n} y_j.$$

These expressions lead to a computationally stable one-pass algorithm for computing the sample variance (see equation (10.7) on page 411).

The Helmert matrix that corresponds to this partitioning has the form

$$H_n = \begin{bmatrix} 1/\sqrt{n} & 1/\sqrt{n} & 1/\sqrt{n} & \cdots & 1/\sqrt{n} \\ 1/\sqrt{2} & -1/\sqrt{2} & 0 & \cdots & 0 \\ 1/\sqrt{6} & 1/\sqrt{6} & -2/\sqrt{6} & \cdots & 0 \\ \vdots & \vdots & \vdots & \ddots & \vdots \\ \frac{1}{\sqrt{n(n-1)}} & \frac{1}{\sqrt{n(n-1)}} & \frac{1}{\sqrt{n(n-1)}} & \cdots & -\frac{(n-1)}{\sqrt{n(n-1)}} \end{bmatrix}$$

$$= \begin{bmatrix} 1/\sqrt{n}\, 1_n^{\mathrm{T}} \\ K_{n-1} \end{bmatrix}, \tag{8.90}$$

where K_{n-1} is the $(n-1) \times n$ matrix below the first row. For the full n-vector y, we have

$$y^{\mathrm{T}} K_{n-1}^{\mathrm{T}} K_{n-1} y = \sum (y_i - \bar{y})^2$$
$$= \sum (y_i - \bar{y})^2$$
$$= (n-1) s_y^2.$$

The rows of the matrix in equation (8.90) correspond to *orthogonal contrasts* in the analysis of linear models (see Section 9.2.2).

Obviously, the sums of squares are never computed by forming the Helmert matrix explicitly and then computing the quadratic form, but the computations in partitioned Helmert matrices are performed indirectly in analysis of variance, and representation of the computations in terms of the matrix is often useful in the analysis of the computations.

8.8.2 Vandermonde Matrices

A *Vandermonde matrix* is an $n \times m$ matrix with columns that are defined by monomials,

$$V_{n \times m} = \begin{bmatrix} 1 & x_1 & x_1^2 & \cdots & x_1^{m-1} \\ 1 & x_2 & x_2^2 & \cdots & x_2^{m-1} \\ \vdots & \vdots & \vdots & \ddots & \vdots \\ 1 & x_n & x_n^2 & \cdots & x_n^{m-1} \end{bmatrix},$$

where $x_i \neq x_j$ if $i \neq j$. The Vandermonde matrix arises in polynomial regression analysis. For the model equation $y_i = \beta_0 + \beta_1 x_i + \cdots + \beta_p x_i^p + \epsilon_i$, given observations on y and x, a Vandermonde matrix is the matrix in the standard representation $y = X\beta + \epsilon$.

Because of the relationships among the columns of a Vandermonde matrix, computations for polynomial regression analysis can be subject to numerical errors, and so sometimes we make transformations based on orthogonal polynomials. (The "condition number", which we define in Section 6.1, for a Vandermonde matrix is large.) A Vandermonde matrix, however, can be used to form simple orthogonal vectors that correspond to orthogonal polynomials. For example, if the xs are chosen over a grid on $[-1, 1]$, a QR factorization (see Section 5.7 on page 188) yields orthogonal vectors that correspond to Legendre polynomials. These vectors are called discrete Legendre polynomials. Although not used in regression analysis so often now, orthogonal vectors are useful in selecting settings in designed experiments.

Vandermonde matrices also arise in the representation or approximation of a probability distribution in terms of its moments.

The determinant of a square Vandermonde matrix has a particularly simple form (see Exercise 8.11).

8.8.3 Hadamard Matrices and Orthogonal Arrays

In a wide range of applications, including experimental design, cryptology, and other areas of combinatorics, we often encounter matrices whose elements are chosen from a set of only a few different elements. In experimental design, the elements may correspond to the levels of the factors; in cryptology, they may represent the letters of an alphabet. In two-level factorial designs, the entries may be either 0 or 1. Matrices all of whose entries are either 1 or -1 can represent the same layouts, and such matrices may have interesting mathematical properties.

An $n \times n$ matrix with $-1, 1$ entries whose determinant is $n^{n/2}$ is called a *Hadamard matrix*. The name comes from the bound derived by Hadamard for the determinant of any matrix A with $|a_{ij}| \leq 1$ for all i, j: $|\det(A)| \leq n^{n/2}$. A Hadamard matrix achieves this upper bound. A maximal determinant is often used as a criterion for a good experimental design. One row and one column of an $n \times n$ Hadamard matrix consist of all 1s; all $n - 1$ other rows and columns consist of $n/2$ 1s and $n/2$ -1s. We often denote an $n \times n$ Hadamard matrix by H_n, which is the same notation often used for a Helmert matrix, but in the case of Hadamard matrices, the matrix is not unique. All rows are orthogonal and so are all columns. The norm of each row or column is n, so $H_n^{\mathrm{T}} H_n = nI$.

A Hadamard matrix is often represented as a mosaic of black and white squares, as in Figure 8.7.

$$\begin{bmatrix} 1 & -1 & -1 & 1 \\ 1 & 1 & -1 & -1 \\ 1 & -1 & 1 & -1 \\ 1 & 1 & 1 & 1 \end{bmatrix}$$

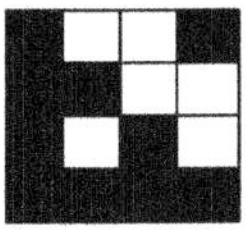

Fig. 8.7. A 4×4 Hadamard Matrix

Hadamard matrices do not exist for all n. Clearly, n must be even because $|H_n| = n^{n/2}$, but some experimentation (or an exhaustive search) quickly shows that there is no Hadamard matrix for $n = 6$. It has been conjectured, but not proven, that Hadamard matrices exist for any n divisible by 4. Given any $n \times n$ Hadamard matrix, H_n, and any $m \times m$ Hadamard matrix, H_m, an $nm \times nm$ Hadamard matrix can be formed as a partitioned matrix in which each 1 in H_n is replaced by the block submatrix H_m and each -1 is replaced by the block submatrix $-H_m$. For example, the 4×4 Hadamard matrix shown in Figure 8.7 is formed using the 2×2 Hadamard matrix

$$\begin{bmatrix} 1 & -1 \\ 1 & 1 \end{bmatrix}$$

as both H_n and H_m. Not all Hadamard matrices can be formed from other Hadamard matrices in this way, however.

A somewhat more general type of matrix corresponds to an $n \times m$ array with the elements in the jth column being members of a set of k_j elements and such that, for some fixed $p \leq m$, in every $n \times p$ submatrix all possible combinations of the elements of the m sets occur equally often as a row. (I make a distinction between the *matrix* and the *array* because often in applications the elements in the array are treated merely as symbols without the assumptions of an algebra of a field. A terminology for orthogonal arrays has evolved that is different from the terminology for matrices; for example, a *symmetric* orthogonal array is one in which $k_1 = \cdots = k_m$. On the other hand, treating the orthogonal arrays as matrices with real elements may provide solutions to combinatorial problems such as may arise in optimal design.)

The 4×4 Hadamard matrix shown in Figure 8.7 is a symmetric orthogonal array with $k_1 = \cdots = k_4 = 2$ and $p = 4$, so in the array each of the possible combinations of elements occurs exactly once. This array is a member of a simple class of symmetric orthogonal arrays that has the property that in any two rows each ordered pair of elements occurs exactly once.

Orthogonal arrays are particularly useful in developing fractional factorial plans. (The robust designs of Taguchi correspond to orthogonal arrays.) Dey and Mukerjee (1999) discuss orthogonal arrays with an emphasis on the applications in experimental design, and Hedayat, Sloane, and Stufken (1999) provide extensive an discussion of the properties of orthogonal arrays.

8.8.4 Toeplitz Matrices

If the elements of the matrix A are such that $a_{i,i+c_k} = d_{c_k}$, where d_{c_k} is constant for fixed c_k, then A is called a *Toeplitz matrix*,

$$
\begin{bmatrix}
d_0 & d_1 & d_2 & \cdots & d_{n-1} \\
d_{-1} & d_0 & d_1 & \cdots & d_{n-2} \\
\vdots & \vdots & \vdots & \ddots & \vdots \\
d_{-n+2} & d_{-n+3} & d_{-n+4} & \ddots & d_1 \\
d_{-n+1} & d_{-n+2} & d_{-n+3} & \cdots & d_0
\end{bmatrix} ;
$$

that is, a Toeplitz matrix is a matrix with constant codiagonals. A Toeplitz matrix may or may not be a band matrix (i.e., have many 0 codiagonals) and it may or may not be symmetric.

Banded Toeplitz matrices arise frequently in time series studies. The covariance matrix in an $\text{ARMA}(p, q)$ process, for example, is a symmetric Toeplitz matrix with $2\max(p, q)$ nonzero off-diagonal bands. See page 364 for an example and further discussion.

Inverses of Toeplitz Matrices and Other Banded Matrices

A Toeplitz matrix that occurs often in stationary time series is the $n \times n$ variance-covariance matrix of the form

$$
V = \sigma^2
\begin{bmatrix}
1 & \rho & \rho^2 & \cdots & \rho^{n-1} \\
\rho & 1 & \rho & \cdots & \rho^{n-2} \\
\vdots & \vdots & \vdots & \vdots & \vdots \\
\rho^{n-1} & \rho^{n-2} & \rho^{n-3} & \cdots & 1
\end{bmatrix}.
$$

It is easy to see that V^{-1} exists if $\sigma \neq 0$ and $\rho \neq 1$, and that it is the type 2 matrix

$$
V^{-1} = \frac{1}{(1-\rho^2)\sigma^2}
\begin{bmatrix}
1 & -\rho & 0 & \cdots & 0 \\
-\rho & 1+\rho^2 & -\rho & \cdots & 0 \\
0 & -\rho & 1+\rho^2 & \cdots & 0 \\
\vdots & \vdots & \vdots & \vdots & \vdots \\
0 & 0 & 0 & \cdots & 1
\end{bmatrix}.
$$

Type 2 matrices also occur as the inverses of other matrices with special patterns that arise in other common statistical applications (see Graybill, 1983, for examples).

The inverses of all banded invertible matrices have off-diagonal submatrices that are zero or have low rank, depending on the bandwidth of the original matrix (see Strang and Nguyen, 2004, for further discussion and examples).

8.8.5 Hankel Matrices

A *Hankel matrix* is an $n \times m$ matrix $H(c, r)$ generated by an n-vector c and an m-vector r such that the (i, j) element is

$$
\begin{aligned}
& c_{i+j-1} \text{ if } i+j-1 \leq n, \\
& r_{i+j-n} \text{ otherwise.}
\end{aligned}
$$

A common form of Hankel matrix is an $n \times n$ skew upper triangular matrix, and it is formed from the c vector only. This kind of matrix occurs in the spectral analysis of time series. If $f(t)$ is a (discrete) time series, for $t = 0, 1, 2, \ldots$, the Hankel matrix of the time series has as the (i, j) element

$$
\begin{aligned}
& f(i+j-2) \text{ if } i+j-1 \leq n, \\
& 0 \qquad\qquad \text{ otherwise.}
\end{aligned}
$$

The L_2 norm of the Hankel matrix of the time series (of the "impulse function", f) is called the *Hankel norm* of the filter frequency response (the Fourier transform).

The simplest form of the square skew upper triangular Hankel matrix is formed from the vector $c = (1, 2, \ldots, n)$:

$$
\begin{bmatrix}
1 & 2 & 3 & \cdots & n \\
2 & 3 & 4 & \cdots & 0 \\
\vdots & \vdots & \vdots & \cdots & \vdots \\
n & 0 & 0 & \cdots & 0
\end{bmatrix}.
\qquad (8.91)
$$

8.8.6 Cauchy Matrices

Another type of special $n \times m$ matrix whose elements are determined by a few n-vectors and m-vectors is a Cauchy-type matrix. The standard Cauchy matrix is built from two vectors, x and y. The more general form defined below uses two additional vectors.

A *Cauchy matrix* is an $n \times m$ matrix $C(x, y, v, w)$ generated by n-vectors x and v and m-vectors y and w of the form

$$
C(x, y, v, w) =
\begin{bmatrix}
\dfrac{v_1 w_1}{x_1 - y_1} & \cdots & \dfrac{v_1 w_m}{x_1 - y_m} \\
\vdots & \cdots & \vdots \\
\dfrac{v_n w_1}{x_n - y_1} & \cdots & \dfrac{v_n w_m}{x_n - y_m}
\end{bmatrix}.
\qquad (8.92)
$$

Cauchy-type matrices often arise in the numerical solution of partial differential equations (PDEs). For Cauchy matrices, the order of the number of computations for factorization or solutions of linear systems can be reduced from a power of three to a power of two. This is a very significant improvement for large matrices. In the PDE applications, the matrices are generally not large, but nevertheless, even in those applications, it it worthwhile to use algorithms that take advantage of the special structure. Fasino and Gemignani (2003) describe such an algorithm.

8.8.7 Matrices Useful in Graph Theory

Many problems in statistics and applied mathematics can be posed as graphs, and various methods of graph theory can be used in their solution.

Graph theory is particularly useful in cluster analysis or classification. These involve the analysis of relationships of objects for the purpose of identifying similar groups of objects. The objects are associated with vertices of the graph, and an edge is generated if the relationship (measured somehow) between two objects is sufficiently great. For example, suppose the question of interest is the authorship of some text documents. Each document is a vertex, and an edge between two vertices exists if there are enough words in common between the two documents. A similar application could be the determination of which computer user is associated with a given computer session. The vertices would correspond to login sessions, and the edges would be established based on the commonality of programs invoked or files accessed. In applications such as these, there would typically be a training dataset consisting of text documents with known authors or consisting of session logs with known users. In both of these types of applications, decisions would have to be made about the extent of commonality of words, phrases, programs invoked, or files accessed in order to establish an edge between two documents or sessions.

Unfortunately, as is often the case for an area of mathematics or statistics that developed from applications in diverse areas or through the efforts of applied mathematicians somewhat outside of the mainstream of mathematics, there are major inconsistencies in the notation and terminology employed in graph theory. Thus, we often find different terms for the same object; for example, adjacency matrix and connectivity matrix. This unpleasant situation, however, is not so disagreeable as a one-to-many inconsistency, such as the designation of the eigenvalues of a graph to be the eigenvalues of one type of matrix in some of the literature and the eigenvalues of different types of matrices in other literature.

Adjacency Matrix; Connectivity Matrix

We discussed adjacency or connectivity matrices on page 265. A matrix, such as an adjacency matrix, that consists of only 1s and 0s is called a *Boolean matrix*.

Two vertices that are not connected and hence correspond to a 0 in a connectivity matrix are said to be *independent*.

If no edges connect a vertex with itself, the adjacency matrix is a hollow matrix.

Because the 1s in a connectivity matrix indicate a strong association, and we would naturally think of a vertex as having a strong association with itself, we sometimes modify the connectivity matrix so as to have 1s along the diagonal. Such a matrix is sometimes called an *augmented connectivity matrix* or *augmented associativity matrix*.

The eigenvalues of the adjacency matrix reveal some interesting properties of the graph and are sometimes called the *eigenvalues of the graph*. The eigenvalues of another matrix, which we discuss below, are more useful, however, and we will refer to them as the eigenvalues of the graph.

Digraphs

The digraph represented in Figure 8.4 on page 266 is a network with five vertices, perhaps representing cities, and directed edges between some of the vertices. The edges could represent airline connections between the cities; for example, there are flights from x to u and from u to x, and from y to z, but not from z to y.

In a digraph, the relationships are directional. (An example of a directional relationship that might be of interest is when each observational unit has a different number of measured features, and a relationship exists from v_i to v_j if a majority of the features of v_i are identical to measured features of v_j.)

Use of the Connectivity Matrix

The analysis of a network may begin by identifying which vertices are connected with others; that is, by construction of the connectivity matrix.

The connectivity matrix can then be used to analyze other levels of association among the data represented by the graph or digraph. For example, from the connectivity matrix in equation (8.2) on page 266, we have

$$
C^2 = \begin{bmatrix}
4 & 1 & 0 & 0 & 1 \\
0 & 1 & 1 & 1 & 1 \\
1 & 1 & 1 & 1 & 2 \\
1 & 2 & 1 & 1 & 1 \\
1 & 1 & 1 & 1 & 1
\end{bmatrix}.
$$

In terms of the application suggested on page 266 for airline connections, the matrix C^2 represents the number of connections between the cities that consist of exactly two flights. From C^2 we see that there are two ways to go from city y to city w in just two flights but only one way to go from w to y in two flights.

A power of a connectivity matrix for a nondirected graph is symmetric.

The Laplacian Matrix of a Graph

Spectral graph theory is concerned with the analysis of the eigenvalues of a graph. As mentioned above, there are two different definitions of the eigenvalues of a graph. The more useful definition, and the one we use here, takes the eigenvalues of a graph to be the eigenvalues of a matrix, called the Laplacian matrix, formed from the adjacency matrix and a diagonal matrix consisting of the degrees of the vertices.

Given the graph $\mathcal{G}$, let $D(\mathcal{G})$ be a diagonal matrix consisting of the degrees of the vertices of $\mathcal{G}$ (that is, $D(\mathcal{G}) = \mathrm{diag}(d(\mathcal{G}))$) and let $C(\mathcal{G})$ be the adjacency matrix of $\mathcal{G}$. If there are no isolated vertices (that is if $d(\mathcal{G}) > 0$), then the *Laplacian matrix* of the graph, $L(\mathcal{G})$ is given by

$$
L(\mathcal{G}) = I - D(\mathcal{G})^{-\frac{1}{2}} C(\mathcal{G}) D(\mathcal{G})^{-\frac{1}{2}}. \tag{8.93}
$$

Some authors define the Laplacian in other ways:

$$
L_a(\mathcal{G}) = I - D(\mathcal{G})^{-1} C(\mathcal{G}) \tag{8.94}
$$

or

$$
L_b(\mathcal{G}) = D(\mathcal{G}) - C(\mathcal{G}). \tag{8.95}
$$

The eigenvalues of the Laplacian matrix are the *eigenvalues of a graph*. The definition of the Laplacian matrix given in equation (8.93) seems to be more useful in terms of bounds on the eigenvalues of the graph. The set of unique eigenvalues (the spectrum of the matrix L) is called the spectrum of the graph.

So long as $d(\mathcal{G}) > 0$, $L(\mathcal{G}) = D(\mathcal{G})^{-\frac{1}{2}} L_a(\mathcal{G}) D(\mathcal{G})^{-\frac{1}{2}}$. Unless the graph is regular, the matrix $L_b(\mathcal{G})$ is not symmetric. Note that if $\mathcal{G}$ is k-regular, $L(\mathcal{G}) = I - C(\mathcal{G})/k$, and $L_b(\mathcal{G}) = L(\mathcal{G})$.

For a digraph, the degrees are replaced by either the indegrees or the outdegrees. (Some authors define it one way and others the other way. The essential properties hold either way.)

The Laplacian can be viewed as an operator on the space of functions $f : V(\mathcal{G}) \rightarrow \mathbb{R}$ such that for the vertex v

$$L(f(v)) = \frac{1}{\sqrt{d_v}} \sum_{w, w \sim v} \left(\frac{f(v)}{\sqrt{d_v}} - \frac{f(w)}{\sqrt{d_w}} \right),$$

where $w \sim v$ means vertices w and v that are adjacent, and d_u is the degree of the vertex u.

The Laplacian matrix is symmetric, so its eigenvalues are all real. We can see that the eigenvalues are all nonnegative by forming the Rayleigh quotient (equation (3.209)) using an arbitrary vector g, which can be viewed as a real-valued function over the vertices,

$$\begin{aligned}
R_L(g) &= \frac{\langle g, Lg \rangle}{\langle g, g \rangle} \\
&= \frac{\langle g, D^{-\frac{1}{2}} L_a D^{-\frac{1}{2}} g \rangle}{\langle g, g \rangle} \\
&= \frac{\langle f, L_a f \rangle}{\langle D^{\frac{1}{2}} f, D^{\frac{1}{2}} f \rangle} \\
&= \frac{\sum_{v \sim w} (f(v) - f(w))^2}{f^{\mathrm{T}} D f},
\end{aligned} \tag{8.96}$$

where $f = D^{-\frac{1}{2}} g$, and $f(u)$ is the element of the vector corresponding to vertex u. Because the Raleigh quotient is nonnegative, all eigenvalues are nonnegative, and because there is an $f \neq 0$ for which the Rayleigh quotient is 0, we see that 0 is an eigenvalue of a graph. Furthermore, using the Cauchy-Schwartz inequality, we see that the spectral radius is less than or equal to 2.

The eigenvalues of a matrix are the basic objects in spectral graph theory. They provide information about the properties of networks and other systems modeled by graphs. We will not explore them further here, and the interested reader is referred to Chung (1997) or other general texts on the subject.

If $\mathcal{G}$ is the graph represented in Figure 8.2 on page 264, with $V(\mathcal{G}) = \{a, b, c, d, e\}$, the degrees of the vertices of the graph are $d(\mathcal{G}) = (4, 2, 2, 3, 3)$. Using the adjacency matrix given in equation (8.1), we have

$$
L(\mathcal{G}) = \begin{bmatrix}
1 & -\frac{\sqrt{2}}{4} & -\frac{\sqrt{2}}{4} & -\frac{\sqrt{3}}{6} & -\frac{\sqrt{3}}{6} \\[1em]
-\frac{\sqrt{2}}{4} & 1 & 0 & 0 & -\frac{\sqrt{6}}{6} \\[1em]
-\frac{\sqrt{2}}{4} & 0 & 1 & -\frac{\sqrt{6}}{6} & 0 \\[1em]
-\frac{\sqrt{3}}{6} & 0 & -\frac{\sqrt{6}}{6} & 1 & -\frac{1}{3} \\[1em]
-\frac{\sqrt{3}}{6} & -\frac{\sqrt{6}}{6} & 0 & -\frac{1}{3} & 1
\end{bmatrix}. \tag{8.97}
$$

This matrix is singular, and the unnormalized eigenvector corresponding to the 0 eigenvalue is $(2\sqrt{14}, 2\sqrt{7}, 2\sqrt{7}, \sqrt{42}, \sqrt{42})$.

8.8.8 M-Matrices

In certain applications in physics and in the solution of systems of nonlinear differential equations, a class of matrices called M-matrices is important.

The matrices in these applications have nonpositive off-diagonal elements. A square matrix all of whose off-diagonal elements are nonpositive is called a Z-matrix.

A Z-matrix that is positive stable (see page 125) is called an M-matrix. A real symmetric M-matrix is positive definite.

In addition to the properties that constitute the definition, M-matrices have a number of remarkable properties, which we state here without proof. If A is a real M-matrix, then

- all principal minors of A are positive;
- all diagonal elements of A are positive;
- all diagonal elements of L and U in the LU decomposition of A are positive;
- for any i, $\sum_j a_{ij} \geq 0$; and
- A is nonsingular and $A^{-1} \geq 0$.

Proofs of these facts can be found in Horn and Johnson (1991).

Exercises

8.1. Ordering of nonnegative definite matrices.
   a) A relation $\bowtie$ on a set is a *partial ordering* if, for elements a, b, and c,
   - it is reflexive: $a \bowtie a$;
   - it is antisymmetric: $a \bowtie b \bowtie a \implies a = b$; and
   - it is transitive: $a \bowtie b \bowtie c \implies a \bowtie c$.
   Show that the relation $\succeq$ (equation (8.19)) is a partial ordering.

b) Show that the relation $\succ$ (equation (8.20)) is transitive.

8.2. Show that a diagonally dominant symmetric matrix with positive diagonals is positive definite.

8.3. Show that the number of positive eigenvalues of an idempotent matrix is the rank of the matrix.

8.4. Show that two idempotent matrices of the same rank are similar.

8.5. Under the given conditions, show that properties (a) and (b) on page 285 imply property (c).

8.6. Projections.

   a) Show that the matrix given in equation (8.42) (page 287) is a projection matrix.

   b) Write out the projection matrix for projecting a vector onto the plane formed by two vectors, x_1 and x_2, as indicated on page 287, and show that it is the same as the hat matrix of equation (8.55).

8.7. Show that the matrix $X^{\mathrm{T}}X$ is symmetric (for any matrix X).

8.8. Correlation matrices.

   A correlation matrix can be defined in terms of a Gramian matrix formed by a centered and scaled matrix, as in equation (8.72). Sometimes in the development of statistical theory, we are interested in the properties of correlation matrices with given eigenvalues or with given ratios of the largest eigenvalue to other eigenvalues.

   Write a program to generate $n \times n$ random correlation matrices R with specified eigenvalues, $c_1, \ldots, c_n$. The only requirements on R are that its diagonals be 1, that it be symmetric, and that its eigenvalues all be positive and sum to n. Use the following method due to Davies and Higham (2000) that uses random orthogonal matrices with the Haar uniform distribution generated using the method described in Exercise 4.7.

   0. Generate a random orthogonal matrix Q; set $k = 0$, and form

$$R^{(0)} = Q\mathrm{diag}(c1, \ldots, c_n)Q^{\mathrm{T}}.$$

   1. If $r_{ii}^{(k)} = 1$ for all i in $\{1, \ldots, n\}$, go to step 3.

   2. Otherwise, choose p and q with $p < j$, such that $r_{pp}^{(k)} < 1 < r_{qq}^{(k)}$ or $r_{pp}^{(k)} > 1 > r_{qq}^{(k)}$, and form $G^{(k)}$ as in equation (5.13), where c and s are as in equations (5.17) and (5.17), with $a = 1$.
   Form $R^{(k+1)} = (G^{(k)})^{\mathrm{T}}R^{(k)}G^{(k)}$.
   Set $k = k + 1$, and go to step 1.

   3. Deliver $R = R^{(k)}$.

8.9. Use the relationship (8.80) to prove properties 1 and 4 on page 304.

8.10. Leslie matrices.

   a) Write the characteristic polynomial of the Leslie matrix, equation (8.88).

   b) Show that the Leslie matrix has a single, unique positive eigenvalue.

8.11. Write out the determinant for an $n \times n$ Vandermonde matrix.

8.12. Write out the determinant for the $n \times n$ skew upper triangular Hankel matrix in (8.91).

9

Selected Applications in Statistics

Data come in many forms. In the broad view, the term "data" embraces all representations of information or knowledge. There is no single structure that can efficiently contain all of these representations. Some data are in free-form text (for example, the Federalist Papers, which was the subject of a famous statistical analysis), other data are in a hierarchical structure (for example, political units and subunits), and still other data are encodings of methods or algorithms. (This broad view is entirely consistent with the concept of a "stored-program computer"; the program is the data.)

Structure in Data and Statistical Data Analysis

Data often have a logical structure as described in Section 8.1.1; that is, a two-dimensional array in which columns correspond to variables or measurable attributes and rows correspond to an observation on all attributes taken together. A matrix is obviously a convenient object for representing numeric data organized this way. An objective in analyzing data of this form is to uncover relationships among the variables, or to characterize the distribution of the sample over $\mathrm{I\!R}^m$. Interesting relationships and patterns are called "structure" in the data. This is a different meaning from that of the word used in the phrase "logical structure" or in the phrase "data structure" used in computer science.

Another type of pattern that may be of interest is a temporal pattern; that is, a set of relationships among the data and the time or the sequence in which the data were observed.

The objective of this chapter is to illustrate how some of the properties of matrices and vectors that were covered in previous chapters relate to statistical models and to data analysis procedures. The field of statistics is far too large for a single chapter on "applications" to cover more than just a small part of the area. Similarly, the topics covered previously are too extensive to give examples of applications of all of them.

A probability distribution is a specification of the stochastic structure of random variables, so we begin with a brief discussion of properties of multivariate probability distributions. The emphasis is on the multivariate normal distribution and distributions of linear and quadratic transformations of normal random variables. We then consider an important structure in multivariate data, a linear model. We discuss some of the computational methods used in analyzing the linear model. We then describe some computational method for identifying more general linear structure and patterns in multivariate data. Next we consider approximation of matrices in the absence of complete data. Finally, we discuss some models of stochastic processes. The special matrices discussed in Chapter 8 play an important role in this chapter.

9.1 Multivariate Probability Distributions

Most methods of statistical inference are based on assumptions about some underlying probability distribution of a random variable. In some cases these assumptions completely specify the form of the distribution, and in other cases, especially in nonparametric methods, the assumptions are more general. Many statistical methods in estimation and hypothesis testing rely on the properties of various transformations of a random variable.

In this section, we do not attempt to develop a theory of probability distribution; rather we assume some basic facts and then derive some important properties that depend on the matrix theory of the previous chapters.

9.1.1 Basic Definitions and Properties

One of the most useful descriptors of a random variable is its probability density function (PDF), or probability function. Various functionals of the PDF define standard properties of the random variable, such as the mean and variance, as we discussed in Section 4.5.3.

If X is a random variable over $\mathrm{I\!R}^d$ with PDF $p_X(\cdot)$ and $f(\cdot)$ is a measurable function (with respect to a dominating measure of $p_X(\cdot)$) from $\mathrm{I\!R}^d$ to $\mathrm{I\!R}^k$, the *expected value* of $f(X)$, which is in $\mathrm{I\!R}^k$ and is denoted by $\mathrm{E}(g(X))$, is defined by

$$\mathrm{E}(f(X)) = \int_{\mathrm{I\!R}^d} f(t) p_X(t)\, \mathrm{d}t.$$

The *mean* of X is the d-vector $\mathrm{E}(X)$, and the *variance* or *variance-covariance* of X, denoted by $\mathrm{V}(X)$, is the $d \times d$ matrix

$$\mathrm{V}(X) = \mathrm{E}\left((X - \mathrm{E}(X))(X - \mathrm{E}(X))^{\mathrm{T}} \right).$$

Given a random variable X, we are often interested in a random variable defined as a function of X, say $Y = g(X)$. To analyze properties of Y, we identify g^{-1}, which may involve another random variable. (For example, if $g(x) = x^2$ and the support of X is $\mathrm{I\!R}$, then $g^{-1}(Y) = (-1)^\alpha \sqrt{Y}$, where $\alpha = 1$ with probability $\Pr(X < 0)$ and $\alpha = 0$ otherwise.) Properties of Y can be evaluated using the Jacobian of $g^{-1}(\cdot)$, as in equation (4.12).

9.1.2 The Multivariate Normal Distribution

The most important multivariate distribution is the multivariate normal, which we denote as $\mathrm{N}_d(\mu, \Sigma)$ for d dimensions; that is, for a random d-vector. The PDF for the d-variate normal distribution is

$$p_X(x) = (2\pi)^{-d/2}|\Sigma|^{-1/2}\mathrm{e}^{-(x-\mu)^\mathrm{T}\Sigma^{-1}(x-\mu)/2}, \tag{9.1}$$

where the normalizing constant is Aitken's integral given in equation (4.39). The multivariate normal distribution is a good model for a wide range of random phenomena.

9.1.3 Derived Distributions and Cochran's Theorem

If X is a random variable with distribution $\mathrm{N}_d(\mu, \Sigma)$, A is a $q \times d$ matrix with rank q (which implies $q \le d$), and $Y = AX$, then the straightforward change-of-variables technique yields the distribution of Y as $\mathrm{N}_d(A\mu, A\Sigma A^\mathrm{T})$.

Useful transformations of the random variable X with distribution $\mathrm{N}_d(\mu, \Sigma)$ are $Y_1 = \Sigma^{-1/2}X$ and $Y_2 = \Sigma_C^{-1}X$, where Σ_C is a Cholesky factor of Σ. In either case, the variance-covariance matrix of the transformed variate Y_1 or Y_2 is I_d.

Quadratic forms involving a Y that is distributed as $\mathrm{N}_d(\mu, I_d)$ have useful properties. For statistical inference it is important to know the distribution of these quadratic forms. The simplest quadratic form involves the identity matrix: $S_d = Y^\mathrm{T}Y$.

We can derive the PDF of S_d by beginning with $d = 1$ and using induction. If $d = 1$, for $t > 0$, we have

$$\Pr(S_1 \le t) = \Pr(Y \le \sqrt{t}) - \Pr(Y \le -\sqrt{t}),$$

where $Y \sim \mathrm{N}_1(\mu, 1)$, and so the PDF of S_1 is

$$p_{S_1}(t) = \frac{1}{2\sqrt{2\pi t}} \left(e^{-(\sqrt{t}-\mu)^2/2} + e^{-(-\sqrt{t}-\mu)^2/2} \right)$$

$$= \frac{e^{-\mu^2/2}e^{-t/2}}{2\sqrt{2\pi t}} \left(e^{\mu\sqrt{t}} + e^{-\mu\sqrt{t}} \right)$$

$$= \frac{e^{-\mu^2/2}e^{-t/2}}{2\sqrt{2\pi t}} \left(\sum_{j=0}^{\infty} \frac{(\mu\sqrt{t})^j}{j!} + \sum_{j=0}^{\infty} \frac{(-\mu\sqrt{t})^j}{j!} \right)$$

$$= \frac{e^{-\mu^2/2}e^{-t/2}}{\sqrt{2t}} \sum_{j=0}^{\infty} \frac{(\mu^2 t)^j}{\sqrt{\pi}(2j)!}$$

$$= \frac{e^{-\mu^2/2}e^{-t/2}}{\sqrt{2t}} \sum_{j=0}^{\infty} \frac{(\mu^2 t)^j}{j!\Gamma(j+1/2)2^{2j}},$$

in which we use the fact that

$$\Gamma(j+1/2) = \frac{\sqrt{\pi}(2j)!}{j!2^{2j}}$$

(see page 484). This can now be written as

$$p_{S_1}(t) = e^{-\mu^2/2} \sum_{j=0}^{\infty} \frac{(\mu^2)^j}{j!2^j} \frac{1}{\Gamma(j+1/2)2^{j+1/2}} t^{j-1/2}e^{-t/2}, \tag{9.2}$$

in which we recognize the PDF of the central chi-squared distribution with $2j+1$ degrees of freedom,

$$p_{\chi^2_{2j+1}}(t) = \frac{1}{\Gamma(j+1/2)2^{j+1/2}} t^{j-1/2}e^{-t/2}.$$

A similar manipulation for $d = 2$ (that is, for $Y \sim \mathrm{N}_2(\mu, 1)$, and maybe $d = 3$, or as far as you need to go) leads us to a general form for the PDF of the $\chi^2_d(\delta)$ random variable S_d:

$$p_{S_d}(t) = e^{-\mu^2/2} \sum_{j=0}^{\infty} \frac{(\mu^2/2)^j}{j!} p_{\chi^2_{2j+1}}(t). \tag{9.3}$$

We can show that equation (9.3) holds for any d by induction. The distribution of S_d is called the noncentral chi-squared distribution with d degrees of freedom and noncentrality parameter $\delta = \mu^{\mathrm{T}}\mu$. We denote this distribution as $\chi^2_d(\delta)$.

The induction method above involves a special case of a more general fact: if X_i for $i = 1, \ldots, k$ are independently distributed as $\chi^2_{n_i}(\delta_i)$, then $\sum_i X_i$ is distributed as $\chi^2_n(\delta)$, where $n = \sum_i n_i$ and $\delta = \sum_i \delta_i$.

In applications of linear models, a quadratic form involving Y is often partitioned into a sum of quadratic forms. Assume that Y is distributed as

$N_d(\mu, I_d)$, and for $i = 1, \ldots k$, let A_i be a $d \times d$ symmetric matrix with rank r_i such that $\sum_i A_i = I_d$. This yields a partition of the total sum of squares $Y^T Y$ into k components:

$$Y^T Y = Y^T A_1 Y + \cdots + Y^T A_k Y. \tag{9.4}$$

One of the most important results in the analysis of linear models states that the $Y^T A_i Y$ have independent noncentral chi-squared distributions $\chi^2_{r_i}(\delta_i)$ with $\delta_i = \mu^T A_i \mu$ if and only if $\sum_i r_i = d$.

This is called Cochran's theorem. On page 283, we discussed a form of Cochran's theorem that applies to properties of idempotent matrices. Those results immediately imply the conclusion above.

9.2 Linear Models

Some of the most important applications of statistics involve the study of the relationship of one variable, often called a "response variable", to other variables. The response variable is usually modeled as a random variable, which we indicate by using a capital letter. A general model for the relationship of a variable, Y, to other variables, x (a vector), is

$$Y \approx f(x). \tag{9.5}$$

In this asymmetric model and others like it, we call Y the *dependent variable* and the elements of x the *independent variables*.

It is often reasonable to formulate the model with a *systematic component* expressing the relationship and an *additive random component* or "additive error". We write

$$Y = f(x) + E, \tag{9.6}$$

where E is a random variable with an expected value of 0; that is,

$$E(E) = 0.$$

(Although this is by far the most common type of model used by data analysts, there are other ways of building a model that incorporates systematic and random components.) The zero expectation of the random error yields the relationship

$$E(Y) = f(x),$$

although this expression is not equivalent to the additive error model above because the random component could just as well be multiplicative (with an expected value of 1) and the same value of $E(Y)$ would result.

Because the functional form f of the relationship between Y and x may contain a *parameter*, we may write the model as

$$Y = f(x; \theta) + E. \tag{9.7}$$

A specific form of this model is

$$Y = \beta^{\mathrm{T}} x + E, \tag{9.8}$$

which expresses the systematic component as a linear combination of the xs using the vector parameter β.

A model is more than an equation; there may be associated statements about the distribution of the random variable or about the nature of f or x. We may assume β (or θ) is a fixed but unknown constant, or we may assume it is a realization of a random variable. Whatever additional assumptions we may make, there are some standard assumptions that go with the model. We assume that Y and x are *observable* and θ and E are *unobservable*.

Models such as these that express an asymmetric relationship between some variables ("dependent variables") and other variables ("independent variables") are called regression models. A model such as equation (9.8) is called a linear regression model. There are many useful variations of the model (9.5) that express other kinds of relationships between the response variable and the other variables.

Notation

In data analysis with regression models, we have a set of observations $\{y_i, x_i\}$ where x_i is an m-vector. One of the primary tasks is to determine a reasonable value of the parameter. That is, in the linear regression model, for example, we think of β as an unknown variable (rather than as a fixed constant or a realization of a random variable), and we want to find a value of it such that the model fits the observations well,

$$y_i = \beta^{\mathrm{T}} x_i + \epsilon_i,$$

where β and x_i are m-vectors. (In the expression (9.8), "E" is an uppercase epsilon. We attempt to use notation consistently; "E" represents a random variable, and "ϵ" represents a realization, though an unobservable one, of the random variable. We will not always follow this convention, however; sometimes it is convenient to use the language more loosely and to speak of ϵ_i as a random variable.) The meaning of the phrase "the model fits the observations well" may vary depending on other aspects of the model, in particular, on any assumptions about the distribution of the random component E. If we make assumptions about the distribution, we have a basis for statistical estimation of β; otherwise, we can define some purely mathematical criterion for "fitting well" and proceed to determine a value of β that optimizes that criterion.

For any choice of β, say b, we have $y_i = b^{\mathrm{T}} x_i + r_i$. The r_is are determined by the observations. An approach that does not depend on any assumptions about the distribution but can nevertheless yield optimal estimators under many distributions is to choose the estimator so as to minimize some measure of the set of r_is.

Given the observations $\{y_i, x_i\}$, we can represent the regression model and the data as

$$y = X\beta + \epsilon, \tag{9.9}$$

where X is the $n \times m$ matrix whose rows are the x_is and ϵ is the vector of deviations ("errors") of the observations from the functional model. Throughout the rest of this section, *we will assume that the number of rows of X (that is, the number of observations n) is greater than the number of columns of X (that is, the number of variables m).*

We will occasionally refer to submatrices of the basic data matrix X using notation developed in Chapter 3. For example, $X_{(i_1,\ldots,i_k)(j_1,\ldots,j_l)}$ refers to the $k \times l$ matrix formed by retaining only the $i_1,\ldots,i_k$ rows and the $j_1,\ldots,j_l$ columns of X, and $X_{-(i_1,\ldots,i_k)(j_1,\ldots,j_l)}$ refers to the matrix formed by deleting the $i_1,\ldots,i_k$ rows and the $j_1,\ldots,j_l$ columns of X. We also use the notation x_{i*} to refer to the ith row of X (the row is a vector, a column vector), and x_{*j} to refer to the jth column of X. See page 487 for a summary of this notation.

9.2.1 Fitting the Model

In a model for a given dataset as in equation (9.9), although the errors are no longer random variables (they are realizations of random variables), they are not observable. To fit the model, we replace the unknowns with variables: β with b and ϵ with r. This yields

$$y = Xb + r.$$

We then proceed by applying some criterion for fitting.

The criteria generally focus on the "residuals" $r = y - Xb$. Two general approaches to fitting are:

- Define a likelihood function of r based on an assumed distribution of E, and determine a value of b that maximizes that likelihood.
- Decide on an appropriate norm on r, and determine a value of b that minimizes that norm.

There are other possible approaches, and there are variations on these two approaches. For the first approach, it must be emphasized that r is not a realization of the random variable E. Our emphasis will be on the second approach, that is, on methods that minimize a norm on r.

Statistical Estimation

The statistical problem is to *estimate* β. (Notice the distinction between the phrases "to *estimate* β" and "to determine a value of β that minimizes ...". The mechanical aspects of the two problems may be the same, of course.) The statistician uses the model and the given observations to explore relationships between the response and the regressors. Considering ϵ to be a realization of a random variable E (a vector) and assumptions about a distribution of the random variable ϵ allow us to make statistical inferences about a "true" β.

Ordinary Least Squares

The r vector contains the distances of the observations on y from the values of the variable y defined by the hyperplane $b^{\mathrm{T}}x$, measured *in the direction of the y axis*. The objective is to determine a value of b that minimizes some norm of r. The use of the L_2 norm is called "least squares". The estimate is the b that minimizes the dot product

$$(y - Xb)^{\mathrm{T}}(y - Xb) = \sum_{i=1}^{n}(y_i - x_{i*}^{\mathrm{T}}b)^2. \tag{9.10}$$

As we saw in Section 6.7 (where we used slightly different notation), using elementary calculus to determine the minimum of equation (9.10) yields the "normal equations"

$$X^{\mathrm{T}}X\widehat{\beta} = X^{\mathrm{T}}y. \tag{9.11}$$

Weighted Least Squares

The elements of the residual vector may be weighted differently. This is appropriate if, for instance, the variance of the residual depends on the value of x; that is, in the notation of equation (9.6), $\mathrm{V}(E) = g(x)$, where g is some function. If the function is known, we can address the problem almost identically as in the use of ordinary least squares, as we saw on page 225. Weighted least squares may also be appropriate if the observations in the sample are not independent. In this case also, if we know the variance-covariance structure, after a simple transformation, we can use ordinary least squares. If the function g or the variance-covariance structure must be estimated, the fitting problem is still straightforward, but formidable complications are introduced into other aspects of statistical inference. We discuss weighted least squares further in Section 9.2.6.

Variations on the Criteria for Fitting

Rather than minimizing a norm of r, there are many other approaches we could use to fit the model to the data. Of course, just the choice of the norm yields different approaches. Some of these approaches may depend on distributional assumptions, which we will not consider here. The point that we want to emphasize here, with little additional comment, is that the standard approach to regression modeling is not the only one. We mentioned some of these other approaches and the computational methods of dealing with them in Section 6.8. Alternative criteria for fitting regression models are sometimes considered in the many textbooks and monographs on data analysis using a linear regression model. This is because the fits may be more "robust" or more resistant to the effects of various statistical distributions.

Regularized Fits

Some variations on the basic approach of minimizing residuals involve a kind of regularization that may take the form of an additive penalty on the objective function. Regularization often results in a shrinkage of the estimator toward 0. One of the most common types of shrinkage estimator is the ridge regression estimator, which for the model $y = X\beta + \epsilon$ is the solution of the modified normal equations $(X^{\mathrm{T}}X + \lambda I)\beta = X^{\mathrm{T}}y$. We discuss this further in Section 9.4.4.

Orthogonal Distances

Another approach is to define an optimal value of β as one that minimizes a norm of the distances of the observed values of y from the vector $X\beta$. This is sometimes called "orthogonal distance regression". The use of the L_2 norm on this vector is sometimes called "total least squares". This is a reasonable approach when it is assumed that the observations in X are realizations of some random variable; that is, an "errors-in-variables" model is appropriate. The model in equation (9.9) is modified to consist of two error terms: one for the errors in the variables and one for the error in the equation. The methods discussed in Section 6.8.3 can be used to fit a model using a criterion of minimum norm of orthogonal residuals. As we mentioned there, weighting of the orthogonal residuals can be easily accomplished in the usual way of handling weights on the different observations.

The weight matrix often is formed as an inverse of a variance-covariance matrix Σ; hence, the modification is to premultiply the matrix $[X|y]$ in equation (6.51) by the Cholesky factor Σ_{C}^{-1}. In the case of errors-in-variables, however, there may be another variance-covariance structure to account for. If the variance-covariance matrix of the columns of X (that is, the independent variables) together with y is T, then we handle the weighting for variances and covariances of the columns of X in the same way, except of course we postmultiply the matrix $[X|y]$ in equation (6.51) by T_{C}^{-1}. This matrix is $(m+1) \times (m+1)$; however, it may be appropriate to assume any error in y is already accounted for, and so the last row and column of T may be 0 except for the $(m+1, m+1)$ element, which would be 1. The appropriate model depends on the nature of the data, of course.

Collinearity

A major problem in regression analysis is collinearity (or "multicollinearity"), by which we mean a "near singularity" of the X matrix. This can be made more precise in terms of a condition number, as discussed in Section 6.1. Ill-conditioning may not only present computational problems, but also may result in an estimate with a very large variance.

9.2.2 Linear Models and Least Squares

The most common estimator of β is one that minimizes the L_2 norm of the vertical distances in equation (9.9); that is, the one that forms a least squares fit. This criterion leads to the normal equations (9.11), whose solution is

$$\widehat{\beta} = (X^\mathrm{T}X)^- X^\mathrm{T}y. \tag{9.12}$$

(As we have pointed out many times, we often write formulas that are not to be used for computing a result; this is the case here.) If X is of full rank, the generalized inverse in equation (9.12) is, of course, the inverse, and $\widehat{\beta}$ is the unique least squares estimator. If X is not of full rank, we generally use the Moore-Penrose inverse, $(X^\mathrm{T}X)^+$, in equation (9.12).

As we saw in equations (6.39) and (6.40), we also have

$$\widehat{\beta} = X^+ y. \tag{9.13}$$

Equation (9.13) indicates the appropriate way to compute $\widehat{\beta}$. As we have seen many times before, however, we often use an expression without computing the individual terms. Instead of computing X^+ in equation (9.13) explicitly, we use either Householder or Givens transformations to obtain the orthogonal decomposition

$$X = QR,$$

or

$$X = QRU^\mathrm{T}$$

if X is not of full rank. As we have seen, the QR decomposition of X can be performed row-wise using Givens transformations. This is especially useful if the data are available only one observation at a time. The equation used for computing $\widehat{\beta}$ is

$$R\widehat{\beta} = Q^\mathrm{T}y, \tag{9.14}$$

which can be solved by back substitution in the triangular matrix R.

Because

$$X^\mathrm{T}X = R^\mathrm{T}R,$$

the quantities in $X^\mathrm{T}X$ or its inverse, which are useful for making inferences using the regression model, can be obtained from the QR decomposition.

If X is not of full rank, the expression (9.13) not only is a least squares solution but the one with minimum length (minimum Euclidean norm), as we saw in equations (6.40) and (6.41).

The vector $\widehat{y} = X\widehat{\beta}$ is the projection of the n-vector y onto a space of dimension equal to the (column) rank of X, which we denote by r_X. The vector of the model, $\mathrm{E}(Y) = X\beta$, is also in the r_X-dimensional space span(X). The projection matrix $I - X(X^\mathrm{T}X)^+X^\mathrm{T}$ projects y onto an $(n - r_X)$-dimensional residual space that is orthogonal to span(X). Figure 9.1 represents these subspaces and the vectors in them.

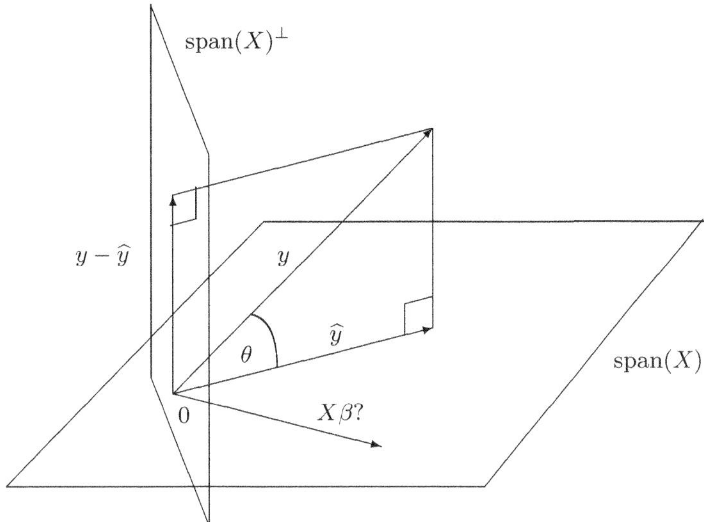

Fig. 9.1. The Linear Least Squares Fit of y with X

In the $(r_X + 1)$-order vector space of the variables, the hyperplane defined by $\widehat{\beta}^{\mathrm{T}} x$ is the estimated model (assuming $\widehat{\beta} \neq 0$; otherwise, the space is of order r_X).

Degrees of Freedom

In the absence of any model, the vector y can range freely over an n-dimensional space. We say the degrees of freedom of y, or the *total degrees of freedom*, is n. If we fix the mean of y, then the *adjusted total degrees of freedom* is $n - 1$.

The model $X\beta$ can range over a space with dimension equal to the (column) rank of X; that is, r_X. We say that the *model degrees of freedom is* r_X. Note that the space of $X\widehat{\beta}$ is the same as the space of $X\beta$.

Finally, the space orthogonal to $X\widehat{\beta}$ (that is, the space of the residuals $y - X\widehat{\beta}$) has dimension $n - r_X$. We say that the *residual (or error) degrees of freedom is* $n - r_X$. (Note that the error vector ϵ can range over an n-dimensional space, but because $\widehat{\beta}$ is a least squares fit, $y - X\widehat{\beta}$ can only range over an $(n - r_X)$-dimensional space.)

The Hat Matrix and Leverage

The projection matrix $H = X(X^{\mathrm{T}}X)^{+}X^{\mathrm{T}}$ is sometimes called the "hat matrix" because

$$\widehat{y} = X\widehat{\beta}$$
$$= X(X^{\mathrm{T}}X)^{+}X^{\mathrm{T}}y$$
$$= Hy, \tag{9.15}$$

that is, it projects y onto $\widehat{y}$ in the span of X. Notice that the hat matrix can be computed without knowledge of the observations in y.

The elements of H are useful in assessing the effect of the particular pattern of the regressors on the predicted values of the response. The extent to which a given point in the row space of X affects the regression fit is called its "leverage". The leverage of the ith observation is

$$h_{ii} = x_{i*}^{\mathrm{T}}(X^{\mathrm{T}}X)^{+}x_{i*}. \tag{9.16}$$

This is just the partial derivative of $\hat{y}_i$ with respect to y_i (Exercise 9.2). A relatively large value of h_{ii} compared with the other diagonal elements of the hat matrix means that the ith observed response, y_i, has a correspondingly relatively large effect on the regression fit.

9.2.3 Statistical Inference

Fitting a model by least squares or by minimizing some other norm of the residuals in the data might be a sensible thing to do without any concern for a probability distribution. "Least squares" per se is not a statistical criterion. Certain statistical criteria, such as maximum likelihood or minimum variance estimation among a certain class of unbiased estimators, however, lead to an estimator that is the solution to a least squares problem for specific probability distributions.

For statistical inference about the parameters of the model $y = X\beta + \epsilon$ in equation (9.9), we must add something to the model. As in statistical inference generally, we must identify the random variables and make some statements (assumptions) about their distribution. The simplest assumptions are that ϵ is a random variable and $\mathrm{E}(\epsilon) = 0$. Whether or not the matrix X is random, our interest is in making inference conditional on the observed values of X.

Estimability

One of the most important questions for statistical inference involves estimating or testing some linear combination of the elements of the parameter β; for example, we may wish to estimate $\beta_1 - \beta_2$ or to test the hypothesis that $\beta_1 - \beta_2 = c_1$ for some constant c_1. In general, we will consider the linear combination $l^{\mathrm{T}}\beta$. Whether or not it makes sense to estimate such a linear combination depends on whether there is a function of the observable random variable Y such that $g(\mathrm{E}(Y)) = l^{\mathrm{T}}\beta$.

We generally restrict our attention to linear functions of $\mathrm{E}(Y)$ and formally define a linear combination $l^{\mathrm{T}}\beta$ to be (linearly) *estimable* if there exists a vector t such that

$$t^{\mathrm{T}}\mathrm{E}(Y) = l^{\mathrm{T}}\beta \tag{9.17}$$

for any β.

It is clear that if X is of full column rank, $l^{\mathrm{T}}\beta$ is linearly estimable for any l or, more generally, $l^{\mathrm{T}}\beta$ is linearly estimable for any $l \in \mathrm{span}(X^{\mathrm{T}})$. (The t vector is just the normalized coefficients expressing l in terms of the columns of X.)

Estimability depends only on the simplest distributional assumption about the model; that is, that $\mathrm{E}(\epsilon) = 0$. Under this assumption, we see that the estimator $\widehat{\beta}$ based on the least squares fit of β is unbiased for the linearly estimable function $l^{\mathrm{T}}\beta$. Because $l \in \mathrm{span}(X^{\mathrm{T}}) = \mathrm{span}(X^{\mathrm{T}}X)$, we can write $l = X^{\mathrm{T}}X\tilde{t}$. Now, we have

$$\begin{aligned}
\mathrm{E}(l^{\mathrm{T}}\widehat{\beta}) &= \mathrm{E}(l^{\mathrm{T}}(X^{\mathrm{T}}X)^{+}X^{\mathrm{T}}y) \\
&= \tilde{t}^{\mathrm{T}}X^{\mathrm{T}}X(X^{\mathrm{T}}X)^{+}X^{\mathrm{T}}X\beta \\
&= \tilde{t}^{\mathrm{T}}X^{\mathrm{T}}X\beta \\
&= l^{\mathrm{T}}\beta.
\end{aligned} \tag{9.18}$$

Although we have been taking $\widehat{\beta}$ to be $(X^{\mathrm{T}}X)^{+}X^{\mathrm{T}}y$, the equations above follow for other least squares fits, $b = (X^{\mathrm{T}}X)^{-}X^{\mathrm{T}}y$, for any generalized inverse. In fact, the estimator of $l^{\mathrm{T}}\beta$ is invariant to the choice of the generalized inverse. This is because if $b = (X^{\mathrm{T}}X)^{-}X^{\mathrm{T}}y$, we have $X^{\mathrm{T}}Xb = X^{\mathrm{T}}y$, and so

$$l^{\mathrm{T}}\widehat{\beta} - l^{\mathrm{T}}b = \tilde{t}^{\mathrm{T}}X^{\mathrm{T}}X(\widehat{\beta} - b) = \tilde{t}^{\mathrm{T}}(X^{\mathrm{T}}y - X^{\mathrm{T}}y) = 0. \tag{9.19}$$

Other properties of the estimators depend on additional assumptions about the distribution of ϵ, and we will consider some of them below.

When X is not of full rank, we often are interested in an orthogonal basis for $\mathrm{span}(X^{\mathrm{T}})$. If X includes a column of 1s, the elements of any vector in the basis must sum to 0. Such vectors are called *contrasts*. The second and subsequent rows of the Helmert matrix (see Section 8.8.1 on page 308) are contrasts that are often of interest because of their regular patterns and their interpretability in applications involving the analysis of levels of factors in experiments.

Testability

We define a linear hypothesis $l^{\mathrm{T}}\beta = c_1$ as *testable* if $l^{\mathrm{T}}\beta$ is estimable. We generally restrict our attention to testable hypotheses.

It is often of interest to test multiple hypotheses concerning linear combinations of the elements of β. For the model (9.9), the *general linear hypothesis* is

$$\mathrm{H}_0: \ L^{\mathrm{T}}\beta = c,$$

where L is $m \times q$, of rank q, and such that $\mathrm{span}(L) \subseteq \mathrm{span}(X)$.

The test for a hypothesis depends on the distributions of the random variables in the model. If we assume that the elements of ϵ are i.i.d. normal with a mean of 0, then the general linear hypothesis is tested using an F statistic whose numerator is the difference in the residual sum of squares from fitting the model with the restriction $L^{\mathrm{T}}\beta = c$ and the residual sum of squares from fitting the unrestricted model. This reduced sum of squares is

$$(L^{\mathrm{T}}\widehat{\beta} - c)^{\mathrm{T}} \, (L^{\mathrm{T}}(X^{\mathrm{T}}X)^{*}L)^{-1} \, (L^{\mathrm{T}}\widehat{\beta} - c), \tag{9.20}$$

where $(X^{\mathrm{T}}X)^{*}$ is any g_2 inverse of $X^{\mathrm{T}}X$. This test is a likelihood ratio test. (See a text on linear models, such as Searle, 1971, for more discussion on this testing problem.)

To compute the quantity in expression (9.20), first observe

$$L^{\mathrm{T}}(X^{\mathrm{T}}X)^{*}L = (X(X^{\mathrm{T}}X)^{*}L)^{\mathrm{T}} \, (X(X^{\mathrm{T}}X)^{*}L). \tag{9.21}$$

Now, if $X(X^{\mathrm{T}}X)^{*}L$, which has rank q, is decomposed as

$$X(X^{\mathrm{T}}X)^{*}L = P \begin{bmatrix} T \\ 0 \end{bmatrix},$$

where P is an $m \times m$ orthogonal matrix and T is a $q \times q$ upper triangular matrix, we can write the reduced sum of squares (9.20) as

$$(L^{\mathrm{T}}\widehat{\beta} - c)^{\mathrm{T}} \, (T^{\mathrm{T}}T)^{-1} \, (L^{\mathrm{T}}\widehat{\beta} - c)$$

or

$$\left(T^{-\mathrm{T}}(L^{\mathrm{T}}\widehat{\beta} - c)\right)^{\mathrm{T}} \, \left(T^{-\mathrm{T}}(L^{\mathrm{T}}\widehat{\beta} - c)\right)$$

or

$$v^{\mathrm{T}}v. \tag{9.22}$$

To compute v, we solve

$$T^{\mathrm{T}}v = L^{\mathrm{T}}\widehat{\beta} - c \tag{9.23}$$

for v, and the reduced sum of squares is then formed as $v^{\mathrm{T}}v$.

The Gauss-Markov Theorem

The Gauss-Markov theorem provides a restricted optimality property for estimators of estimable functions of β under the condition that $\mathrm{E}(\epsilon) = 0$ and $\mathrm{V}(\epsilon) = \sigma^2 I$; that is, in addition to the assumption of zero expectation, which we have used above, we also assume that the elements of ϵ have constant variance and that their covariances are zero. (We are not assuming independence or normality, as we did in order to develop tests of hypotheses.)

Given $y = X\beta + \epsilon$ and $\mathrm{E}(\epsilon) = 0$ and $\mathrm{V}(\epsilon) = \sigma^2 I$, the Gauss-Markov theorem states that $l^{\mathrm{T}}\widehat{\beta}$ is the unique *best linear unbiased estimator* (BLUE) of the estimable function $l^{\mathrm{T}}\beta$.

"Linear" estimator in this context means a linear combination of y; that is, an estimator in the form $a^{\mathrm{T}}y$. It is clear that $l^{\mathrm{T}}\widehat{\beta}$ is linear, and we have already seen that it is unbiased for $l^{\mathrm{T}}\beta$. "Best" in this context means that its variance is no greater than any other estimator that fits the requirements. Hence, to prove the theorem, first let $a^{\mathrm{T}}y$ be any unbiased estimator of $l^{\mathrm{T}}\beta$, and write $l = X^{\mathrm{T}}X\tilde{t}$ as above. Because $a^{\mathrm{T}}y$ is unbiased for any β, as we saw above, it must be the case that $a^{\mathrm{T}}X = l^{\mathrm{T}}$. Recalling that $X^{\mathrm{T}}X\widehat{\beta} = X^{\mathrm{T}}y$, we have

$$\begin{aligned}
\mathrm{V}(a^{\mathrm{T}}y) &= \mathrm{V}(a^{\mathrm{T}}y - l^{\mathrm{T}}\widehat{\beta} + l^{\mathrm{T}}\widehat{\beta}) \\
&= \mathrm{V}(a^{\mathrm{T}}y - \tilde{t}^{\mathrm{T}}X^{\mathrm{T}}y + l^{\mathrm{T}}\widehat{\beta}) \\
&= \mathrm{V}(a^{\mathrm{T}}y - \tilde{t}^{\mathrm{T}}X^{\mathrm{T}}y) + \mathrm{V}(l^{\mathrm{T}}\widehat{\beta}) + 2\mathrm{Cov}(a^{\mathrm{T}}y - \tilde{t}^{\mathrm{T}}X^{\mathrm{T}}y, \ \tilde{t}^{\mathrm{T}}X^{\mathrm{T}}y).
\end{aligned}$$

Now, under the assumptions on the variance-covariance matrix of ϵ, which is also the (conditional, given X) variance-covariance matrix of y, we have

$$\begin{aligned}
\mathrm{Cov}(a^{\mathrm{T}}y - \tilde{t}^{\mathrm{T}}X^{\mathrm{T}}y, \ l^{\mathrm{T}}\widehat{\beta}) &= (a^{\mathrm{T}} - \tilde{t}^{\mathrm{T}}X^{\mathrm{T}})\sigma^2 I X\tilde{t} \\
&= (a^{\mathrm{T}}X - \tilde{t}^{\mathrm{T}}X^{\mathrm{T}}X)\sigma^2 I\tilde{t} \\
&= (l^{\mathrm{T}} - l^{\mathrm{T}})\sigma^2 I\tilde{t} \\
&= 0;
\end{aligned}$$

that is,

$$\mathrm{V}(a^{\mathrm{T}}y) = \mathrm{V}(a^{\mathrm{T}}y - \tilde{t}^{\mathrm{T}}X^{\mathrm{T}}y) + \mathrm{V}(l^{\mathrm{T}}\widehat{\beta}).$$

This implies that

$$\mathrm{V}(a^{\mathrm{T}}y) \geq \mathrm{V}(l^{\mathrm{T}}\widehat{\beta});$$

that is, $l^{\mathrm{T}}\widehat{\beta}$ has minimum variance among the linear unbiased estimators of $l^{\mathrm{T}}\beta$. To see that it is unique, we consider the case in which $\mathrm{V}(a^{\mathrm{T}}y) = \mathrm{V}(l^{\mathrm{T}}\widehat{\beta})$; that is, $\mathrm{V}(a^{\mathrm{T}}y - \tilde{t}^{\mathrm{T}}X^{\mathrm{T}}y) = 0$. For this variance to equal 0, it must be the case that $a^{\mathrm{T}} - \tilde{t}^{\mathrm{T}}X^{\mathrm{T}} = 0$ or $a^{\mathrm{T}}y = \tilde{t}^{\mathrm{T}}X^{\mathrm{T}}y = l^{\mathrm{T}}\widehat{\beta}$; that is, $l^{\mathrm{T}}\widehat{\beta}$ is the unique linear unbiased estimator that achieves the minimum variance.

If we assume further that $\epsilon \sim \mathrm{N}_n(0, \sigma^2 I)$, we can show that $l^{\mathrm{T}}\widehat{\beta}$ is the uniformly minimum variance unbiased estimator (UMVUE) for $l^{\mathrm{T}}\beta$. This is because $(X^{\mathrm{T}}y, \ (y - X\widehat{\beta})^{\mathrm{T}}(y - X\widehat{\beta}))$ is complete and sufficient for (β, σ^2). This line of reasoning also implies that $(y - X\widehat{\beta})^{\mathrm{T}}(y - X\widehat{\beta})/(n - r)$, where $r = \mathrm{rank}(X)$, is UMVUE for σ^2. We will not go through the details here. The interested reader is referred to a text on mathematical statistics, such as Shao (2003).

9.2.4 The Normal Equations and the Sweep Operator

The coefficient matrix in the normal equations, $X^{\mathrm{T}}X$, or the adjusted version $X_c^{\mathrm{T}}X_c$, where X_c is the centered matrix as in equation (8.67) on page 293, is often of interest for reasons other than just to compute the least squares estimators. The condition number of $X^\mathrm{T}X$ is the square of the condition number of X, however, and so any ill-conditioning is exacerbated by formation of the sums of squares and cross products matrix. The adjusted sums of squares and cross products matrix, $X_c^\mathrm{T}X_c$, tends to be better conditioned, so it is usually the one used in the normal equations, but of course the condition number of $X_c^\mathrm{T}X_c$ is the square of the condition number of X_c.

A useful matrix can be formed from the normal equations:

$$\begin{bmatrix} X^\mathrm{T}X & X^\mathrm{T}y \\ y^\mathrm{T}X & y^\mathrm{T}y \end{bmatrix}. \tag{9.24}$$

Applying m elementary operations on this matrix, we can get

$$\begin{bmatrix} (X^\mathrm{T}X)^+ & X^+y \\ y^\mathrm{T}X^{+\mathrm{T}} & y^\mathrm{T}y - y^\mathrm{T}X(X^\mathrm{T}X)^+X^\mathrm{T}y \end{bmatrix}.$$

(If X is not of full rank, in order to get the Moore-Penrose inverse in this expression, the elementary operations must be applied in a fixed manner.) The matrix in the upper left of the partition is related to the estimated variance-covariance matrix of the particular solution of the normal equations, and it can be used to get an estimate of the variance-covariance matrix of estimates of any independent set of linearly estimable functions of β. The vector in the upper right of the partition is the unique minimum-length solution to the normal equations, $\widehat{\beta}$. The scalar in the lower right partition, which is the Schur complement of the full inverse (see equations (3.145) and (3.165)), is the square of the residual norm. The squared residual norm provides an estimate of the variance of the residuals in equation (9.9) after proper scaling.

The elementary operations can be grouped into a larger operation, called the "sweep operation", which is performed for a given row. The sweep operation on row i, S_i, of the nonnegative definite matrix A to yield the matrix B, which we denote by

$$\mathrm{S}_i(A) = B,$$

is defined in Algorithm 9.1.

Algorithm 9.1 Sweep of the ith Row

1. If $a_{ii} = 0$, skip the following operations.
2. Set $b_{ii} = a_{ii}^{-1}$.
3. For $j \neq i$, set $b_{ij} = a_{ii}^{-1}a_{ij}$.
4. For $k \neq i$, set $b_{kj} = a_{kj} - a_{ki}a_{ii}^{-1}a_{ij}$.

Skipping the operations if $a_{ii} = 0$ allows the sweep operator to handle non-full rank problems. The sweep operator is its own inverse:

$$\mathrm{S}_i(\mathrm{S}_i(A)) = A.$$

The sweep operator applied to the matrix (9.24) corresponds to adding or removing the ith variable (column) of the X matrix to the regression equation.

9.2.5 Linear Least Squares Subject to Linear Equality Constraints

In the regression model (9.9), it may be known that β satisfies certain constraints, such as that all the elements be nonnegative. For constraints of the form $g(\beta) \in C$, where C is some m-dimensional space, we may estimate β by the *constrained least squares estimator*; that is, the vector $\widehat{\beta}_C$ that minimizes the dot product (9.10) among all b that satisfy $g(b) \in C$.

The nature of the constraints may or may not make drastic changes to the computational problem. (The constraints also change the statistical inference problem in various ways, but we do not address that here.) If the constraints are nonlinear, or if the constraints are inequality constraints (such as that all the elements be nonnegative), there is no general closed-form solution.

It is easy to handle linear equality constraints of the form

$$g(\beta) = L\beta$$
$$= c,$$

where L is a $q \times m$ matrix of full rank. The solution is, analogous to equation (9.12),

$$\widehat{\beta}_C = (X^{\mathrm{T}}X)^+ X^{\mathrm{T}}y + (X^{\mathrm{T}}X)^+ L^{\mathrm{T}}(L(X^{\mathrm{T}}X)^+ L^{\mathrm{T}})^+(c - L(X^{\mathrm{T}}X)^+ X^{\mathrm{T}}y).$$
$$(9.25)$$

When X is of full rank, this result can be derived by using Lagrange multipliers and the derivative of the norm (9.10) (see Exercise 9.4 on page 365). When X is not of full rank, it is slightly more difficult to show this, but it is still true. (See a text on linear regression, such as Draper and Smith, 1998.)

The restricted least squares estimate, $\widehat{\beta}_C$, can be obtained (in the (1, 2) block) by performing $m + q$ sweep operations on the matrix,

$$\begin{bmatrix} X^{\mathrm{T}}X & X^{\mathrm{T}}y & L^{\mathrm{T}} \\ y^{\mathrm{T}}X & y^{\mathrm{T}}y & c^{\mathrm{T}} \\ L & c & 0 \end{bmatrix},$$
$$(9.26)$$

analogous to matrix (9.24).

9.2.6 Weighted Least Squares

In fitting the regression model $y \approx X\beta$, it is often desirable to weight the observations differently, and so instead of minimizing equation (9.10), we minimize

$$\sum w_i(y_i - x_{i*}^{\mathrm{T}}b)^2,$$

where w_i represents a nonnegative weight to be applied to the ith observation. One purpose of the weight may be to control the effect of a given observation on the overall fit. If a model of the form of equation (9.9),

$$y = X\beta + \epsilon,$$

is assumed, and ϵ is taken to be a random variable such that ϵ_i has variance σ_i^2, an appropriate value of w_i may be $1/\sigma_i^2$. (Statisticians almost always naturally assume that ϵ is a random variable. Although usually it is modeled this way, here we are allowing for more general interpretations and more general motives in fitting the model.)

The normal equations can be written as

$$\left(X^{\mathrm{T}}\mathrm{diag}((w_1, w_2, \ldots, w_n))X\right)\widehat{\beta} = X^{\mathrm{T}}\mathrm{diag}((w_1, w_2, \ldots, w_n))y.$$

More generally, we can consider W to be a weight matrix that is not necessarily diagonal. We have the same set of normal equations:

$$(X^{\mathrm{T}}WX)\widehat{\beta}_W = X^{\mathrm{T}}Wy. \tag{9.27}$$

When W is a diagonal matrix, the problem is called "weighted least squares". Use of a nondiagonal W is also called weighted least squares but is sometimes called "generalized least squares". The weight matrix is symmetric and generally positive definite, or at least nonnegative definite. The weighted least squares estimator is

$$\widehat{\beta}_W = (X^{\mathrm{T}}WX)^{+}X^{\mathrm{T}}Wy.$$

As we have mentioned many times, an expression such as this is not necessarily a formula for computation. The matrix factorizations discussed above for the unweighted case can also be used for computing weighted least squares estimates.

In a model $y = X\beta + \epsilon$, where ϵ is taken to be a random variable with variance-covariance matrix Σ, the choice of W as Σ^{-1} yields estimators with certain desirable statistical properties. (Because this is a natural choice for many models, statisticians sometimes choose the weighting matrix without fully considering the reasons for the choice.) As we pointed out on page 225, weighted least squares can be handled by premultiplication of both y and X by the Cholesky factor of the weight matrix. In the case of an assumed variance-covariance matrix Σ, we transform each side by Σ_{C}^{-1}, where Σ_{C} is the Cholesky factor of Σ. The residuals whose squares are to be minimized are $\Sigma_{\mathrm{C}}^{-1}(y - Xb)$. Under the assumptions, the variance-covariance matrix of the residuals is I.

9.2.7 Updating Linear Regression Statistics

In Section 6.7.4 on page 228, we discussed the general problem of updating a least squares solution to an overdetermined system when either the number of equations (rows) or the number of variables (columns) is changed. In the linear regression problem these correspond to adding or deleting observations and adding or deleting terms in the linear model, respectively.

Adding More Variables

Suppose first that more variables are added, so the regression model is

$$y \approx \begin{bmatrix} X & X_+ \end{bmatrix} \theta,$$

where X_+ represents the observations on the additional variables. (We use θ to represent the parameter vector; because the model is different, it is not just β with some additional elements.)

If $X^{\mathrm{T}}X$ has been formed and the sweep operator is being used to perform the regression computations, it can be used easily to add or delete variables from the model, as we mentioned above. The Sherman-Morrison-Woodbury formulas (6.24) and (6.26) and the Hemes formula (6.27) (see page 221) can also be used to update the solution.

In regression analysis, one of the most important questions is the identification of independent variables from a set of potential explanatory variables that should be in the model. This aspect of the analysis involves adding and deleting variables. We discuss this further in Section 9.4.2.

Adding More Observations

If we have obtained more observations, the regression model is

$$\begin{bmatrix} y \\ y_+ \end{bmatrix} \approx \begin{bmatrix} X \\ X_+ \end{bmatrix} \beta,$$

where y_+ and X_+ represent the additional observations.

If the QR decomposition of X is available, we simply augment it as in equation (6.42):

$$\begin{bmatrix} R & c_1 \\ 0 & c_2 \\ X_+ & y_+ \end{bmatrix} = \begin{bmatrix} Q^{\mathrm{T}} & 0 \\ 0 & I \end{bmatrix} \begin{bmatrix} X & y \\ X_+ & y_+ \end{bmatrix}.$$

We now apply orthogonal transformations to this to zero out the last rows and produce

$$\begin{bmatrix} R_* & c_{1*} \\ 0 & c_{2*} \end{bmatrix},$$

where R_* is an $m \times m$ upper triangular matrix and c_{1*} is an m-vector as before, but c_{2*} is an $(n - m + k)$-vector. We then have an equation of the form (9.14) and we use back substitution to solve it.

Adding More Observations Using Weights

Another way of approaching the problem of adding or deleting observations is by viewing the problem as weighted least squares. In this approach, we also have more general results for updating regression statistics. Following Escobar and Moser (1993), we can consider two weighted least squares problems: one with weight matrix W and one with weight matrix V. Suppose we have the solutions $\widehat{\beta}_W$ and $\widehat{\beta}_V$. Now let

$$\Delta = V - W,$$

and use the subscript $*$ on any matrix or vector to denote the subarray that corresponds only to the nonnull rows of Δ. The symbol Δ_*, for example, is the square subarray of Δ consisting of all of the nonzero rows and columns of Δ, and X_* is the subarray of X consisting of all the columns of X and only the rows of X that correspond to Δ_*. From the normal equations (9.27) using W and V, and with the solutions $\widehat{\beta}_W$ and $\widehat{\beta}_V$ plugged in, we have

$$(X^{\mathrm{T}}WX)\widehat{\beta}_W + (X^{\mathrm{T}}\Delta X)\widehat{\beta}_V = X^{\mathrm{T}}Wy + X^{\mathrm{T}}\Delta y,$$

and so

$$\widehat{\beta}_V - \widehat{\beta}_W = (X^{\mathrm{T}}WX)^+ X_*^{\mathrm{T}}\Delta_*(y - X\widehat{\beta}_V)_*.$$

This gives

$$(y - X\widehat{\beta}_V)_* = (I + X(X^{\mathrm{T}}WX)^+ X_*^{\mathrm{T}}\Delta_*)^+(y - X\widehat{\beta}_W)_*,$$

and finally

$$\widehat{\beta}_V = \widehat{\beta}_W + (X^{\mathrm{T}}WX)^+ X_*^{\mathrm{T}}\Delta_* \left(I + X_*(X^{\mathrm{T}}WX)^+ X_*^{\mathrm{T}}\Delta_* \right)^+ (y - X\widehat{\beta}_W)_*.$$

If Δ_* can be written as $\pm GG^{\mathrm{T}}$, using this equation and the equations (3.133) on page 93 (which also apply to pseudoinverses), we have

$$\widehat{\beta}_V = \widehat{\beta}_W \pm (X^{\mathrm{T}}WX)^+ X_*^{\mathrm{T}}G(I \pm G^{\mathrm{T}}X_*(X^{\mathrm{T}}WX)^+ X_*^{\mathrm{T}}G)^+ G^{\mathrm{T}}(y - X\widehat{\beta}_W)_*. \tag{9.28}$$

The sign of GG^{T} is positive when observations are added and negative when they are deleted.

Equation (9.28) is particularly simple in the case where W and V are identity matrices (of different sizes, of course). Suppose that we have obtained more observations in y_+ and X_+. (In the following, the reader must be careful to distinguish "+" as a subscript to represent more data and "+" as a superscript with its usual meaning of a Moore-Penrose inverse.) Suppose we already have the least squares solution for $y \approx X\beta$, say $\widehat{\beta}_W$. Now $\widehat{\beta}_W$ is the weighted least squares solution to the model with the additional data and with weight matrix

$$W = \begin{bmatrix} I & 0 \\ 0 & 0 \end{bmatrix}.$$

We now seek the solution to the same system with weight matrix V, which is a larger identity matrix. From equation (9.28), the solution is

$$\widehat{\beta} = \widehat{\beta}_W + (X^{\mathrm{T}}X)^+ X_+^{\mathrm{T}}(I + X_+(X^{\mathrm{T}}X)^+ X_+^{\mathrm{T}})^+(y - X\widehat{\beta}_W)_*. \tag{9.29}$$

9.2.8 Linear Smoothing

The interesting reasons for doing regression analysis are to understand relationships and to predict a value of the dependent value given a value of the independent variable. As a side benefit, a model with a smooth equation $f(x)$ "smoothes" the observed responses; that is, the elements in $\hat{y} = \widehat{f(x)}$ exhibit less variation than the elements in y, meaning the model sum of squares is less than the total sum of squares. (Of course, the important fact for our purposes is that $\|y - \hat{y}\|$ is smaller than $\|y\|$ or $\|y - \bar{y}\|$.)

The use of the hat matrix emphasizes the smoothing perspective:

$$\hat{y} = Hy.$$

The concept of a smoothing matrix was discussed in Section 8.6.2. From this perspective, using H, we project y onto a vector in $\mathrm{span}(H)$, and that vector has a smaller variation than y; that is, H has *smoothed* y. It does not matter what the specific values in the vector y are so long as they are associated with the same values of the independent variables.

We can extend this idea to a general $n \times n$ *smoothing matrix* H_λ:

$$\tilde{y} = H_\lambda y.$$

The smoothing matrix depends only on the kind and extent of smoothing to be performed and on the observed values of the independent variables. The extent of the smoothing may be indicated by the indexing parameter λ. Once the smoothing matrix is obtained, it does not matter how the independent variables are related to the model.

In Section 6.8.2, we discussed regularized solutions of overdetermined systems of equations, which in the present case is equivalent to solving

$$\min_b \left((y - Xb)^{\mathrm{T}}(y - Xb) + \lambda b^{\mathrm{T}} b \right).$$

The solution of this yields the smoothing matrix

$$S_\lambda = X(X^{\mathrm{T}} X + \lambda I)^{-1} X^{\mathrm{T}}$$

(see equation (8.62)). This has the effect of shrinking the $\hat{y}$ of equation (8.56) toward 0. (In regression analysis, this is called "ridge regression".)

We discuss ridge regression and general shrinkage estimation in Section 9.4.4. Loader (2004) provides additional background and discusses more general issues in smoothing.

9.3 Principal Components

The analysis of multivariate data involves various linear transformations that help in understanding the relationships among the features that the data represent. The second moments of the data are used to accommodate the differences in the scales of the individual variables and the covariances among pairs of variables.

If X is the matrix containing the data stored in the usual way, a useful statistic is the sums of squares and cross products matrix, $X^T X$, or the "adjusted" squares and cross products matrix, $X_c^T X_c$, where X_c is the centered matrix formed by subtracting from each element of X the mean of the column containing that element. The sample variance-covariance matrix, as in equation (8.70), is the Gramian matrix

$$S_X = \frac{1}{n-1} X_c^T X_c, \tag{9.30}$$

where n is the number of observations (the number of rows in X).

In data analysis, the sample variance-covariance matrix S_X in equation (9.30) plays an important role. In more formal statistical inference, it is a consistent estimator of the population variance-covariance matrix (if it is positive definite), and under assumptions of independent sampling from a normal distribution, it has a known distribution. It also has important numerical properties; it is symmetric and positive definite (or, at least, nonnegative definite; see Section 8.6). Other estimates of the variance-covariance matrix or the correlation matrix of the underlying distribution may not be positive definite, however, and in Section 9.4.6 and Exercise 9.14 we describe possible ways of adjusting a matrix to be positive definite.

9.3.1 Principal Components of a Random Vector

It is often of interest to transform a given random vector into a vector whose elements are independent. We may also be interested in which of those elements of the transformed random vector have the largest variances. The transformed vector may be more useful in making inferences about the population. In more informal data analysis, it may allow use of smaller observational vectors without much loss in information.

Stating this more formally, if Y is a random d-vector with variance-covariance matrix Σ, we seek a transformation matrix A such that $\widetilde{Y} = AY$ has a diagonal variance-covariance matrix. We are additionally interested in a transformation $a^T Y$ that has maximal variance for a given $\|a\|$.

Because the variance of $a^T Y$ is $V(a^T Y) = a^T \Sigma a$, we have already obtained the solution in equation (3.208). The vector a is the eigenvector corresponding to the maximum eigenvalue of Σ, and if a is normalized, the variance of $a^T Y$ is the maximum eigenvalue.

Because Σ is symmetric, it is orthogonally diagonalizable and the properties discussed in Section 3.8.7 on page 119 not only provide the transformation immediately but also indicate which elements of $\widetilde{Y}$ have the largest variances. We write the orthogonal diagonalization of Σ as (see equation (3.197))

$$\Sigma = \Gamma \Lambda \Gamma^{\mathrm{T}}, \tag{9.31}$$

where $\Gamma \Gamma^{\mathrm{T}} = \Gamma^{\mathrm{T}} \Gamma = I$, and Λ is diagonal with elements $\lambda_1 \geq \cdots \geq \lambda_m \geq 0$ (because a variance-covariance matrix is nonnegative definite). Choosing the transformation as

$$\widetilde{Y} = \Gamma Y, \tag{9.32}$$

we have $\mathrm{V}(\widetilde{Y}) = \Lambda$; that is, the ith element of $\widetilde{Y}$ has variance λ_i, and

$$\mathrm{Cov}(\widetilde{Y}_i, \widetilde{Y}_j) = 0 \quad \text{if } i \neq j.$$

The elements of $\widetilde{Y}$ are called the *principal components* of Y. The *first principal component*, $\widetilde{Y}_1$, which is the signed magnitude of the projection of Y in the direction of the eigenvector corresponding to the maximum eigenvalue, has the maximum variance of any of the elements of $\widetilde{Y}$, and $\mathrm{V}(\widetilde{Y}_1) = \lambda_1$. (It is, of course, possible that the maximum eigenvalue is not simple. In that case, there is no one-dimensional first principal component. If m_1 is the multiplicity of λ_1, all one-dimensional projections within the m_1-dimensional eigenspace corresponding to λ_1 have the same variance, and m_1 projections can be chosen as mutually independent.)

The second and third principal components, and so on, are determined directly from the spectral decomposition.

9.3.2 Principal Components of Data

The same ideas of principal components carry over to observational data. Given an $n \times d$ data matrix X, we seek a transformation as above that will yield the linear combination of the columns that has maximum sample variance, and other linear combinations that are independent. This means that we work with the centered matrix X_{c} (equation (8.67)) and the variance-covariance matrix S_X, as above, or the centered and scaled matrix X_{cs} (equation (8.68)) and the correlation matrix R_X (equation (8.72)). See Section 3.3 in Jolliffe (2002) for discussions of the differences in using the centered but not scaled matrix and using the centered and scaled matrix.

In the following, we will use S_X, which plays a role similar to Σ for the random variable. (This role could be stated more formally in terms of statistical estimation. Additionally, the scaling may require more careful consideration. The issue of scaling naturally arises from the arbitrariness of units of measurement in data. Random variables have no units of measurement.)

In data analysis, we seek a normalized transformation vector a to apply to any centered observation x_c, so that the sample variance of $a^{\mathrm{T}} x_c$, that is,

$$a^{\mathrm{T}} S_X a, \tag{9.33}$$

is maximized.

From equation (3.208) or the spectral decomposition equation (3.200), we know that the solution to this maximization problem is the eigenvector,

v_1, corresponding to the largest eigenvalue, c_1, of S_X, and the value of the expression (9.33); that is, $v_1^{\mathrm{T}} S_X v_1$ at the maximum is the largest eigenvalue. In applications, this vector is used to transform the rows of X_c into scalars. If we think of a generic row of X_c as the vector x, we call $v_1^{\mathrm{T}} x$ the *first principal component* of x. There is some ambiguity about the precise meaning of "principal component". The definition just given is a scalar; that is, a combination of values of a vector of variables. This is consistent with the definition that arises in the population model in Section 9.3.1. Sometimes, however, the eigenvector v_1 itself is referred to as the first principal component. More often, the vector $X_c v_1$ of linear combinations of the columns of X_c is called the first principal component. We will often use the term in this latter sense.

If the largest eigenvalue, c_1, is of algebraic multiplicity $m_1 > 1$, we have seen that we can choose m_1 orthogonal eigenvectors that correspond to c_1 (because S_X, being symmetric, is simple). Any one of these vectors may be called a first principal component of X.

The second and third principal components, and so on, are determined directly from the nonzero eigenvalues in the spectral decomposition of S_X.

The full set of principal components of X_c, analogous to equation (9.32) except that here the random vectors correspond to the rows in X_c, is

$$Z = X_c V, \tag{9.34}$$

where V has r_X columns. (As before, r_X is the rank of X.)

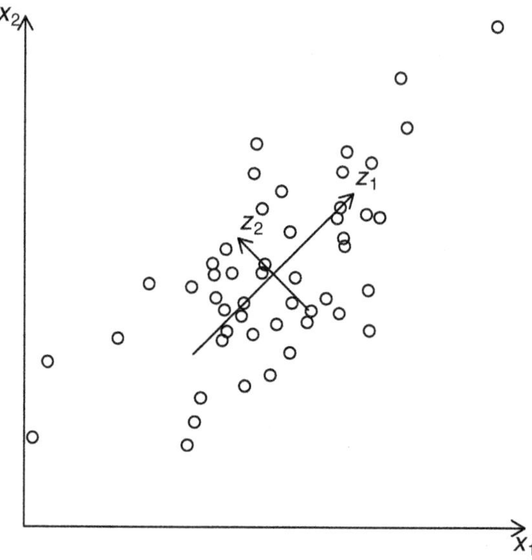

Fig. 9.2. Principal Components

Principal Components Directly from the Data Matrix

Formation of the S_X matrix emphasizes the role that the sample covariances play in principal component analysis. However, there is no reason to form a matrix such as $X_c^T X_c$, and indeed we may introduce significant rounding errors by doing so. (Recall our previous discussions of the condition numbers of $X^T X$ and X.)

The singular value decomposition of the $n \times m$ matrix X_c yields the square roots of the eigenvalues of $X_c^T X_c$ and the same eigenvectors. (The eigenvalues of $X_c^T X_c$ are $(n-1)$ times the eigenvalues of S_X.) We will assume that there are more observations than variables (that is, that $n > m$). In the SVD of the centered data matrix $X_c = UAV^T$, U is an $n \times r_X$ matrix with orthogonal columns, V is an $m \times r_X$ matrix whose first r_X columns are orthogonal and the rest are 0, and A is an $r_X \times r_X$ diagonal matrix whose entries are the nonnegative singular values of $X - \overline{X}$. (As before, r_X is the column rank of X.)

The spectral decomposition in terms of the singular values and outer products of the columns of the factor matrices is

$$X_c = \sum_i^{r_X} \sigma_i u_i v_i^T. \tag{9.35}$$

The vectors u_i are the same as the eigenvectors of S_X.

Dimension Reduction

If the columns of a data matrix X are viewed as variables or features that are measured for each of several observational units, which correspond to rows in the data matrix, an objective in principal components analysis may be to determine some small number of linear combinations of the columns of X that contain almost as much information as the full set of columns. (Here we are not using "information" in a precise sense; in a general sense, it means having similar statistical properties.) Instead of a space of dimension equal to the (column) rank of X (that is, r_X), we seek a subspace of span(X) with rank less than r_X that approximates the full space (in some sense). As we discussed on page 138, the best approximation in terms of the usual norm (the Frobenius norm) of X_c by a matrix of rank p is

$$\widetilde{X}_p = \sum_i^p \sigma_i u_i v_i^T \tag{9.36}$$

for some $p < \min(n, m)$.

Principal components analysis is often used for "dimension reduction" by using the first few principal components in place of the original data. There are various ways of choosing the number of principal components (that is, p in equation (9.36)). There are also other approaches to dimension reduction. A general reference on this topic is Mizuta (2004).

9.4 Condition of Models and Data

In Section 6.1, we describe the concept of "condition" of a matrix for certain kinds of computations. In Section 6.4, we discuss how a large condition number may indicate the level of numerical accuracy in the solution of a system of linear equations, and on page 225 we extend this discussion to overdetermined systems such as those encountered in regression analysis. (We return to the topic of condition in Section 11.2 with even more emphasis on the numerical computations.) The condition of the X matrices has implications for the accuracy we can expect in the numerical computations for regression analysis.

There are other connections between the condition of the data and statistical analysis that go beyond just the purely computational issues. Analysis involves more than just computations. Ill-conditioned data also make interpretation of relationships difficult because we may be concerned with both conditional and marginal relationships. In ill-conditioned data, the relationships between any two variables may be quite different depending on whether or not the relationships are conditioned on relationships with other variables in the dataset.

9.4.1 Ill-Conditioning in Statistical Applications

We have described ill-conditioning heuristically as a situation in which small changes in the input data may result in large changes in the solution. Ill-conditioning in statistical modeling is often the result of high correlations among the independent variables. When such correlations exist, the computations may be subject to severe rounding error. This was a problem in using computer software many years ago, as Longley (1967) pointed out. When there are large correlations among the independent variables, the model itself must be examined, as Beaton, Rubin, and Barone (1976) emphasize in reviewing the analysis performed by Longley. Although the work of Beaton, Rubin, and Barone was criticized for not paying proper respect to high-accuracy computations, ultimately it is the utility of the fitted model that counts, not the accuracy of the computations.

Large correlations are reflected in the condition number of the X matrix. A large condition number may indicate the possibility of harmful numerical errors. Some of the techniques for assessing the accuracy of a computed result may be useful. In particular, the analyst may try the suggestion of Mullet and Murray (1971) to regress $y + dx_j$ on $x_1, \ldots, x_m$, and compare the results with the results obtained from just using y.

Other types of ill-conditioning may be more subtle. Large variations in the leverages may be the cause of ill-conditioning.

Often, numerical problems in regression computations indicate that the linear model may not be entirely satisfactory for the phenomenon being studied. Ill-conditioning in statistical data analysis often means that the approach or the model is wrong.

9.4.2 Variable Selection

Starting with a model such as equation (9.8),

$$Y = \beta^{\mathrm{T}} x + E,$$

we are ignoring the most fundamental problem in data analysis: which variables *are really related* to Y, and *how are they related?*

We often begin with the premise that a linear relationship is at least a good approximation locally; that is, with restricted ranges of the variables. This leaves us with one of the most important tasks in linear regression analysis: selection of the variables to include in the model. There are many statistical issues that must be taken into consideration. We will not discuss these issues here; rather we refer the reader to a comprehensive text on regression analysis, such as Draper and Smith (1998), or to a text specifically on this topic, such as Miller (2002). Some aspects of the statistical analysis involve tests of linear hypotheses, such as discussed in Section 9.2.3. There is a major difference, however; those tests were based on knowledge of the *correct* model. The basic problem in variable selection is that we do not know the correct model. Most reasonable procedures to determine the correct model yield biased statistics. Some people attempt to circumvent this problem by recasting the problem in terms of a "full" model; that is, one that includes all independent variables that the data analyst has looked at. (Looking at a variable and then making a decision to exclude that variable from the model can bias further analyses.)

We generally approach the variable selection problem by writing the model with the data as

$$y = X_{\mathrm{i}}\beta_{\mathrm{i}} + X_{\mathrm{o}}\beta_{\mathrm{o}} + \epsilon, \tag{9.37}$$

where X_{i} and X_{o} are matrices that form some permutation of the columns of X, $X_{\mathrm{i}}|X_{\mathrm{o}} = X$, and β_{i} and β_{o} are vectors consisting of corresponding elements from β. We then consider the model

$$y = X_{\mathrm{i}}\beta_{\mathrm{i}} + \epsilon_{\mathrm{i}}. \tag{9.38}$$

It is interesting to note that the least squares estimate of β_{i} in the model (9.38) is the same as the least squares estimate in the model

$$\hat{y}_{\mathrm{io}} = X_{\mathrm{i}}\beta_{\mathrm{i}} + \epsilon_{\mathrm{i}},$$

where $\hat{y}_{\mathrm{io}}$ is the vector of predicted values obtained by fitting the full model (9.37). An interpretation of this fact is that fitting the model (9.38) that includes only a subset of the variables is the same as using that subset to *approximate* the predictions of the full model. The fact itself can be seen from the normal equations associated with these two models. We have

$$X_{\mathrm{i}}^{\mathrm{T}} X (X^{\mathrm{T}} X)^{-1} X^{\mathrm{T}} = X_{\mathrm{i}}^{\mathrm{T}}. \tag{9.39}$$

This follows from the fact that $X(X^{\mathrm{T}}X)^{-1}X^{\mathrm{T}}$ is a projection matrix, and X_{i} consists of a set of columns of X (see Section 8.5 and Exercise 9.11 on page 368).

As mentioned above, there are many difficult statistical issues in the variable selection problem. The exact methods of statistical inference generally do not apply (because they are based on a model, and we are trying to choose a model). In variable selection, as in any statistical analysis that involves the choice of a model, the effect of the given dataset may be greater than warranted, resulting in overfitting. One way of dealing with this kind of problem is to use part of the dataset for fitting and part for validation of the fit. There are many variations on exactly how to do this, but in general, "cross validation" is an important part of any analysis that involves building a model.

The computations involved in variable selection are the same as those discussed in Sections 9.2.3 and 9.2.7.

9.4.3 Principal Components Regression

A somewhat different approach to the problem of variable selection involves selecting some linear combinations of all of the variables. The first p principal components of X cover the space of $\mathrm{span}(X)$ optimally (in some sense), and so these linear combinations themselves may be considered as the "best" variables to include in a regression model. If V_p is the first p columns from V in the full set of principal components of X, equation (9.34), we use the regression model

$$y \approx Z_p\gamma, \tag{9.40}$$

where

$$Z_p = XV_p. \tag{9.41}$$

This is the idea of principal components regression.

In principal components regression, even if $p < m$ (which is the case, of course; otherwise principal components regression would make no sense), all of the original variables are included in the model. Any linear combination forming a principal component may include all of the original variables. The weighting on the original variables tends to be such that the coefficients of the original variables that have extreme values in the ordinary least squares regression are attenuated in the principal components regression using only the first p principal components.

The principal components do not involve y, so it may not be obvious that a model using only a set of principal components selected without reference to y would yield a useful regression model. Indeed, sometimes important independent variables do not get sufficient weight in principal components regression.

9.4.4 Shrinkage Estimation

As mentioned in the previous section, instead of selecting specific independent variables to include in the regression model, we may take the approach of shrinking the coefficient estimates toward zero. This of course has the effect of introducing a bias into the estimates (in the case of a true model being used), but in the process of reducing the inherent instability due to collinearity in the independent variables, it may also reduce the mean squared error of linear combinations of the coefficient estimates. This is one approach to the problem of overfitting.

The shrinkage can also be accomplished by a regularization of the fitting criterion. If the fitting criterion is minimization of a norm of the residuals, we add a norm of the coefficient estimates to minimize

$$\|r(b)\|_f + \lambda\|b\|_b, \tag{9.42}$$

where λ is a tuning parameter that allows control over the relative weight given to the two components of the objective function. This regularization is also related to the variable selection problem by the association of superfluous variables with the individual elements of the optimal b that are close to zero.

Ridge Regression

If the fitting criterion is least squares, we may also choose an L_2 norm on b, and we have the fitting problem

$$\min_b \left((y - Xb)^{\mathrm{T}}(y - Xb) + \lambda b^{\mathrm{T}}b\right). \tag{9.43}$$

This is called Tikhonov regularization (from A. N. Tikhonov), and it is by far the most commonly used regularization. This minimization problem yields the modified normal equations

$$(X^{\mathrm{T}}X + \lambda I)b = X^{\mathrm{T}}y, \tag{9.44}$$

obtained by adding λI to the sums of squares and cross products matrix. This is the ridge regression we discussed on page 291, and as we saw in Section 6.1, the addition of this positive definite matrix has the effect of reducing numerical ill-conditioning.

Interestingly, these normal equations correspond to a least squares approximation for

$$\begin{pmatrix} y \\ 0 \end{pmatrix} \approx \begin{bmatrix} X \\ \sqrt{\lambda}I \end{bmatrix} \beta.$$

The shrinkage toward 0 is evident in this formulation. Because of this, we say the "effective" degrees of freedom of a ridge regression model decreases with increasing λ. In Equation (8.64), we formally defined the *effective model degrees of freedom* of any linear fit

$$\widehat{y} = S_\lambda y$$

as

$$\mathrm{tr}(S_\lambda).$$

Even if all variables are left in the model, the ridge regression approach may alleviate some of the deleterious effects of collinearity in the independent variables.

Lasso Regression

The norm for the regularization in expression (9.42) does not have to be the same as the norm applied to the model residuals. An alternative fitting criterion, for example, is to use an L_1 norm,

$$\min_b (y - Xb)^{\mathrm{T}}(y - Xb) + \lambda\|b\|_1.$$

Rather than strictly minimizing this expression, we can formulate a constrained optimization problem

$$\min_{\|b\|_1 < t} (y - Xb)^{\mathrm{T}}(y - Xb), \tag{9.45}$$

for some tuning constant t. The solution of this quadratic programming problem yields a b with some elements identically 0, depending on t. As t decreases, more elements of the optimal b are identically 0, and thus this is an effective method for variable selection. The use of expression (9.45) is called lasso regression. ("Lasso" stands for "least absolute shrinkage and selection operator".)

Lasso regression is computationally expensive if several values of t are explored. Efron et al. (2004) propose "least angle regression" (LAR) to compute the entire lasso regularization path simultaneously.

9.4.5 Testing the Rank of a Matrix

The rank of a matrix is not a continuous function of the elements of the matrix. For this reason, among others, it is difficult to compute the rank of a matrix. Numerical analysts refer to computations to estimate the rank of a matrix. (See Section 11.4, where we discuss the rank-revealing QR (or LU) method for estimating the rank of a matrix.) "Estimation" in that sense refers to "approximation" rather than to statistical estimation. This is an important distinction that is often lost. Estimation and testing in a statistical sense do not apply to a given entity; these methods of inference apply to properties of a random variable. A statistical test is a decision rule for rejection of a hypothesis about which empirical evidence is available. The empirical evidence consists of observations on some random variable, and the hypothesis is a statement about the distribution of the random variable. In simple cases of hypothesis testing, the distribution is assumed to be characterized by a parameter, and the hypothesis merely specifies the value of that parameter. The statistical test is based on the distribution of the underlying random variable if the hypothesis is true.

Gill and Lewbel (1992) describe a statistical test of the rank of a matrix. The test uses factors from an LDU factorization. The rows and/or columns of the matrices are permuted so that the values of D that are larger in magnitude occur in the earlier positions. The $n \times m$ matrix A (with $n \geq m$ without loss of generality) can be decomposed as

$$
\begin{aligned}
PAQ &= LDU \\
&= \begin{bmatrix} L_{11} & 0 & 0 \\ L_{21} & L_{22} & 0 \\ L_{31} & L_{32} & I_{n-m} \end{bmatrix} \begin{bmatrix} D_1 & 0 & 0 \\ 0 & D_2 & 0 \\ 0 & 0 & 0 \end{bmatrix} \begin{bmatrix} U_{11} & U_{12} \\ 0 & U_{22} \\ 0 & 0 \end{bmatrix},
\end{aligned} \tag{9.46}
$$

where the matrices L_{11}, U_{11}, and D_1 are $r \times r$, and the elements of the diagonal submatrices D_1 and D_2 are arranged in nonincreasing order. If the rank of A is r, the entries in D_2 are 0.

Now assume A is some parameter characterizing a distribution, and realizations of the underlying random variable can be used to estimate A. (The random variable is vector-valued.) Now let $\widehat{A}$ be an estimate of A based on k such realizations, and assume the central limit property,

$$
\sqrt{k} \operatorname{vec}(\widehat{A} - A) \to_{\mathrm{d}} \mathrm{N}(0, V), \tag{9.47}
$$

where V is $nm \times nm$ and positive definite. Now if $D_2 = 0$ (that is, if A has rank r) and $\widehat{A}$ is decomposed in the same way as A in equation (9.46), then

$$
\sqrt{k} \operatorname{diag}(\widehat{D}_2) \to_{\mathrm{d}} \mathrm{N}(0, W)
$$

for some positive definite matrix W, and the quantity

$$
n \widehat{d}_2^{\mathrm{T}} W^{-1} \widehat{d}_2, \tag{9.48}
$$

where

$$
\widehat{d}_2 = \operatorname{diag}(\widehat{D}_2),
$$

has an asymptotic chi-squared distribution with $(m - r)$ degrees of freedom. If a consistent estimator of W, say $\widehat{W}$, is used in place of W in the expression (9.48), this would be a test statistic for the hypothesis that the rank of A is r. (Note that W is $m - r \times m - r$.)

Gill and Lewbel (1992) derive a consistent estimator to use in expression (9.48) as a test statistic. Following their derivation, first let $\widehat{V}$ be a consistent estimator of V. (It would typically be a sample variance-covariance matrix.) Then

$$
\left(\widehat{Q}^{\mathrm{T}} \otimes \widehat{P} \right) \widehat{V} \left(\widehat{Q} \otimes \widehat{P}^{\mathrm{T}} \right)
$$

is a consistent estimator of the variance-covariance of $\mathrm{vec}(\widehat{P}(\widehat{A} - A)\widehat{Q})$. Next, define the matrices

$$\widehat{H} = \left[-\widehat{L}_{22}^{-1}\widehat{L}_{21}\widehat{L}_{11}^{-1} \mid \widehat{L}_{22}^{-1} \mid 0 \right],$$

$$\widehat{K} = \begin{bmatrix} -\widehat{U}_{11}^{-1}\widehat{U}_{12}\widehat{U}_{22}^{-1} \\ \widehat{U}_{22}^{-1} \end{bmatrix},$$

and T such that

$$\mathrm{vec}(\widehat{D}_2) = T\widehat{d}_2.$$

The matrix T is $(m - r)^2 \times (m - r)$, consisting of a stack of square matrices with 0s in all positions except for a 1 in one diagonal element. The matrix is orthogonal; that is,

$$T^{\mathrm{T}}T = I_{m-r}.$$

The matrix

$$(\widehat{K} \otimes \widehat{H}^{\mathrm{T}})T$$

transforms $\mathrm{vec}(\widehat{P}(\widehat{A} - A)\widehat{Q})$ into $\widehat{d}_2$; hence the variance-covariance estimator, $(\widehat{Q}^{\mathrm{T}} \otimes \widehat{P})\widehat{V}(\widehat{Q} \otimes \widehat{P}^{\mathrm{T}})$, is adjusted by this matrix. The estimator $\widehat{W}$ therefore is given by

$$\widehat{W} = T^{\mathrm{T}}(\widehat{K}^{\mathrm{T}} \otimes \widehat{H})(\widehat{Q}^{\mathrm{T}} \otimes \widehat{P})\widehat{V}(\widehat{Q} \otimes \widehat{P}^{\mathrm{T}})(\widehat{K} \otimes \widehat{H}^{\mathrm{T}})T.$$

The test statistic is

$$n\widehat{d}_2^{\mathrm{T}}\widehat{W}^{-1}\widehat{d}_2, \tag{9.49}$$

with an approximate chi-squared distribution with $(m - r)$ degrees of freedom.

The decomposition in equation (9.46) can affect the limiting distribution of $\sqrt{k}\,\mathrm{vec}(\widehat{A} - A)$. The effect can be exacerbated by complete pivoting, and Gill and Lewbel (1992) recommend that pivoting be limited to row pivoting.

A test for the rank of matrices has many applications, especially in time series, and Gill and Lewbel give examples of several.

9.4.6 Incomplete Data

Missing values in a dataset can not only result in ill-conditioned problems but can cause some matrix statistics to lack their standard properties, such as covariance or correlation matrices not being positive definite.

In the standard flat data file represented in Figure 8.1, where a row holds data from a given observation and a column represents a specific variable or feature, it is often the case that some values are missing for some observation/variable combination. This can occur for various reasons, such as a failure of a measuring device, refusal to answer a question in a survey, or an indeterminate or infinite value for a derived variable (for example, a coefficient of variation when the mean is 0). This causes problems for our standard storage of data in a matrix. The values for some cells are not available.

The need to make provisions for missing data is one of the important differences between statistical numerical processing and ordinary numerical analysis. First of all, we need a method for representing a "not available" (NA) value, and then we need a mechanism for avoiding computations with this NA value. There are various ways of doing this, including the use of a special computer number (see page 386).

The layout of the data may be of the form

$$
X = \begin{bmatrix} X & X & NA \\ X & NA & NA \\ X & NA & X \\ X & X & X \end{bmatrix}. \tag{9.50}
$$

In the data matrix of equation (9.50), all rows could be used for summary statistics relating to the first variable, but only two rows could be used for summary statistics relating to the second and third variables. For summary statistics such as the mean or variance for any one variable, it would seem to make sense to use all of the available data.

The picture is not so clear, however, for statistics on two variables, such as the covariance. If all observations that contain data on both variables are used for computing the covariance, then the covariance matrix may not be positive definite. If the correlation matrix is computed using covariances computed in this way but variances computed on all of the data, some off-diagonal elements may be larger than 1. If the correlation matrix is computed using covariances from all available pairs and variances computed only from the data in complete pairs (that is, the variances used in computing correlations involving a given variable are different for different variables), then no off-diagonal element can be larger than 1, but the correlation matrix may not be nonnegative definite.

An alternative, of course, is to use only data in records that are complete. This is called "casewise deletion", whereas use of all available data for bivariate statistics is called "pairwise deletion". One must be very careful in computing bivariate statistics from data with missing values; see Exercise 9.13 (and a solution on page 499).

Estimated or approximate variance-covariance or correlation matrices that are not positive definite can arise in other ways in applications. For example, the data analyst may have an estimate of the correlation matrix that was not based on a single sample.

Various approaches to handling an approximate correlation matrix that is not positive definite have been considered. Devlin, Gnanadesikan, and Kettenring (1975) describe a method of shrinking the given R toward a chosen positive definite matrix, R_1, which may be an estimator of a correlation matrix computed in other ways (perhaps a robust estimator) or may just be chosen arbitrarily; for example, R_1 may just be the identity matrix. The method is to choose the largest value α in $[0, 1]$ such that the matrix

$$\widetilde{R} = \alpha R + (1 - \alpha)R_1 \tag{9.51}$$

is positive definite. This optimization problem can be solved iteratively starting with $\alpha = 1$ and decreasing α in small steps while checking whether $\widetilde{R}$ is positive definite. (The checks may require several computations.) A related method is to use a modified Cholesky decomposition. If the symmetric matrix S is not positive definite, a diagonal matrix D can be determined so that $S + D$ is positive definite. Eskow and Schnabel (1991) describe a method to determine D with values near zero and to compute a Cholesky decomposition of $S + D$.

Devlin, Gnanadesikan, and Kettenring (1975) also describe nonlinear shrinking methods in which all of the off-diagonal elements r_{ij} are replaced iteratively, beginning with $r_{ij}^{(0)} = r_{ij}$ and proceeding with

$$
r_{ij}^{(k)} = \begin{cases}
f^{-1}\left(f\left(r_{ij}^{(k-1)}\right) + \delta\right) & \text{if } r_{ij}^{(k-1)} < -f^{-1}(\delta) \\[2ex]
0 & \text{if } \left|r_{ij}^{(k-1)}\right| \leq f^{-1}(\delta) \\[2ex]
f^{-1}\left(f\left(r_{ij}^{(k-1)}\right) - \delta\right) & \text{if } r_{ij}^{(k-1)} > f^{-1}(\delta)
\end{cases}
\tag{9.52}
$$

for some invertible positive-valued function f and some small positive constant δ (for example, 0.05). The function f may be chosen in various ways; one suggested function is the hyperbolic tangent, which makes f^{-1} Fisher's variance-stabilizing function for a correlation coefficient; see Exercise 9.18b.

Rousseeuw and Molenberghs (1993) suggest a method in which some approximate correlation matrices can be adjusted to a nearby correlation matrix, where closeness is determined by the Frobenius norm. Their method applies to pseudo-correlation matrices. Recall that any symmetric nonnegative definite matrix with ones on the diagonal is a correlation matrix. A *pseudo-correlation matrix* is a symmetric matrix R with positive diagonal elements (but not necessarily 1s) and such that $r_{ij}^2 \leq r_{ii}r_{jj}$. (This is inequality (8.12), which is a necessary but not sufficient condition for the matrix to be nonnegative definite.)

The method of Rousseeuw and Molenberghs adjusts an $m \times m$ pseudo-correlation matrix R to the closest correlation matrix $\widetilde{R}$, where closeness is determined by the Frobenius norm; that is, we seek $\widetilde{R}$ such that

$$\|R - \widetilde{R}\|_{\mathrm{F}} \tag{9.53}$$

is minimum over all choices of $\widetilde{R}$ that are correlation matrices (that is, matrices with 1s on the diagonal that are positive definite). The solution to this optimization problem is not as easy as the solution to the problem we consider on page 138 of finding the best approximate matrix of a given rank. Rousseeuw and Molenberghs describe a computational method for finding $\widetilde{R}$ to minimize expression (9.53). A correlation matrix $\widetilde{R}$ can be formed as a Gramian matrix formed from a matrix U whose columns, $u_1, \ldots, u_m$, are normalized vectors, where

$$\tilde{r}_{ij} = u_i^{\mathrm{T}} u_j.$$

If we choose the vector u_i so that only the first i elements are nonzero, then they form the Cholesky factor elements of $\widetilde{R}$ with nonnegative diagonal elements,

$$\widetilde{R} = U^{\mathrm{T}} U,$$

and each u_i can be completely represented in $\mathrm{I\!R}^i$. We can associate the $m(m-1)/2$ unknown elements of U with the angles in their spherical coordinates. In u_i, the jth element is 0 if $j > i$ and otherwise is

$$\sin(\theta_{i1}) \cdots \sin(\theta_{i,i-j}) \cos(\theta_{i,i-j+1}),$$

where $\theta_{i1}, \ldots, \theta_{i,i-j}, \theta_{i,i-j+1}$ are the unknown angles that are the variables in the optimization problem for the Frobenius norm (9.53). The problem now is to solve

$$\min \sum_{i=1}^{m} \sum_{j=1}^{i} (r_{ij} - \sin(\theta_{i1}) \cdots \sin(\theta_{i,i-j}) \cos(\theta_{i,i-j+1}))^2. \tag{9.54}$$

This optimization problem is well-behaved and can be solved by steepest descent (see page 158). Rousseeuw and Molenberghs (1993) also mention that a weighted least squares problem in place of equation (9.54) may be more appropriate if the elements of the pseudo-correlation matrix R result from different numbers of observations.

In Exercise 9.14, we describe another way of converting an approximate correlation matrix that is not positive definite into a correlation matrix by iteratively replacing negative eigenvalues with positive ones.

9.5 Optimal Design

When an experiment is designed to explore the effects of some variables (usually called "factors") on another variable, the settings of the factors (independent variables) should be determined so as to yield a maximum amount of information from a given number of observations. The basic problem is to determine from a set of candidates the best rows for the data matrix X. For example, if there are six factors and each can be set at three different levels, there is a total of $3^6 = 729$ combinations of settings. In many cases, because of the expense in conducting the experiment, only a relatively small number of runs can be made. If, in the case of the 729 possible combinations, only 30 or so runs can be made, the scientist must choose the subset of combinations that will be most informative. A row in X may contain more elements than just the number of factors (because of interactions), but the factor settings completely determine the row.

We may quantify the information in terms of variances of the estimators. If we assume a linear relationship expressed by

$$y = \beta_0 1 + X\beta + \epsilon$$

and make certain assumptions about the probability distribution of the residuals, the variance-covariance matrix of estimable linear functions of the least squares solution (9.12) is formed from

$$(X^{\mathrm{T}}X)^{-}\sigma^2.$$

(The assumptions are that the residuals are independently distributed with a constant variance, σ^2. We will not dwell on the statistical properties here, however.) If the emphasis is on estimation of β, then X should be of full rank. In the following, we assume X is of full rank; that is, that $(X^{\mathrm{T}}X)^{-1}$ exists.

An objective is to minimize the variances of estimators of linear combinations of the elements of β. We may identify three types of relevant measures of the variance of the estimator $\widehat{\beta}$: the average variance of the elements of $\widehat{\beta}$, the maximum variance of any elements, and the "generalized variance" of the vector $\widehat{\beta}$. The property of the design resulting from maximizing the information by reducing these measures of variance is called, respectively, A-optimality, E-optimality, and D-optimality. They are achieved when X is chosen as follows:

- A-optimality: minimize $\mathrm{tr}((X^{\mathrm{T}}X)^{-1})$.
- E-optimality: minimize $\rho((X^{\mathrm{T}}X)^{-1})$.
- D-optimality: minimize $\det((X^{\mathrm{T}}X)^{-1})$.

Using the properties of eigenvalues and determinants that we discussed in Chapter 3, we see that E-optimality is achieved by maximizing $\rho(X^{\mathrm{T}}X)$ and D-optimality is achieved by maximizing $\det(X^{\mathrm{T}}X)$.

D-Optimal Designs

The D-optimal criterion is probably used most often. If the residuals have a normal distribution (and the other distributional assumptions are satisfied), the D-optimal design results in the smallest volume of confidence ellipsoids for β. (See Titterington, 1975; Nguyen and Miller, 1992; and Atkinson and Donev, 1992. Identification of the D-optimal design is related to determination of a minimum-volume ellipsoid for multivariate data.) The computations required for the D-optimal criterion are the simplest, and this may be another reason it is used often.

To construct an optimal X with a given number of rows, n, from a set of N potential rows, one usually begins with an initial choice of rows, perhaps random, and then determines the effect on the determinant by exchanging a selected row with a different row from the set of potential rows. If the matrix X has n rows and the row vector x^{T} is appended, the determinant of interest is

$$\det(X^{\mathrm{T}}X + xx^{\mathrm{T}})$$

or its inverse. Using the relationship $\det(AB) = \det(A)\det(B)$, it is easy to see that

$$\det(X^{\mathrm{T}}X + xx^{\mathrm{T}}) = \det(X^{\mathrm{T}}X)(1 + x^{\mathrm{T}}(X^{\mathrm{T}}X)^{-1}x). \qquad (9.55)$$

Now, if a row x_{+}^{T} is exchanged for the row x_{-}^{T}, the effect on the determinant is given by

$$\det(X^{\mathrm{T}}X + x_{+}x_{+}^{\mathrm{T}} - x_{-}x_{-}^{\mathrm{T}}) = \det(X^{\mathrm{T}}X) \times$$
$$\left(1 + x_{+}^{\mathrm{T}}(X^{\mathrm{T}}X)^{-1}x_{+} - \right.$$
$$x_{-}^{\mathrm{T}}(X^{\mathrm{T}}X)^{-1}x_{-}(1 + x_{+}^{\mathrm{T}}(X^{\mathrm{T}}X)^{-1}x_{+}) +$$
$$\left. (x_{+}^{\mathrm{T}}(X^{\mathrm{T}}X)^{-1}x_{-})^{2} \right) \qquad (9.56)$$

(see Exercise 9.7).

Following Miller and Nguyen (1994), writing $X^{\mathrm{T}}X$ as $R^{\mathrm{T}}R$ from the QR decomposition of X, and introducing z_{+} and z_{-} as

$$Rz_{+} = x_{+}$$

and

$$Rz_{-} = x_{-},$$

we have the right-hand side of equation (9.56):

$$z_{+}^{\mathrm{T}}z_{+} - z_{-}^{\mathrm{T}}z_{-}(1 + z_{+}^{\mathrm{T}}z_{+}) + (z_{-}^{\mathrm{T}}z_{+})^{2}. \qquad (9.57)$$

Even though there are $n(N - n)$ possible pairs (x_{+}, x_{-}) to consider for exchanging, various quantities in (9.57) need be computed only once. The corresponding (z_{+}, z_{-}) are obtained by back substitution using the triangular matrix R. Miller and Nguyen use the Cauchy-Schwarz inequality (2.10) (page 16) to show that the quantity (9.57) can be no larger than

$$z_{+}^{\mathrm{T}}z_{+} - z_{-}^{\mathrm{T}}z_{-}; \qquad (9.58)$$

hence, when considering a pair (x_{+}, x_{-}) for exchanging, if the quantity (9.58) is smaller than the largest value of (9.57) found so far, then the full computation of (9.57) can be skipped. Miller and Nguyen also suggest not allowing the last point added to the design to be considered for removal in the next iteration and not allowing the last point removed to be added in the next iteration.

The procedure begins with an initial selection of design points, yielding the $n \times m$ matrix $X^{(0)}$ that is of full rank. At the kth step, each row of $X^{(k)}$

is considered for exchange with a candidate point, subject to the restrictions mentioned above. Equations (9.57) and (9.58) are used to determine the best exchange. If no point is found to improve the determinant, the process terminates. Otherwise, when the optimal exchange is determined, $R^{(k+1)}$ is formed using the updating methods discussed in the previous sections. (The programs of Gentleman, 1974, referred to in Section 6.7.4 can be used.)

9.6 Multivariate Random Number Generation

The need to simulate realizations of random variables arises often in statistical applications, both in the development of statistical theory and in applied data analysis. In this section, we will illustrate only a couple of problems in multivariate random number generation. These make use of some of the properties we have discussed previously.

Most methods for random number generation assume an underlying source of realizations of a uniform $(0, 1)$ random variable. If U is a uniform $(0, 1)$ random variable, and F is the cumulative distribution function of a continuous random variable, then the random variable

$$X = F^{-1}(U)$$

has the cumulative distribution function F. (If the support of X is finite, $F^{-1}(0)$ and $F^{-1}(1)$ are interpreted as the limits of the support.) This same idea, the basis of the so-called inverse CDF method, can also be applied to discrete random variables.

The Multivariate Normal Distribution

If Z has a multivariate normal distribution with the identity as variance-covariance matrix, then for a given positive definite matrix Σ, both

$$Y_1 = \Sigma^{1/2} Z \tag{9.59}$$

and

$$Y_2 = \Sigma_C Z, \tag{9.60}$$

where Σ_C is a Cholesky factor of Σ, have a multivariate normal distribution with variance-covariance matrix Σ (see page 323).

This leads to a very simple method for generating a multivariate normal random d-vector: generate into a d-vector z d independent $N_1(0, 1)$. Then form a vector from the desired distribution by the transformation in equation (9.59) or (9.60) together with the addition of a mean vector if necessary.

Random Correlation Matrices

Occasionally we wish to generate random numbers but do not wish to specify the distribution fully. We may want a "random" matrix, but we do not know an exact distribution that we wish to simulate. (There are only a few "standard" distributions of matrices. The Wishart distribution and the Haar distribution (page 169) are the only two common ones. We can also, of course, specify the distributions of the individual elements.)

We may want to simulate random correlation matrices. Although we do not have a specific distribution, we may want to specify some characteristics, such as the eigenvalues. (All of the eigenvalues of a correlation matrix, not just the largest and smallest, determine the condition of data matrices that are realizations of random variables with the given correlation matrix.)

Any nonnegative definite (symmetric) matrix with 1s on the diagonal is a correlation matrix. A correlation matrix is diagonalizable, so if the eigenvalues are $c_1, \ldots, c_d$, we can represent the matrix as

$$V \operatorname{diag}(c_1, \ldots, c_d) V^{\mathrm{T}}$$

for an orthogonal matrix V. (For a $d \times d$ correlation matrix, we have $\sum c_i = d$; see page 295.) Generating a random correlation matrix with given eigenvalues becomes a problem of generating the random orthogonal eigenvectors and then forming the matrix V from them. (Recall from page 119 that the eigenvectors of a symmetric matrix can be chosen to be orthogonal.) In the following, we let $C = \operatorname{diag}(c_1, \ldots, c_d)$ and begin with $E = I$ (the $d \times d$ identity) and $k = 1$. The method makes use of deflation in step 6 (see page 243). The underlying randomness is that of a normal distribution.

Algorithm 9.2 Random Correlation Matrices with Given Eigenvalues

1. Generate a d-vector w of i.i.d. standard normal deviates, form $x = Ew$, and compute $a = x^{\mathrm{T}}(I - C)x$.
2. Generate a d-vector z of i.i.d. standard normal deviates, form $y = Ez$, and compute $b = x^{\mathrm{T}}(I - C)y$, $c = y^{\mathrm{T}}(I - C)y$, and $e^2 = b^2 - ac$.
3. If $e^2 < 0$, then go to step 2.
4. Choose a random sign, $s = -1$ or $s = 1$. Set $r = \dfrac{b + se}{a}x - y$.
5. Choose another random sign, $s = -1$ or $s = 1$, and set $v_k = \dfrac{sr}{(r^{\mathrm{T}}r)^{\frac{1}{2}}}$.
6. Set $E = E - v_k v_k^{\mathrm{T}}$, and set $k = k + 1$.
7. If $k < d$, then go to step 1.
8. Generate a d-vector w of i.i.d. standard normal deviates, form $x = Ew$, and set $v_d = \dfrac{x}{(x^{\mathrm{T}}x)^{\frac{1}{2}}}$.
9. Construct the matrix V using the vectors v_k as its rows. Deliver VCV^{T} as the random correlation matrix. ∎

9.7 Stochastic Processes

Many stochastic processes are modeled by a "state vector" and rules for updating the state vector through a sequence of discrete steps. At time t, the elements of the state vector x_t are values of various characteristics of the system. A model for the stochastic process is a probabilistic prescription for x_{t_a} in terms of x_{t_b}, where $t_a > t_b$; that is, given observations on the state vector prior to some point in time, the model gives probabilities for, or predicts values of, the state vector at later times.

A stochastic process is distinguished in terms of the countability of the space of states, $\mathcal{X}$, and the index of the state (that is, the parameter space, $\mathcal{T}$); either may or may not be countable. If the parameter space is continuous, the process is called a *diffusion process*. If the parameter space is countable, we usually consider it to consist of the nonnegative integers.

If the properties of a stochastic process do not depend on the index, the process is said to be *stationary*. If the properties also do not depend on any initial state, the process is said to be *time homogeneous* or *homogeneous with respect to the parameter space*. (We usually refer to such processes simply as "homogeneous".)

9.7.1 Markov Chains

The *Markov* (or Markovian) *property* in a stochastic process is the condition where the current state does not depend on any states prior to the immediately previous state; that is, the process is *memoryless*. If the parameter space is countable, the Markov property is the condition where the probability distribution of the state at time $t + 1$ depends only on the state at time t.

In what follows, we will briefly consider some Markov processes in which both the state space and the parameter space (time) are countable. Such a process is called a *Markov chain*. (Some authors' use of the term "Markov chain" allows only the state space to be continuous, and others' allows only time to be continuous; here we are not defining the term. We will be concerned with only a subclass of Markov chains, whichever way they are defined. The models for this subclass are easily formulated in terms of vectors and matrices.)

If the state space is countable, it is equivalent to $\mathcal{X} = \{1, 2, \ldots\}$. If X is a random variable from some sample space to $\mathcal{X}$, and

$$\pi_i = \Pr(X = i),$$

then the vector π defines a distribution of X on $\mathcal{X}$. (A vector of nonnegative numbers that sum to 1 is a *distribution*.) Formally, we define a Markov chain (of random variables) $X_0, X_1, \ldots$ in terms of an initial distribution π and a conditional distribution for X_{t+1} given X_t. Let X_0 have distribution π, and given $X_t = i$, let X_{t+1} have distribution $(p_{ij}; j \in \mathcal{X})$; that is, p_{ij} is the probability of a transition from state i at time t to state j at time $t + 1$. Let

$$P = (p_{ij}).$$

This square matrix is called the *transition matrix* of the chain. The initial distribution π and the transition matrix P characterize the chain, which we sometimes denote as *Markov*(π, P). It is clear that P is a stochastic matrix, and hence $\rho(P) = \|P\|_\infty = 1$, and $(1, 1)$ is an eigenpair of P (see page 306).

If P does not depend on the time (and our notation indicates that we are assuming this), the Markov chain is stationary.

If the state space is countably infinite, the vectors and matrices have infinite order; that is, they have "infinite dimension". (Note that this use of "dimension" is different from our standard definition that is based on linear independence.)

We denote the distribution at time t by $\pi(t)$ and hence often write the initial distribution as $\pi(0)$. A distribution at time t can be expressed in terms of π and P if we extend the definition of (Cayley) matrix multiplication in equation (3.34) in the obvious way so that

$$(P^2)_{ij} = \sum_{k \in \mathcal{X}} p_{ik} p_{kj}.$$

We see immediately that

$$\pi(t) = (P^t)^\mathrm{T} \pi(0). \tag{9.61}$$

(The somewhat awkward notation here results from the historical convention in Markov chain theory of expressing distributions as row vectors.) Because of equation (9.61), P^t is often called the *t-step transition matrix*.

The transition matrix determines various relationships among the states of a Markov chain. State j is said to be *accessible* from state i if it can be reached from state i in a finite number of steps. This is equivalent to $(P^t)_{ij} > 0$ for some t. If state j is *accessible* from state i and state i is *accessible* from state j, states j and i are said to *communicate*. Communication is clearly an equivalence relation. (A binary relation $\sim$ is an *equivalence relation* over some set S if for $x, y, z \in S$, (1) $x \sim x$, (2) $x \sim y \Rightarrow x \sim y$, and (3) $x \sim y \wedge y \sim z \Rightarrow x \sim z$; that is, it is reflexive, symmetric, and transitive.) The set of all states that communicate with each other is an *equivalence class*. States belonging to different equivalence classes do not communicate, although a state in one class may be accessible from a state in a different class. If all states in a Markov chain are in a single equivalence class, the chain is said to be *irreducible*. Reducibility of Markov chains is clearly related to the reducibility in graphs that we discussed in Section 8.1.2, and reducibility in both cases is related to properties of a nonnegative matrix; in the case of graphs, it is the connectivity matrix, and for Markov chains it is the transition matrix.

The limiting behavior of the Markov chain is of interest. This of course can be analyzed in terms of $\lim_{t \to \infty} P^t$. Whether or not this limit exists depends

on the properties of P. If P is primitive and irreducible, we can make use of
the results in Section 8.7.2. In particular, because 1 is an eigenvalue and the
vector 1 is the eigenvector associated with 1, from equation (8.82), we have

$$\lim_{t \to \infty} P^k = 1\pi_s^{\mathrm{T}}, \tag{9.62}$$

where π_s is the Perron vector of P^{T}.

This also gives us the *limiting distribution* for an irreducible, primitive
Markov chain,

$$\lim_{t \to \infty} \pi(t) = \pi_s.$$

The Perron vector has the property $\pi_s = P^{\mathrm{T}}\pi_s$ of course, so this distribution
is the *invariant distribution* of the chain.

There are many other interesting properties of Markov chains that follow
from various properties of nonnegative matrices that we discuss in Section 8.7,
but rather than continuing the discussion here, we refer the interested reader
to a text on Markov chains, such as Norris (1997).

9.7.2 Markovian Population Models

A simple but useful model for population growth measured at discrete points
in time, $t, t+1, \ldots$, is constructed as follows. We identify k age groupings for
the members of the population; we determine the number of members in each
age group at time t, calling this $p^{(t)}$,

$$p^{(t)} = \left(p_1^{(t)}, \ldots, p_k^{(t)}\right);$$

determine the reproductive rate in each age group, calling this α,

$$\alpha = (\alpha_1, \ldots, \alpha_k);$$

and determine the survival rate in each of the first $k-1$ age groups, calling
this σ,

$$\sigma = (\sigma_1, \ldots, \sigma_{k-1}).$$

It is assumed that the reproductive rate and the survival rate are constant
in time. (There are interesting statistical estimation problems here that are
described in standard texts in demography or in animal population models.)
The survival rate σ_i is the proportion of members in age group i at time t
who survive to age group $i+1$. (It is assumed that the members in the last
age group do not survive from time t to time $t+1$.) The total size of the
population at time t is $N^{(t)} = 1^{\mathrm{T}}p^{(t)}$. (The use of the capital letter N for
a scalar variable is consistent with the notation used in the study of finite
populations.)

If the population in each age group is relatively large, then given the sizes
of the population age groups at time t, the approximate sizes at time $t+1$
are given by

$$p^{(t+1)} = Ap^{(t)},$$ (9.63)

where A is a Leslie matrix as in equation (8.88),

$$A = \begin{bmatrix} \alpha_1 & \alpha_2 & \cdots & \alpha_{m-1} & \alpha_m \\ \sigma_1 & 0 & \cdots & 0 & 0 \\ 0 & \sigma_2 & \cdots & 0 & 0 \\ \vdots & \vdots & \vdots & \ddots & \vdots \\ 0 & 0 & \cdots & \sigma_{m-1} & 0 \end{bmatrix},$$ (9.64)

where $0 \leq \alpha_i$ and $0 \leq \sigma_i \leq 1$.

The Leslie population model can be useful in studying various species of plants or animals. The parameters in the model determine the vitality of the species. For biological realism, at least one α_i and all σ_i must be positive. This model provides a simple approach to the study and simulation of population dynamics. The model depends critically on the eigenvalues of A.

As we have seen (Exercise 8.10), the Leslie matrix has a single unique positive eigenvalue. If that positive eigenvalue is strictly greater in modulus than any other eigenvalue, then given some initial population size, $p^{(0)}$, the model yields a few damping oscillations and then an exponential growth,

$$p^{(t_0+t)} = p^{(t_0)}e^{rt},$$ (9.65)

where r is the *rate constant*. The vector $p^{(t_0)}$ (or any scalar multiple) is called the *stable age distribution*. (You are asked to show this in Exercise 9.21a.) If 1 is an eigenvalue and all other eigenvalues are strictly less than 1 in modulus, then the population eventually becomes constant; that is, there is a *stable population*. (You are asked to show this in Exercise 9.21b.)

The survival rates and reproductive rates constitute an age-dependent *life table*, which is widely used in studying population growth. The age groups in life tables for higher-order animals are often defined in years, and the parameters often are defined only for females. The first age group is generally age 0, and so $\alpha_1 = 0$. The *net reproductive rate*, r_0, is the average number of (female) offspring born to a given (female) member of the population over the lifetime of that member; that is,

$$r_0 = \sum_{i=2}^{m} \alpha_i \sigma_{i-1}.$$ (9.66)

The *average generation time*, T, is given by

$$T = \sum_{i=2}^{m} i\alpha_i \sigma_{i-1}/r_0.$$ (9.67)

The net reproductive rate, average generation time, and exponential growth rate constant are related by

$$r = \log(r_0)/T. \tag{9.68}$$

(You are asked to show this in Exercise 9.21c.)

Because the process being modeled is continuous in time and this model is discrete, there are certain averaging approximations that must be made. There are various refinements of this basic model to account for continuous time. There are also refinements to allow for time-varying parameters and for the intervention of exogenous events. Of course, from a statistical perspective, the most interesting questions involve the estimation of the parameters. See Cullen (1985), for example, for further discussions of this modeling problem.

Various starting age distributions can be used in this model to study the population dynamics.

9.7.3 Autoregressive Processes

Another type of application arises in the pth-order autoregressive time series defined by the stochastic difference equation

$$x_t + \alpha_1 x_{t-1} + \cdots + \alpha_p x_{t-p} = e_t,$$

where the e_t are mutually independent normal random variables with mean 0, and $\alpha_p \neq 0$. If the roots of the associated polynomial $m^p + \alpha_1 m^{p-1} + \cdots + \alpha_p = 0$ are less than 1 in absolute value, we can express the parameters of the time series as

$$R\alpha = -\rho, \tag{9.69}$$

where α is the vector of the α_is, the ith element of the vector; ρ is the auto-covariance of lag i; and the $(i, j)^{\text{th}}$ element of the $p \times p$ matrix $R(h)$ is the autocorrelation between x_i and x_j. Equation (9.69) is called the Yule-Walker equation. Because the autocorrelation depends only on the difference $|i - j|$, the diagonals of R are constant,

$$R = \begin{bmatrix} 1 & \rho_1 & \rho_2 & \cdots & \rho_{p-1} \\ \rho_1 & 1 & \rho_1 & \cdots & \rho_{p-2} \\ \rho_2 & \rho_1 & 1 & \cdots & \rho_{p-3} \\ \vdots & & & \ddots & \vdots \\ \rho_{p-1} & \rho_{p-2} & \rho_{p-3} & \cdots & 1 \end{bmatrix};$$

that is, R is a Toeplitz matrix (see Section 8.8.4). Algorithm 9.3 can be used to solve the system (9.69).

Algorithm 9.3 Solution of the Yule-Walker System (9.69)

1. Set $k = 0$; $\alpha_1^{(k)} = -\rho_1$; $b^{(k)} = 1$; and $a^{(k)} = -\rho_1$.
2. Set $k = k + 1$.
3. Set $b^{(k)} = \left(1 - \left(a^{(k-1)}\right)^2\right) b^{(k-1)}$.

4. Set $a^{(k)} = - \left(\rho_{k+1} + \sum_{i=1}^{k} \rho_{k+1-i} \alpha_1^{(k-1)} \right) / b^{(k)}$.

5. For $i = 1, 2, \ldots, k$
   set $y_i = \alpha_i^{(k-1)} + a^{(k)} \alpha_{k+1-i}^{(k-1)}$.

6. For $i = 1, 2, \ldots, k$
   set $\alpha_i^{(k)} = y_i$.

7. Set $\alpha_{k+1}^{(k)} = a^{(k)}$.

8. If $k < p - 1$, go to step 1; otherwise terminate.

This algorithm is $O(p)$ (see Golub and Van Loan, 1996).

The Yule-Walker equations arise in many places in the analysis of stochastic processes. Multivariate versions of the equations are used for a vector time series (see Fuller, 1995, for example).

Exercises

9.1. Let X be an $n \times m$ matrix with $n > m$ and with entries sampled independently from a continuous distribution (of a real-valued random variable). What is the probability that $X^{\mathrm{T}} X$ is positive definite?

9.2. From equation (9.15), we have $\hat{y}_i = y^{\mathrm{T}} X (X^{\mathrm{T}} X)^+ x_{i*}$. Show that h_{ii} in equation (9.16) is $\partial \hat{y}_i / \partial y_i$.

9.3. Formally prove from the definition that the sweep operator is its own inverse.

9.4. Consider the regression model

$$y = X\beta + \epsilon \tag{9.70}$$

subject to the linear equality constraints

$$L\beta = c, \tag{9.71}$$

and assume that X is of full column rank.

a) Let λ be the vector of Lagrange multipliers. Form

$$(b^{\mathrm{T}} L^{\mathrm{T}} - c^{\mathrm{T}}) \lambda$$

and

$$(y - Xb)^{\mathrm{T}} (y - Xb) + (b^{\mathrm{T}} L^{\mathrm{T}} - c^{\mathrm{T}}) \lambda.$$

Now differentiate these two expressions with respect to λ and b, respectively, set the derivatives equal to zero, and solve to obtain

$$\widehat{\beta}_C = (X^{\mathrm{T}} X)^{-1} X^{\mathrm{T}} y - \frac{1}{2} (X^{\mathrm{T}} X)^{-1} L^{\mathrm{T}} \widehat{\lambda}_C$$

$$= \widehat{\beta} - \frac{1}{2} (X^{\mathrm{T}} X)^{-1} L^{\mathrm{T}} \widehat{\lambda}_C$$

and

$$\widehat{\lambda}_C = -2(L(X^{\mathrm{T}}X)^{-1}L^{\mathrm{T}})^{-1}(c - L\widehat{\beta}).$$

Now combine and simplify these expressions to obtain expression (9.25) (on page 337).

b) Prove that the stationary point obtained in Exercise 9.4a actually minimizes the residual sum of squares subject to the equality constraints.

Hint: First express the residual sum of squares as

$$(y - X\widehat{\beta})^{\mathrm{T}}(y - X\widehat{\beta}) + (\widehat{\beta} - b)^{\mathrm{T}}X^{\mathrm{T}}X(\widehat{\beta} - b),$$

and show that is equal to

$$(y - X\widehat{\beta})^{\mathrm{T}}(y - X\widehat{\beta}) + (\widehat{\beta} - \widehat{\beta}_C)^{\mathrm{T}}X^{\mathrm{T}}X(\widehat{\beta} - \widehat{\beta}_C) + (\widehat{\beta}_C - b)^{\mathrm{T}}X^{\mathrm{T}}X(\widehat{\beta}_C - b),$$

which is minimized when $b = \widehat{\beta}_C$.

c) Show that sweep operations applied to the matrix (9.26) on page 337 yield the restricted least squares estimate in the (1,2) block.

d) For the weighting matrix W, derive the expression, analogous to equation (9.25), for the generalized or weighted least squares estimator for β in equation (9.70) subject to the equality constraints (9.71).

9.5. Derive a formula similar to equation (9.29) to update $\widehat{\beta}$ due to the deletion of the ith observation.

9.6. When data are used to fit a model such as $y = X\beta + \epsilon$, a large leverage of an observation is generally undesirable. If an observation with large leverage just happens not to fit the "true" model well, it will cause $\widehat{\beta}$ to be farther from β than a similar observation with smaller leverage.

a) Use artificial data to study influence. There are two main aspects to consider in choosing the data: the pattern of X and the values of the residuals in ϵ. The true values of β are not too important, so β can be chosen as 1. Use 20 observations. First, use just one independent variable ($y_i = \beta_0 + \beta_1 x_i + \epsilon_i$). Generate 20 x_is more or less equally spaced between 0 and 10, generate 20 ϵ_is, and form the corresponding y_is. Fit the model, and plot the data and the model. Now, set $x_{20} = 20$, set ϵ_{20} to various values, form the y_i's and fit the model for each value. Notice the influence of x_{20}.

Now, do similar studies with three independent variables. (Do not plot the data, but perform the computations and observe the effect.) Carefully write up a clear description of your study with tables and plots.

b) Heuristically, the leverage of a point arises from the distance from the point to a fulcrum. In the case of a linear regression model, the measure of the distance of observation i is

$$\Delta(x_i, X1/n) = \|x_i, X1/n\|.$$

(This is not the same quantity from the hat matrix that is defined as the leverage on page 332, but it should be clear that the influence of a point for which $\Delta(x_i, X1/n)$ is large is greater than that of a point for which the quantity is small.) It may be possible to overcome some of the undesirable effects of differential leverage by using weighted least squares to fit the model. The weight w_i would be a decreasing function of $\Delta(x_i, X1/n)$.

Now, using datasets similar to those used in the previous part of this exercise, study the use of various weighting schemes to control the influence. Weight functions that may be interesting to try include

$$w_i = e^{-\Delta(x_i, X1/n)}$$

and

$$w_i = \max(w_{\max}, \|\Delta(x_i, X1/n)\|^{-p})$$

for some $w_{\max}$ and some $p > 0$. (Use your imagination!)

Carefully write up a clear description of your study with tables and plots.

c) Now repeat Exercise 9.6b except use a decreasing function of the leverage, h_{ii} from the hat matrix in equation (9.15) instead of the function $\Delta(x_i, X1/n)$.

Carefully write up a clear description of this study, and compare it with the results from Exercise 9.6b.

9.7. Formally prove the relationship expressed in equation (9.56) on page 357.

Hint: Use equation (9.55) twice.

9.8. On page 161, we used Lagrange multipliers to determine the normalized vector x that maximized $x^{\mathrm{T}} Ax$. If A is S_X, this is the first principal component. We also know the principal components from the spectral decomposition. We could also find them by sequential solutions of Lagrangians. After finding the first principal component, we would seek the linear combination z such that $X_c z$ has maximum variance among all normalized z that are orthogonal to the space spanned by the first principal component; that is, that are $X_c^{\mathrm{T}} X_c$-*conjugate* to the first principal component (see equation (3.65) on page 71). If V_1 is the matrix whose columns are the eigenvectors associated with the largest eigenvector, this is equivalent to finding z so as to maximize $z^{\mathrm{T}} Sz$ subject to $V_1^{\mathrm{T}} z = 0$. Using the method of Lagrange multipliers as in equation (4.29), we form the Lagrangian corresponding to equation (4.31) as

$$z^{\mathrm{T}} Sz - \lambda(z^{\mathrm{T}} z - 1) - \phi V_1^{\mathrm{T}} z,$$

where λ is the Lagrange multiplier associated with the normalization requirement $z^{\mathrm{T}} z = 1$, and ϕ is the Lagrange multiplier associated with the orthogonality requirement. Solve this for the second principal component, and show that it is the same as the eigenvector corresponding to the second-largest eigenvalue.

9.9. Obtain the "Longley data". (It is a dataset in R, and it is also available from `statlib`.) Each observation is for a year from 1947 to 1962 and consists of the number of people employed, five other economic variables, and the year itself. Longley (1967) fitted the number of people employed to a linear combination of the other variables, including the year.

a) Use a regression program to obtain the fit.

b) Now consider the year variable. The other variables are measured (estimated) at various times of the year, so replace the year variable with a "midyear" variable (i.e., add $\frac{1}{2}$ to each year). Redo the regression. How do your estimates compare?

c) Compute the L_2 condition number of the matrix of independent variables. Now add a ridge regression diagonal matrix, as in the matrix (9.72), and compute the condition number of the resulting matrix. How do the two condition numbers compare?

9.10. Consider the least squares regression estimator (9.12) for full rank $n \times m$ matrix X $(n > m)$:

$$\widehat{\beta} = (X^\mathrm{T} X)^{-1} X^\mathrm{T} y.$$

a) Compare this with the ridge estimator

$$\widehat{\beta}_{R(d)} = (X^\mathrm{T} X + dI_m)^{-1} X^\mathrm{T} y$$

for $d \geq 0$. Show that

$$\|\widehat{\beta}_{R(d)}\| \leq \|\widehat{\beta}\|.$$

b) Show that $\widehat{\beta}_{R(d)}$ is the least squares solution to the regression model similar to $y = X\beta + \epsilon$ except with some additional artificial data; that is, y is replaced with

$$\begin{pmatrix} y \\ 0 \end{pmatrix},$$

where 0 is an m-vector of 0s, and X is replaced with

$$\begin{bmatrix} X \\ dI_m \end{bmatrix}. \tag{9.72}$$

Now explain why $\widehat{\beta}_{R(d)}$ is shorter than $\widehat{\beta}$.

9.11. Use the Schur decomposition (equation (3.145), page 95) of the inverse of $(X^\mathrm{T} X)$ to prove equation (9.39).

9.12. Given the matrix

$$A = \begin{bmatrix} 2 & 1 & 3 \\ 1 & 2 & 3 \\ 1 & 1 & 1 \\ 1 & 0 & 1 \end{bmatrix},$$

assume the random 3×2 matrix X is such that

$$\mathrm{vec}(X - A)$$

has a $\mathrm{N}(0, V)$ distribution, where V is block diagonal with the matrix

$$\begin{bmatrix} 2 & 1 & 1 & 1 \\ 1 & 2 & 1 & 1 \\ 1 & 1 & 2 & 1 \\ 1 & 1 & 1 & 2 \end{bmatrix}$$

along the diagonal. Generate ten realizations of X matrices, and use them to test that the rank of A is 2. Use the test statistic (9.49) on page 352.

9.13. Construct a 9×2 matrix X with some missing values, such that S_X computed using all available data for the covariance or correlation matrix is not nonnegative definite.

9.14. Consider an $m \times m$, symmetric nonsingular matrix, R, with 1s on the diagonal and with all off-diagonal elements less than 1 in absolute value. If this matrix is positive definite, it is a correlation matrix. Suppose, however, that some of the eigenvalues are negative. Iman and Davenport (1982) describe a method of adjusting the matrix to a "nearby" matrix that is positive definite. (See Ronald L. Iman and James M. Davenport, 1982, *An Iterative Algorithm to Produce a Positive Definite Correlation Matrix from an "Approximate Correlation Matrix"*, Sandia Report SAND81-1376, Sandia National Laboratories, Albuquerque, New Mexico.) For their method, they assumed the eigenvalues are unique, but this is not necessary in the algorithm.

Before beginning the algorithm, choose a small positive quantity, ϵ, to use in the adjustments, set $k = 0$, and set $R^{(k)} = R$.

1. Compute the eigenvalues of $R^{(k)}$,

$$c_1 \geq c_2 \geq \ldots \geq c_m,$$

and let p be the number of eigenvalues that are negative. If $p = 0$, stop. Otherwise, set

$$c_i^* = \begin{cases} \epsilon & \text{if } c_i < \epsilon \\ c_i & \text{otherwise} \end{cases} \qquad \text{for } i = p_1, \ldots, m - p, \qquad (9.73)$$

where $p_1 = \max(1, m - 2p)$.

2. Let

$$\sum_i c_i v_i v_i^{\mathrm{T}}$$

be the spectral decomposition of R (equation (3.200), page 120), and form the matrix R^*:

$$R^* = \sum_{i=1}^{p_1} c_i v_i v_i^{\mathrm{T}} + \sum_{i=p_1+1}^{m-p} c_i^* v_i v_i^{\mathrm{T}} + \sum_{i=m-p+1}^{m} \epsilon v_i v_i^{\mathrm{T}}.$$

3. Form $R^{(k)}$ from R^* by setting all diagonal elements to 1.

4. Set $k = k + 1$, and go to step 1. (The algorithm iterates on k until $p = 0$.)

Write a program to implement this adjustment algorithm. Write your program to accept any size matrix and a user-chosen value for ϵ. Test your program on the correlation matrix from Exercise 9.13.

9.15. Consider some variations of the method in Exercise 9.14. For example, do not make the adjustments as in equation (9.73), or make different ones. Consider different adjustments of R^*; for example, adjust any off-diagonal elements that are greater than 1 in absolute value.
Compare the performance of the variations.

9.16. Investigate the convergence of the method in Exercise 9.14. Note that there are several ways the method could converge.

9.17. Suppose the method in Exercise 9.14 converges to a positive definite matrix $R^{(n)}$. Prove that all off-diagonal elements of $R^{(n)}$ are less than 1 in absolute value. (This is true for any positive definite matrix with 1s on the diagonal.)

9.18. Shrinkage adjustments of approximate correlation matrices.

a) Write a program to implement the linear shrinkage adjustment of equation (9.51). Test your program on the correlation matrix from Exercise 9.13.

b) Write a program to implement the nonlinear shrinkage adjustment of equation (9.52). Let $\delta = 0.05$ and

$$f(x) = \tanh(x).$$

Test your program on the correlation matrix from Exercise 9.13.

c) Write a program to implement the scaling adjustment of equation (9.53). Recall that this method applies to an approximate correlation matrix that is a pseudo-correlation matrix. Test your program on the correlation matrix from Exercise 9.13.

9.19. Show that the matrices generated in Algorithm 9.2 are correlation matrices. (They are clearly nonnegative definite, but how do we know that they have 1s on the diagonal?)

9.20. Consider a two-state Markov chain with transition matrix

$$P = \begin{bmatrix} 1 - \alpha & \alpha \\ \beta & 1 - \beta \end{bmatrix}$$

for $0 < \alpha < 1$ and $0 < \beta < 1$. Does an invariant distribution exist, and if so what is it?

9.21. Recall from Exercise 8.10 that a Leslie matrix has a single unique positive eigenvalue.

a) What are the conditions on a Leslie matrix A that allow a stable age distribution? Prove your assertion.

Hint: Review the development of the power method in equations (7.8) and (7.9).

b) What are the conditions on a Leslie matrix A that allow a stable population, that is, for some x_t, $x_{t+1} = x_t$?

c) Derive equation (9.68). (Recall that there are approximations that result from the use of a discrete model of a continuous process.)

Part III

Numerical Methods and Software

10

Numerical Methods

The computer is a tool for storage, manipulation, and presentation of data. The data may be numbers, text, or images, but no matter what the data are, they must be coded into a sequence of 0s and 1s. For each type of data, there are several ways of coding that can be used to store the data and specific ways the data may be manipulated.

How much a computer user needs to know about the way the computer works depends on the complexity of the use and the extent to which the necessary operations of the computer have been encapsulated in software that is oriented toward the specific application. This chapter covers many of the basics of how digital computers represent data and perform operations on the data. Although some of the specific details we discuss will not be important for the computational scientist or for someone doing statistical computing, the consequences of those details are important, and the serious computer user must be at least vaguely aware of the consequences. The fact that multiplying two positive numbers on the computer can yield a negative number should cause anyone who programs a computer to take care.

Data of whatever form are represented by groups of 0s and 1s, called *bits* from the words "binary" and "digits". (The word was coined by John Tukey.) For representing simple text (that is, strings of characters with no special representation), the bits are usually taken in groups of eight, called *bytes*, and associated with a specific character according to a fixed coding rule. Because of the common association of a byte with a character, those two words are often used synonymously.

The most widely used code for representing characters in bytes is "ASCII" (pronounced "askey", from American Standard Code for Information Interchange). Because the code is so widely used, the phrase "ASCII data" is sometimes used as a synonym for text or character data. The ASCII code for the character "A", for example, is 01000001; for "a" it is 01100001; and for "5" it is 00110101. Humans can more easily read shorter strings with several different characters than they can longer strings, even if those longer strings consist of only two characters. Bits, therefore, are often grouped into strings of fours; a four-bit string is equivalent to a hexadecimal digit, 1, 2, ..., 9, A, B, ..., or F. Thus, the ASCII codes just shown could be written in hexadecimal notation as 41 ("A"), 61 ("a"), and 35 ("5").

Because the common character sets differ from one language to another (both natural languages and computer languages), there are several modifications of the basic ASCII code set. Also, when there is a need for more different characters than can be represented in a byte (2^8), codes to associate characters with larger groups of bits are necessary. For compatibility with the commonly used ASCII codes using groups of 8 bits, these codes usually are for groups of 16 bits. These codes for "16-bit characters" are useful for representing characters in some Oriental languages, for example. The Unicode Consortium (1990, 1992) has developed a 16-bit standard, called Unicode, that is widely used for representing characters from a variety of languages. For any ASCII character, the Unicode representation uses eight leading 0s and then the same eight bits as the ASCII representation.

A standard scheme for representing data is very important when data are moved from one computer system to another or when researchers at different sites want to share data. Except for some bits that indicate how other bits are to be formed into groups (such as an indicator of the end of a file, or the delimiters of a record within a file), a set of data in ASCII representation is the same on different computer systems. Software systems that process documents either are specific to a given computer system or must have some standard coding to allow portability. The Java system, for example, uses Unicode to represent characters so as to ensure that documents can be shared among widely disparate platforms.

In addition to standard schemes for representing the individual data elements, there are some standard formats for organizing and storing sets of data. Although most of these formats are defined by commercial software vendors, two that are open and may become more commonly used are the Common Data Format (CDF), developed by the National Space Science Data Center, and the Hierarchical Data Format (HDF), developed by the National Center for Supercomputing Applications. Both standards allow a variety of types and structures of data; the standardization is in the descriptions that accompany the datasets.

Types of Data

Bytes that correspond to characters are often concatenated to form *character string data* (or just "strings"). Strings represent text without regard to the appearance of the text if it were to be printed. Thus, a string representing "ABC" does not distinguish between "ABC", "*ABC*", and "**ABC**". The appearance of the printed character must be indicated some other way, perhaps by additional bit strings designating a font.

The appearance of characters or other visual entities such as graphs or pictures is often represented more directly as a "bitmap". Images on a display medium such as paper or a CRT screen consist of an arrangement of small dots, possibly of various colors. The dots must be coded into a sequence of bits, and there are various coding schemes in use, such as JPEG (for Joint Photographic Experts Group). Image representations of "ABC", "*ABC*", and "**ABC**" would all be different. The computer's internal representation may correspond directly to the dots that are displayed or may be a formula to generate the dots, but in each case, the data are represented as a set of dots located with respect to some coordinate system. More dots would be turned on to represent "**ABC**" than to represent "ABC". The location of the dots and the distance between the dots depend on the coordinate system; thus the image can be repositioned or rescaled.

Computers initially were used primarily to process numeric data, and numbers are still the most important type of data in statistical computing. There are important differences between the numerical quantities with which the computer works and the numerical quantities of everyday experience. The fact that numbers in the computer must have a finite representation has very important consequences.

10.1 Digital Representation of Numeric Data

For representing a number in a finite number of digits or bits, the two most relevant things are the magnitude of the number and the precision with which the number is to be represented. Whenever a set of numbers are to be used in the same context, we must find a method of representing the numbers that will accommodate their full range and will carry enough precision for all of the numbers in the set.

Another important aspect in the choice of a method to represent data is the way data are communicated within a computer and between the computer and peripheral components such as data storage units. Data are usually treated as a fixed-length sequence of bits. The basic grouping of bits in a computer is sometimes called a "word" or a "storage unit". The lengths of words or storage units commonly used in computers are 32 or 64 bits.

Unlike data represented in ASCII (in which the representation is actually of the characters, which in turn represent the data themselves), the same numeric data will very often have different representations on different computer systems. It is also necessary to have different kinds of representations for different sets of numbers, even on the same computer. Like the ASCII standard for characters, however, there are some standards for representation of, and operations on, numeric data. The Institute of Electrical and Electronics Engineers (IEEE) and, subsequently, the International Electrotechnical Commission (IEC) have been active in promulgating these standards, and the standards themselves are designated by an IEEE number and/or an IEC number.

The two mathematical models that are often used for numeric data are the ring of integers, $\mathbb{Z}$, and the field of reals, $\mathbb{R}$. We use two computer models, $\mathbb{I}$ and $\mathbb{F}$, to simulate these mathematical entities. (Unfortunately, neither $\mathbb{I}$ nor $\mathbb{F}$ is a simple mathematical construct such as a ring or field.)

10.1.1 The Fixed-Point Number System

Because an important set of numbers is a finite set of reasonably sized integers, efficient schemes for representing these special numbers are available in most computing systems. The scheme is usually some form of a base 2 representation and may use one storage unit (this is most common), two storage units, or one half of a storage unit. For example, if a storage unit consists of 32 bits and one storage unit is used to represent an integer, the integer 5 may be represented in binary notation using the low-order bits, as shown in Figure 10.1.

| 0 | 0 | 0 | 0 | 0 | 0 | 0 | 0 | 0 | 0 | 0 | 0 | 0 | 0 | 0 | 0 | 0 | 0 | 0 | 0 | 0 | 0 | 0 | 0 | 0 | 0 | 0 | 0 | 0 | 1 | 0 | 1 |

Fig. 10.1. The Value 5 in a Binary Representation

The sequence of bits in Figure 10.1 represents the value 5, using one storage unit. The character "5" is represented in the ASCII code shown previously, 00110101.

If the set of integers includes the negative numbers also, some way of indicating the sign must be available. The first bit in the bit sequence (usually one storage unit) representing an integer is usually used to indicate the sign; if it is 0, a positive number is represented; if it is 1, a negative number is represented. In a common method for representing negative integers, called "twos-complement representation", the sign bit is set to 1 and the remaining bits are set to their opposite values (0 for 1; 1 for 0), and then 1 is added to the result. If the bits for 5 are ...00101, the bits for -5 would be ...11010 + 1, or ...11011. If there are k bits in a storage unit (and one storage unit is used to represent a single integer), the integers from 0 through $2^{k-1} - 1$ would be represented in ordinary binary notation using $k - 1$ bits. An integer i in the interval $[-2^{k-1}, -1]$ would be represented by the same bit pattern by which the nonnegative integer $2^{k-1} - i$ is represented, except the sign bit would be 1.

The sequence of bits in Figure 10.2 represents the value -5 using twos-complement notation in 32 bits, with the leftmost bit being the sign bit and the rightmost bit being the least significant bit; that is, the 1 position. The ASCII code for "-5" consists of the codes for "$-$" and "5"; that is, 00101101 00110101.

| 1 | 1 | 1 | 1 | 1 | 1 | 1 | 1 | 1 | 1 | 1 | 1 | 1 | 1 | 1 | 1 | 1 | 1 | 1 | 1 | 1 | 1 | 1 | 1 | 1 | 1 | 1 | 1 | 0 | 1 | 1 |

Fig. 10.2. The Value -5 in a Twos-Complement Representation

The special representations for numeric data are usually chosen so as to facilitate manipulation of data. The twos-complement representation makes arithmetic operations particularly simple. It is easy to see that the largest integer that can be represented in the twos-complement form is $2^{k-1} - 1$ and that the smallest integer is -2^{k-1}.

A representation scheme such as that described above is called *fixed-point* representation or *integer* representation, and the set of such numbers is denoted by $\mathbb{I}$. The notation $\mathbb{I}$ is also used to denote the system built on this set. This system is similar in some ways to a ring, which is what the integers $\mathbb{Z}$ are.

There are several variations of the fixed-point representation. The number of bits used and the method of representing negative numbers are two aspects that generally vary from one computer to another. Even within a single computer system, the number of bits used in fixed-point representation may vary; it is typically one storage unit or half of a storage unit.

We discuss the operations with numbers in the fixed-point system in Section 10.2.1.

10.1.2 The Floating-Point Model for Real Numbers

In a fixed-point representation, all bits represent values greater than or equal to 1; the *base point* or *radix point* is at the far right, before the first bit. In a fixed-point representation scheme using k bits, the range of representable numbers is of the order of 2^k, usually from approximately -2^{k-1} to 2^{k-1}. Numbers outside of this range cannot be represented directly in the fixed-point scheme. Likewise, nonintegral numbers cannot be represented. Large numbers and fractional numbers are generally represented in a scheme similar to what is sometimes called "scientific notation" or in a type of logarithmic notation. Because within a fixed number of digits the radix point is not fixed, this scheme is called *floating-point* representation, and the set of such numbers is denoted by $\mathbb{F}$. The notation $\mathbb{F}$ is also used to denote the system built on this set.

In a misplaced analogy to the real numbers, a floating-point number is also called "real". Both computer "integers", $\mathbb{I}$, and "reals", $\mathbb{F}$, represent useful subsets of the corresponding mathematical entities, $\mathbb{Z}$ and $\mathbb{R}$, but while the computer numbers called "integers" do constitute a fairly simple subset of the integers, the computer numbers called "real" do not correspond to the real numbers in a natural way. In particular, the floating-point numbers do not occur uniformly over the real number line.

Within the allowable range, a mathematical integer is exactly represented by a computer fixed-point number, but a given real number, even a rational, of any size may or may not have an exact representation by a floating-point number. This is the familiar situation where fractions such as $\frac{1}{3}$ have no finite representation in base 10. The simple rule, of course, is that the number must be a rational number whose denominator in reduced form factors into only primes that appear in the factorization of the base. In base 10, for example, only rational numbers whose factored denominators contain only 2s and 5s have an exact, finite representation; and in base 2, only rational numbers whose factored denominators contain only 2s have an exact, finite representation.

For a given real number x, we will occasionally use the notation

$$[x]_c$$

to indicate the floating-point number used to approximate x, and we will refer to the exact value of a floating-point number as a *computer number*. We will also use the phrase "computer number" to refer to the value of a computer fixed-point number. It is important to understand that computer numbers are members of proper finite subsets, $\mathbb{I}$ and $\mathbb{F}$, of the corresponding sets $\mathbb{Z}$ and $\mathbb{R}$.

Our main purpose in using computers, of course, is not to evaluate functions of the set of computer floating-point numbers or the set of computer integers; the main immediate purpose usually is to perform operations in the field of real (or complex) numbers or occasionally in the ring of integers. Doing computations on the computer, then, involves using the sets of computer numbers to simulate the sets of reals or integers.

The Parameters of the Floating-Point Representation

The parameters necessary to define a floating-point representation are the *base* or *radix*, the range of the *mantissa* or *significand*, and the range of the *exponent*. Because the number is to be represented in a fixed number of bits, such as one storage unit or word, the ranges of the significand and exponent must be chosen judiciously so as to fit within the number of bits available. If the radix is b and the integer digits d_i are such that $0 \leq d_i < b$, and there are enough bits in the significand to represent p digits, then a real number is approximated by

$$\pm 0.d_1 d_2 \cdots d_p \times b^e, \tag{10.1}$$

where e is an integer. This is the standard model for the floating-point representation. (The d_i are called "digits" from the common use of base 10.)

The number of bits allocated to the exponent e must be sufficient to represent numbers within a reasonable range of magnitudes; that is, so that the smallest number in magnitude that may be of interest is approximately $b^{e_{\min}}$ and the largest number of interest is approximately $b^{e_{\max}}$, where $e_{\min}$ and $e_{\max}$

are, respectively, the smallest and the largest allowable values of the exponent. Because $e_{\min}$ is likely negative and $e_{\max}$ is positive, the exponent requires a sign. In practice, most computer systems handle the sign of the exponent by defining a *bias* and then subtracting the bias from the value of the exponent evaluated without regard to a sign.

The parameters b, p, and $e_{\min}$ and $e_{\max}$ are so fundamental to the operations of the computer that on most computers they are fixed, except for a choice of two or three values for p and maybe two choices for the range of e.

In order to ensure a unique representation for all numbers (except 0), most floating-point systems require that the leading digit in the significand be nonzero unless the magnitude is less than $b^{e_{\min}}$. A number with a nonzero leading digit in the significand is said to be *normalized*.

The most common value of the base b is 2, although 16 and even 10 are sometimes used. If the base is 2, in a normalized representation, the first digit in the significand is always 1; therefore, it is not necessary to fill that bit position, and so we effectively have an extra bit in the significand. The leading bit, which is not represented, is called a "hidden bit". This requires a special representation for the number 0, however.

In a typical computer using a base of 2 and 64 bits to represent one floating-point number, 1 bit may be designated as the sign bit, 52 bits may be allocated to the significand, and 11 bits allocated to the exponent. The arrangement of these bits is somewhat arbitrary, and of course the physical arrangement on some kind of storage medium would be different from the "logical" arrangement. A common logical arrangement assigns the first bit as the sign bit, the next 11 bits as the exponent, and the last 52 bits as the significand. (Computer engineers sometimes label these bits as $0, 1, \ldots$, and then get confused as to which is the ith bit. When we say "first", we mean "first", whether an engineer calls it the "0th" or the "1st".) The range of exponents for the base of 2 in this typical computer would be 2,048. If this range is split evenly between positive and negative values, the range of orders of magnitude of representable numbers would be from -308 to 308. The bits allocated to the significand would provide roughly 16 decimal places of precision.

Figure 10.3 shows the bit pattern to represent the number 5, using $b = 2$, $p = 24$, $e_{\min} = -126$, and a bias of 127, in a word of 32 bits. The first bit on the left is the sign bit, the next 8 bits represent the exponent, 129, in ordinary base 2 with a bias, and the remaining 23 bits represent the significand beyond the leading bit, known to be 1. (The binary point is to the right of the leading bit that is not represented.) The value is therefore $+1.01 \times 2^2$ in binary notation.

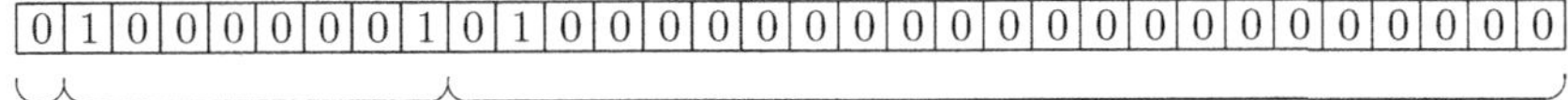

Fig. 10.3. The Value 5 in a Floating-Point Representation

While in fixed-point twos-complement representations there are considerable differences between the representation of a given integer and the negative of that integer (see Figures 10.1 and 10.2), the only difference between the floating-point representation of a number and its additive inverse is usually just in one bit. In the example of Figure 10.3, only the first bit would be changed to represent the number -5.

As mentioned above, the set of floating-point numbers is not uniformly distributed over the ordered set of the reals. There are the same number of floating-point numbers in the interval $[b^i, b^{i+1}]$ as in the interval $[b^{i+1}, b^{i+2}]$, even though the second interval is b times as long as the first. Figures 10.4 through 10.6 illustrate this. The fixed-point numbers, on the other hand, are uniformly distributed over their range, as illustrated in Figure 10.7.

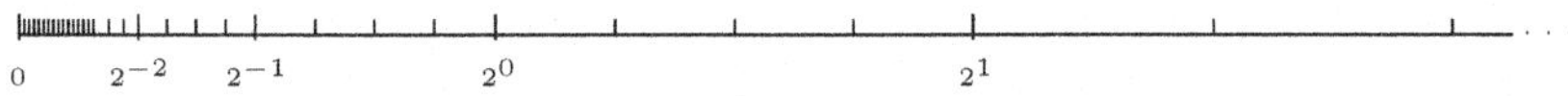

Fig. 10.4. The Floating-Point Number Line, Nonnegative Half

Fig. 10.5. The Floating-Point Number Line, Nonpositive Half

Fig. 10.6. The Floating-Point Number Line, Nonnegative Half; Another View

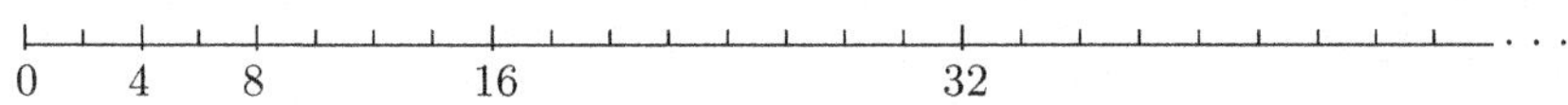

Fig. 10.7. The Fixed-Point Number Line, Nonnegative Half

The density of the floating-point numbers is generally greater closer to zero. Notice that if floating-point numbers are all normalized, the spacing between 0 and $b^{e_{\min}}$ is $b^{e_{\min}}$ (that is, there is no floating-point number in that open interval), whereas the spacing between $b^{e_{\min}}$ and $b^{e_{\min}+1}$ is $b^{e_{\min}-p+1}$. Most systems do not require floating-point numbers less than $b^{e_{\min}}$ in magnitude to be normalized. This means that the spacing between 0 and $b^{e_{\min}}$

can be $b^{e_{\min}-p}$, which is more consistent with the spacing just above $b^{e_{\min}}$. When these nonnormalized numbers are the result of arithmetic operations, the result is called "graceful" or "gradual" underflow.

The spacing between floating-point numbers has some interesting (and, for the novice computer user, surprising!) consequences. For example, if 1 is repeatedly added to x, by the recursion

$$x^{(k+1)} = x^{(k)} + 1,$$

the resulting quantity does not continue to get larger. Obviously, it could not increase without bound because of the finite representation. It does not even approach the largest number representable, however! (This is assuming that the parameters of the floating-point representation are reasonable ones.) In fact, if x is initially smaller in absolute value than $b^{e_{\max}-p}$ (approximately), the recursion

$$x^{(k+1)} = x^{(k)} + c$$

will converge to a stationary point for any value of c smaller in absolute value than $b^{e_{\max}-p}$.

The way the arithmetic is performed would determine these values precisely; as we shall see below, arithmetic operations may utilize more bits than are used in the representation of the individual operands.

The spacings of numbers just smaller than 1 and just larger than 1 are particularly interesting. This is because we can determine the *relative spacing* at any point by knowing the spacing around 1. These spacings at 1 are sometimes called the "machine epsilons", denoted $\epsilon_{\min}$ and $\epsilon_{\max}$ (not to be confused with $e_{\min}$ and $e_{\max}$ defined earlier). It is easy to see from the model for floating-point numbers on page 380 that

$$\epsilon_{\min} = b^{-p}$$

and

$$\epsilon_{\max} = b^{1-p};$$

see Figure 10.8. The more conservative value, $\epsilon_{\max}$, sometimes called "the machine epsilon", ϵ or ϵ_{mach}, provides an upper bound on the rounding that occurs when a floating-point number is chosen to represent a real number. A floating-point number near 1 can be chosen within $\epsilon_{\max}/2$ of a real number that is near 1. This bound, $\frac{1}{2}b^{1-p}$, is called the *unit roundoff.*

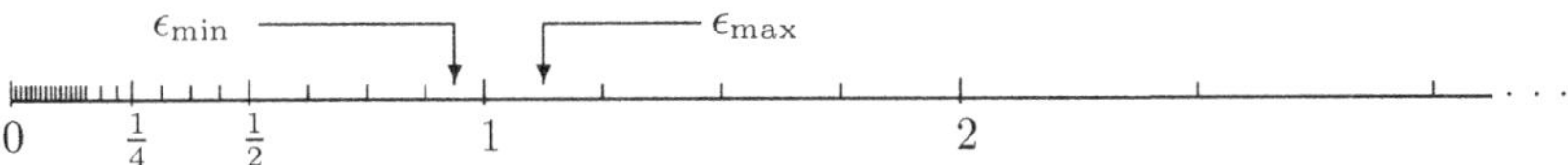

Fig. 10.8. Relative Spacings at 1: "Machine Epsilons"

These machine epsilons are also called the "smallest relative spacing" and the "largest relative spacing" because they can be used to determine the relative spacing at the point x (Figure 10.8).

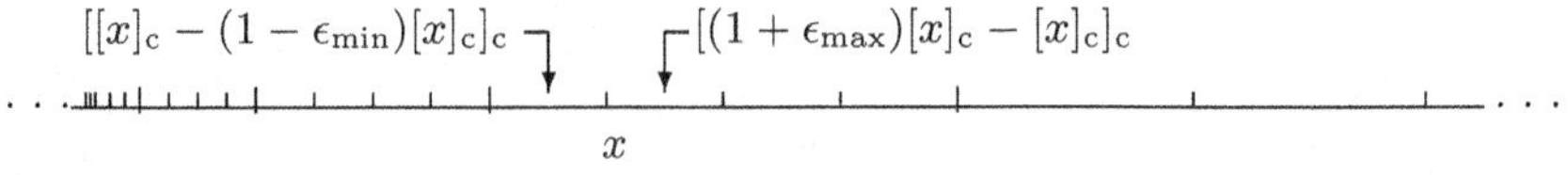

$$[[x]_{\mathrm{c}} - (1 - \epsilon_{\min})[x]_{\mathrm{c}}]_{\mathrm{c}} \qquad [(1 + \epsilon_{\max})[x]_{\mathrm{c}} - [x]_{\mathrm{c}}]_{\mathrm{c}}$$

Fig. 10.9. Relative Spacings

If x is not zero, the relative spacing at x is approximately

$$\frac{x - (1 - \epsilon_{\min})x}{x}$$

or

$$\frac{(1 + \epsilon_{\max})x - x}{x}.$$

Notice that we say "approximately". First of all, we do not even know that x is representable. Although $(1 - \epsilon_{\min})$ and $(1 + \epsilon_{\max})$ are members of the set of floating-point numbers by definition, that does not guarantee that the product of either of these numbers and $[x]_{\mathrm{c}}$ is also a member of the set of floating-point numbers. However, the quantities $[(1 - \epsilon_{\min})[x]_{\mathrm{c}}]_{\mathrm{c}}$ and $[(1 + \epsilon_{\max})[x]_{\mathrm{c}}]_{\mathrm{c}}$ are representable (by the definition of $[\cdot]_{\mathrm{c}}$ as a floating point number approximating the quantity within the brackets); and, in fact, they are respectively the next smallest number than $[x]_{\mathrm{c}}$ (if $[x]_{\mathrm{c}}$ is positive, or the next largest number otherwise) and the next largest number than $[x]_{\mathrm{c}}$ (if $[x]_{\mathrm{c}}$ is positive). The spacings at $[x]_{\mathrm{c}}$ therefore are

$$[x]_{\mathrm{c}} - [(1 - \epsilon_{\min})[x]_{\mathrm{c}}]_{\mathrm{c}}$$

and

$$[(1 + \epsilon_{\max})[x]_{\mathrm{c}} - [x]_{\mathrm{c}}]_{\mathrm{c}}.$$

As an aside, note that this implies it is probable that

$$[(1 - \epsilon_{\min})[x]_{\mathrm{c}}]_{\mathrm{c}} = [(1 + \epsilon_{\min})[x]_{\mathrm{c}}]_{\mathrm{c}}.$$

In practice, to compare two numbers x and y, we must compare $[x]_{\mathrm{c}}$ and $[y]_{\mathrm{c}}$. We consider x and y different if

$$[|y|]_{\mathrm{c}} < [|x|]_{\mathrm{c}} - [\epsilon_{\min}[|x|]_{\mathrm{c}}]_{\mathrm{c}}$$

or if

$$[|y|]_{\mathrm{c}} > [|x|]_{\mathrm{c}} + [\epsilon_{\max}[|x|]_{\mathrm{c}}]_{\mathrm{c}}.$$

The relative spacing at any point obviously depends on the value represented by the least significant digit in the significand. This digit (or bit) is called the "unit in the last place", or "ulp". The magnitude of an ulp depends of course on the magnitude of the number being represented. Any real number within the range allowed by the exponent can be approximated within $\frac{1}{2}$ ulp by a floating-point number.

The subsets of numbers that we need in the computer depend on the kinds of numbers that are of interest for the problem at hand. Often, however, the kinds of numbers of interest change dramatically within a given problem. For example, we may begin with integer data in the range from 1 to 50. Most simple operations, such as addition, squaring, and so on, with these data would allow a single paradigm for their representation. The fixed-point representation should work very nicely for such manipulations.

Something as simple as a factorial, however, immediately changes the paradigm. It is unlikely that the fixed-point representation would be able to handle the resulting large numbers. When we significantly change the range of numbers that must be accommodated, another change that occurs is the ability to represent the numbers exactly. If the beginning data are integers between 1 and 50, and no divisions or operations leading to irrational numbers are performed, one storage unit would almost surely be sufficient to represent all values exactly. If factorials are evaluated, however, the results cannot be represented exactly in one storage unit and so must be approximated (even though the results are integers). When data are not integers, it is usually obvious that we must use approximations, but it may also be true for integer data.

Standardization of Floating-Point Representation

As we have indicated, different computers represent numeric data in different ways. There has been some attempt to provide standards, at least in the range representable and in the precision of floating-point quantities. There are two IEEE standards that specify characteristics of floating-point numbers (IEEE, 1985). The IEEE Standard 754, which became the IEC 60559 standard, is a binary standard that specifies the exact layout of the bits for two different precisions, "single" and "double". In both cases, the standard requires that the radix be 2. For single precision, p must be 24, $e_{\max}$ must be 127, and $e_{\min}$ must be -126. For double precision, p must be 53, $e_{\max}$ must be 1023, and $e_{\min}$ must be -1022.

The IEEE Standard 754, or IEC 60559, also defines two additional precisions, "single extended" and "double extended". For each of the extended precisions, the standard sets bounds on the precision and exponent ranges rather than specifying them exactly. The extended precisions have larger exponent ranges and greater precision than the corresponding precision that is not "extended".

The IEEE Standard 854 requires that the radix be either 2 or 10 and defines ranges for floating-point representations. Formerly, the most widely used computers (IBM System 360 and derivatives) used base 16 representation; and some computers still use this base. Additional information about the IEEE standards for floating-point numbers can be found in Overton (2001). Both IEEE Standards 754 and 854 are undergoing modest revisions, and it is likely that 854 will be merged into 754.

Most of the computers developed in the past few years comply with the standards, but it is up to the computer manufacturers to conform voluntarily to these standards. We would hope that the marketplace would penalize the manufacturers who do not conform.

Special Floating-Point Numbers

It is convenient to be able to represent certain special numeric entities, such as infinity or "indeterminate" $(0/0)$, which do not have ordinary representations in any base-digit system. Although 8 bits are available for the exponent in the single-precision IEEE binary standard, $e_{max} = 127$ and $e_{min} = -126$. This means there are two unused possible values for the exponent; likewise, for the double-precision standard, there are two unused possible values for the exponent. These extra possible values for the exponent allow us to represent certain special floating-point numbers. An exponent of $e_{min} - 1$ allows us to handle 0 and the numbers between 0 and $b^{e_{min}}$ unambiguously even though there is a hidden bit (see the discussion above about normalization and gradual underflow). The special number 0 is represented with an exponent of $e_{min} - 1$ and a significand of $00\ldots0$.

An exponent of $e_{max} + 1$ allows us to represent $\pm\infty$ or the indeterminate value. A floating-point number with this exponent and a significand of 0 represents $\pm\infty$ (the sign bit determines the sign, as usual). A floating-point number with this exponent and a nonzero significand represents an indeterminate value such as $\frac{0}{0}$. This value is called "not-a-number", or NaN. In statistical data processing, a NaN is sometimes used to represent a missing value. Because a NaN is indeterminate, if a variable x has a value of NaN, $x \neq x$. Also, because a NaN can be represented in different ways, however, a programmer must be careful in testing for NaNs. Some software systems provide explicit functions for testing for a NaN. The IEEE binary standard recommended that a function `isnan` be provided to test for a NaN. Cody and Coonen (1993) provide C programs for `isnan` and other functions useful in working with floating-point numbers. We discuss computations with floating-point numbers in Section 10.2.2

10.1.3 Language Constructs for Representing Numeric Data

Most general-purpose computer programming languages, such as Fortran and C, provide constructs for the user to specify the type of representation for numeric quantities. These specifications are made in declaration statements that are made at the beginning of some section of the program for which they apply.

The difference between fixed-point and floating-point representations has a conceptual basis that may correspond to the problem being addressed. The differences between other kinds of representations often are not because of conceptual differences; rather, they are the results of increasingly irrelevant limitations of the computer. The reasons there are "short" and "long", or "signed" and "unsigned", representations do not arise from the problem the user wishes to solve; the representations are to allow more efficient use of computer resources. The wise software designer nowadays eschews the space-saving constructs that apply to only a relatively small proportion of the data. In some applications, however, the short representations of numeric data still have a place.

In C, the types of all variables must be specified with a basic declarator, which may be qualified further. For variables containing numeric data, the possible types are shown in Table 10.1.

Table 10.1. Numeric Data Types in C

Basic type	Basic declarator	Fully qualified declarator

fixed-point	`int`	`signed short int`
		`unsigned short int`
		`signed long int`
		`unsigned long int`

floating-point	`float` `double`	`double` `long double`

Exactly what these types mean is not specified by the language but depends on the specific implementation, which associates each type with some natural type supported by the specific computer. Common storage for a fixed-point variable of type `short int` uses 16 bits and for type `long int` uses 32 bits. An `unsigned` quantity of either type specifies that no bit is to be used as a sign bit, which effectively doubles the largest representable number. Of course, this is essentially irrelevant for scientific computations, so `unsigned` integers are generally just a nuisance. If neither `short` nor `long` is specified, there is a default interpretation that is implementation-dependent. The default always favors `signed` over `unsigned`. There is a movement toward standardization of the meanings of these types. The American National Standards Institute (ANSI) and its international counterpart, the International Organization for Standardization (ISO), have specified standard definitions of several programming languages. ANSI (1989) is a specification of the C language. ANSI C requires that `short int` use at least 16 bits, that `long int` use at least 32 bits, and that `long int` be at least as long as `int`, which in turn must be least as long as `short int`. The `long double` type may or may not have more precision and a larger range than the `double` type.

C does not provide a complex data type. This deficiency can be overcome to some extent by means of a user-defined data type. The user must write functions for all the simple arithmetic operations on complex numbers, just as is done for the simple exponentiation for floats.

The object-oriented hybrid language built on C, C++ (ANSI, 1998), provides the user with the ability also to define operator functions, so that the four simple arithmetic operations can be implemented by the operators "+", "−", "∗", and "/". There is no good way of defining an exponentiation operator, however, because the user-defined operators are limited to extended versions of the operators already defined in the language.

In Fortran, variables have a default numeric type that depends on the first letter in the name of the variable. The type can be explicitly declared (and, in fact, should be in careful programming). The `signed` and `unsigned` qualifiers of C, which have very little use in scientific computing, are missing in Fortran. Fortran has a fixed-point type that corresponds to integers and two floating-point types that correspond to reals and complex numbers. For one standard version of Fortran, called Fortran 77, the possible types for variables containing numeric data are shown in Table 10.2.

Table 10.2. Numeric Data Types in Fortran 77

Basic type	Basic declarator	Default variable name
fixed-point	`integer`	begin with i–n or I–N
floating-point	`real`	begin with a–h or o–z or with A–H or O–Z
	`double precision`	no default, although d or D is sometimes used
complex	`complex`	no default, although c or C is sometimes used

Although the standards organizations have defined these constructs for the Fortran 77 language (ANSI, 1978), just as is the case with C, exactly what these types mean is not specified by the language but depends on the specific implementation. Some extensions to the language allow the number of bytes to use for a type to be specified (e.g., `real*8`) and allow the type `double complex`.

The complex type is not so much a data type as a data structure composed of two floating-point numbers that has associated operations that simulate the operations defined on the field of complex numbers.

The Fortran 90/95 language and subsequent versions of Fortran support the same types as Fortran 77 but also provide much more flexibility in selecting the number of bits to use in the representation of any of the basic types. (There are only small differences between Fortran 90 and Fortran 95. There is also a version called Fortran 2003. Most of the features I discuss are in all of these versions, and since the version I currently use is Fortran 95, I will generally just refer to "Fortran 95" or "Fortran 90 and subsequent versions".) A fundamental concept for the numeric types in Fortran 95 is called *"kind"*. The kind is a qualifier for the basic type; thus a fixed-point number may be an `integer` of kind 1 or kind 2, for example. The actual value of the qualifier kind may differ from one compiler to another, so the user defines a program parameter to be the kind that is appropriate to the range and precision required for a given variable. Fortran 95 provides the functions `selected_int_kind` and `selected_real_kind` to do this. Thus, to declare some fixed-point variables that have at least three decimal digits and some more fixed-point variables that have at least eight decimal digits, the user may write the following statements:

```
integer, parameter  :: little = selected_int_kind(3)
integer, parameter  :: big    = selected_int_kind(8)
integer (little)    :: ismall, jsmall
integer (big)       :: itotal_accounts, igain
```

The variables `little` and `big` would have integer values, chosen by the compiler designer, that could be used in the program to qualify integer types to ensure that range of numbers could be handled. Thus, `ismall` and `jsmall` would be fixed-point numbers that could represent integers between -999 and 999, and `itotal_accounts` and `igain` would be fixed-point numbers that could represent integers between $-99{,}999{,}999$ and $99{,}999{,}999$. Depending on the basic hardware, the compiler may assign two bytes as kind = `little`, meaning that integers between $-32{,}768$ and $32{,}767$ could probably be accommodated by any variable, such as `ismall`, that is declared as `integer (little)`. Likewise, it is probable that the range of variables declared as `integer (big)` could handle numbers in the range $-2{,}147{,}483{,}648$ and $2{,}147{,}483{,}647$. For declaring floating-point numbers, the user can specify a minimum range and precision with the function `selected_real_kind`, which takes two arguments, the number of decimal digits of precision and the exponent of 10 for the range. Thus, the statements

```
integer, parameter  :: real4 = selected_real_kind(6,37)
integer, parameter  :: real8 = selected_real_kind(15,307)
```

would yield designators of floating-point types that would have either six decimals of precision and a range up to 10^{37} or fifteen decimals of precision and a range up to 10^{307}. The statements

```
real (real4)   :: x, y
real (real8)   :: dx, dy
```

declare x and y as variables corresponding roughly to `real` on most systems and dx and dy as variables corresponding roughly to `double precision`.

If the system cannot provide types matching the requirements specified in `selected_int_kind` or `selected_real_kind`, these functions return -1. Because it is not possible to handle such an error situation in the declaration statements, the user should know in advance the available ranges. Fortran 90 and subsequent versions of Fortran provide a number of intrinsic functions, such as `epsilon`, `rrspacing`, and `huge`, to use in obtaining information about the fixed- and floating-point numbers provided by the system.

Fortran 90 and subsequent versions also provide a number of intrinsic functions for dealing with bits. These functions are essentially those specified in the MIL-STD-1753 standard of the U.S. Department of Defense. These bit functions, which have been a part of many Fortran implementations for years, provide for shifting bits within a string, extracting bits, exclusive or inclusive oring of bits, and so on. (See ANSI, 1992; Lemmon and Schafer 2005; or Metcalf, Reid, and Cohen, 2004, for more extensive discussions of the types and intrinsic functions provided in Fortran 90 and subsequent versions.)

Many higher-level languages and application software packages do not give the user a choice of how to represent numeric data. The software system may consistently use a type thought to be appropriate for the kinds of applications addressed. For example, many statistical analysis application packages choose to use a floating-point representation with about 64 bits for all numeric data. Making a choice such as this yields more comparable results across a range of computer platforms on which the software system may be implemented.

Whenever the user chooses the type and precision of variables, it is a good idea to use some convention to name the variable in such a way as to indicate the type and precision. Books or courses on elementary programming suggest using mnemonic names, such as "`time`", for a variable that holds the measure of time. If the variable takes fixed-point values, a better name might be "`itime`". It still has the mnemonic value of "time", but it also helps us to remember that, in the computer, `itime/length` may not be the same thing as `time/xlength`. Although the variables are declared in the program to be of a specific type, the programmer can benefit from a reminder of the type. Even as we "humanize" computing, we must remember that there are details about the computer that matter. (The operator "/" is said to be "overloaded": in a general way, it means "divide", but it means different things depending on the contexts of the two expressions above.) Whether a quantity is a member of II or IF may have major consequences for the computations, and a careful choice of notation can help to remind us of that, even if the notation may look old-fashioned.

Numerical analysts sometimes use the phrase "full precision" to refer to a precision of about sixteen decimal digits and the phrase "half precision" to refer to a precision of about seven decimal digits. These terms are not defined precisely, but they do allow us to speak of the precision in roughly equivalent ways for different computer systems without specifying the precision exactly. Full precision is roughly equivalent to Fortran `double precision` on the common 32-bit workstations and to Fortran `real` on "supercomputers" and other 64-bit machines. Half precision corresponds roughly to Fortran `real` on the common 32-bit workstations. Full and half precision can be handled in a portable way in Fortran 90 and subsequent versions of Fortran. The following statements declare a variable x to be one with full precision:

```
integer, parameter  :: full = selected\_real\_kind(15,307)
real (full)         :: x
```

In a construct of this kind, the user can define "full" or "half" as appropriate.

Determining the Numerical Characteristics of a Particular Computer

The environmental inquiry program MACHAR by Cody (1988) can be used to determine the characteristics of a specific computer's floating-point representation and its arithmetic. The program, which is available in CALGO from `netlib` (see page 505 in the Bibliography), was written in Fortran 77 and has been translated into C and R. In R, the results on a given system are stored in the variable `.Machine`. Other R objects that provide information on a computer's characteristics are the variable `.Platform` and the function `capabilities`.

10.1.4 Other Variations in the Representation of Data; Portability of Data

As we have indicated already, computer designers have a great deal of latitude in how they choose to represent data. The ASCII standards of ANSI and ISO have provided a common representation for individual characters. The IEEE standard 754 referred to previously (IEEE, 1985) has brought some standardization to the representation of floating-point data but does not specify how the available bits are to be allocated among the sign, exponent, and significand.

Because the number of bits used as the basic storage unit has generally increased over time, some computer designers have arranged small groups of bits, such as bytes, together in strange ways to form words. There are two common schemes of organizing bits into bytes and bytes into words. In one scheme, called "big end" or "big endian", the bits are indexed from the "left", or most significant, end of the byte, and bytes are indexed within words and words are indexed within groups of words in the same direction.

In another scheme, called "little end" or "little endian", the bytes are indexed within the word in the opposite direction. Figures 10.11 through 10.13 illustrate some of the differences, using the program shown in Figure 10.10.

```
          character a
          character*4 b
          integer i, j
          equivalence (b,i), (a,j)
          print '(10x, a7 , a8)', ' Bits  ', '   Value'
          a = 'a'
          print '(1x, a10, z2, 7x, a1)', 'a:          ', a, a
          print '(1x, a10, z8, 1x, i12)', 'j (=a):   ', j, j
          b = 'abcd'
          print '(1x, a10, z8, 1x, a4)', 'b:          ', b, b
          print '(1x, a10, z8, 1x, i12)', 'i (=b):   ', i, i
          end
```

Fig. 10.10. A Fortran Program Illustrating Bit and Byte Organization

```
           Bits       Value
a:         61         a
j (=a):          61              97
b:         64636261 abcd
i (=b):    64636261   1684234849
```

Fig. 10.11. Output from a Little Endian System (VAX Running Unix or VMS)

```
           Bits       Value
a:         61         a
j (=a):    00000061             97
b:         61626364 abcd
i (=b):    64636261   1684234849
```

Fig. 10.12. Output from a Little Endian System (Intel x86, Pentium, or AMD, Running Microsoft Windows)

```
           Bits       Value
a:         61         a
j (=a):    61000000   1627389952
b:         61626364 abcd
i (=b):    61626364   1633837924
```

Fig. 10.13. Output from a Big Endian System (Sun SPARC or Silicon Graphics, Running Unix)

The R function .Platform provides information on the type of endian of the given machine on which the program is running.

These differences are important only when accessing the individual bits and bytes, when making data type transformations directly, or when moving data from one machine to another without interpreting the data in the process ("binary transfer"). One lesson to be learned from observing such subtle differences in the way the same quantities are treated in different computer systems is that programs should rarely rely on the inner workings of the computer. A program that does will not be *portable*; that is, it will not give the same results on different computer systems. Programs that are not portable may work well on one system, and the developers of the programs may never intend for them to be used anywhere else. As time passes, however, systems change or users change systems. When that happens, the programs that were not portable may cost more than they ever saved by making use of computer-specific features.

The *external data representation*, or XDR, standard format, developed by Sun Microsystems for use in remote procedure calls, is a widely used machine-independent standard for binary data structures.

10.2 Computer Operations on Numeric Data

As we have emphasized above, the numerical quantities represented in the computer are used to simulate or approximate more interesting quantities, namely the real numbers or perhaps the integers. Obviously, because the sets (computer numbers and real numbers) are not the same, we could not define operations on the computer numbers that would yield the same field as the familiar field of the reals. In fact, because of the nonuniform spacing of floating-point numbers, we would suspect that some of the fundamental properties of a field may not hold. Depending on the magnitudes of the quantities involved, it is possible, for example, that if we compute ab and ac and then $ab + ac$, we may not get the same thing as if we compute $(b + c)$ and then $a(b + c)$. Just as we use the computer quantities to simulate real quantities, we define operations on the computer quantities to simulate the familiar operations on real quantities. Designers of computers attempt to define computer operations so as to correspond closely to operations on real numbers, but we must not lose sight of the fact that the computer uses a different arithmetic system.

The basic operational objective in numerical computing, of course, is that a computer operation, when applied to computer numbers, yield computer numbers that approximate the number that would be yielded by a certain mathematical operation applied to the numbers approximated by the original computer numbers. Just as we introduced the notation

$$[x]_{\mathrm{c}}$$

on page 380 to denote the computer floating-point number approximation to the real number x, we occasionally use the notation

$$[\circ]_{\mathrm{c}}$$

to refer to a computer operation that simulates the mathematical operation $\circ$. Thus,

$$[+]_{\mathrm{c}}$$

represents an operation similar to addition but that yields a result in a set of computer numbers. (We use this notation only where necessary for emphasis, however, because it is somewhat awkward to use it consistently.) The failure of the familiar laws of the field of the reals, such as the distributive law cited above, can be anticipated by noting that

$$[[a]_{\mathrm{c}} \ [+]_{\mathrm{c}} \ [b]_{\mathrm{c}}]_{\mathrm{c}} \neq [a + b]_{\mathrm{c}},$$

or by considering the simple example in which all numbers are rounded to one decimal and so $\frac{1}{3} + \frac{1}{3} \neq \frac{2}{3}$ (that is, $.3 + .3 \neq .7$).

The three familiar laws of the field of the reals (commutativity of addition and multiplication, associativity of addition and multiplication, and distribution of multiplication over addition) result in the independence of the order in which operations are performed; the failure of these laws implies that the order of the operations may make a difference. When computer operations are performed sequentially, we can usually define and control the sequence fairly easily. If the computer performs operations in parallel, the resulting differences in the orders in which some operations may be performed can occasionally yield unexpected results.

Because the operations are not closed, special notice may need to be taken when the operation would yield a number not in the set. Adding two numbers, for example, may yield a number too large to be represented well by a computer number, either fixed-point or floating-point. When an operation yields such an anomalous result, an *exception* is said to exist.

The computer operations for the two different types of computer numbers are different, and we discuss them separately.

10.2.1 Fixed-Point Operations

The operations of addition, subtraction, and multiplication for fixed-point numbers are performed in an obvious way that corresponds to the similar operations on the ring of integers. Subtraction is addition of the additive inverse. (In the usual twos-complement representation we described earlier, all fixed-point numbers have additive inverses except -2^{k-1}.) Because there is no multiplicative inverse, however, division is not multiplication by the inverse. The result of division with fixed-point numbers is the result of division with the corresponding real numbers rounded toward zero. This is not considered an exception.

As we indicated above, the set of fixed-point numbers together with addition and multiplication is not the same as the ring of integers, if for no other reason than that the set is finite. Under the ordinary definitions of addition and multiplication, the set is not closed under either operation. The computer operations of addition and multiplication, however, are defined so that the set is closed. These operations occur as if there were additional higher-order bits and the sign bit were interpreted as a regular numeric bit. The result is then whatever would be in the standard number of lower-order bits. If the lost higher-order bits are necessary, the operation is said to *overflow*. If fixed-point overflow occurs, the result is not correct under the usual interpretation of the operation, so an error situation, or an exception, has occurred. Most computer systems allow this error condition to be detected, but most software systems do not take note of the exception. The result, of course, depends on the specific computer architecture. On many systems, aside from the interpretation of the sign bit, the result is essentially the same as would result from a modular reduction. There are some special-purpose algorithms that actually use this modified modular reduction, although such algorithms would not be portable across different computer systems.

10.2.2 Floating-Point Operations

As we have seen, real numbers within the allowable range may or may not have an exact floating-point operation, and the computer operations on the computer numbers may or may not yield numbers that represent exactly the real number that would result from mathematical operations on the numbers. If the true result is r, the best we could hope for would be $[r]_c$. As we have mentioned, however, the computer operation may not be exactly the same as the mathematical operation being simulated, and furthermore, there may be several operations involved in arriving at the result. Hence, we expect some error in the result.

Errors

If the computed value is $\tilde{r}$ (for the true value r), we speak of the *absolute error*,

$$|\tilde{r} - r|,$$

and the *relative error*,

$$\frac{|\tilde{r} - r|}{|r|}$$

(so long as $r \neq 0$). An important objective in numerical computation obviously is to ensure that the error in the result is small.

We will discuss error in floating-point computations further in Section 10.3.1.

Guard Digits and Chained Operations

Ideally, the result of an operation on two floating-point numbers would be the same as if the operation were performed exactly on the two operands (considering them to be exact also) and the result was then rounded. Attempting to do this would be very expensive in both computational time and complexity of the software. If care is not taken, however, the relative error can be very large. Consider, for example, a floating-point number system with $b = 2$ and $p = 4$. Suppose we want to add 8 and -7.5. In the floating-point system, we would be faced with the problem

$$8 : 1.000 \times 2^3$$
$$7.5 : 1.111 \times 2^2.$$

To make the exponents the same, we have

$$8 : 1.000 \times 2^3 \quad \text{or} \quad 8 : 1.000 \times 2^3$$
$$7.5 : 0.111 \times 2^3 \qquad\qquad 7.5 : 1.000 \times 2^3.$$

The subtraction will yield either 0.000_2 or $1.000_2 \times 2^0$, whereas the correct value is $1.000_2 \times 2^{-1}$. Either way, the absolute error is 0.5_{10}, and the relative error is 1. Every bit in the significand is wrong. The magnitude of the error is the same as the magnitude of the result. This is not acceptable. (More generally, we could show that the relative error in a similar computation could be as large as $b - 1$ for any base b.) The solution to this problem is to use one or more *guard digits*. A guard digit is an extra digit in the significand that participates in the arithmetic operation. If one guard digit is used (and this is the most common situation), the operands each have $p + 1$ digits in the significand. In the example above, we would have

$$8 : 1.0000 \times 2^3$$
$$7.5 : 0.1111 \times 2^3,$$

and the result is exact. In general, one guard digit can ensure that the relative error is less than $2\epsilon_{\max}$. The use of guard digits requires that the operands be stored in special storage units. Whenever multiple operations are to be performed together, the operands and intermediate results can all be kept in the special registers to take advantage of the guard digits or even longer storage units. This is called chaining of operations.

Addition of Several Numbers

When several numbers x_i are to be summed, it is likely that as the operations proceed serially, the magnitudes of the partial sum and the next summand will be quite different. In such a case, the full precision of the next summand is lost. This is especially true if the numbers are of the same sign. As we mentioned earlier, a computer program to implement serially the algorithm implied by $\sum_{i=1}^{\infty} i$ will converge to some number much smaller than the largest floating-point number.

If the numbers to be summed are not all the same constant (and if they are constant, just use multiplication!), the accuracy of the summation can be increased by first sorting the numbers and summing them in order of increasing magnitude. If the numbers are all of the same sign and have roughly the same magnitude, a pairwise "fan-in" method may yield good accuracy. In the fan-in method, the n numbers to be summed are added two at a time to yield $\lceil n/2 \rceil$ partial sums. The partial sums are then added two at a time, and so on, until all sums are completed. The name "fan-in" comes from the tree diagram of the separate steps of the computations:

$$
\begin{array}{cccc}
s_1^{(1)} = x_1 + x_2 & \left| s_2^{(1)} = x_3 + x_4 \right| \ldots \left| s_{2m-1}^{(1)} = x_{4m-3} + x_{4m-2} \right| s_{2m}^{(1)} = \right| \ldots \\
\searrow \qquad\qquad \swarrow \qquad\qquad \ldots \qquad\qquad \searrow \qquad\qquad \swarrow \qquad \ldots \\
s_1^{(2)} = s_1^{(1)} + s_2^{(1)} \qquad\qquad \ldots \qquad\qquad s_m^{(2)} = s_{2m-1}^{(1)} + s_{2m}^{(1)} \qquad \ldots \\
\searrow \qquad\qquad\qquad \ldots \qquad\qquad\qquad \downarrow \qquad\qquad\qquad \ldots \\
s_1^{(3)} = s_1^{(2)} + s_2^{(2)} \qquad \ldots \qquad\qquad\qquad \ldots \qquad\qquad\qquad \ldots
\end{array}
$$

It is likely that the numbers to be added will be of roughly the same magnitude at each stage. Remember we are assuming they have the same sign initially; this would be the case, for example, if the summands are squares.

Another way that is even better is due to W. Kahan:

$$
\begin{aligned}
&s = x_1 \\
&a = 0 \\
&\text{for } i = 2, \ldots, n \\
&\{ \\
&\qquad y = x_i - a \\
&\qquad t = s + y \\
&\qquad a = (t - s) - y \\
&\qquad s = t \\
&\}.
\end{aligned}
\tag{10.2}
$$

Catastrophic Cancellation

Another kind of error that can result because of the finite precision used for floating-point numbers is *catastrophic cancellation*. This can occur when two rounded values of approximately equal magnitude and opposite signs are added. (If the values are exact, cancellation can also occur, but it is *benign*.) After catastrophic cancellation, the digits left are just the digits that represented the rounding. Suppose $x \approx y$ and that $[x]_c = [y]_c$. The computed result will be zero, whereas the correct (rounded) result is $[x - y]_c$. The relative error is 100%. This error is caused by rounding, but it is different from the "rounding error" discussed above. Although the loss of information arising from the rounding error is the culprit, the rounding would be of little consequence were it not for the cancellation.

To avoid catastrophic cancellation, watch for possible additions of quantities of approximately equal magnitude and opposite signs, and consider rearranging the computations. Consider the problem of computing the roots of a quadratic polynomial, $ax^2 + bx + c$ (see Rice, 1993). In the quadratic formula

$$x = \frac{-b \pm \sqrt{b^2 - 4ac}}{2a}, \tag{10.3}$$

the square root of the discriminant, $(b^2 - 4ac)$, may be approximately equal to b in magnitude, meaning that one of the roots is close to zero and, in fact, may be computed as zero. The solution is to compute only one of the roots, x_1, by the formula (the "$-$" root if b is positive and the "$+$" root if b is negative) and then compute the other root, x_2 by the relationship $x_1 x_2 = c/a$.

Standards for Floating-Point Operations

The IEEE Binary Standard 754 (IEEE, 1985) applies not only to the representation of floating-point numbers but also to certain operations on those numbers. The standard requires correct rounded results for addition, subtraction, multiplication, division, remaindering, and extraction of the square root. It also requires that conversion between fixed-point numbers and floating-point numbers yield correct rounded results.

The standard also defines how exceptions should be handled. The exceptions are divided into five types: overflow, division by zero, underflow, invalid operation, and inexact operation. If an operation on floating-point numbers would result in a number beyond the range of representable floating-point numbers, the exception, called *overflow*, is generally very serious. (It is serious in fixed-point operations also if it is unplanned. Because we have the alternative of using floating-point numbers if the magnitude of the numbers is likely to exceed what is representable in fixed-point numbers, the user is expected to use this alternative. If the magnitude exceeds what is representable in floating-point numbers, however, the user must resort to some indirect means, such as scaling, to solve the problem.)

Division by zero does not cause overflow; it results in a special number if the dividend is nonzero. The result is either ∞ or $-\infty$, and these have special representations, as we have seen.

Underflow occurs whenever the result is too small to be represented as a normalized floating-point number. As we have seen, a nonnormalized representation can be used to allow a gradual underflow.

An invalid operation is one for which the result is not defined because of the value of an operand. The invalid operations are addition of ∞ to $-\infty$, multiplication of $\pm\infty$ and 0, 0 divided by 0 or by $\pm\infty$, $\pm\infty$ divided by 0 or by $\pm\infty$, extraction of the square root of a negative number (some systems, such as Fortran, have a special type for complex numbers and deal correctly with them), and remaindering any quantity with 0 or remaindering $\pm\infty$ with any quantity. An invalid operation results in a NaN. Any operation with a NaN also results in a NaN. Some systems distinguish two types of NaN: a "quiet NaN" and a "signaling NaN".

An inexact operation is one for which the result must be rounded. For example, if all p bits of the significand are required to represent both the multiplier and multiplicand, approximately $2p$ bits would be required to represent the product. Because only p are available, however, the result must be rounded.

Conformance to the IEEE Binary Standard 754 does not ensure that the results of multiple floating-point computations will be the same on all computers. The standard does not specify the order of the computations, and differences in the order can change the results. The slight differences are usually unimportant, but Blackford et al. (1997a) describe some examples of problems that occurred when computations were performed in parallel using a heterogeneous network of computers all of which conformed to the IEEE standard. See also Gropp (2005) for further discussion of some of these issues.

Comparison of Reals and Floating-Point Numbers

For most applications, the system of floating-point numbers simulates the field of the reals very well. It is important, however, to be aware of some of the differences in the two systems. There is a very obvious useful measure for the reals, namely the Lebesgue measure, μ, based on lengths of open intervals. An approximation of this measure is appropriate for floating-point numbers, even though the set is finite. The finiteness of the set of floating-point numbers means that there is a difference in the cardinality of an open interval and a closed interval with the same endpoints. The uneven distribution of floating-point values relative to the reals (Figures 10.4 and 10.5) means that the cardinalities of two interval-bounded sets with the same interval length may be different. On the other hand, a counting measure does not work well at all.

Some general differences in the two systems are exhibited in Table 10.3. The last four properties in Table 10.3 are properties of a field (except for the divergence of $\sum_{x=1}^{\infty} x$). The important facts are that $\mathbb{R}$ is an uncountable field and that $\mathbb{F}$ is a more complicated finite mathematical structure.

10.2.3 Exact Computations; Rational Fractions

If the input data can be represented exactly as rational fractions, it may be possible to preserve exact values of the results of computations. Using rational fractions allows avoidance of reciprocation, which is the operation that most commonly yields a nonrepresentable value from one that is representable. Of course, any addition or multiplication that increases the magnitude of an integer in a rational fraction beyond a value that can be represented exactly (that is, beyond approximately 2^{23}, 2^{31}, or 2^{53}, depending on the computing system) may break the error-free chain of operations. Exact computations with integers can be carried out using *residue arithmetic*, in which each quantity is as a vector of residues, all from a vector of relatively prime moduli. (See Szabó and Tanaka, 1967, for a discussion of the use of residue arithmetic in numerical

Table 10.3. Differences in Real Numbers and Floating-Point Numbers

	$\mathbb{R}$	$\mathbb{F}$
cardinality:	uncountable	finite
measure:	$\mu((x,y)) = \lvert x - y \rvert$ $\mu((x,y)) = \mu([x,y])$	$\nu((\mathtt{x},\mathtt{y})) = \nu([\mathtt{x},\mathtt{y}]) = \lvert \mathtt{x} - \mathtt{y} \rvert$ $\exists\, \mathtt{x},\mathtt{y},\mathtt{z},\mathtt{w} \ni \lvert \mathtt{x} - \mathtt{y} \rvert = \lvert \mathtt{z} - \mathtt{w} \rvert$, but $\#(\mathtt{x},\mathtt{y}) \neq \#(\mathtt{z},\mathtt{w})$
continuity:	if $x < y$, $\exists z \ni x < z < y$ and $\mu([x,y]) = \mu((x,y))$	$\mathtt{x} < \mathtt{y}$, but no $\mathtt{z} \ni \mathtt{x} < \mathtt{z} < \mathtt{y}$ and $\#[\mathtt{x},\mathtt{y}] > \#(\mathtt{x},\mathtt{y})$
closure:	$x, y \in \mathbb{R} \Rightarrow x + y \in \mathbb{R}$ $x, y \in \mathbb{R} \Rightarrow xy \in \mathbb{R}$	not closed wrt addition not closed wrt multiplication (exclusive of infinities)
operations with an identity, a or a:	$a = 0$, unique $a + x = x$, for any x $x - x = a$, for any x	$\mathtt{a} + \mathtt{x} = \mathtt{b} + \mathtt{x}$, but $\mathtt{b} \neq \mathtt{a}$ $\mathtt{a} + \mathtt{x} = \mathtt{x}$, but $\mathtt{a} + \mathtt{y} \neq \mathtt{y}$ $\mathtt{a} + \mathtt{x} = \mathtt{x}$, but $\mathtt{x} - \mathtt{x} \neq \mathtt{a}$
convergence	$\sum_{x=1}^{\infty} x$ diverges	$\sum_{\mathtt{x}=1}^{\infty} \mathtt{x}$ converges, if interpreted as $(\cdots((1 + 2) + 3)\cdots)$
associativity:	$x, y, z \in \mathbb{R} \Rightarrow$ $(x + y) + z = x + (y + z)$ $(xy)z = x(yz)$	not associative not associative
distributivity:	$x, y, z \in \mathbb{R} \Rightarrow$ $x(y + z) = xy + xz$	not distributive

computations; and see Stallings and Boullion, 1972, and Keller-McNulty and Kennedy, 1986, for applications of this technology in matrix computations.)

Computations with rational fractions are sometimes performed using a fixed-point representation. Gregory and Krishnamurthy (1984) discuss in detail these and other methods for performing error-free computations.

10.2.4 Language Constructs for Operations on Numeric Data

Most general-purpose computer programming languages, such as Fortran and C, provide constructs for operations that correspond to the common operations on scalar numeric data, such as "+", "−", "∗" (multiplication), and "/". These operators *simulate* the corresponding mathematical operations. As we mentioned on page 393, we will occasionally use notation such as

$$[+]_c$$

to indicate the computer operator. The operators have slightly different meanings depending on the operand objects; that is, the operations are "*overloaded*". Most of these operators are *binary infix* operators, meaning that the operator is written between the two operands.

Some languages provide operations beyond the four basic scalar arithmetic operations. C provides some specialized operations, such as the unary postfix increment "++" and decrement "−−" operators, for trivial common operations but does not provide an operator for exponentiation. (Exponentiation is handled by a function provided in a standard supplemental library in C, `<math.h>`.) C also overloads the basic multiplication operator so that it can indicate a change of meaning of a variable in addition to indicating the multiplication of two scalar numbers. A standard library in C (`<signal.h>`) allows easy handling of arithmetic exceptions. With this facility, for example, the user can distinguish a quiet NaN from a signaling NaN.

The C language does not directly provide for operations on special data structures. For operations on complex data, for example, the user must define the type and its operations in a header file (or else, of course, just do the operations as if they were operations on an array of length 2).

Fortran provides the four basic scalar numeric operators plus an exponentiation operator ("∗∗"). (Exactly what this operator means may be slightly different in different versions of Fortran. Some versions interpret the operator always to mean

1. take log
2. multiply by power
3. exponentiate

if the base and the power are both floating-point types. This, of course, will not work if the base is negative, even if the power is an integer. Most versions of Fortran will determine at run time if the power is an integer and use repeated multiplication if it is.)

Fortran also provides the usual five operators for complex data (the basic four plus exponentiation). Fortran 90 and subsequent versions of Fortran provide the same set of scalar numeric operators plus a basic set of array and vector/matrix operators. The usual vector/matrix operators are implemented as functions or prefix operators in Fortran 95.

In addition to the basic arithmetic operators, both Fortran and C, as well as other general programming languages, provide several other types of operators, including relational operators and operators for manipulating structures of data.

Multiple Precision

Software packages have been built on Fortran and C to extend their accuracy. Two ways in which this is done are by using *multiple precision* and by using *interval arithmetic*.

Multiple-precision operations are performed in the software by combining more than one computer-storage unit to represent a single number. For example, to operate on x and y, we may represent x as $a \cdot 10^p + b$ and y as $c \cdot 10^p + d$. The product xy then is formed as $ac \cdot 10^{2p} + (ad + bc) \cdot 10^p + bd$. The representation is chosen so that any of the coefficients of the scaling factors (in this case powers of 10) can be represented to within the desired accuracy.

Multiple precision is different from "extended precision", discussed earlier; extended precision is implemented at the hardware level or at the microcode level. Brent (1978) and Smith (1991) have produced Fortran packages for multiple-precision computations, and Bailey (1993, 1995) gives software for instrumenting Fortran code to use multiple-precision operations.

A multiple-precision package may allow the user to specify the number of digits to use in representing data and performing computations. The software packages for symbolic computations, such as Maple, generally provide multiple precision capabilities.

Interval Arithmetic

Interval arithmetic maintains intervals in which the exact data and solution are known to lie. Instead of working with single-point approximations, for which we used notation such as

$$[x]_c$$

on page 380 for the value of the floating-point approximation to the real number x and

$$[\circ]_c$$

on page 393 for the simulated operation $\circ$, we can approach the problem by identifying a closed interval in which x lies and a closed interval in which the result of the operation $\circ$ lies. We denote the interval operation as

$$[\circ]_I.$$

For the real number x, we identify two floating-point numbers, x_l and x_u, such that $x_l \leq x \leq x_u$. (This relationship also implies $x_l \leq [x]_c \leq x_u$.) The real number x is then considered to be the interval $[x_l, x_u]$. For this approach to be useful, of course, we seek tight bounds. If $x = [x]_c$, the best interval is degenerate. In other cases, either x_l or x_c is $[x]_c$ and the length of the interval is the floating-point spacing from $[x]_c$ in the appropriate direction.

Addition and multiplication in interval arithmetic yield intervals

$$x \; [+]_I \; y = [x_l + y_l, \; x_u + y_u]$$

and

$$x \; [*]_I \; y = [\min(x_l y_l, x_l y_u, x_u y_l, x_u y_u), \; \max(x_l y_l, x_l y_u, x_u y_l, x_u y_u)].$$

A change of sign results in $[-x_u, \; -x_l]$ and if $0 \notin [x_l, \; x_u]$, reciprocation results in $[1/x_u, \; 1/x_l]$. See Moore (1979) or Alefeld and Herzberger (1983) for discussions of these kinds of operations and an extensive treatment of interval arithmetic. The journal *Reliable Computing* is devoted to interval computations. The book edited by Kearfott and Kreinovich (1996) addresses various aspects of interval arithmetic. One chapter in that book, by Walster (1996), discusses how both hardware and system software could be designed to implement interval arithmetic.

Most software support for interval arithmetic is provided through subroutine libraries. The `ACRITH` package of IBM (see Jansen and Weidner, 1986) is a library of Fortran subroutines that perform computations in interval arithmetic and also in extended precision. Kearfott et al. (1994) have produced a portable Fortran library of basic arithmetic operations and elementary functions in interval arithmetic, and Kearfott (1996) gives a Fortran 90 module defining an interval data type. Jaulin et al. (2001) give additional sources of software. Sun Microsystems Inc. has provided full intrinsic support for interval data types in their Fortran compiler Sun^{TM} ONE Studio Fortran 95; see Walster (2005) for a description of the compiler extensions.

10.3 Numerical Algorithms and Analysis

We will use the term "algorithm" rather loosely but always in the general sense of a *method* or a *set of instructions* for doing something. (Formally, an "algorithm" must terminate; however, respecting that definition would not allow us to refer to a method as an algorithm until it has been proven to terminate.) Algorithms are sometimes distinguished as "numerical", "seminumerical", and "nonnumerical", depending on the extent to which operations on real numbers are simulated.

Algorithms and Programs

Algorithms are expressed by means of a flowchart, a series of steps, or in a computer language or pseudolanguage. The expression in a computer language is a source program or module; hence, we sometimes use the words "algorithm" and "program" synonymously.

The program is the set of computer instructions that implement the algorithm. A poor implementation can render a good algorithm useless. A good implementation will preserve the algorithm's accuracy and efficiency and will detect data that are inappropriate for the algorithm. Robustness is more a property of the program than of the algorithm.

The exact way an algorithm is implemented in a program depends of course on the programming language, but it also may depend on the computer and associated system software. A program that will run on most systems without modification is said to be *portable*.

The two most important aspects of a computer algorithm are its accuracy and its efficiency. Although each of these concepts appears rather simple on the surface, each is actually fairly complicated, as we shall see.

10.3.1 Error in Numerical Computations

An "accurate" algorithm is one that gets the "right" answer. Knowing that the right answer may not be representable and that rounding within a set of operations may result in variations in the answer, we often must settle for an answer that is "close". As we have discussed previously, we measure error, or closeness, as either the absolute error or the relative error of a computation.

Another way of considering the concept of "closeness" is by looking backward from the computed answer and asking what perturbation of the original problem would yield the computed answer exactly. This approach, developed by Wilkinson (1963), is called *backward error analysis*. The backward analysis is followed by an assessment of the effect of the perturbation on the solution. Although backward error analysis may not seem as natural as "forward" analysis (in which we assess the difference between the computed and true solutions), it is easier to perform because all operations in the backward analysis are performed in $\mathbb{IF}$ instead of in $\mathbb{IR}$. Each step in the backward analysis involves numbers in the set $\mathbb{IF}$, that is, numbers that could actually have participated in the computations that were performed. Because the properties of the arithmetic operations in $\mathbb{IR}$ do not hold and, at any step in the sequence of computations, the result in $\mathbb{IF}$ may not exist in $\mathbb{IR}$, it is very difficult to carry out a forward error analysis.

There are other complications in assessing errors. Suppose the answer is a vector, such as a solution to a linear system. What norm do we use to compare the closeness of vectors? Another, more complicated situation for which assessing correctness may be difficult is random number generation. It would be difficult to assign a meaning to "accuracy" for such a problem.

The basic source of error in numerical computations is the inability to work with the reals. The field of reals is simulated with a finite set. This has several consequences. A real number is rounded to a floating-point number; the result of an operation on two floating-point numbers is rounded to another floating-point number; and passage to the limit, which is a fundamental concept in the field of reals, is not possible in the computer.

Rounding errors that occur just because the result of an operation is not representable in the computer's set of floating-point numbers are usually not too bad. Of course, if they accumulate through the course of many operations, the final result may have an unacceptably large accumulated rounding error.

A natural approach to studying errors in floating-point computations is to define random variables for the rounding at all stages, from the initial representation of the operands, through any intermediate computations, to the final result. Given a probability model for the rounding error in the representation of the input data, a statistical analysis of rounding errors can be performed. Wilkinson (1963) introduced a uniform probability model for rounding of input and derived distributions for computed results based on that model. Linnainmaa (1975) discusses the effects of accumulated errors in floating-point computations based on a more general model of the rounding for the input. This approach leads to a forward error analysis that provides a probability distribution for the error in the final result. Analysis of errors in fixed-point computations presents altogether different problems because, for values near 0, the relative errors cannot approach 0 in any realistic manner.

The obvious probability model for floating-point representations is that the reals within an interval between any two floating-point numbers have a uniform distribution (see Figure 10.4 on page 382 and Calvetti, 1991). A probability model for the real line can be built up as a mixture of the uniform distributions (see Exercise 10.9 on page 424). The density is obviously 0 in the tails. While a model based on simple distributions may be appropriate for the rounding error due to the finite-precision *representation* of real numbers, probability models for rounding errors in floating point *computations* are not so simple. This is because the rounding errors in computations are not random. See Chaitin-Chatelin and Frayssé (1996) for a further discussion of probability models for rounding errors. Dempster and Rubin (1983) discuss the application of statistical methods for dealing with grouped data to the data resulting from rounding in floating-point computations.

Another, more pernicious, effect of rounding can occur in a single operation, resulting in catastrophic cancellation, as we have discussed previously (see page 397).

Measures of Error and Bounds for Errors

For the simple case of representing the real number r by an approximation $\tilde{r}$, we define absolute error, $|\tilde{r} - r|$, and relative error, $|\tilde{r} - r|/|r|$ (so long as $r \neq 0$). These same types of measures are used to express the errors in

numerical computations. As we indicated above, however, the result may not
be a simple real number; it may consist of several real numbers. For example,
in statistical data analysis, the numerical result, $\tilde{r}$, may consist of estimates
of several regression coefficients, various sums of squares and their ratio, and
several other quantities. We may then be interested in some more general
measure of the difference of $\tilde{r}$ and r,

$$\Delta(\tilde{r}, r),$$

where $\Delta(\cdot, \cdot)$ is a nonnegative, real-valued function. This is the absolute error,
and the relative error is the ratio of the absolute error to $\Delta(r, r_0)$, where r_0
is a baseline value, such as 0. When r, instead of just being a single number,
consists of several components, we must measure error differently. If r is a
vector, the measure may be based on some norm, and in that case, $\Delta(\tilde{r}, r)$
may be denoted by $\|(\tilde{r} - r)\|$. A norm tends to become larger as the number
of elements increases, so instead of using a raw norm, it may be appropriate
to scale the norm to reflect the number of elements being computed.

However the error is measured, for a given algorithm, we would like to have
some knowledge of the amount of error to expect or at least some bound on the
error. Unfortunately, almost any measure contains terms that depend on the
quantity being evaluated. Given this limitation, however, often we can develop
an upper bound on the error. In other cases, we can develop an estimate of an
"average error" based on some assumed probability distribution of the data
comprising the problem. In a Monte Carlo method, we estimate the solution
based on a "random" sample, so just as in ordinary statistical estimation,
we are concerned about the variance of the estimate. We can usually derive
expressions for the variance of the estimator in terms of the quantity being
evaluated, and of course we can estimate the variance of the estimator using
the realized random sample. The standard deviation of the estimator provides
an indication of the distance around the computed quantity within which we
may have some confidence that the true value lies. The standard deviation is
sometimes called the "standard error", and nonstatisticians speak of it as a
"probabilistic error bound".

It is often useful to identify the "order of the error" whether we are concerned about error bounds, average expected error, or the standard deviation
of an estimator. In general, we speak of the order of one function in terms of
another function as a common argument of the functions approaches a given
value. A function $f(t)$ is said to be of order $g(t)$ at t_0, written $\mathrm{O}(g(t))$ ("big O
of $g(t)$"), if there exists a positive constant M such that

$$|f(t)| \leq M|g(t)| \quad \text{as } t \to t_0.$$

This is the *order of convergence* of one function to another function at a given
point.

If our objective is to compute $f(t)$ and we use an approximation $\tilde{f}(t)$, the
order of the *error due to the approximation* is the order of the convergence.

In this case, the argument of the order of the error may be some variable that defines the approximation. For example, if $\tilde{f}(t)$ is a finite series approximation to $f(t)$ using, say, k terms, we may express the error as $O(h(k))$ for some function $h(k)$. Typical orders of errors due to the approximation may be $O(1/k)$, $O(1/k^2)$, or $O(1/k!)$. An approximation with order of error $O(1/k!)$ is to be preferred over one order of error $O(1/k)$ because the error is decreasing more rapidly. The order of error due to the approximation is only one aspect to consider; roundoff error in the representation of any intermediate quantities must also be considered.

We will discuss the order of error in iterative algorithms further in Section 10.3.3 beginning on page 417. (We will discuss order also in measuring the speed of an algorithm in Section 10.3.2.)

The special case of convergence to the constant zero is often of interest. A function $f(t)$ is said to be "little o of $g(t)$" at t_0, written $o(g(t))$, if

$$f(t)/g(t) \to 0 \quad \text{as } t \to t_0.$$

If the function $f(t)$ approaches 0 at t_0, $g(t)$ can be taken as a constant and $f(t)$ is said to be $o(1)$.

Big O and little o convergences are defined in terms of dominating functions. In the analysis of algorithms, it is often useful to consider analogous types of convergence in which the function of interest dominates another function. This type of relationship is similar to a lower bound. A function $f(t)$ is said to be $\Omega(g(t))$ ("big omega of $g(t)$") if there exists a positive constant m such that

$$|f(t)| \geq m|g(t)| \quad \text{as } t \to t_0.$$

Likewise, a function $f(t)$ is said to be "little omega of $g(t)$" at t_0, written $\omega(g(t))$, if

$$g(t)/f(t) \to 0 \quad \text{as } t \to t_0.$$

Usually the limit on t in order expressions is either 0 or ∞, and because it is obvious from the context, mention of it is omitted. The order of the error in numerical computations usually provides a measure in terms of something that can be controlled in the algorithm, such as the point at which an infinite series is truncated in the computations. The measure of the error usually also contains expressions that depend on the quantity being evaluated, however.

Error of Approximation

Some algorithms are exact, such as an algorithm to multiply two matrices that just uses the definition of matrix multiplication. Other algorithms are approximate because the result to be computed does not have a finite closed-form expression. An example is the evaluation of the normal cumulative distribution function. One way of evaluating this is by using a rational polynomial approximation to the distribution function. Such an expression may be evaluated with very little rounding error, but the expression has an *error of approximation*.

When solving a differential equation on the computer, the differential equation is often approximated by a difference equation. Even though the differences used may not be constant, they are finite and the passage to the limit can never be effected. This kind of approximation leads to a *discretization error*. The amount of the discretization error has nothing to do with rounding error. If the last differences used in the algorithm are δt, then the error is usually of order $O(\delta t)$, even if the computations are performed exactly.

Another type of error of approximation occurs when the algorithm uses a series expansion. The series may be exact, and in principle the evaluation of all terms would yield an exact result. The algorithm uses only a smaller number of terms, and the resulting error is *truncation error*. This is the type of error we discussed in connection with Fourier expansions on pages 30 and 76. Often the exact expansion is an infinite series, and we approximate it with a finite series. When a truncated Taylor series is used to evaluate a function at a given point x_0, the order of the truncation error is the derivative of the function that would appear in the first unused term of the series, evaluated at x_0.

We need to have some knowledge of the magnitude of the error. For algorithms that use approximations, it is often useful to express the order of the error in terms of some quantity used in the algorithm or in terms of some aspect of the problem itself. We must be aware, however, of the limitations of such measures of the errors or error bounds. For an oscillating function, for example, the truncation error may never approach zero over any nonzero interval.

Algorithms and Data

The performance of an algorithm may depend on the data. We have seen that even the simple problem of computing the roots of a quadratic polynomial, $ax^2 + bx + c$, using the quadratic formula, equation (10.3), can lead to severe cancellation. For many values of a, b, and c, the quadratic formula works perfectly well. Data that are likely to cause computational problems are referred to as ill-conditioned data, and, more generally, we speak of the "condition" of data. The concept of condition is understood in the context of a particular set of operations. Heuristically, data for a given problem are ill-conditioned if small changes in the data may yield large changes in the solution.

Consider the problem of finding the roots of a high-degree polynomial, for example. Wilkinson (1959) gave an example of a polynomial that is very simple on the surface yet whose solution is very sensitive to small changes of the values of the coefficients:

$$f(x) = (x - 1)(x - 2) \cdots (x - 20)$$
$$= x^{20} - 210x^{19} + \cdots + 20!.$$

While the solution is easy to see from the factored form, the solution is very sensitive to perturbations of the coefficients. For example, changing the coefficient 210 to $210 + 2^{-23}$ changes the roots drastically; in fact, ten of them are now complex. Of course, the extreme variation in the magnitudes of the coefficients should give us some indication that the problem may be ill-conditioned.

Condition of Data

We attempt to quantify the condition of a set of data for a particular set of operations by means of a *condition number*. Condition numbers are defined to be positive and in such a way that large values of the numbers mean that the data or problems are ill-conditioned. A useful condition number for the problem of finding roots of a function can be defined to be increasing as the reciprocal of the absolute value of the derivative of the function in the vicinity of a root.

In the solution of a linear system of equations, the coefficient matrix determines the condition of this problem. The most commonly used condition number is the number associated with a matrix with respect to the problem of solving a linear system of equations. This is the number we discuss in Section 6.4 on page 218.

Condition numbers are only indicators of possible numerical difficulties for a given problem. They must be used with some care. For example, according to the condition number for finding roots based on the derivative, Wilkinson's polynomial is well-conditioned.

Robustness of Algorithms

The ability of an algorithm to handle a wide range of data and either to solve the problem as requested or to determine that the condition of the data does not allow the algorithm to be used is called the *robustness* of the algorithm.

Stability of Algorithms

Another concept that is quite different from robustness is *stability*. An algorithm is said to be *stable* if it always yields a solution that is an *exact* solution to a perturbed problem; that is, for the problem of computing $f(x)$ using the input data x, an algorithm is stable if the result it yields, $\tilde{f}(x)$, is

$$f(x + \delta x)$$

for some (bounded) perturbation δx of x. Stated another way, an algorithm is stable if small perturbations in the input or in intermediate computations do not result in large differences in the results.

The concept of stability for an algorithm should be contrasted with the concept of condition for a problem or a dataset. If a problem is ill-conditioned, a stable algorithm (a "good algorithm") will produce results with large differences for small differences in the specification of the problem. This is because the exact results have large differences. An algorithm that is not stable, however, may produce large differences for small differences in the computer description of the problem, which may involve rounding, truncation, or discretization, or for small differences in the intermediate computations performed by the algorithm.

The concept of stability arises from backward error analysis. The stability of an algorithm may depend on how continuous quantities are discretized, such as when a range is gridded for solving a differential equation. See Higham (2002) for an extensive discussion of stability.

Reducing the Error in Numerical Computations

An objective in designing an algorithm to evaluate some quantity is to avoid accumulated rounding error and to avoid catastrophic cancellation. In the discussion of floating-point operations above, we have seen two examples of how an algorithm can be constructed to mitigate the effect of accumulated rounding error (using equations (10.2) on page 397 for computing a sum) and to avoid possible catastrophic cancellation in the evaluation of the expression (10.3) for the roots of a quadratic equation.

Another example familiar to statisticians is the computation of the sample sum of squares:

$$\sum_{i=1}^{n}(x_i - \bar{x})^2 = \sum_{i=1}^{n} x_i^2 - n\bar{x}^2. \tag{10.4}$$

This quantity is $(n-1)s^2$, where s^2 is the sample variance.

Either expression in equation (10.4) can be thought of as describing an algorithm. The expression on the left-hand side implies the "two-pass" algorithm:

$$
\begin{aligned}
&a = x_1 \\
&\text{for } i = 2, \ldots, n \\
&\{ \\
&\quad a = x_i + a \\
&\} \\
&a = a/n \\
&b = (x_1 - a)^2 \\
&\text{for } i = 2, \ldots, n \\
&\{ \\
&\quad b = (x_i - a)^2 + b \\
&\}.
\end{aligned}
\tag{10.5}
$$

This algorithm yields $\bar{x} = a$ and $(n-1)s^2 = b$. Each of the sums computed in this algorithm may be improved by using equations (10.2). A problem with this algorithm is the fact that it requires two passes through the data. Because the quantities in the second summation are squares of residuals, they are likely to be of relatively equal magnitude. They are of the same sign, so there will be no catastrophic cancellation in the early stages when the terms being accumulated are close in size to the current value of b. There will be some accuracy loss as the sum b grows, but the addends $(x_i - a)^2$ remain roughly the same size. The accumulated rounding error, however, may not be too bad.

The expression on the right-hand side of equation (10.4) implies the "one-pass" algorithm:

$$
\begin{aligned}
&a = x_1 \\
&b = x_1^2 \\
&\text{for } i = 2, \ldots, n \\
&\{ \\
&\quad a = x_i + a \\
&\quad b = x_i^2 + b \\
&\} \\
&a = a/n \\
&b = b - na^2.
\end{aligned}
\tag{10.6}
$$

This algorithm requires only one pass through the data, but if the x_is have magnitudes larger than 1, the algorithm has built up two relatively large quantities, b and na^2. These quantities may be of roughly equal magnitudes; subtracting one from the other may lead to catastrophic cancellation (see Exercise 10.16, page 426).

Another algorithm is shown in equations (10.7). It requires just one pass through the data, and the individual terms are generally accumulated fairly accurately. Equations (10.7) are a form of the Kalman filter (see, for example, Grewal and Andrews, 1993).

$$
\begin{aligned}
&a = x_1 \\
&b = 0 \\
&\text{for } i = 2, \ldots, n \\
&\{ \\
&\quad d = (x_i - a)/i \\
&\quad a = d + a \\
&\quad b = i(i - 1)d^2 + b \\
&\}.
\end{aligned}
\tag{10.7}
$$

Chan and Lewis (1979) propose a condition number to quantify the sensitivity in s, the sample standard deviation, to the data, the x_is. Their condition number is

$$
\kappa = \frac{\sum_{i=1}^{n} x_i^2}{\sqrt{n-1}\, s}.
\tag{10.8}
$$

This is a measure of the "stiffness" of the data. It is clear that if the mean is large relative to the variance, this condition number will be large. (Recall that large condition numbers imply ill-conditioning, and also recall that condition numbers must be interpreted with some care.) Notice that this condition number achieves its minimum value of 1 for the data $x_i - \bar{x}$, so if the computations for $\bar{x}$ and $x_i - \bar{x}$ were exact, the data in the last part of the algorithm in equations (10.5) would be perfectly conditioned. A dataset with a large mean relative to the variance is said to be *stiff*.

Often when a finite series is to be evaluated, it is necessary to accumulate a set of terms of the series that have similar magnitudes, and then combine this with similar partial sums. It may also be necessary to scale the individual terms by some very large or very small multiplicative constant while the terms are being accumulated and then remove the scale after some computations have been performed.

Chan, Golub, and LeVeque (1982) propose a modification of the algorithm in equations (10.7) to use pairwise accumulations (as in the fan-in method discussed previously). Chan, Golub, and LeVeque (1983) make extensive comparisons of the methods and give error bounds based on the condition number.

10.3.2 Efficiency

The *efficiency* of an algorithm refers to its usage of computer resources. The two most important resources are the processing units and memory. The amount of time the processing units are in use and the amount of memory required are the key measures of efficiency. A limiting factor for the time the processing units are in use is the number and type of operations required. Some operations take longer than others; for example, the operation of adding floating-point numbers may take more time than the operation of adding fixed-point numbers. This, of course, depends on the computer system and on what kinds of floating-point or fixed-point numbers we are dealing with. If we have a measure of the size of the problem, we can characterize the performance of a given algorithm by specifying the number of operations of each type or just the number of operations of the slowest type.

High-Performance Computing

In "high-performance" computing, major emphasis is placed on computational efficiency. The architecture of the computer becomes very important, and the programs are designed to take advantage of the particular characteristics of the computer on which they are to run. The three main architectural elements are memory, processing units, and communication paths. A controlling unit oversees how these elements work together. There are various ways memory can be organized. There is usually a hierarchy of types of memory with different speeds of access. The various levels can also be organized into banks with separate communication links to the processing units. There are various types of processing units. The unit may be distributed and consist of multiple central processing units. The units may consist of multiple processors within the same core. The processing units may include vector processors. Dongarra et al. (1998) provide a good overview of the various designs and their relevance to high-performance computing.

If more than one processing unit is available, it may be possible to perform operations simultaneously. In this case, the amount of time required may be drastically smaller for an efficient parallel algorithm than it would for the most efficient serial algorithm that utilizes only one processor at a time. An analysis of the efficiency must take into consideration how many processors are available, how many computations can be performed in parallel, and how often they can be performed in parallel.

Measuring Efficiency

Often, instead of the exact number of operations, we use the *order* of the number of operations in terms of the measure of problem size. If n is some measure of the size of the problem, an algorithm has order $O(f(n))$ if, as $n \to \infty$, the number of computations $\to cf(n)$, where c is some constant that does not depend on n. For example, to multiply two $n \times n$ matrices in the obvious way requires $O(n^3)$ multiplications and additions; to multiply an $n \times m$ matrix and an $m \times p$ matrix requires $O(nmp)$ multiplications and additions. In the latter case, n, m, and p are all measures of the size of the problem.

Notice that in the definition of order there is a constant c. Two algorithms that have the same order may have different constants and in that case are said to "differ only in the constant". The order of an algorithm is a measure of how well the algorithm "scales"; that is, the extent to which the algorithm can deal with truly large problems.

Let n be a measure of the problem size, and let b and q be constants. An algorithm of order $O(b^n)$ has *exponential order*, one of order $O(n^q)$ has *polynomial order*, and one of order $O(\log n)$ has *log order*. Notice that for log order it does not matter what the base is. Also, notice that $O(\log n^q) = O(\log n)$. For a given task with an obvious algorithm that has polynomial order, it is often possible to modify the algorithm to address parts of the problem so that in the order of the resulting algorithm one n factor is replaced by a factor of $\log n$.

Although it is often relatively easy to determine the order of an algorithm, an interesting question in algorithm design involves the *order of the problem*; that is, the order of the most efficient algorithm possible. A problem of polynomial order is usually considered tractable, whereas one of exponential order may require a prohibitively excessive amount of time for its solution. An interesting class of problems are those for which a solution can be verified in polynomial time yet for which no polynomial algorithm is known to exist. Such a problem is called a *nondeterministic polynomial*, or NP, problem. "Nondeterministic" does not imply any randomness; it refers to the fact that no polynomial algorithm for determining the solution is known. Most interesting NP problems can be shown to be equivalent to each other in order by reductions that require polynomial time. Any problem in this subclass of NP problems is equivalent in some sense to all other problems in the subclass and so such a problem is said to be *NP-complete.*

For many problems it is useful to measure the size of a *problem* in some standard way and then to identify the order of an *algorithm* for the problem with separate components. A common measure of the size of a problem is L, the length of the stream of data elements. An $n \times n$ matrix would have length proportional to $L = n^2$, for example. To multiply two $n \times n$ matrices in the obvious way requires $O(L^{3/2})$ multiplications and additions, as we mentioned above.

In analyzing algorithms for more complicated problems, we may wish to determine the order in the form

$$O(f(n)g(L))$$

because L is an essential measure of the problem size and n may depend on how the computations are performed. For example, in the linear programming problem, with n variables and m constraints with a dense coefficient matrix, there are order nm data elements. Algorithms for solving this problem generally depend on the limit on n, so we may speak of a linear programming algorithm as being $O(n^3 L)$, for example, or of some other algorithm as being $O(\sqrt{n}L)$. (In defining L, it is common to consider the magnitudes of the data elements or the precision with which the data are represented, so that L is the order of the total number of bits required to represent the data. This level of detail can usually be ignored, however, because the limits involved in the order are generally not taken on the magnitude of the data but only on the number of data elements.)

The order of an algorithm (or, more precisely, the "order of *operations* of an algorithm") is an asymptotic measure of the operation count as the size of the problem goes to infinity. The order of an algorithm is important, but in practice the actual count of the operations is also important. In practice, an algorithm whose operation count is approximately n^2 may be more useful than one whose count is $1000(n \log n + n)$, although the latter would have order $O(n \log n)$, which is much better than that of the former, $O(n^2)$. When an algorithm is given a fixed-size task many times, the finite efficiency of the algorithm becomes very important.

The number of computations required to perform some tasks depends not only on the size of the problem but also on the data. For example, for most sorting algorithms, it takes fewer computations (comparisons) to sort data that are already almost sorted than it does to sort data that are completely unsorted. We sometimes speak of the *average* time and the *worst-case* time of an algorithm. For some algorithms, these may be very different, whereas for other algorithms or for some problems these two may be essentially the same.

Our main interest is usually not in how many computations occur but rather in how long it takes to perform the computations. Because some computations can take place simultaneously, even if all kinds of computations

required the same amount of time, the *order of time* could be different from
the order of the number of computations.

The actual number of floating-point operations divided by the time required
to perform the operations is called the FLOPS (floating-point operations per
second) rate. Confusingly, "FLOP" also means "floating-point operation", and
"FLOPs" is the plural of "FLOP". Of course, as we tend to use lowercase more
often, we must use the context to distinguish "flops" as a rate from "flops"
the plural of "flop".

In addition to the actual processing, the data may need to be copied from
one storage position to another. Data movement slows the algorithm and may
cause it not to use the processing units to their fullest capacity. When groups
of data are being used together, blocks of data may be moved from ordinary
storage locations to an area from which they can be accessed more rapidly. The
efficiency of a program is enhanced if all operations that are to be performed
on a given block of data are performed one right after the other. Sometimes a
higher-level language prevents this from happening. For example, to add two
arrays (matrices) in Fortran 95, a single statement is sufficient:

```
A = B + C
```

Now, if we also want to add B to the array E, we may write

```
A = B + C
D = B + E
```

These two Fortran 95 statements together may be less efficient than writing
a traditional loop in Fortran or in C because the array B may be accessed
a second time needlessly. (Of course, this is relevant only if these arrays are
very large.)

Improving Efficiency

There are many ways to attempt to improve the efficiency of an algorithm.
Often the best way is just to look at the task from a higher level of detail and
attempt to construct a new algorithm. Many obvious algorithms are serial
methods that would be used for hand computations, and so are not the best
for use on the computer.

An effective general method of developing an efficient algorithm is called
divide and conquer. In this method, the problem is broken into subproblems,
each of which is solved, and then the subproblem solutions are combined into
a solution for the original problem. In some cases, this can result in a net
savings either in the number of computations, resulting in an improved order
of computations, or in the number of computations that must be performed
serially, resulting in an improved order of time.

Let the time required to solve a problem of size n be $t(n)$, and consider
the recurrence relation

$$t(n) = pt(n/p) + cn$$

for p positive and c nonnegative. Then $t(n) = O(n \log n)$ (see Exercise 10.18, page 426). Divide and conquer strategies can sometimes be used together with a simple method that would be $O(n^2)$ if applied directly to the full problem to reduce the order to $O(n \log n)$.

The "fan-in algorithm" is an example of a divide and conquer strategy that allows $O(n)$ operations to be performed in $O(\log n)$ time if the operations can be performed simultaneously. The number of operations does not change materially; the improvement is in the time.

Although there have been orders of magnitude improvements in the speed of computers because the hardware is better, the order of time required to solve a problem is almost entirely dependent on the algorithm. The improvements in efficiency resulting from hardware improvements are generally differences only in the constant. The practical meaning of the order of the time must be considered, however, and so the constant may be important. In the fan-in algorithm, for example, the improvement in order is dependent on the unrealistic assumption that as the problem size increases without bound, the number of processors also increases without bound. Divide and conquer strategies do not require multiple processors for their implementation, of course.

Some algorithms are designed so that each step is as efficient as possible, without regard to what future steps may be part of the algorithm. An algorithm that follows this principle is called a *greedy algorithm*. A greedy algorithm is often useful in the early stages of computation for a problem or when a problem lacks an understandable structure.

Bottlenecks and Limits

There is a maximum FLOPS rate possible for a given computer system. This rate depends on how fast the individual processing units are, how many processing units there are, and how fast data can be moved around in the system. The more efficient an algorithm is, the closer its achieved FLOPS rate is to the maximum FLOPS rate.

For a given computer system, there is also a maximum FLOPS rate possible for a given problem. This has to do with the nature of the tasks within the given problem. Some kinds of tasks can utilize various system resources more easily than other tasks. If a problem can be broken into two tasks, T_1 and T_2, such that T_1 must be brought to completion before T_2 can be performed, the total time required for the problem depends more on the task that takes longer. This tautology has important implications for the limits of efficiency of algorithms. It is the basis of "Amdahl's law" or "Ware's law", (Amdahl, 1967) which puts limits on the speedup of problems that consist of both tasks that must be performed sequentially and tasks that can be performed in parallel. It is also the basis of the following childhood riddle:

> You are to make a round trip to a city 100 miles away. You want
> to average 50 miles per hour. Going, you travel at a constant rate
> of 25 miles per hour. How fast must you travel coming back?

The efficiency of an algorithm may depend on the organization of the computer, the implementation of the algorithm in a programming language, and the way the program is compiled.

Computations in Parallel

The most effective way of decreasing the time required for solving a computational problem is to perform the computations in parallel if possible. There are some computations that are essentially serial, but in almost any problem there are subtasks that are independent of each other and can be performed in any order. Parallel computing remains an important research area. See Nakano (2004) for a summary discussion.

10.3.3 Iterations and Convergence

Many numerical algorithms are iterative; that is, groups of computations form successive approximations to the desired solution. In a program, this usually means a loop through a common set of instructions in which each pass through the loop changes the initial values of operands in the instructions.

We will generally use the notation $x^{(k)}$ to refer to the computed value of x at the kth iteration.

An iterative algorithm terminates when some *convergence criterion* or *stopping criterion* is satisfied. An example is to declare that an algorithm has converged when

$$\Delta(x^{(k)}, x^{(k-1)}) \leq \epsilon,$$

where $\Delta(x^{(k)}, x^{(k-1)})$ is some measure of the difference of $x^{(k)}$ and $x^{(k-1)}$ and ϵ is a small positive number. Because x may not be a single number, we must consider general measures of the difference of $x^{(k)}$ and $x^{(k-1)}$. For example, if x is a vector, the measure may be some metric, such as we discuss in Chapter 2. In that case, $\Delta(x^{(k)}, x^{(k-1)})$ may be denoted by $\|x^{(k)} - x^{(k-1)}\|$.

An iterative algorithm may have more than one stopping criterion. Often, a maximum number of iterations is set so that the algorithm will be sure to terminate whether it converges or not. (Some people define the term "algorithm" to refer only to methods that converge. Under this definition, whether or not a method is an "algorithm" may depend on the input data unless a stopping rule based on something independent of the data, such as the number of iterations, is applied. In any event, it is always a good idea, in addition to stopping criteria based on convergence of the solution, to have a stopping criterion that is independent of convergence and that limits the number of operations.)

The *convergence ratio* of the sequence $x^{(k)}$ to a constant x_0 is

$$\lim_{k \to \infty} \frac{\Delta(x^{(k+1)}, x_0)}{\Delta(x^{(k)}, x_0)}$$

if this limit exists. If the convergence ratio is greater than 0 and less than 1, the sequence is said to converge *linearly*. If the convergence ratio is 0, the sequence is said to converge *superlinearly*.

Other measures of the rate of convergence are based on

$$\lim_{k \to \infty} \frac{\Delta(x^{(k+1)}, x_0)}{(\Delta(x^{(k)}, x_0))^r} = c \tag{10.9}$$

(again, assuming the limit exists; i.e., $c < \infty$). In equation (10.9), the exponent r is called the *rate of convergence*, and the limit c is called the *rate constant*. If $r = 2$ (and c is finite), the sequence is said to converge *quadratically*. It is clear that for any $r > 1$ (and finite c), the convergence is superlinear.

Convergence defined in terms of equation (10.9) is sometimes referred to as "Q-convergence" because the criterion is a quotient. Types of convergence may then be referred to as "Q-linear", "Q-quadratic", and so on.

The convergence rate is often a function of k, say $h(k)$. The convergence is then expressed as an order in k, $O(h(k))$.

Extrapolation

As we have noted, many numerical computations are performed on a discrete set that approximates the reals or $\mathrm{I\!R}^d$, resulting in *discretization errors*. By "discretization error", we do not mean a rounding error resulting from the computer's finite representation of numbers. The discrete set used in computing some quantity such as an integral is often a grid. If h is the interval width of the grid, the computations may have errors that can be expressed as a function of h. For example, if the true value is x and, because of the discretization, the *exact value* that would be computed is x_h, then we can write

$$x = x_h + e(h).$$

For a given algorithm, suppose the error $e(h)$ is proportional to some power of h, say h^n, and so we can write

$$x = x_h + ch^n \tag{10.10}$$

for some constant c. Now, suppose we use a different discretization, with interval length rh having $0 < r < h$. We have

$$x = x_{rh} + c(rh)^n$$

and, after subtracting from equation (10.10),

$$0 = x_h - x_{rh} + c(h^n - (rh)^n)$$

or

$$ch^n = \frac{(x_h - x_{rh})}{r^n - 1}. \tag{10.11}$$

This analysis relies on the assumption that the error in the discrete algorithm is proportional to h^n. Under this assumption, ch^n in equation (10.11) is the discretization error in computing x, using exact computations, and is an estimate of the error due to discretization in actual computations. A more realistic regularity assumption is that the error is $O(h^n)$ as $h \to 0$; that is, instead of (10.10), we have

$$x = x_h + ch^n + O(h^{n+\alpha})$$

for $\alpha > 0$.

Whenever this regularity assumption is satisfied, equation (10.11) provides us with an inexpensive improved estimate of x:

$$x_R = \frac{x_{rh} - r^n x_h}{1 - r^n}. \tag{10.12}$$

It is easy to see that $|x - x_R|$ is less than the absolute error using an interval size of either h or rh.

The process described above is called *Richardson extrapolation*, and the value in equation (10.12) is called the Richardson extrapolation estimate. Richardson extrapolation is also called "Richardson's deferred approach to the limit". It has general applications in numerical analysis, but is most widely used in numerical quadrature. Bickel and Yahav (1988) use Richardson extrapolation to reduce the computations in a bootstrap. Extrapolation can be extended beyond just one step, as in the presentation above.

Reducing the computational burden by using extrapolation is very important in higher dimensions. In many cases, for example in direct extensions of quadrature rules, the computational burden grows exponentially with the number of dimensions. This is sometimes called "the curse of dimensionality" and can render a fairly straightforward problem in one or two dimensions unsolvable in higher dimensions.

A direct extension of Richardson extrapolation in higher dimensions would involve extrapolation in each direction, with an exponential increase in the amount of computation. An approach that is particularly appealing in higher dimensions is splitting extrapolation, which avoids independent extrapolations in all directions. See Liem, Lü, and Shih (1995) for an extensive discussion of splitting extrapolation, with numerous applications.

10.3.4 Other Computational Techniques

In addition to techniques to improve the efficiency and the accuracy of computations, there are also special methods that relate to the way we build programs or store data.

Recursion

The algorithms for many computations perform some operation, update the operands, and perform the operation again.

1. perform operation
2. test for exit
3. update operands
4. go to 1

If we give this algorithm the name `doit` and represent its operands by x, we could write the algorithm as

Algorithm $\mathtt{doit}(x)$
1. operate on x
2. test for exit
3. update x: x'
4. $\mathtt{doit}(x')$

The algorithm for computing the mean and the sum of squares (10.7) on page 411 can be derived as a recursion. Suppose we have the mean a_k and the sum of squares s_k for k elements $x_1, x_2, \ldots, x_k$, and we have a new value x_{k+1} and wish to compute a_{k+1} and s_{k+1}. The obvious solution is

$$a_{k+1} = a_k + \frac{x_{k+1} - a_k}{k+1}$$

and

$$s_{k+1} = s_k + \frac{k(x_{k+1} - a_k)^2}{k+1}.$$

These are the same computations as in equations (10.7) on page 411.

Another example of how viewing the problem as an update problem can result in an efficient algorithm is in the evaluation of a polynomial of degree d,

$$p_d(x) = c_d x^d + c_{d-1} x^{d-1} + \cdots + c_1 x + c_0.$$

Doing this in a naive way would require $d-1$ multiplications to get the powers of x, d additional multiplications for the coefficients, and d additions. If we write the polynomial as

$$p_d(x) = x(c_d x^{d-1} + c_{d-1} x^{d-2} + \cdots + c_1) + c_0,$$

we see a polynomial of degree $d-1$ from which our polynomial of degree d can be obtained with but one multiplication and one addition; that is, the number of multiplications is equal to the increase in the degree — not two times the increase in the degree. Generalizing, we have

$$p_d(x) = x(\cdots x(x(c_d x + c_{d-1}) + \cdots) + c_1) + c_0, \tag{10.13}$$

which has a total of d multiplications and d additions. The method for evaluating polynomials in equation (10.13) is called *Horner's method.*

A computer subprogram that implements recursion invokes itself. Not only must the programmer be careful in writing the recursive subprogram, but the programming system must maintain call tables and other data properly to allow for recursion. Once a programmer begins to understand recursion, there may be a tendency to overuse it. To compute a factorial, for example, the inexperienced C programmer may write

```
float Factorial(int n)
   {
    if(n==0)
      return 1;
    else
      return n*Factorial(n-1);
   }
```

The problem is that this is implemented by storing a stack of statements. Because n may be relatively large, the stack may become quite large and inefficient. It is just as easy to write the function as a simple loop, and it would be a much better piece of code.

Both C and Fortran 95 allow for recursion. Many versions of Fortran have supported recursion for years, but it was not part of the Fortran standards before Fortran 90.

Computations without Storing Data

For computations involving large sets of data, it is desirable to have algorithms that sequentially use a single data record, update some cumulative data, and then discard the data record. Such an algorithm is called a *real-time* algorithm, and operation of such an algorithm is called *online* processing. An algorithm that has all of the data available throughout the computations is called a *batch* algorithm.

An algorithm that generally processes data sequentially in a similar manner as a real-time algorithm but may have subsequent access to the same data is called an *online* algorithm or an *"out-of-core"* algorithm. (This latter name derives from the erstwhile use of "core" to refer to computer memory.) Any real-time algorithm is an online or out-of-core algorithm, but an online or out-of-core algorithm may make more than one pass through the data. (Some people restrict "online" to mean "real-time" as we have defined it above.)

If the quantity t is to be computed from the data $x_1, x_2, \ldots, x_n$, a real-time algorithm begins with a quantity $t^{(0)}$ and from $t^{(0)}$ and x_1 computes $t^{(1)}$. The algorithm proceeds to compute $t^{(2)}$ using x_2 and so on, never retaining more than just the current value, $t^{(k)}$. The quantities $t^{(k)}$ may of course consist of multiple elements. The point is that the number of elements in $t^{(k)}$ is independent of n.

Many summary statistics can be computed in online processes. For example, the algorithms discussed beginning on page 411 for computing the sample sum of squares are real-time algorithms. The algorithm in equations (10.5) requires two passes through the data so it is not a real-time algorithm, although it is out-of-core. There are stable online algorithms for other similar statistics, such as the sample variance-covariance matrix. The least squares linear regression estimates can also be computed by a stable one-pass algorithm that, incidentally, does not involve computation of the variance-covariance matrix (or the sums of squares and cross products matrix). There is no real-time algorithm for finding the median. The number of data records that must be retained and reexamined depends on n.

In addition to the reduced storage burden, a real-time algorithm allows a statistic computed from one sample to be updated using data from a new sample. A real-time algorithm is necessarily $O(n)$.

Exercises

10.1. An important attitude in the computational sciences is that the computer is to be used as a tool for exploration and discovery. The computer should be used to check out "hunches" or conjectures, which then later should be subjected to analysis in the traditional manner. There are limits to this approach, however. An example is in limiting processes. Because the computer deals with finite quantities, the results of a computation may be misleading. Explore each of the situations below using C or Fortran. A few minutes or even seconds of computing should be enough to give you a feel for the nature of the computations.

In these exercises, you may write computer programs in which you perform tests for equality. A word of warning is in order about such tests. If a test involving a quantity x is executed soon after the computation of x, the test may be invalid within the set of floating-point numbers with which the computer nominally works. This is because the test may be performed using the extended precision of the computational registers.

a) Consider the question of the convergence of the series

$$\sum_{i=1}^{\infty} i.$$

Obviously, this series does not converge in $\rm I\!R$. Suppose, however, that we begin summing this series using floating-point numbers. Will the computations overflow? If so, at what value of i (approximately)? Or will the series converge in $\rm I\!F$? If so, to what value, and at what value of i (approximately)? In either case, state your answer in terms of the standard parameters of the floating-point model, b, p, $e_{\min}$, and $e_{\max}$ (page 380).

b) Consider the question of the convergence of the series

$$\sum_{i=1}^{\infty} 2^{-2i}$$

and answer the same questions as in Exercise 10.1a.

c) Consider the question of the convergence of the series

$$\sum_{i=1}^{\infty} \frac{1}{i}$$

and answer the same questions as in Exercise 10.1a.

d) Consider the question of the convergence of the series

$$\sum_{i=1}^{\infty} \frac{1}{i^x},$$

for $x \geq 1$. Answer the same questions as in Exercise 10.1a, except address the variable x.

10.2. We know, of course, that the harmonic series in Exercise 10.1c does not converge (although the naive program to compute it does). It is, in fact, true that

$$H_n = \sum_{i=1}^{n} \frac{1}{i}$$
$$= f(n) + \gamma + o(1),$$

where f is an increasing function and γ is Euler's constant. For various n, compute H_n. Determine a function f that provides a good fit and obtain an approximation of Euler's constant.

10.3. Machine characteristics.

a) Write a program to determine the smallest and largest relative spacings. Use it to determine them on the machine you are using.

b) Write a program to determine whether your computer system implements gradual underflow.

c) Write a program to determine the bit patterns of $+\infty$, $-\infty$, and NaN on a computer that implements the IEEE binary standard. (This may be more difficult than it seems.)

d) Obtain the program MACHAR (Cody, 1988) and use it to determine the smallest positive floating-point number on the computer you are using. (MACHAR is included in CALGO, which is available from `netlib`. See the Bibliography.)

10.4. Write a program in Fortran or C to determine the bit patterns of fixed-point numbers, floating-point numbers, and character strings. Run your program on different computers, and compare your results with those shown in Figures 10.1 through 10.3 and Figures 10.11 through 10.13.

10.5. What is the numerical value of the rounding unit ($\frac{1}{2}$ ulp) in the IEEE Standard 754 double precision?

10.6. Consider the standard model (10.1) for the floating-point representation,

$$\pm 0.d_1 d_2 \cdots d_p \times b^e,$$

with $e_{\min} \leq e \leq e_{\max}$. Your answers to the following questions may depend on an additional assumption or two. Either choice of (standard) assumptions is acceptable.

a) How many floating-point numbers are there?

b) What is the smallest positive number?

c) What is the smallest number larger than 1?

d) What is the smallest number X such that $X + 1 = X$?

e) Suppose $p = 4$ and $b = 2$ (and $e_{\min}$ is very small and $e_{\max}$ is very large). What is the next number after 20 in this number system?

10.7. a) Define parameters of a floating-point model so that the number of numbers in the system is less than the largest number in the system.

b) Define parameters of a floating-point model so that the number of numbers in the system is greater than the largest number in the system.

10.8. Suppose that a certain computer represents floating-point numbers in base 10 using eight decimal places for the mantissa, two decimal places for the exponent, one decimal place for the sign of the exponent, and one decimal place for the sign of the number.

a) What are the "smallest relative spacing" and the "largest relative spacing"? (Your answer may depend on certain additional assumptions about the representation; state any assumptions.)

b) What is the largest number g such that $417 + g = 417$?

c) Discuss the associativity of addition using numbers represented in this system. Give an example of three numbers, a, b, and c, such that using this representation $(a + b) + c \neq a + (b + c)$ unless the operations are chained. Then show how chaining could make associativity hold for some more numbers but still not hold for others.

d) Compare the maximum rounding error in the computation $x + x + x + x$ with that in $4 * x$. (Again, you may wish to mention the possibilities of chaining operations.)

10.9. Consider the same floating-point system as in Exercise 10.8.

a) Let X be a random variable uniformly distributed over the interval

$$[1 - .000001, \ 1 + .000001].$$

Develop a probability model for the representation $[X]_c$. (This is a discrete random variable with 111 mass points.)

    b) Let X and Y be random variables uniformly distributed over the same interval as above. Develop a probability model for the representation $[X + Y]_c$. (This is a discrete random variable with 121 mass points.)

    c) Develop a probability model for $[X]_c \, [+]_c \, [Y]_c$. (This is also a discrete random variable with 121 mass points.)

10.10. Give an example to show that the sum of three floating-point numbers can have a very large relative error.

10.11. Write a single program in Fortran or C to compute the following

    a)
$$\sum_{i=0}^{5} \binom{10}{i} 0.25^i 0.75^{20-i}.$$

    b)
$$\sum_{i=0}^{10} \binom{20}{i} 0.25^i 0.75^{20-i}.$$

    c)
$$\sum_{i=0}^{50} \binom{100}{i} 0.25^i 0.75^{20-i}.$$

10.12. In standard mathematical libraries, there are functions for $\log(x)$ and $\exp(x)$ called `log` and `exp` respectively. There is a function in the IMSL Libraries to evaluate $\log(1 + x)$ and one to evaluate $(\exp(x) - 1)/x$. (The names in Fortran for single precision are `alnrel` and `exprl`.)

    a) Explain why the designers of the libraries included those functions, even though `log` and `exp` are available.

    b) Give an example in which the standard log loses precision. Evaluate it using `log` in the standard math library of Fortran or C. Now evaluate it using a Taylor series expansion of $\log(1 + x)$.

10.13. Suppose you have a program to compute the cumulative distribution function for the chi-squared distribution. The input for the program is x and df, and the output is $\Pr(X \le x)$. Suppose you are interested in probabilities in the extreme upper range and high accuracy is very important. What is wrong with the design of the program for this problem?

10.14. Write a program in Fortran or C to compute e^{-12} using a Taylor series directly, and then compute e^{-12} as the reciprocal of e^{12}, which is also computed using a Taylor series. Discuss the reasons for the differences in the results. To what extent is truncation error a problem?

10.15. Errors in computations.

    a) Explain the difference in truncation and cancellation.

    b) Why is cancellation not a problem in multiplication?

10.16. Assume we have a computer system that can maintain seven digits of precision. Evaluate the sum of squares for the dataset $\{9000, 9001, 9002\}$.

  a) Use the algorithm in equations (10.5) on page 410.
  b) Use the algorithm in equations (10.6) on page 411.
  c) Now assume there is one guard digit. Would the answers change?

10.17. Develop algorithms similar to equations (10.7) on page 411 to evaluate the following.
  a) The weighted sum of squares

$$\sum_{i=1}^{n} w_i(x_i - \bar{x})^2.$$

  b) The third central moment

$$\sum_{i=1}^{n}(x_i - \bar{x})^3.$$

  c) The sum of cross products

$$\sum_{i=1}^{n}(x_i - \bar{x})(y_i - \bar{y}).$$

  Hint: Look at the difference in partial sums,

$$\sum_{i=1}^{j}(\cdot) - \sum_{i=1}^{j-1}(\cdot).$$

10.18. Given the recurrence relation

$$t(n) = pt(n/p) + cn$$

for p positive and c nonnegative, show that $t(n)$ is $O(n \log n)$.

  Hint: First assume n is a power of p.

10.19. In statistical data analysis, it is common to have some missing data. This may be because of nonresponse in a survey questionnaire or because an experimental or observational unit dies or discontinues participation in the study. When the data are recorded, some form of missing-data indicator must be used. Discuss the use of NaN as a missing-value indicator. What are some of its advantages and disadvantages?

10.20. Consider the four properties of a dot product listed on page 15. For each one, state whether the property holds in computer arithmetic. Give examples to support your answers.

10.21. Assuming the model (10.1) on page 380 for the floating-point number system, give an example of a nonsingular 2×2 matrix that is algorithmically singular.

10.22. A Monte Carlo study of condition number and size of the matrix.
For $n = 5, 10, \ldots, 30$, generate 100 $n \times n$ matrices whose elements have independent $N(0, 1)$ distributions. For each, compute the L_2 condition number and plot the mean condition number versus the size of the matrix. At each point, plot error bars representing the sample "standard error" (the standard deviation of the sample mean at that point). How would you describe the relationship between the condition number and the size?
In any such Monte Carlo study we must consider the extent to which the random samples represent situations of interest. (How often do we have matrices whose elements have independent $N(0, 1)$ distributions?)

11

Numerical Linear Algebra

Most scientific computational problems involve vectors and matrices. It is necessary to work with either the elements of vectors and matrices individually or with the arrays themselves. Programming languages such as Fortran 77 and C provide the capabilities for working with the individual elements but not directly with the arrays. Fortran 95 and higher-level languages such as Octave or Matlab and R allow direct manipulation with vectors and matrices.

The distinction between the set of real numbers, $\mathbb{R}$, and the set of floating-point numbers, $\mathbb{F}$, that we use in the computer has important implications for numerical computations. As we discussed in Section 10.2, beginning on page 393, an element x of a vector or matrix is approximated by $[x]_c$, and a mathematical operation $\circ$ is simulated by a computer operation $[\circ]_c$. The familiar laws of algebra for the field of the reals do not hold in $\mathbb{F}$, especially if uncontrolled parallel operations are allowed. These distinctions, of course, carry over to arrays of floating-point numbers that represent real numbers, and the properties of vectors and matrices that we discussed in earlier chapters may not hold for their computer counterparts. For example, the dot product of a nonzero vector with itself is positive (see page 15), but $\langle x_c, x_c \rangle_c = 0$ does not imply $x_c = 0$.

A good general reference on the topic of numerical linear algebra is Čížková and Čížek (2004).

11.1 Computer Representation of Vectors and Matrices

The elements of vectors and matrices are represented as ordinary numeric data, as we described in Section 10.1, in either fixed-point or floating-point representation.

Storage Modes

The elements are generally stored in a logically contiguous area of the computer's memory. What is logically contiguous may not be physically contiguous, however.

Accessing data from memory in a single pipeline may take more computer time than the computations themselves. For this reason, computer memory may be organized into separate modules, or *banks*, with separate paths to the central processing unit. Logical memory is *interleaved* through the banks; that is, two consecutive logical memory locations are in separate banks. In order to take maximum advantage of the computing power, it may be necessary to be aware of how many interleaved banks the computer system has. We will not consider these issues further but rather refer the interested reader to Dongarra et al. (1998).

There are no convenient mappings of computer memory that would allow matrices to be stored in a logical rectangular grid, so matrices are usually stored either as columns strung end-to-end (a "column-major" storage) or as rows strung end-to-end (a "row-major" storage). In using a computer language or a software package, sometimes it is necessary to know which way the matrix is stored.

For some software to deal with matrices of varying sizes, the user must specify the length of one dimension of the array containing the matrix. (In general, the user must specify the lengths of all dimensions of the array except one.) In Fortran subroutines, it is common to have an argument specifying the leading dimension (number of rows), and in C functions it is common to have an argument specifying the column dimension. (See the examples in Figure 12.1 on page 459 and Figure 12.2 on page 460 for illustrations of the leading dimension argument.)

Strides

Sometimes in accessing a partition of a given matrix, the elements occur at fixed distances from each other. If the storage is row-major for an $n \times m$ matrix, for example, the elements of a given column occur at a fixed distance of m from each other. This distance is called the "stride", and it is often more efficient to access elements that occur with a fixed stride than it is to access elements randomly scattered.

Just accessing data from the computer's memory contributes significantly to the time it takes to perform computations. A stride that is not a multiple of the number of banks in an interleaved bank memory organization can measurably increase the computational time in high-performance computing.

Sparsity

If a matrix has many elements that are zeros, and if the positions of those zeros are easily identified, many operations on the matrix can be speeded up.

Matrices with many zero elements are called *sparse matrices*. They occur often in certain types of problems; for example in the solution of differential equations and in statistical designs of experiments. The first consideration is how to represent the matrix and to store the matrix and the location information. Different software systems may use different schemes to store sparse matrices. The method used in the IMSL Libraries, for example, is described on page 458. An important consideration is how to preserve the sparsity during intermediate computations.

11.2 General Computational Considerations for Vectors and Matrices

All of the computational methods discussed in Chapter 10 apply to vectors and matrices, but there are some additional general considerations for vectors and matrices.

11.2.1 Relative Magnitudes of Operands

One common situation that gives rise to numerical errors in computer operations is when a quantity x is transformed to $t(x)$ but the value computed is unchanged:

$$[t(x)]_{\mathrm{c}} = [x]_{\mathrm{c}}; \tag{11.1}$$

that is, the operation actually accomplishes nothing. A type of transformation that has this problem is

$$t(x) = x + \epsilon, \tag{11.2}$$

where $|\epsilon|$ is much smaller than $|x|$. If all we wish to compute is $x + \epsilon$, the fact that we get x is probably not important. Usually, of course, this simple computation is part of some larger set of computations in which ϵ was computed. This, therefore, is the situation we want to anticipate and avoid.

Another type of problem is the addition to x of a computed quantity y that overwhelms x in magnitude. In this case, we may have

$$[x + y]_{\mathrm{c}} = [y]_{\mathrm{c}}. \tag{11.3}$$

Again, this is a situation we want to anticipate and avoid.

Condition

A measure of the worst-case numerical error in numerical computation involving a given mathematical entity is the "condition" of that entity for the particular computations. The *condition number* of a matrix is the most generally useful such measure. For the matrix A, we denote the condition number as $\kappa(A)$. We discussed the condition number in Section 6.1 and illustrated it in the toy example of equation (6.1). The condition number provides a bound on the relative norms of a "correct" solution to a linear system and a solution to a nearby problem. A specific condition number therefore depends on the norm, and we defined κ_1, κ_2, and κ_∞ condition numbers (and saw that they are generally roughly of the same magnitude). We saw in equation (6.10) that the L_2 condition number, $\kappa_2(A)$, is the ratio of magnitudes of the two extreme eigenvalues of A.

The condition of data depends on the particular computations to be performed. The relative magnitudes of other eigenvalues (or singular values) may be more relevant for some types of computations. Also, we saw in Section 10.3.1 that the "stiffness" measure in equation (10.8) is a more appropriate measure of the extent of the numerical error to be expected in computing variances.

Pivoting

Pivoting, discussed on page 209, is a method for avoiding a situation like that in equation (11.3). In Gaussian elimination, for example, we do an addition, $x+y$, where the y is the result of having divided some element of the matrix by some other element and x is some other element in the matrix. If the divisor is very small in magnitude, y is large and may overwhelm x as in equation (11.3).

"Modified" and "Classical" Gram-Schmidt Transformations

Another example of how to avoid a situation similar to that in equation (11.1) is the use of the correct form of the Gram-Schmidt transformations.

The orthogonalizing transformations shown in equations (2.34) on page 27 are the basis for Gram-Schmidt transformations of matrices. These transformations in turn are the basis for other computations, such as the QR factorization. (Exercise 5.9 required you to apply Gram-Schmidt transformations to develop a QR factorization.)

As mentioned on page 27, there are two ways we can extend equations (2.34) to more than two vectors, and the method given in Algorithm 2.1 is the correct way to do it. At the kth stage of the Gram-Schmidt method, the vector $x_k^{(k)}$ is taken as $x_k^{(k-1)}$ and the vectors $x_{k+1}^{(k)}, x_{k+2}^{(k)}, \ldots, x_m^{(k)}$ are all made orthogonal to $x_k^{(k)}$. After the first stage, all vectors have been transformed. This method is sometimes called "modified Gram-Schmidt" because some people have performed the basic transformations in a different way, so that at the kth iteration, starting at $k = 2$, the first $k - 1$ vectors are unchanged (i.e., $x_i^{(k)} = x_i^{(k-1)}$ for $i = 1, 2, \ldots, k-1$), and $x_k^{(k)}$ is made orthogonal to the $k - 1$ previously orthogonalized vectors $x_1^{(k)}, x_2^{(k)}, \ldots, x_{k-1}^{(k)}$. This method is called "classical Gram-Schmidt" for no particular reason. The "classical" method is not as stable, and should not be used; see Rice (1966) and Björck (1967) for discussions. In this book, "Gram-Schmidt" is the same as what is sometimes called "modified Gram-Schmidt". In Exercise 11.1, you are asked to experiment with the relative numerical accuracy of the "classical Gram-Schmidt" and the correct Gram-Schmidt. The problems with the former method show up with the simple set of vectors $x_1 = (1, \epsilon, \epsilon)$, $x_2 = (1, \epsilon, 0)$, and $x_3 = (1, 0, \epsilon)$, with ϵ small enough that

$$[1 + \epsilon^2]_c = 1.$$

11.2.2 Iterative Methods

As we saw in Chapter 6, we often have a choice between direct methods (that is, methods that compute a closed-form solution) and iterative methods. Iterative methods are usually to be favored for large, sparse systems.

Iterative methods are based on a sequence of approximations that (it is hoped) converge to the correct solution. The fundamental trade-off in iterative methods is between the amount of work expended in getting a good approximation at each step and the number of steps required for convergence.

Preconditioning

In order to achieve acceptable rates of convergence for iterative algorithms, it is often necessary to precondition the system; that is, to replace the system $Ax = b$ by the system

$$M^{-1}Ax = M^{-1}b$$

for some suitable matrix M. As we indicated in Chapters 6 and 7, the choice of M involves some art, and we will not consider any of the results here. Benzi (2002) provides a useful survey of the general problem and work up to that time, but this is an area of active research.

Restarting and Rescaling

In many iterative methods, not all components of the computations are updated in each iteration. An approximation to a given matrix or vector may be adequate during some sequence of computations without change, but then at some point the approximation is no longer close enough, and a new approximation must be computed. An example of this is in the use of quasi-Newton methods in optimization in which an approximate Hessian is updated, as indicated in equation (4.24) on page 159. We may, for example, just compute an approximation to the Hessian every few iterations, perhaps using second differences, and then use that approximate matrix for a few subsequent iterations.

Another example of the need to restart or to rescale is in the use of fast Givens rotations. As we mentioned on page 185 when we described the fast Givens rotations, the diagonal elements in the accumulated C matrices in the fast Givens rotations can become widely different in absolute values, so to avoid excessive loss of accuracy, it is usually necessary to rescale the elements periodically. Anda and Park (1994, 1996) describe methods of doing the rescaling dynamically. Their methods involve adjusting the first diagonal element by multiplication by the square of the cosine and adjusting the second diagonal element by division by the square of the cosine. Bindel et al. (2002) discuss in detail techniques for performing Givens rotations efficiently while still maintaining accuracy. (The BLAS routines (see Section 12.1.4) `rotmg` and `rotm`, respectively, set up and apply fast Givens rotations.)

Preservation of Sparsity

In computations involving large sparse systems, we may want to preserve the sparsity, even if that requires using approximations, as discussed in Section 5.10. Fill-in (when a zero position in a sparse matrix becomes nonzero) would cause loss of the computational and storage efficiencies of software for sparse matrices.

In forming a preconditioner for a sparse matrix A, for example, we may choose a matrix $M = \widetilde{L}\widetilde{U}$, where $\widetilde{L}$ and $\widetilde{U}$ are approximations to the matrices in an LU decomposition of A, as in equation (5.43). These matrices are constructed as indicated in equation (5.44) so as to have zeros everywhere A has, and $A \approx \widetilde{L}\widetilde{U}$. This is called incomplete factorization, and often, instead of an exact factorization, an approximate factorization may be more useful because of computational efficiency.

Iterative Refinement

Even if we are using a direct method, it may be useful to refine the solution by one step computed in extended precision. A method for iterative refinement of a solution of a linear system is given in Algorithm 6.3.

11.2.3 Assessing Computational Errors

As we discuss in Section 10.2.2 on page 395, we measure error by a scalar quantity, either as *absolute error*, $|\tilde{r} - r|$, where r is the true value and $\tilde{r}$ is the computed or rounded value, or as *relative error*, $|\tilde{r} - r|/r$ (as long as $r \neq 0$). We discuss general ways of reducing them in Section 10.3.1.

Errors in Vectors and Matrices

The errors in vectors or matrices are generally expressed in terms of norms; for example, the relative error in the representation of the vector v, or as a result of computing v, may be expressed as $\|\tilde{v} - v\|/\|v\|$ (as long as $\|v\| \neq 0$), where $\tilde{v}$ is the computed vector. We often use the notation $\tilde{v} = v + \delta v$, and so $\|\delta v\|/\|v\|$ is the relative error. The choice of which vector norm to use may depend on practical considerations about the errors in the individual elements. The L_∞ norm, for example, gives weight only to the element with the largest single error, while the L_1 norm gives weights to all magnitudes equally.

Assessing Errors in Given Computations

In real-life applications, the correct solution is not known, but we would still like to have some way of assessing the accuracy using the data themselves. Sometimes a convenient way to do this in a given problem is to perform internal consistency tests. An internal consistency test may be an assessment of the agreement of various parts of the output. Relationships among the output are exploited to ensure that the individually computed quantities satisfy these relationships. Other internal consistency tests may be performed by comparing the results of the solutions of two problems with a known relationship.

The solution to the linear system $Ax = b$ has a simple relationship to the solution to the linear system $Ax = b + ca_j$, where a_j is the jth column of A and c is a constant. A useful check on the accuracy of a computed solution to $Ax = b$ is to compare it with a computed solution to the modified system. Of course, if the expected relationship does not hold, we do not know which solution is incorrect, but it is probably not a good idea to trust either. Mullet and Murray (1971) describe this kind of consistency test for regression software. To test the accuracy of the computed regression coefficients for regressing y on $x_1, \ldots, x_m$, they suggest comparing them to the computed regression coefficients for regressing $y + dx_j$ on $x_1, \ldots, x_m$. If the expected relationships do not obtain, the analyst has strong reason to doubt the accuracy of the computations.

Another simple modification of the problem of solving a linear system with a known exact effect is the permutation of the rows or columns. Although this perturbation of the problem does not change the solution, it does sometimes result in a change in the computations, and hence it may result in a different computed solution. This obviously would alert the user to problems in the computations.

Another simple internal consistency test that is applicable to many problems is to use two levels of precision in the computations. In using this test, one must be careful to make sure that the input data are the same. Rounding of the input data may cause incorrect output to result, but that is not the fault of the computational algorithm.

Internal consistency tests cannot confirm that the results are correct; they can only give an indication that the results are incorrect.

11.3 Multiplication of Vectors and Matrices

Arithmetic on vectors and matrices involves arithmetic on the individual elements. The arithmetic on the individual elements is performed as we have discussed in Section 10.2.

The way the storage of the individual elements is organized is very important for the efficiency of computations. Also, the way the computer memory is organized and the nature of the numerical processors affect the efficiency and may be an important consideration in the design of algorithms for working with vectors and matrices.

The best methods for performing operations on vectors and matrices in the computer may not be the methods that are suggested by the definitions of the operations.

In most numerical computations with vectors and matrices, there is more than one way of performing the operations on the scalar elements. Consider the problem of evaluating the matrix times vector product, $c = Ab$, where A is $n \times m$. There are two obvious ways of doing this:

- compute each of the n elements of c, one at a time, as an inner product of m-vectors, $c_i = a_i^{\mathrm{T}} b = \sum_j a_{ij} b_j$, or
- update the computation of all of the elements of c simultaneously as

1. For $i = 1, \ldots, n$, let $c_i^{(0)} = 0$.
2. For $j = 1, \ldots, m$,
   {
       for $i = 1, \ldots, n$,
       {
           let $c_i^{(i)} = c_i^{(i-1)} + a_{ij} b_j$.
       }
   }

If there are p processors available for parallel processing, we could use a fan-in algorithm (see page 397) to evaluate Ax as a set of inner products:

$$
\begin{array}{llll}
c_1^{(1)} = & c_2^{(1)} = & \cdots \quad c_{2m-1}^{(1)} = & c_{2m}^{(1)} = \cdots \\
\quad a_{i1}b_1 + a_{i2}b_2 & \quad a_{i3}b_3 + a_{i4}b_4 & \cdots \quad a_{i,4m-3}b_{4m-3} + a_{i,4m-2}b_{4m-2} & \cdots \\
\qquad \searrow & \qquad \swarrow & \qquad \searrow & \qquad \swarrow \\
c_1^{(2)} = & & \cdots \quad c_m^{(2)} = & \cdots \\
\quad c_1^{(1)} + c_2^{(1)} & & \qquad c_{2m-1}^{(1)} + c_{2m}^{(1)} & \cdots \\
\qquad \searrow & & \qquad \downarrow & \\
\quad c_1^{(3)} = c_1^{(2)} + c_2^{(2)} & \cdots & \cdots & \cdots
\end{array}
$$

The order of the computations is nm (or n^2).

Multiplying two matrices A and B can be considered as a problem of multiplying several vectors b_i by a matrix A, as described above. In the following we will assume A is $n \times m$ and B is $m \times p$, and we will use the notation a_i to represent the ith column of A, a_i^{T} to represent the ith row of A, b_i to represent the ith column of B, c_i to represent the ith column of $C = AB$, and so on. (This notation is somewhat confusing because here we are not using a_i^{T} to represent the transpose of a_i as we normally do. The notation should be clear in the context of the diagrams below, however.) Using the inner product method above results in the first step of the matrix multiplication forming

$$
\begin{bmatrix} a_1^{\mathrm{T}} \\ \hline \cdots \\ \ddots \\ \cdots \end{bmatrix}
\begin{bmatrix} b_1 & \cdots \\ & \cdots \\ & \ddots \\ & \cdots \end{bmatrix}
\longrightarrow
\begin{bmatrix} c_{11} = a_1^{\mathrm{T}} b_1 & \cdots \\ \hline & \cdots \\ \vdots & \ddots \\ & \cdots \end{bmatrix} .
$$

Using the second method above, in which the elements of the product vector are updated all at once, results in the first step of the matrix multiplication forming

$$
\begin{bmatrix} a_1 & \cdots \\ & \cdots \\ & \ddots \\ & \cdots \end{bmatrix}
\begin{bmatrix} b_{11} & \cdots \\ \hline & \cdots \\ \vdots & \ddots \\ & \cdots \end{bmatrix}
\longrightarrow
\begin{bmatrix} c_{11}^{(1)} = a_{11}b_{11} & \cdots \\ c_{21}^{(1)} = a_{21}b_{11} & \cdots \\ \vdots & \ddots \\ c_{n1}^{(1)} = a_{n1}b_{11} & \cdots \end{bmatrix} .
$$

The next and each successive step in this method are axpy operations:

$$
c_1^{(k+1)} = b_{(k+1),1} a_1 + c_1^{(k)},
$$

for k going to $m - 1$.

Another method for matrix multiplication is to perform axpy operations using all of the elements of b_1^{T} before completing the computations for any of the columns of C. In this method, the elements of the product are built as the sum of the outer products $a_i b_i^{\mathrm{T}}$. In the notation used above for the other methods, we have

$$
\begin{bmatrix} a_1 & \cdots \\ & \cdots \\ & \ddots \\ & \cdots \end{bmatrix}
\begin{bmatrix} b_1^{\mathrm{T}} \\ \hline \cdots \\ \ddots \\ \cdots \end{bmatrix}
\longrightarrow
\begin{bmatrix} c_{ij}^{(1)} = a_1 b_1^{\mathrm{T}} \end{bmatrix} ,
$$

and the update is

$$
c_{ij}^{(k+1)} = a_{k+1} b_{k+1}^{\mathrm{T}} + c_{ij}^{(k)} .
$$

The order of computations for any of these methods is $\mathrm{O}(nmp)$, or just $\mathrm{O}(n^3)$, if the dimensions are all approximately the same. Strassen's method, discussed next, reduces the order of the computations.

Strassen's Algorithm

Another method for multiplying matrices that can be faster for large matrices is the so-called *Strassen algorithm* (from Strassen, 1969). Suppose A and B are square matrices with equal and even dimensions. Partition them into submatrices of equal size, and consider the block representation of the product,

$$\begin{bmatrix} C_{11} & C_{12} \\ C_{21} & C_{22} \end{bmatrix} = \begin{bmatrix} A_{11} & A_{12} \\ A_{21} & A_{22} \end{bmatrix} \begin{bmatrix} B_{11} & B_{12} \\ B_{21} & B_{22} \end{bmatrix},$$

where all blocks are of equal size. Form

$$P_1 = (A_{11} + A_{22})(B_{11} + B_{22}),$$
$$P_2 = (A_{21} + A_{22})B_{11},$$
$$P_3 = A_{11}(B_{12} - B_{22}),$$
$$P_4 = A_{22}(B_{21} - B_{11}),$$
$$P_5 = (A_{11} + A_{12})B_{22},$$
$$P_6 = (A_{21} - A_{11})(B_{11} + B_{12}),$$
$$P_7 = (A_{12} - A_{22})(B_{21} + B_{22}).$$

Then we have (see the discussion on partitioned matrices in Section 3.1)

$$C_{11} = P_1 + P_4 - P_5 + P_7,$$
$$C_{12} = P_3 + P_5,$$
$$C_{21} = P_2 + P_4,$$
$$C_{22} = P_1 + P_3 - P_2 + P_6.$$

Notice that the total number of multiplications of matrices is seven instead of the eight it would be in forming

$$\begin{bmatrix} A_{11} & A_{12} \\ A_{21} & A_{22} \end{bmatrix} \begin{bmatrix} B_{11} & B_{12} \\ B_{21} & B_{22} \end{bmatrix}$$

directly. Whether the blocks are matrices or scalars, the same analysis holds. Of course, in either case there are more additions. The addition of two $k \times k$ matrices is $O(k^2)$, so for a large enough value of n the total number of operations using the Strassen algorithm is less than the number required for performing the multiplication in the usual way.

The partitioning of the matrix factors can also be used recursively; that is, in the formation of the P matrices. If the dimension, n, contains a factor 2^e, the algorithm can be used directly e times, and then conventional matrix multiplication can be used on any submatrix of dimension $\leq n/2^e$.) If the dimension of the matrices is not even, or if the matrices are not square, it may be worthwhile to pad the matrices with zeros, and the use the Strassen algorithm recursively.

The order of computations of the Strassen algorithm is $O(n^{\log_2 7})$, instead of $O(n^3)$ as in the ordinary method ($\log_2 7 = 2.81$). The algorithm can be implemented in parallel (see Bailey, Lee, and Simon, 1990).

11.4 Other Matrix Computations

Rank Determination

It is often easy to determine that a matrix is of full rank. If the matrix is not of full rank, however, or if it is very ill-conditioned, it is difficult to determine its rank. This is because the computations to determine the rank eventually approximate 0. It is difficult to approximate 0; the relative error (if defined) would be either 0 or infinite. The rank-revealing QR factorization (equation (5.36), page 190) is the preferred method for estimating the rank.

When this decomposition is used to estimate the rank, it is recommended that complete pivoting be used in computing the decomposition. The LDU decomposition, described on page 186, can be modified the same way we used the modified QR to estimate the rank of a matrix. Again, it is recommended that complete pivoting be used in computing the decomposition.

The singular value decomposition (SVD) shown in equation (3.218) on page 127 also provides an indication of the rank of the matrix. For the $n \times m$ matrix A, the SVD is

$$A = UDV^{\mathrm{T}},$$

where U is an $n \times n$ orthogonal matrix, V is an $m \times m$ orthogonal matrix, and D is a diagonal matrix of the singular values. The number of nonzero singular values is the rank of the matrix. Of course, again, the question is whether or not the singular values are zero. It is unlikely that the values computed are exactly zero.

A problem related to rank determination is to approximate the matrix A with a matrix A_r of rank $r \leq \mathrm{rank}(A)$. The singular value decomposition provides an easy way to do this,

$$A_r = UD_rV^{\mathrm{T}},$$

where D_r is the same as D, except with zeros replacing all but the r largest singular values. A result of Eckart and Young (1936) guarantees A_r is the rank r matrix closest to A as measured by the Frobenius norm,

$$\|A - A_r\|_{\mathrm{F}},$$

(see Section 3.10). This kind of matrix approximation is the basis for dimension reduction by principal components.

Computing the Determinant

The determinant of a square matrix can be obtained easily as the product of the diagonal elements of the triangular matrix in any factorization that yields an orthogonal matrix times a triangular matrix. As we have stated before, it is not often that the determinant need be computed, however.

One application in statistics is in optimal experimental designs. The D-optimal criterion, for example, chooses the design matrix, X, such that $|X^{\mathrm{T}}X|$ is maximized (see Section 9.2.2).

Computing the Condition Number

The computation of a condition number of a matrix can be quite involved. Clearly, we would not want to use the definition, $\kappa(A) = \|A\| \, \|A^{-1}\|$, directly. Although the choice of the norm affects the condition number, recalling the discussion in Section 6.1, we choose whichever condition number is easiest to compute or estimate.

Various methods have been proposed to estimate the condition number using relatively simple computations. Cline et al. (1979) suggest a method that is easy to perform and is widely used. For a given matrix A and some vector v, solve

$$A^{\mathrm{T}}x = v$$

and then

$$Ay = x.$$

By tracking the computations in the solution of these systems, Cline et al. conclude that

$$\frac{\|y\|}{\|x\|}$$

is approximately equal to, but less than, $\|A^{-1}\|$. This estimate is used with respect to the L_1 norm in the LINPACK software library (Dongarra et al., 1979), but the approximation is valid for any norm. Solving the two systems above probably does not require much additional work because the original problem was likely to solve $Ax = b$, and solving a system with multiple right-hand sides can be done efficiently using the solution to one of the right-hand sides. The approximation is better if v is chosen so that $\|x\|$ is as large as possible relative to $\|v\|$.

Stewart (1980) and Cline and Rew (1983) investigated the validity of the approximation. The LINPACK estimator can underestimate the true condition number considerably, although generally not by an order of magnitude. Cline, Conn, and Van Loan (1982) give a method of estimating the L_2 condition number of a matrix that is a modification of the L_1 condition number used in LINPACK. This estimate generally performs better than the L_1 estimate, but the Cline/Conn/Van Loan estimator still can have problems (see Bischof, 1990).

Hager (1984) gives another method for an L_1 condition number. Higham (1988) provides an improvement of Hager's method, given as Algorithm 11.1 below, which is used in the LAPACK software library (Anderson et al., 2000).

Algorithm 11.1 The Hager/Higham LAPACK Condition Number Estimator γ of the $n \times n$ Matrix A

Assume $n > 1$; otherwise set $\gamma = \|A\|$. (All norms are L_1 unless specified otherwise.)

0. Set $k = 1$; $v^{(k)} = \frac{1}{n}A1$; $\gamma^{(k)} = \|v^{(k)}\|$; and $x^{(k)} = A^{\mathrm{T}}\mathrm{sign}(v^{(k)})$.

1. Set $j = \min\{i,\ \text{s.t.}\ |x_i^{(k)}| = \|x^{(k)}\|_\infty\}$.
2. Set $k = k + 1$.
3. Set $v^{(k)} = Ae_j$.
4. Set $\gamma^{(k)} = \|v^{(k)}\|$.
5. If $\text{sign}(v^{(k)}) = \text{sign}(v^{(k-1)})$ or $\gamma^{(k)} \leq \gamma^{(k-1)}$, then go to step 8.
6. Set $x^{(k)} = A^{\mathrm{T}}\text{sign}(v^{(k)})$.
7. If $\|x^{(k)}\|_\infty \neq x_j^{(k)}$ and $k \leq k_{\max}$, then go to step 1.
8. For $i = 1, 2, \ldots, n$, set $x_i = (-1)^{i+1}\left(1 + \frac{i-1}{n-1}\right)$.
9. Set $x = Ax$.
10. If $\frac{2\|x\|}{(3n)} > \gamma^{(k)}$, set $\gamma^{(k)} = \frac{2\|x\|}{(3n)}$.
11. Set $\gamma = \gamma^{(k)}$. ∎

Higham (1987) compares Hager's condition number estimator with that of Cline et al. (1979) and finds that the Hager LAPACK estimator is generally more useful. Higham (1990) gives a survey and comparison of the various ways of estimating and computing condition numbers. You are asked to study the performance of the LAPACK estimate using Monte Carlo methods in Exercise 11.5 on page 442.

Exercises

11.1. Gram-Schmidt orthonormalization.
   a) Write a program module (in Fortran, C, R or S-Plus, Octave or Matlab, or whatever language you choose) to implement Gram-Schmidt orthonormalization using Algorithm 2.1. Your program should be for an arbitrary order and for an arbitrary set of linearly independent vectors.
   b) Write a program module to implement Gram-Schmidt orthonormalization using equations (2.34) and (2.35).
   c) Experiment with your programs. Do they usually give the same results? Try them on a linearly independent set of vectors all of which point "almost" in the same direction. Do you see any difference in the accuracy? Think of some systematic way of forming a set of vectors that point in almost the same direction. One way of doing this would be, for a given x, to form $x + \epsilon e_i$ for $i = 1, \ldots, n - 1$, where e_i is the ith unit vector and ϵ is a small positive number. The difference can even be seen in hand computations for $n = 3$. Take $x_1 = (1, 10^{-6}, 10^{-6})$, $x_2 = (1, 10^{-6}, 0)$, and $x_3 = (1, 0, 10^{-6})$.
11.2. Given the $n \times k$ matrix A and the k-vector b (where n and k are large), consider the problem of evaluating $c = Ab$. As we have mentioned, there are two obvious ways of doing this: (1) compute each element of c, one at a time, as an inner product $c_i = a_i^{\mathrm{T}}b = \sum_j a_{ij}b_j$, or (2) update the computation of all of the elements of c in the inner loop.

a) What is the order of computation of the two algorithms?

b) Why would the relative efficiencies of these two algorithms be different for different programming languages, such as Fortran and C?

c) Suppose there are p processors available and the fan-in algorithm on page 436 is used to evaluate Ax as a set of inner products. What is the order of time of the algorithm?

d) Give a heuristic explanation of why the computation of the inner products by a fan-in algorithm is likely to have less roundoff error than computing the inner products by a standard serial algorithm. (This does not have anything to do with the parallelism.)

e) Describe how the following approach could be parallelized. (This is the second general algorithm mentioned above.)

$$
\begin{aligned}
&\text{for } i = 1, \ldots, n \\
&\{ \\
&\quad c_i = 0 \\
&\quad \text{for } j = 1, \ldots, k \\
&\quad \{ \\
&\quad\quad c_i = c_i + a_{ij}b_j \\
&\quad \} \\
&\}
\end{aligned}
$$

f) What is the order of time of the algorithms you described?

11.3. Consider the problem of evaluating $C = AB$, where A is $n \times m$ and B is $m \times q$. Notice that this multiplication can be viewed as a set of matrix/vector multiplications, so either of the algorithms in Exercise 11.2d above would be applicable. There is, however, another way of performing this multiplication, in which all of the elements of C could be evaluated simultaneously.

a) Write pseudocode for an algorithm in which the nq elements of C could be evaluated simultaneously. Do not be concerned with the parallelization in this part of the question.

b) Now suppose there are nmq processors available. Describe how the matrix multiplication could be accomplished in $O(m)$ steps (where a step may be a multiplication and an addition).

Hint: Use a fan-in algorithm.

11.4. Write a Fortran or C program to compute an estimate of the L_1 LAPACK condition number γ using Algorithm 11.1 on page 440.

11.5. Design and conduct a Monte Carlo study to assess the performance of the LAPACK estimator of the L_1 condition number using your program from Exercise 11.4. Consider a few different sizes of matrices, say 5×5, 10×10, and 20×20, and consider a range of condition numbers, say 10, 10^4, and 10^8. In order to assess the accuracy of the condition number estimator, the random matrices in your study must have known condition numbers. It is easy to construct a diagonal matrix with a given condition number. The condition number of the diagonal matrix D, with nonzero elements $d_1, \ldots, d_n$, is $\max |d_i| / \min |d_i|$. It is not so clear how to construct a general (square) matrix with a given condition number. The L_2 condition number of the matrix UDV, where U and V are orthogonal matrices is the same as the L_2 condition number of U. We can therefore construct a wide range of matrices with given L_2 condition numbers. In your Monte Carlo study, use matrices with known L_2 condition numbers. The next question is what kind of random matrices to generate. Again, make a choice of convenience. Generate random diagonal matrices D, subject to fixed $\kappa(D) = \max |d_i| / \min |d_i|$. Then generate random orthogonal matrices as described in Exercise 4.7 on page 171. Any conclusions made on the basis of a Monte Carlo study, of course, must be restricted to the domain of the sampling of the study. (See Stewart, 1980, for a Monte Carlo study of the performance of the LINPACK condition number estimator.)

12

Software for Numerical Linear Algebra

There is a variety of computer software available to perform the operations on vectors and matrices discussed in Chapter 11. We can distinguish the software based on the kinds of applications it emphasizes, the level of the objects it works with directly, and whether or not it is interactive. Some software is designed only to perform certain functions, such as eigenanalysis, while other software provides a wide range of computations for linear algebra. Some software supports only real matrices and real associated values, such as eigenvalues. In some software systems, the basic units must be scalars, and so operations on matrices or vectors must be performed on individual elements. In these systems, higher-level functions to work directly on the arrays are often built and stored in libraries. In other software systems, the array itself is a fundamental operand. Finally, some software for linear algebra is interactive and computations are performed immediately in response to the user's input.

There are many software systems that provide capabilities for numerical linear algebra. Some of these grew out of work at universities and government labs. Others are commercial products. These include the IMSL^{TM} Libraries, MATLAB$^{®}$, S-PLUS$^{®}$, the GAUSS Mathematical and Statistical System^{TM}, IDL$^{®}$, PV-Wave$^{®}$, Maple$^{®}$, Mathematica$^{®}$, and SAS IML$^{®}$. In this chapter, we briefly discuss some of these systems and give some of the salient features from the user's point of view. We also occasionally refer to two standard software packages for linear algebra, LINPACK (Dongarra et al., 1979) and LAPACK. (Anderson et al., 2000).

The Guide to Available Mathematical Software (GAMS) is a good source of information about software. This guide is organized by types of computations. Computations for linear algebra are in Class D. The web site is

```
http://gams.nist.gov/serve.cgi/Class/D/
```

Much of the software is available through `statlib` or `netlib` (see page 505 in the Bibliography).

For some types of software, it is important to be aware of the way the data are stored in the computer, as we discussed in Section 11.1 beginning on page 429. This may include such things as whether the storage is row-major or column-major, which will determine the stride and may determine the details of an algorithm so as to enhance the efficiency. Software written in a language such as Fortran or C often requires the specification of the number of rows (in Fortran) or columns (in C) that have been allocated for the storage of a matrix. As we have indicated before, the amount of space allocated for the storage of a matrix may not correspond exactly to the size of the matrix.

There are many issues to consider in evaluating software or to be aware of when developing software. The portability of the software is an important consideration because a user's programs are often moved from one computing environment to another.

Some situations require special software that is more efficient than general-purpose software would be. Software for sparse matrices, for example, is specialized to take advantage of the zero entries. For sparse matrices it is necessary to have a scheme for identifying the locations of the nonzeros and for specifying their values. The nature of storage schemes varies from one software package to another. The reader is referred to GAMS as a resource for information about software for sparse matrices.

Occasionally we need to operate on vectors or matrices whose elements are variables. Software for symbolic manipulation, such as Maple, can perform vector/matrix operations on variables. See Exercise 12.6 on page 476.

Operations on matrices are often viewed from the narrow perspective of the numerical analyst rather than from the broader perspective of a user with a task to perform. For example, the user may seek a solution to the linear system $Ax = b$. Most software to solve a linear system requires A to be square and of full rank. If this is not the case, then there are three possibilities: the system has no solution, the system has multiple solutions, or the system has a unique solution. A program to solve a linear system that requires A to be square and of full rank does not distinguish among these possibilities but rather always refuses to provide any solution. This can be quite annoying to a user who wants to solve a large number of systems using the same code.

Writing Mathematics and Writing Programs

In writing either mathematics or programs, it is generally best to think of objects at the highest level that is appropriate for the problem at hand. The details of some computational procedure may be of the form

$$\sum_i \sum_j \sum_k a_{ki} x_{kj}. \tag{12.1}$$

We sometimes think of the computations in this form because we have pro-grammed them in some low-level language at some time. In some cases, it is important to look at the computations in this form, but usually it is better to think of the computations at a higher level, say

$$A^{\mathrm{T}}X. \tag{12.2}$$

The compactness of the expression is not the issue (although it certainly is more pleasant to read). The issue is that expression (12.1) leads us to think of some nested computational loops, while expression (12.2) leads us to look for more efficient computational modules, such as the BLAS, which we discuss below. In a higher-level language system such as R, the latter expression is more likely to cause us to use the system more efficiently.

12.1 Fortran and C

Fortran and C are the most commonly used procedural languages for scientific computation. The American National Standards Institute (ANSI) and its international counterpart, the International Organization for Standardization (ISO), have specified standard definitions of these languages. Whenever ANSI and ISO both have a standard for a given version of a language, the standards are the same. There are various dialects of these languages, most of which result from "extensions" provided by writers of compilers. While these extensions may make program development easier and occasionally provide modest enhancements to execution efficiency, a major effect of the extensions is to lock the user into a specific compiler. Because users usually outlive compilers, it is best to eschew the extensions and to program according to the ANSI/ISO standards. Several libraries of program modules for numerical linear algebra are available both in Fortran and in C.

C began as a low-level language that provided many of the capabilities of a higher-level language together with more direct access to the operating system. It lacks some of the facilities that are very useful in scientific computation, such as complex data types, an exponentiation operator, and direct manipulation of arrays as vectors or matrices.

C++ is an object-oriented programming language built on C. The object-oriented features make it much more useful in computing with vectors and matrices or other arrays and more complicated data structures. Class libraries can be built in C++ to provide capabilities similar to those available in Fortran. There are ANSI standard versions of both C and C++.

An advantage of C is that it provides for easier communication between program units, so it is often used when larger program systems are being put together. Another advantage of C is that inexpensive compilers are readily available, and it is widely taught as a programming language in beginning courses in computer science.

Fortran has evolved over many years of use by scientists and engineers. There are two related families of Fortran languages, which we will call "Fortran 77" and "Fortran 95" or "Fortran 90 and subsequent versions", after the model ISO/ANSI standards. Both ANSI and ISO have specified standard definitions of various versions of Fortran. A version called FORTRAN was defined in 1977 (see ANSI, 1978). We refer to this version along with a modest number of extensions as Fortran 77. If we meant to exclude any extensions or modifications, we refer to it as ANSI Fortran 77. A new standard (not a replacement standard) was adopted in 1990 by ANSI, at the insistence of ISO. This standard language is called ANSI Fortran 90 or ISO Fortran 90 (see ANSI, 1992). It has a number of features that extend its usefulness, especially in numerical linear algebra. There have been a few revisions of Fortran 90 in the past several years. There are only small differences between Fortran 90 and subsequent versions, which are called Fortran 95, Fortran 2000, and Fortran 2003. Most of the features I discuss are in all of these versions, and since the version I currently use is Fortran 95, I will generally just refer to "Fortran 95", or to "Fortran 90 and subsequent versions".

Fortran 95 provides additional facilities for working directly with arrays. For example, to add matrices A and B we can write the Fortran expression A+B (see Lemmon and Schafer, 2005; Metcalf, Reid, and Cohen, 2004; or Press et al., 1996).

Compilers for Fortran are often more expensive and less widely available than compilers for C/C++. An open-source compiler for Fortran 95 is available at

```
http://www.g95.org/
```

Another disadvantage of Fortran is that fewer people outside of the numerical computing community know the language.

12.1.1 Programming Considerations

Both users and developers of Fortran and C software need to be aware of a number of programming details.

Indexing Arrays

Neither Fortran 77 nor C allow vectors and matrices to be treated as atomic units. Numerical operations on vectors and matrices are performed either within loops of operations on the individual elements or by invocation of a separate program module.

The natural way of representing vectors and matrices in the earlier versions of Fortran and in C is as array variables with indexes. Fortran handles arrays as multiply indexed memory locations, consistent with the nature of the object. Indexes start at 1, just as in the mathematical notation used throughout this book. The storage of two-dimensional arrays in Fortran is column-major; that is, the array A is stored as vec(A). To reference the contiguous memory locations, the first subscript varies fastest. In general-purpose software consisting of Fortran subprograms, it is often necessary to specify the lengths of all dimensions of a Fortran array except the last one.

An array in C is an ordered set of memory locations referenced by a pointer or by a name and an index. Indexes start at 0. The indexes are enclosed in rectangular brackets following the variable name. An element of a multidimensional array in C is indexed by multiple indexes, each within rectangular brackets. If the 3×4 matrix A is as stored in the C array A, the $(2, 3)$ element $A_{2,3}$ is referenced as A[1][2]. This disconnect between the usual mathematical representations and the C representations results from the historical development of C by computer scientists, who deal with arrays, rather than by mathematical scientists, who deal with matrices and vectors.

Multidimensional arrays in C are arrays of arrays, in which the array constructors operate from right to left. This results in two-dimensional C arrays being stored in row-major order, that is, the array A is stored as $\text{vec}(A^{\mathrm{T}})$. To reference the contiguous memory locations, the last subscript varies fastest. In general-purpose software consisting of C functions, it is often necessary to specify the lengths of all dimensions of a C array except the first one.

Reverse Communication in Iterative Algorithms

Sometimes within the execution of an iterative algorithm it is necessary to perform some operation outside of the basic algorithm itself. The simplest example of this is in an online algorithm, in which more data must be brought in between the operations of the online algorithm. The simplest example of this is perhaps the online computation of a correlation matrix using an algorithm similar to equations (10.7) on page 411. When the first observation is passed to the program doing the computations, that program must be told that this is the first observation (or, more generally, the first n_1 observations). Then, for each subsequent observation (or set of observations), the program must be told that these are intermediate observations. Finally, when the last observation (or set of observations, or even a null set of observations) is passed to the computational program, the program must be told that these are the last observations, and wrap-up computations must be performed (computing correlations from sums of squares). Between the first and last invocations of the computational program, the computational program may preserve intermediate results that are not passed back to the calling program. In this simple example, the communication is one-way, from calling routine to called routine.

In more complicated cases using an iterative algorithm, the computational routine may need more general input or auxiliary computations, and hence there may be two-way communication between the calling routine and the called routine. This is sometimes called reverse communication. An example is the repetition of a preconditioning step in a routine using a conjugate gradient method; as the computations proceed, the computational routine may detect a need for rescaling and so return to a calling routine to perform those services.

Barrett et al. (1994) and Dongarra and Eijkhout (2000) describe a variety of uses of reverse communication in software for numerical linear algebra.

Computational Efficiency

Two seemingly trivial things can have major effects on computational efficiency. One is movement of data from the computer's memory into the computational unit. How quickly this movement occurs depends, among other things, on the organization of the data in the computer. Multiple elements of an array can be retrieved from memory more quickly if they are in contiguous memory locations. (Location in computer memory does not necessarily refer to a physical place; in fact, memory is often divided into banks, and adjacent "locations" are in alternate banks. Memory is organized to optimize access.) The main reason that storage of data in contiguous memory locations affects efficiency involves the different levels of computer memory. A computer often has three levels of randomly accessible memory, ranging from "cache" memory, which is very fast, to "disk" memory, which is relatively slower. When data are used in computations, they may be moved in blocks, or pages, from contiguous locations in one level of memory to a higher level. This allows faster subsequent access to other data in the same page. When one block of data is moved into the higher level of memory, another block is moved out. The movement of data (or program segments, which are also data) from one level of memory to another is called "paging".

In Fortran, a column of a matrix occupies contiguous locations, so when paging occurs, elements in the same column are moved. Hence, a column of a matrix can often be operated on more quickly in Fortran than a row of a matrix. In C, a row can be operated on more quickly for similar reasons.

Some computers have array processors that provide basic arithmetic operations for vectors. The processing units are called vector registers and typically hold 128 or 256 full-precision floating-point numbers (see Section 10.1). For software to achieve high levels of efficiency, computations must be organized to match the length of the vector processors as often as possible.

Another thing that affects the performance of software is the execution of loops. In the simple loop

```
do i = 1, n
   sx(i) = sin(x(i))
end do
```

it may appear that the only computing is just the evaluation of the sine of the elements in the vector x. In fact, a nonnegligible amount of time may be spent in keeping track of the loop index and in accessing memory. A compiler on a vector computer may organize the computations so that they are done in groups corresponding to the length of the vector registers. On a computer that does not have vector processors, a technique called "unrolling do-loops" is sometimes used. For the code segment above, unrolling the do-loop to a depth of 7, for example, would yield the following code:

```fortran
do i = 1, n, 7
    sx(i)   = sin(x(i))
    sx(i+1) = sin(x(i+1))
    sx(i+2) = sin(x(i+2))
    sx(i+3) = sin(x(i+3))
    sx(i+4) = sin(x(i+4))
    sx(i+5) = sin(x(i+5))
    sx(i+6) = sin(x(i+6))
end do
```

plus a short loop for any additional elements in x beyond $7\lfloor n/7 \rfloor$. Obviously, this kind of programming effort is warranted only when n is large and when the code segment is expected to be executed many times. The extra programming is definitely worthwhile for programs that are to be widely distributed and used, such as the BLAS that we discuss later.

Matrix Storage Modes

Matrices that have multiple elements with the same value can often be stored in the computer in such a way that the individual elements do not all have separate locations. Symmetric matrices and matrices with many zeros, such as the upper or lower triangular matrices of the various factorizations we have discussed, are examples of matrices that do not require full rectangular arrays for their storage.

A special indexing method for storing symmetric matrices, called *symmetric storage mode*, uses a linear array to store only the unique elements. Symmetric storage mode is a much more efficient and useful method of storing a symmetric matrix than would be achieved by a vech($\cdot$) operator because with symmetric storage mode, the size of the matrix affects only the elements of the vector near the end. If the number of rows and columns of the matrix is increased, the length of the vector is increased, but the elements are not rearranged. For example, the symmetric matrix

$$\begin{bmatrix} 1\ 2\ 4 \cdots \\ 2\ 3\ 5 \cdots \\ 4\ 5\ 6 \cdots \\ \cdots \end{bmatrix}$$

in symmetric storage mode is represented by the array

$$(1, 2, 3, 4, 5, 6, \cdots).$$

By comparison, the vech($\cdot$) operator yields $(1, 2, 4, \cdots, 3, 5, \cdots, 6, \cdots, \cdots)$. For an $n \times n$ symmetric matrix A, the correspondence with the $n(n + 1)/2$-vector v is $v_{i(i-1)/2+j} = a_{i,j}$ for $i \geq j$. Notice that the relationship does not involve n. For $i \geq j$, in Fortran, it is

```
v(i*(i-1)/2+j) = a(i,j)
```

and in C it is

```
v[i*(i+1)/2+j] = a[i][j]
```

Although the amount of space saved by not storing the full symmetric matrix is only about one half of the amount of space required, the use of rank 1 arrays rather than rank 2 arrays can yield some reference efficiencies. (Recall that in discussions of computer software objects, "rank" usually means the number of dimensions.) For band matrices and other sparse matrices, the savings in storage can be much larger.

12.1.2 Fortran 95

For the scientific programmer, one of the most useful features of Fortran 95 and other versions in that family of Fortran languages is the provision of primitive constructs for vectors and matrices. Whereas all of the Fortran 77 intrinsics are scalar-valued functions, Fortran 95 provides array-valued functions. For example, if **aa** and **bb** represent matrices conformable for multiplication, the statement

```
cc = matmul(aa, bb)
```

yields the Cayley product in **cc**. The **matmul** function also allows multiplication of vectors and matrices.

Indexing of arrays starts at 1 by default (any starting value can be specified, however), and storage is column-major.

Space must be allocated for arrays in Fortran 95, but this can be done at run time. An array can be initialized either in the statement allocating the space or in a regular assignment statement. A vector can be initialized by listing the elements between "(/" and "/)". This list can be generated in various ways. The **reshape** function can be used to initialize matrices.

For example, a Fortran 95 statement to declare that the variable **aa** is to be used as a 3×4 array and to allocate the necessary space is

```
real, dimension(3,4) :: aa
```

A Fortran 95 statement to initialize **aa** with the matrix

$$\begin{bmatrix} 1 & 4 & 7 & 10 \\ 2 & 5 & 8 & 11 \\ 3 & 6 & 9 & 12 \end{bmatrix}$$

is

```
aa = reshape( (/ 1., 2., 3., &
                 4., 5., 6., &
                 7., 8., 9., &
                10.,11.,12./), &
              (/3,4/) )
```

Fortran 95 has an intuitive syntax for referencing subarrays, shown in Table 12.1.

Table 12.1. Subarrays in Fortran 95

`aa(2:3,1:3)`	the 2×3 submatrix in rows 2 and 3 and columns 1 to 3 of **aa**
`aa(:,1:4:2)`	refers to the submatrix with all three rows and the first and third columns of **aa**
`aa(:,4)`	refers to the column vector that is the fourth column of **aa**

Notice that because the indexing starts with 1 (instead of 0) the correspondence between the computer objects and the mathematical objects is a natural one. The subarrays can be used directly in functions. For example, if **bb** is the matrix

$$\begin{bmatrix} 1 & 5 \\ 2 & 6 \\ 3 & 7 \\ 4 & 8 \end{bmatrix},$$

the Fortran 95 function reference

```
matmul(aa(1:2,2:3), bb(3:4,:))
```

yields the Cayley product

$$\begin{bmatrix} 4 & 7 \\ 5 & 8 \end{bmatrix} \begin{bmatrix} 3 & 7 \\ 4 & 8 \end{bmatrix}. \tag{12.3}$$

Libraries built on Fortran 95 allow some of the basic operations of linear algebra to be implemented as operators whose operands are vectors or matrices.

Fortran 95 also contains some of the constructs, such as `forall`, that have evolved to support parallel processing.

More extensive later revisions (Fortran 2000 and subsequent versions) include such features as exception handling, interoperability with C, allocatable components, parameterized derived types, and object-oriented programming.

12.1.3 Matrix and Vector Classes in C++

In an object-oriented language such as C++, it is useful to define classes corresponding to matrices and vectors. Operators and/or functions corresponding to the usual operations in linear algebra can be defined so as to allow use of simple expressions to perform these operations.

A class library in C++ can be defined in such a way that the computer code corresponds more closely to mathematical code. The indexes to the arrays can be defined to start at 1, and the double index of a matrix can be written within a single pair of parentheses. For example, in a C++ class defined for use in scientific computations, the $(10, 10)$ element of the matrix A (that is, $a_{10,10}$) could be referenced as

```
aa(10,10)
```

instead of as

```
aa[9][9]
```

as it would be in ordinary C. Many computer engineers prefer the latter notation, however.

There are various C++ class libraries or templates for matrix and vector computations; for example, those of *Numerical Recipes* (Press et al., 2000). The Template Numerical Toolkit

```
http://math.nist.gov/tnt/
```

and the Matrix Template Library

```
http://www.osl.iu.edu/research/mtl/
```

are templates based on the design approach of the C++ Standard Template Library

```
http://www.sgi.com/tech/stl/
```

The class library in *Numerical Recipes* comes with wrapper classes for use with the Template Numerical Toolkit or the Matrix Template Library.

Use of a C++ class library for linear algebra computations may carry a computational overhead that is unacceptable for large arrays. Both the Template Numerical Toolkit and the Matrix Template Library are designed to be computationally efficient (see Siek and Lumsdaine, 2000).

12.1.4 Libraries

There are a number of libraries of Fortran and C subprograms. The libraries vary in several ways: free or with licensing costs or user fees; low-level computational modules or higher-level, more application-oriented programs; specialized or general purpose; and quality, from high to low.

BLAS

There are several basic computations for vectors and matrices that are very common across a wide range of scientific applications. Computing the dot product of two vectors, for example, is a task that may occur in such diverse areas as fitting a linear model to data or determining the maximum value of a function. While the dot product is relatively simple, the details of how the computations are performed and the order in which they are performed can have effects on both the efficiency and the accuracy. See the discussion beginning on page 396 about the order of summing a list of numbers.

The sets of routines called "basic linear algebra subprograms" (BLAS) implement many of the standard operations for vectors and matrices. The BLAS represent a very significant step toward software standardization because the definitions of the tasks and the user interface are the same on all computing platforms. The actual coding, however, may be quite different to take advantage of special features of the hardware or underlying software, such as compilers.

The level 1 BLAS or BLAS-1, the original set of the BLAS, are for vector operations. They were defined by Lawson et al. (1979). Matrix operations, such as multiplying two matrices, were built using the BLAS-1. Later, a set of the BLAS, called level 2 or the BLAS-2, for operations involving a matrix and a vector was defined by Dongarra et al. (1988), a set called the level 3 BLAS or the BLAS-3, for operations involving two dense matrices, was defined by Dongarra et al. (1990), and a set of the level 3 BLAS for sparse matrices was proposed by Duff et al. (1997). An updated set of BLAS is described by Blackford et al. (2002).

The operations performed by the BLAS often cause an input variable to be updated. For example, in a Givens rotation, two input vectors are rotated into two new vectors. In this case, it is natural and efficient just to replace the input values with the output values (see below). A natural implementation of such an operation is to use an argument that is both input and output. In some programming paradigms, such a "side effect" can be somewhat confusing, but the value of this implementation outweighs the undesirable properties.

There is a consistency of the interface among the BLAS routines. The nature of the arguments and their order in the reference are similar from one routine to the next. The general order of the arguments is:

1. the size or shape of the vector or matrix,
2. the array itself, which may be either input or output,
3. the stride, and
4. other input arguments.

The first and second types of arguments are repeated as necessary for each of the operand arrays and the resultant array.

A BLAS routine is identified by a root character string that indicates the operation, for example, `dot` or `axpy`. The name of the BLAS program module may depend on the programming language. In Fortran, the root may be prefixed by `s` to indicate single precision, by `d` to indicate double precision, or by `c` to indicate complex, for example. If the language allows generic function and subroutine references, just the root of the name is used.

The axpy operation we referred to on page 10 multiplies one vector by a constant and then adds another vector ($ax + y$). The BLAS routine `axpy` performs this operation. The interface is

```
axpy(n, a, x, incx, y, incy)
```

where

n is the number of elements in each vector,

a is the scalar constant,

x is the input/output one-dimensional array that contains the elements of the
    vector x,

incx is the stride in the array x that defines the vector,

y is the input/output one-dimensional array that contains the elements of the
    vector y, and

incy is the stride in the array y that defines the vector.

Another example, the routine **rot** to apply a Givens rotation (similar to
the routine **rotm** for Fast Givens that we referred to earlier), has the interface


```
rot(n, x, incx, y, incy, c, s)
```

where

n is the number of elements in each vector,

x is the input/output one-dimensional array that contains the elements of the
    vector x,

incx is the stride in the array x that defines the vector,

y is the input/output one-dimensional array that contains the elements of the
    vector y,

incy is the stride in the array y that defines the vector,

c is the cosine of the rotation, and

s is the sine of the rotation.

This routine is invoked after **rotg** has been called to determine the cosine and
the sine of the rotation (see Exercise 12.3, page 476).

Source programs and additional information about the BLAS can be obtained at

```
http://www.netlib.org/blas/
```

There is a software suite called ATLAS (Automatically Tuned Linear Algebra Software) that provides Fortran and C interfaces to a portable BLAS
binding as well as to other software for linear algebra for various processors.
Information about the ATLAS software can be obtained at

```
http://math-atlas.sourceforge.net/
```

Other Fortran and C Libraries

When work was being done on the BLAS-1 in the 1970s, those lower-level routines were being incorporated into a higher-level set of Fortran routines for matrix eigensystem analysis called EISPACK (Smith et al., 1976) and into a higher-level set of Fortran routines for solutions of linear systems called LINPACK (Dongarra et al., 1979). As work progressed on the BLAS-2 and BLAS-3 in the 1980s and later, a unified set of Fortran routines for both eigenvalue problems and solutions of linear systems was developed, called LAPACK (Anderson et al., 2000). A Fortran 95 version, LAPACK95, is described by Barker et al. (2001). Information about LAPACK is available at

```
http://www.netlib.org/lapack/
```

There is a graphical user interface to help the user navigate the LAPACK site and download LAPACK routines.

ARPACK is a collection of Fortran 77 subroutines to solve large-scale eigenvalue problems. It is designed to compute a few eigenvalues and corresponding eigenvectors of a general matrix, but it also has special abilities for large sparse or structured matrices. See Lehoucq, Sorensen, and Yang (1998) for a more complete description and for the software itself.

Two of the most widely used Fortran and C libraries are the IMSL Libraries and the Nag Library. The GNU Scientific Library (GSL) is a widely used and freely distributed C library. See Galassi et al., (2002) and the web site

```
http://www.gnu.org/gsl/
```

All of these libraries provide large numbers of routines for numerical linear algebra, ranging from very basic computations as provided in the BLAS through complete routines for solving various types of systems of equations and for performing eigenanalysis.

12.1.5 The IMSL Libraries

The IMSL^{TM} libraries are available in both Fortran and C versions and in both single and double precisions. These libraries use the BLAS and other software from LAPACK.

Matrix Storage Modes

The BLAS and the IMSL Libraries implement a wide range of matrix storage modes:

Symmetric mode. A full matrix is used for storage, but only the upper or lower triangular portion of the matrix is used. Some library routines allow the user to specify which portion is to be used, and others require that it be the upper portion.

Hermitian mode. This is the same as the symmetric mode, except for the obvious changes for the Hermitian transpose.

Triangular mode. This is the same as the symmetric mode (with the obvious changes in the meanings).

Band mode. For the $n \times m$ band matrix A with lower band width w_l and upper band width w_u, an $w_l + w_u \times m$ array is used to store the elements. The elements are stored in the same column of the array, say **aa**, as they are in the matrix; that is,

$$\mathbf{aa}(i - j + w_u + 1, \, j) = a_{i,j}$$

for $i = 1, 2, \ldots, w_l + w_u + 1$.

Band symmetric, band Hermitian, and band triangular modes are all defined similarly. In each case, only the upper or lower bands are referenced.

Sparse storage mode. There are several different schemes for representing sparse matrices. The IMSL Libraries use three arrays, each of rank 1 and with length equal to the number of nonzero elements. The integer array **i** contains the row indicator, the integer array **j** contains the column indicator, and the floating-point array **a** contains the corresponding values; that is, the $(\mathbf{i}(k), \mathbf{j}(k))$ element of the matrix is stored in $\mathbf{a}(k)$. The level 3 BLAS for sparse matrices proposed by Duff et al. (1997) have an argument to allow the user to specify the type of storage mode.

Examples of Use of the IMSL Libraries

There are separate IMSL routines for single and double precisions. The names of the Fortran routines share a common root; the double-precision version has a D as its first character, usually just placed in front of the common root. Functions that return a floating-point number but whose mnemonic root begins with an I through an N have an A in front of the mnemonic root for the single-precision version and have a D in front of the mnemonic root for the double-precision version. Likewise, the names of the C functions share a common root. The function name is of the form **imsl_f_**root_name for single precision and **imsl_d_**root_name for double precision.

Consider the problem of solving the system of linear equations

$$x_1 + 4x_2 + 7x_3 = 10,$$
$$2x_1 + 5x_2 + 8x_3 = 11,$$
$$3x_1 + 6x_2 + 9x_3 = 12.$$

Write the system as $Ax = b$. The coefficient matrix A is real (not necessarily REAL) and square. We can use various IMSL subroutines to solve this problem. The two simplest basic routines are **LSLRG/DLSLRG** and **LSARG/DLSARG**. Both have the same set of arguments:

N, the problem size;
A, the coefficient matrix;
LDA, the leading dimension of A (A can be defined to be bigger than it actually
     is in the given problem);
B, the right-hand sides;
IPATH, an indicator of whether $Ax = b$ or $A^{T}x = b$ is to be solved; and
X, the solution.

The difference in the two routines is whether or not they do iterative refinement. A program to solve the system using LSARG (without iterative refinement) is shown in Figure 12.1.

```
C  Fortran 77 program
      parameter (ida=3)
      integer   n, ipath
      real      a(ida, ida), b(ida), x(ida)
C  Storage is by column;
C  nonblank character in column 6 indicates continuation
      data      a/1.0,   2.0,   3.0,
     +                4.0,   5.0,   6.0,
     +                7.0,   8.0,   9.0/
      data      b/10.0, 11.0, 12.0/
      n     = 3
      ipath = 1
      call lsarg (n, a, lda, b, ipath, x)
      print *, 'The solution is', x
      end
```

Fig. 12.1. IMSL Fortran Program to Solve the System of Linear Equations

The IMSL C function to solve this problem is lin_sol_gen, which is available as *float* *imsl_f_lin_sol_gen or *double* *imsl_d_lin_sol_gen. The only required arguments for *imsl_f_lin_sol_gen are:

int n, the problem size;
float a[], the coefficient matrix; and
float b[], the right-hand sides.

Either function will allow the array a to be larger than n, in which case the number of columns in a must be supplied in an optional argument. Other optional arguments allow the specification of whether $Ax = b$ or $A^{T}x = b$ is to be solved (corresponding to the argument IPATH in the Fortran subroutines LSLRG/DLSLRG and LSARG/DLSARG), the storage of the LU factorization, the storage of the inverse, and so on. A program to solve the system is shown in Figure 12.2. Note the difference between the column orientation of Fortran and the row orientation of C.

```
\*  C program *\
#include <imsl.h>
#include <stdio.h>
main()
{
    int   n = 3;
    float *x;
\* Storage is by row;
   statements are delimited by ';',
   so statements continue automatically. */
   float a[] = {1.0,   4.0,   7.0,
                2.0,   5.0,   8.0,
                3.0,   6.0,   9.0};
   float b[] = {10.0, 11.0, 12.0};
   x = imsl_f_lin_sol_gen (n, a, IMSL_A_COL_DIM, 3, b, 0);
   printf ("The solution is %10.4f%10.4f%10.4f\n",
                            x[0], x[1], x[2]);
}
```

Fig. 12.2. IMSL C Program to Solve the System of Linear Equations

The argument IMSL_A_COL_DIM is optional, taking the value of n, the number of equations, if it is not specified. It is used in Figure 12.2 only for illustration.

12.1.6 Libraries for Parallel Processing

Another standard set of routines, called the BLACS (Basic Linear Algebra Communication Subroutines), provides a portable message-passing interface primarily for linear algebra computations with a user interface similar to that of the BLAS. A slightly higher-level set of routines, the PBLAS, combine both the data communication and computation into one routine, also with a user interface similar to that of the BLAS. Filippone and Colajanni (2000) provide a set of parallel BLAS for sparse matrices. Their system, called PSBLAS, shares the general design of the PBLAS for dense matrices and the design of the level 3 BLAS for sparse matrices proposed by Duff et al. (1997).

A distributed memory version of LAPACK, called ScaLAPACK (see Blackford et al., 1997a), has been built on the BLACS and the PBLAS modules.

A parallel version of the ARPACK library is also available. The message-passing layers currently supported are BLACS and MPI. Parallel ARPACK (PARPACK) is provided as an extension to the current ARPACK library (Release 2.1).

Standards for message passing in a distributed-memory parallel processing environment are evolving. The MPI (message-passing interface) standard being developed primarily at Argonne National Laboratories allows for standardized message passing across languages and systems. See Gropp, Lusk, and

Skjellum (1999) for a description of the MPI system. IBM has built the Message Passing Library (MPL) in both Fortran and C, which provides message-passing kernels. PLAPACK is a package for linear algebra built on MPI (see Van de Geijn, 1997).

Trilinos is a collection of compatible software packages that support parallel linear algebra computations, solution of linear and nonlinear equations and eigensystems of equations and related capabilities. The majority of packages are written in C++ using object-oriented techniques. All packages are self-contained, with the Trilinos top layer providing a common look and feel and infrastructure.

The main Trilinos web site is

```
http://software.sandia.gov/trilinos/
```

All of these packages are available on a range of platforms, especially on high-performance computers.

General references that describe parallel computations and software for linear algebra include Nakano (2004), Quinn (2003), and Roosta (2000).

12.2 Interactive Systems for Array Manipulation

Many of the computations for linear algebra are implemented as simple operators on vectors and matrices in some interactive systems. Some of the more common interactive systems that provide for direct array manipulation are Octave or Matlab, R or S-Plus, SAS IML, APL, Lisp-Stat, Gauss, IDL, and PV-Wave. There is no need to allocate space for the arrays in these systems as there is for arrays in Fortran and C.

Mathematical Objects and Computer Objects

Some difficult design decisions must be made when building systems that provide objects that simulate mathematical objects. One issue is how to treat scalars, vectors, and matrices when their sizes happen to coincide.

- Is a vector with one element a scalar?
- Is a 1×1 matrix a scalar?
- Is a $1 \times n$ matrix a row vector?
- Is an $n \times 1$ matrix a column vector?
- Is a column vector the same as a row vector?

While the obvious answer to all these questions is "no", it is often convenient to design software systems as if the answer, at least to some questions some of the time, is "yes". The answer to any such software design question always must be made in the context of the purpose and intended use (and users) of the software. The issue is not the purity of a mathematical definition. We have already seen that most computer objects and operators do not behave exactly like the mathematical entities they simulate.

The experience of most people engaged in scientific computations over many years has shown that the convenience resulting from the software's equivalent treatment of such different objects as a 1×1 matrix and a scalar outweighs the programming error detection that could be possible if the objects were made to behave as nearly as possible to the way the mathematical entities they simulate behave.

Consider, for example, the following arrays of numbers:

$$A = [1\ 2], \quad B = \begin{bmatrix} 1 \\ 2 \end{bmatrix}, \quad C = \begin{bmatrix} 1 \\ 2 \\ 3 \end{bmatrix}. \tag{12.4}$$

If these arrays are matrices with the usual matrix algebra, then ABC, where juxtaposition indicates Cayley multiplication, is not a valid expression. (Under Cayley multiplication, of course, we do not need to indicate the order of the operations because the operation is associative.)

If, however, we are willing to allow mathematical objects to change types, we come up with a reasonable interpretation of ABC. If the 1×1 matrix AB is interpreted as the scalar 5, then the expression $(AB)C$ can be interpreted as $5C$, that is, a scalar times a matrix.

There is no (reasonable) interpretation that would make the expression $A(BC)$ valid.

If A is a row vector and B is a column vector, it hardly makes sense to define an operation on them that would yield another vector. A vector space cannot consist of such mixtures. Under a strict interpretation of the operations, $(AB)C$ is not a valid expression.

We often think of the "transpose" of a vector (although this is not a viable concept in a vector space), and we denote a dot product in a vector space as $x^{\mathrm{T}}y$. If we therefore interpret a row vector such as A in (12.4) as x^{T} for some x in the vector space of which B is a member, then AB can be interpreted as a dot product (that is, as a scalar) and again $(AB)C$ is a valid expression.

The software systems discussed in this section treat the arrays in (12.4) as different kinds of objects when they evaluate expressions involving the arrays. The possible objects are scalars, row vectors, column vectors, and matrices, corresponding to ordinary mathematical objects, and arrays, for which there is no common corresponding mathematical object. The systems provide different subsets of these objects; some may have only one class of object (matrix would be the most general), while some distinguish all five types. Some systems enforce the mathematical properties of the corresponding objects, and some systems take a more pragmatic approach and coerce the object types to ones that allow an expression to be valid if there is an unambiguous interpretation.

In the next two sections we briefly describe the facilities for linear algebra in Matlab and R. The purpose is to give a very quick comparative introduction.

12.2.1 MATLAB and Octave

MATLAB®, or Matlab®, is a proprietary software package distributed by The Mathworks, Inc. It is built on an interactive, interpretive expression language. The package also has a graphical user interface.

Octave is a freely available package that provides essentially the same core functionality in the same language as Matlab. The graphical interfaces for Octave are more primitive than those for Matlab and do not interact as seamlessly with the operating system.

General Properties

The basic object in Matlab is a rectangular array of numbers (possibly complex). Scalars (even indices) are 1×1 matrices; equivalently, a 1×1 matrix can be treated as a scalar.

Statements in Matlab are line-oriented. A statement is assumed to end at the end of the line, unless the last three characters on the line are periods (...). If an assignment statement in Matlab is not terminated with a semicolon, the matrix on the left-hand side of the assignment is printed. If a statement consists only of the name of a matrix, the object is printed to the standard output device (which is likely to be the monitor).

A comment statement in Matlab begins with a percent sign, "%".

Basic Operations with Vectors and Matrices and for Subarrays

The indexing of arrays in Matlab starts with 1.

A matrix is initialized in Matlab by listing the elements row-wise within brackets and with a semicolon marking the end of a row. (Matlab also has a `reshape` function similar to that of Fortran 95 that treats the matrix in a column-major fashion.)

In general, the operators in Matlab refer to the common vector/matrix operations. For example, Cayley multiplication is indicated by the usual multiplication symbol, "*". The meaning of an operator can often be changed to become the corresponding element-by-element operation by preceding the operator with a period; for example, the symbol ".*" indicates the Hadamard product of two matrices. The expression

```
aa * bb
```

indicates the Cayley product of the matrices, where the number of columns of `aa` must be the same as the number of rows of `bb`; and the expression

```
aa .* bb
```

indicates the Hadamard product of the matrices, where the number of rows and columns of `aa` must be the same as the number of rows and columns of `bb`. The transpose of a vector or matrix is obtained by using a postfix operator "'", which is the same ASCII character as the apostrophe:

```
aa'
```

Figure 12.3 below shows Matlab code that initializes the same matrix **aa** that we used as an example for Fortran 95 above. The code in Figure 12.3 also initializes a vector **xx** and a 4 × 2 matrix **bb** and then forms and prints some products.

```
% Matlab program fragment
xx = [1 2 3 4];
% Storage is by rows; continuation is indicated by ...
aa = [1 4 7 10; ...
      2 5 8 11; ...
      3 6 9 12];
bb = [1 5; 2 6; 3 7; 4 8];
% Printing occurs automatically unless ';' is used
yy = a*xx'
yy = xx(1:3)*aa
cc = aa*bb
```

Fig. 12.3. Matlab Code to Define and Initialize Two Matrices and a Vector and Then Form and Print Their Product

Matlab distinguishes between row vectors and column vectors. A row vector is a matrix whose first dimension is 1, and a column vector is a matrix whose second dimension is 1. In either case, an element of the vector is referenced by a single index.

Subarrays in Matlab are defined in much the same way as in Fortran 95, except for one major difference: the upper limit and the stride are reversed in the triplet used in identifying the row or column indices. Examples of subarray references in Matlab are shown in Table 12.2. Compare these with the Fortran 95 references shown in Table 12.1.

Table 12.2. Subarrays in Matlab

aa(2:3,1:3) the 2 × 3 submatrix in rows 2 and 3   and columns 1 to 3 of **aa**
aa(:,1:2:4) the submatrix with all three rows   and the first and third columns of **aa**
aa(:,4)   the column vector that is the fourth column of **aa**

The subarrays can be used directly in expressions. For example, the expression

```
aa(1:2,2:3) * bb(3:4,:)
```

yields the product

$$\begin{bmatrix} 4 & 7 \\ 5 & 8 \end{bmatrix} \begin{bmatrix} 3 & 7 \\ 4 & 8 \end{bmatrix}$$

as on page 453.

Functions of Vectors and Matrices

Matlab has functions for many of the basic operations on vectors and matrices, some of which are shown in Table 12.3.

Table 12.3. Some Matlab Functions for Vector/Matrix Computations

`norm`	Matrix or vector norm.
	For vectors, all L_p norms are available.
	For matrices, the L_1, L_2, L_∞, and Frobenius norms are available.
`rank`	Number of linearly independent rows or columns.
`det`	Determinant.
`trace`	Trace.
`cond`	Matrix condition number.
`null`	Null space.
`orth`	Orthogonalization.
`inv`	Matrix inverse.
`pinv`	Pseudoinverse.
`lu`	LU decomposition.
`qr`	QR decomposition.
`chol`	Cholesky factorization.
`svd`	Singular value decomposition.
`linsolve`	Solve system of linear equations.
`lscov`	Weighted least squares.
	The operator "\" can be used for ordinary least squares.
`nnls`	Nonnegative least squares.
`eig`	Eigenvalues and eigenvectors.
`poly`	Characteristic polynomial.
`hess`	Hessenberg form.
`schur`	Schur decomposition.
`balance`	Diagonal scaling to improve eigenvalue accuracy.
`expm`	Matrix exponential.
`logm`	Matrix logarithm.
`sqrtm`	Matrix square root.
`funm`	Evaluate general matrix function.

In addition to these functions, Matlab has special operators "\" and "/" for solving linear systems or for multiplying one matrix by the inverse of another. While the statement

```
aa\bb
```

refers to a quantity that has the same value as the quantity indicated by

```
inv(aa)*bb
```

the computations performed are different (and, hence, the values produced may be different). The second expression is evaluated by performing the two operations indicated: `aa` is inverted, and the inverse is used as the left factor in matrix or matrix/vector multiplication. The first expression, `aa\bb`, indicates that the appropriate computations to evaluate x in $Ax = b$ should be performed to evaluate the expression. (Here, x and b may be matrices or vectors.) Another difference between the two expressions is that `inv(aa)` requires `aa` to be square algorithmically nonsingular, whereas `aa\bb` produces a value that simulates $A^- b$.

References

There are a number of books on Matlab, including, for example, Hanselman and Littlefield (2004). The book by Coleman and Van Loan (1988) is not specifically on Matlab but shows how to perform matrix computations in Matlab.

12.2.2 R and S-PLUS

The software system called S was developed at Bell Laboratories in the mid-1970s. S is both a data analysis system and an object-oriented programming language.

S-PLUS® is an enhancement of S, developed by StatSci, Inc. (now a part of Insightful Corporation). The enhancements include graphical interfaces with menus for common analyses, more statistical analysis functionality, and support.

There is a freely available open source system called R that provides generally the same functionality in the same language as S. This system, as well as additional information about it, is available at

```
http://www.r-project.org/
```

There are graphical interfaces for installation and maintenance of R that interact well with the operating system. The menus for analyses provided in S-Plus are not available in R.

In the following, rather than continuing to refer to each of the systems, I will generally refer only to R, but most of the discussion applies to either of the systems. There are some functions that are available in S-Plus and not in R and some available in R and not in S-Plus.

General Properties

The most important R entity is the function. In R, all actions are "functions", and R has an extensive set of functions (that is, verbs). Many functions are provided through packages that although not part of the core R can be easily installed.

Assignment is made by "<-" or "_". (The symbol "_" should *never* be used for assignment, in my opinion. It is not mnemonic, and it is often used as a connective. I have seen students use a variable L_p, with "_" being used as a connective, and then use a statement such as `norm_L_p`, in which the first "_" is an assignment. Using this symbol instead of <- saves exactly one unshifted keystroke!)

A comment statement in R begins with a pound sign, "#".

R has a natural syntax and powerful functions for dealing with vectors and matrices, which are objects in the base language. R has functions for printing, but if a statement consists of just the name of an object, the object is printed to the standard output device (which is likely to be the monitor).

Basic Operations with Vectors and Matrices and for Subarrays

Indexing of arrays starts at 1, and storage is column-major. Indexes are indicated by "[ ]"; for example, `xx[1]` refers to the first element of the one-dimensional array `xx`.

A list is constructed by the `c` function. A list can be treated as a vector without modification. A matrix is constructed from a list by the `matrix` function. A matrix can also be constructed by binding vectors as the columns of the matrix (the `cbind` function) or by binding vectors as the rows of the matrix (the `rbind` function).

Cayley multiplication is indicated by the symbol "%*%". Most operators with array operands are applied elementwise; for example, the symbol "*" indicates the Hadamard product of two matrices. The expression

```
aa %*% bb
```

indicates the Cayley product of the matrices, where the number of columns of `aa` must be the same as the number of rows of `bb` and the expression

```
aa * bb
```

indicates the Hadamard product of the matrices, where the number of rows and columns of `aa` must be the same as the number of rows and columns of `bb`. The transpose of a vector or matrix is obtained by using the function "t":

```
t(aa)
```

Figure 12.4 below shows R code that does the same thing as the Matlab code in Figure 12.3; that is, initialize two matrices and a vector, and then form and print their products.

```
#   R program fragment
xx <- c(1 2 3 4)
#   Storage is by column, but a matrix can be constructed by rows;
#   the form of a statement indicates when it is complete, so
#   statements continue automatically.
aa <- matrix(c( 1,  4,  7, 10,
                2,  5,  8, 11,
                3,  6,  9, 12),
             nrow=3, byrow=T)
bb <- matrix(seq(1,8), nrow=4)
yy <- aa %*% xx
#   Printing is performed by entering the name of the object
yy
yy <- xx[c(1,2,3)] %*% aa
yy
cc <- aa %*% bb
cc
```

Fig. 12.4. R Code to Define and Initialize Two Matrices and a Vector and Then Form and Print Their Product

To the extent that R distinguishes between row vectors and column vectors, a vector is considered to be a column vector. In many cases, however, it does not distinguish. For example, if

```
xx <- c(1,2)
yy <- c(1,2)
```

the expression xx %*% yy is the dot product; that is, xx %*% yy is the same as t(xx) %*% yy; that is, the transpose operator is not required.

The outer product, however, requires either explicit transposition or use of a special binary operator. The outer product is formed by xx %*% t(yy) or by using the special outer product operator %o%; thus, xx %o% yy=xx %*% t(yy). There is also a useful function, outer, that allows more general combinations of the elements of two vectors. For example, if func is a scalar function of two scalar variables, outer(xx,yy,FUN=func) forms a matrix with the rows corresponding to xx and the columns corresponding to yy, and whose $(ij)^{\text{th}}$ element corresponds to func(xx[i],yy[j]). Strings can be used as the argument FUN; thus, outer(xx,yy,FUN="*") = xx %o% yy.

In the expressions

```
yy <- aa %*% xx
```

and

```
yy <- xx[c(1,2,3)] %*% aa
```

in Figure 12.4, the vector is interpreted as a row or column as appropriate for the multiplication to be defined. Compare the similar expressions in the

Matlab code in Figure 12.3 in which a distinction is made between column and row vectors.

Like many other software systems for array manipulation, R usually does not distinguish between scalars and arrays of size 1. For example, if

```
xx <- c(1,2)
yy <- c(1,2)
zz <- c(1,2,3)
```

the expression xx %*% yy %*% zz yields the same value as 5*zz because the expression xx %*% yy %*% zz is interpreted as (xx %*% yy) %*% zz and (xx %*% yy) is a scalar. The expression xx %*% (yy %*% zz) is invalid because yy and zz are not conformable for multiplication.

Examples of subarray references in R are shown in Table 12.4. Compare these with the Fortran 95 references shown in Table 12.1 and the Matlab references shown in Table 12.2. In R, a missing index indicates that the entire corresponding dimension is to be used. Groups of indices can be formed by the c function or the seq function, which is similar to the i:j:k notation of Fortran 95.

Table 12.4. Subarrays in R

`aa[c(2,3),c(1,3)]`	the 2×3 submatrix in rows 2 and 3 and columns 1 to 3 of **aa**
`aa[,seq(1,4,2)]`	the submatrix with all 3 rows and the 1st and 3rd columns of **aa**
`aa[,4]`	the column vector that is the 4th column of **aa**

The subarrays can be used directly in expressions. For example, the expression

```
aa[c(1,2),c(2,3)] %*% bb[c(3,4),]
```

yields the product

$$\begin{bmatrix} 4 & 7 \\ 5 & 8 \end{bmatrix} \begin{bmatrix} 3 & 7 \\ 4 & 8 \end{bmatrix}$$

as on page 453.

Functions of Vectors and Matrices

R has functions for many of the basic operations on vectors and matrices. Some of the R functions are shown in Table 12.5.

Table 12.5. Some R Functions for Vector/Matrix Computations

`norm`	Matrix norm.
	The L_1, L_2, L_∞, and Frobenius norms are available.
`vecnorm`	Vector L_p norm.
`det`	Determinant.
`rcond.Matrix`	Matrix condition number.
`solve.Matrix`	Matrix inverse or pseudoinverse.
`lu`	LU decomposition.
`qr`	QR decomposition.
`chol`	Cholesky factorization.
`svd`	Singular value decomposition.
`solve.Matrix`	Solve system of linear equations.
`lsfit`	Ordinary or weighted least squares.
`nnls.fit`	Nonnegative least squares.
`eigen`	Eigenvalues and eigenvectors.

References

Chambers (1998) provides a basic description of the S language. (John Chambers was the principal designer of S.) There are several texts that describe the use of R in statistical data analysis, such as Maindonald and Braun (2003), Venables and Ripley (2003), and Everitt and Nothorn (2006).

12.3 High-Performance Software

Because computations for linear algebra are so pervasive in scientific applications, it is important to have very efficient software for carrying out these computations. We have discussed several considerations for software efficiency in previous chapters. Goedecker and Hoisie (2001) discuss some of these issues more extensively.

Parallel Processing

It is important that software for numerical linear algebra take full advantage of vector or parallel computer architecture. We discussed some of the issues on page 460. Surveys of specialized software for vector architectures and parallel processors are available in Dongarra and Walker (1995) and Dongarra et al. (2002).

ScaLAPACK, described by Blackford et al. (1997b), is a distributed memory version of LAPACK that uses the BLACS and the PBLAS modules. The computations in ScaLAPACK are organized as if performed in a "distributed linear algebra machine" (DLAM), which is constructed by interconnecting BLAS with a BLACS network. The BLAS perform the usual basic computations and the BLACS network exhanges data using primitive message-passing operations. The DLAM can be constructed either with or without a host process. If a host process is present, it would act like a server in receiving a user request, creating the BLACS network, distributing the data, starting the BLAS processes, and collecting the results. ScaLAPACK has routines for LU, Cholesky, and QR decompositions and for computing eigenvalues of a symmetric matrix. The routines are similar to the corresponding routines in LAPACK. Even the names are similar, for example, in Fortran:

LAPACK	ScaLAPACK	
dgetrf	pdgetrf	LU factorization
dpotrf	pdpotrf	Cholesky factorization
dgeqrf	pdgeqrf	QR factorization
dsyevx	pdsyevx	eigenvalues/vectors of symmetric matrix

The constructs of Fortran 95 are helpful in thinking of operations in such a way that they are naturally parallelized. While the addition of arrays in Fortran 77 or C is an operation that leads to loops of sequential scalar operations, in Fortran 95 it is thought of as a single higher-level operation. How to perform operations in parallel efficiently is still not a natural activity, however. For example, the two Fortran 95 statements to add the arrays **aa** and **bb** and then to add **aa** and **cc**

```
dd = aa + bb
ee = aa + cc
```

may be less efficient than loops because the array **aa** may be accessed twice.

Clusters of Computers

The software package PVM, or Parallel Virtual Machine, which was developed at Oak Ridge National Laboratory, the University of Tennessee, and Emory University, provides a set of C functions or Fortran subroutines that allow a heterogeneous collection of Unix or Linux computers to operate smoothly as a multicomputer (see Geist et al., 1994). Likewise, the libraries built on the MPI standard provide functions that effectively build a multicomputer from a heterogeneous collection of Unix computers.

A cluster of computers is a very cost-effective method for high-performance computing. A standard technology for building a cluster of Unix or Linux computers is called Beowulf (see Gropp, Lusk, and Sterling, 2003). A system called Pooch is available for linking Apple computers into clusters (see Dauger and Decyk, 2005).

Processing Sparse Matrices

If the matrices in large-scale problems are sparse, it is important to take advantage of that sparsity both in the storage and in all computations. We discussed storage schemes on page 451. It is also important to preserve the sparsity during intermediate computations.

Duff, Heroux, and Pozo (2002) discusses special software for sparse matrices. Duff and Vömel (2002) provide a set of Fortran BLAS for sparse matrices.

12.4 Software for Statistical Applications

Statistical applications have needs that go beyond simple linear algebra. The two most common additional requirements are for

- handling metadata and
- accommodating missing data.

Software packages designed for data analysis, such as SAS/IML and R, generally provide for metadata and missing values. Fortran/C libraries generally do not provide for metadata or for handling missing data.

Two other needs that often arise in statistical analysis but often are not dealt with adequately in available software, are the

- graceful handling of nonfull rank matrices and
- working with nonsquare matrices.

Aside from these general capabilities, of course, software packages for statistical applications, even if they are designed for some specific type of analysis, should provide the common operations such as computation of simple univariate statistics, linear regression computations, and some simple graphing capabilities.

12.5 Test Data

Testbeds for software consist of test datasets that vary in condition but have known solutions or for which there is an easy way of verifying the solution. Test data maybe fixed datasets or randomly generated datasets over some population with known and controllable properties.

For testing software for matrix computations, a very common matrix is the *Hilbert matrix*, which has elements

$$h_{ij} = \frac{1}{i + j - 1}.$$

Hilbert matrices have large condition numbers; for example, the 10×10 Hilbert matrix has a generates an $n \times n$ Hilbert matrix.

Randomly generated test data can provide general information about the performance of a computational method over a range of datasets with specified characteristics. Examples of studies using randomly generated datasets are the paper by Birkhoff and Gulati (1979) on the accuracy of computed solutions x_c of the linear system $Ax = b$, where A is $n \times n$ from a BMvN distribution, and the paper by Stewart (1980) using random matrices from a Haar distribution to study approximations to condition numbers (see page 169 and Exercise 4.7). As it turns out, matrices from the BMvN distribution are not sufficiently ill-conditioned often enough to be useful in studies of the accuracy of solutions of linear systems. Birkhoff and Gulati developed a procedure to construct arbitrarily ill-conditioned matrices from ones with a BMvN distribution and then used these matrices in their empirical studies.

Ericksen (1985) describes how to generate matrices with known inverses in such a way that the condition numbers vary widely. To generate an $n \times n$ matrix A, choose $x_1, x_2, \ldots, x_n$ arbitrarily, except such that $x_1 \neq 0$, and take

$$
\begin{aligned}
a_{1j} &= x_1 && \text{for } j = 1, \ldots, n, \\
a_{i1} &= x_i && \text{for } i = 2, \ldots, n, \\
a_{ij} &= a_{i,j-1} + a_{i-1,j-1} && \text{for } i, j = 2, \ldots, n.
\end{aligned}
$$

To represent the elements of the inverse, first define $y_1 = x_1^{-1}$, and for $i = 2, \ldots, n$,

$$
y_i = -y_1 \sum_{k=0}^{i-1} x_{i-k} y_k.
$$

Then the elements of the inverse of A, $B = (b_{ij})$, are given by

$$
b_{in} = (-1)^{i+k} \binom{n-1}{i-1} y_1 \qquad \text{for } i = 1, \ldots, n,
$$

$$
b_{nj} = y_{n+1-j} \qquad \text{for } j = 1, \ldots, n-1,
$$

$$
b_{ij} = x_1 b_{in} b_{nj} + \sum_{k=i+1}^{n} b_{k,j+1} \quad \text{for } i, j = 1, \ldots, n-1,
$$

where the binomial coefficient, $\binom{k}{m}$, is defined to be 0 if $k < m$ or $m < 0$.

The nonzero elements of L and U in the LU decomposition of A are easily seen to be $l_{ij} = x_{i+1-j}$ and $u_{ij} = \binom{j-1}{i-1}$. The nonzero elements of the inverses of L and U are then seen to have (i, j) elements y_{i+1-j} and $(-1)^{i-j} \binom{j-1}{i-1}$. The determinant of A is x_1^n. For some choices of $x_1, \ldots, x_n$, it is easy to determine the condition numbers, especially with respect to the L_1 norm, of the matrices A generated in this way. Ericksen (1985) suggests that the xs be chosen as

$$
x_1 = 2^m \quad \text{for } m \leq 0
$$

and

$$x_i = \begin{pmatrix} k \\ i-1 \end{pmatrix} \quad \text{for } i = 2, \ldots, n \quad \text{and } k \geq 2,$$

in which case the L_1 condition number of 10×10 matrices will range from about 10^7 to 10^{17} as n ranges from 2 to 20 for $m = 0$ and will range from about 10^{11} to 10^{23} as n ranges from 2 to 20 for $m = -1$.

For testing algorithms for computing eigenvalues, a useful matrix is a *Wilkinson matrix*, which is a symmetric, tridiagonal matrix with 1s on the off-diagonals. For an $n \times n$ Wilkinson matrix, the diagonal elements are

$$\frac{n-1}{2}, \frac{n-3}{2}, \frac{n-5}{2}, \ldots, \frac{n-5}{2}, \frac{n-3}{2}, \frac{n-1}{2}.$$

If n is odd, the diagonal includes 0, otherwise all of the diagonal elements are positive. The 5×5 Wilkinson matrix, for example, is

$$\begin{bmatrix} 2 & 1 & 0 & 0 & 0 \\ 1 & 1 & 1 & 0 & 0 \\ 0 & 1 & 0 & 1 & 0 \\ 0 & 0 & 1 & 1 & 1 \\ 0 & 0 & 0 & 1 & 2 \end{bmatrix}.$$

The two largest eigenvalues of a Wilkinson matrix are very nearly equal. Other pairs are likewise almost equal to each other: the third and fourth largest eigenvalues are also close in size, the fifth and sixth largest are likewise, and so on. The largest pair is closest in size, and each smaller pair is less close in size.

The Matlab function `wilkinson(n)` generates an $n \times n$ Wilkinson matrix. Another test matrix available in Matlab is the Rosser test matrix, which is an 8×8 matrix with an eigenvalue of multiplicity 2 and three nearly equal eigenvalues. It is constructed by the Matlab function `rosser`.

A well-known, large, and wide-ranging set of test matrices for computational algorithms for various problems in linear algebra was compiled and described by Gregory and Karney (1969). Higham (1991, 2002) describes a set of test matrices and provides Matlab programs to generate the matrices.

Another set of test matrices is available through the "Matrix Market", designed and developed by R. Boisvert, R. Pozo, and K. Remington of the U.S. National Institute of Standards and Technology with contributions by various other people. The test matrices can be accessed at

`http://math.nist.gov/MatrixMarket`

The database can be searched by specifying characteristics of the test matrix, such as size, symmetry, and so on. Once a particular matrix is found, its sparsity pattern can be viewed at various levels of detail, and other pertinent data can be reviewed. If the matrix seems to be what the user wants, it can be downloaded. The initial database for the Matrix Market is the approximately 300 problems from the Harwell-Boeing Sparse Matrix Collection.

A set of test datasets for statistical analyses has been developed by the National Institute of Standards and Technology. This set, called "statistical reference datasets" (StRD), includes test datasets for linear regression, analysis of variance, nonlinear regression, Markov chain Monte Carlo estimation, and univariate summary statistics. It is available at

```
http://www.itl.nist.gov/div898/strd/
```

Assessing the Accuracy of a Computed Result

In real-life applications, the correct solution is not known, and this may also be the case for randomly generated test datasets. If the correct solution is not known, internal consistency tests as discussed in Section 11.2.3 may be used to assess the accuracy of the computations in a given problem.

Software Reviews

Reviews of available software play an important role in alerting the user to both good software to be used and bad software to be avoided. Software reviews also often have the salutary effect of causing the software producers to improve their products.

Exercises

12.1. Write a recursive function in Fortran, C, Octave or Matlab, R or S-Plus, or PV-Wave to multiply two square matrices using the Strassen algorithm (page 437). Write the function so that it uses an ordinary multiplication method if the size of the matrices is below a threshold that is supplied by the user.

12.2. There are various ways to evaluate the efficiency of a program: counting operations, checking the "wall time", using a shell level timer, and using a call within the program. In C, the timing routine is `ctime`, and in Fortran 95 it is the subroutine `system_clock`. Fortran 77 does not have a built-in timing routine, but the IMSL Fortran Library provides one. For this assignment, you are to write six short C programs and six short Fortran programs. The programs in all cases are to initialize an $n \times m$ matrix so that the entries are equal to the column numbers; that is, all elements in the first column are 1s, all in the second column are 2s, etc. The six programs arise from three matrices of different sizes $10,000 \times 10,000$, $100 \times 1,000,000$, and $1,000,000 \times 100$; and from two different ways of nesting the loops: for each size matrix, first nest the row loop within the column loop and then reverse the loops. The number of operations is the same for all programs. For each program, use both a shell level timer (e.g., in Unix, use `time`) and a timer called from within your program. Make a table of the times:

		10000×10000	100×1000000	1000000×100
Fortran	column-in-row	—	—	—
	row-in-column	—	—	—
C	column-in-row	—	—	—
	row-in-column	—	—	—

12.3. Obtain the BLAS routines `rotg` and `rot` for constructing and applying a Givens rotation. These routines exist in both Fortran and C; they are available in the IMSL Libraries or from *CALGO* (*Collected Algorithms of the ACM*; see the Bibliography).

   a) Using these two routines, apply a Givens rotation to the matrix used in Exercise 5.8 in Chapter 5,

$$A = \begin{bmatrix} 3 & 5 & 6 \\ 6 & 1 & 2 \\ 8 & 6 & 7 \\ 2 & 3 & 1 \end{bmatrix},$$

   so that the second column becomes $(5, \tilde{a}_{22}, 6, 0)$.

   b) Write a routine in Fortran or C that accepts as input a matrix and its dimensions and uses the BLAS routines `rotg` and `rot` to produce its QR decomposition. There are several design issues you should address: how the output is returned (for purposes of this exercise, just return two arrays or pointers to the arrays in full storage mode), how to handle nonfull rank matrices (for this exercise, assume that the matrix is of full rank, so return an error message in this case), how to handle other input errors (what do you do if the user inputs a negative number for a dimension?), and others.

12.4. Using the BLAS routines `rotg` and `rot` for constructing and applying a Givens rotation and the program you wrote in Exercise 12.3, write a Fortran or C routine that accepts a simple symmetric matrix and computes its eigenvalues using the mobile Jacobi scheme. The outer loop of your routine consists of the steps shown on page 249, and the multiple actions of each of those steps can be implemented in a loop in serial mode. The importance of this algorithm, however, is realized when the actions in the individual steps on page 249 are performed in parallel.

12.5. Compute the two largest eigenvalues of the 21×21 Wilkinson matrix to 15 digits.

12.6. Use a symbolic manipulation software package such as Maple to determine the inverse of the matrix:

$$\begin{bmatrix} a & b & c \\ d & e & f \\ g & h & i \end{bmatrix}.$$

Determine conditions for which the matrix would be singular. (You can use the `solve()` function in Maple on certain expressions in the symbolic solution you obtained.)

12.7. Consider the 3×3 symmetric Toeplitz matrix with elements a, b, and c; that is, the matrix that looks like this:

$$\begin{bmatrix} a & b & c \\ b & a & b \\ c & b & a \end{bmatrix}.$$

  a) Invert this matrix.

  b) Determine conditions for which the matrix would be singular.

12.8. Develop a class library in C++ for matrix and vector operations. Discuss carefully the issues you consider in designing the class constructors. Design them in such a way that the references

```
xx(1)
YY(1,1)
```

refer to the implied mathematical entities. Design the operators "+" and "*" so that the references

```
aa + bb
aa * bb
```

will determine whether a and b are matrices and/or vectors conformable for the implied mathematical operations and, if so, will produce the object corresponding to the implied mathematical entity represented by the expression.

A

Notation and Definitions

All notation used in this work is "standard". I have opted for simple notation, which, of course, results in a one-to-many map of notation to object classes. Within a given context, however, the overloaded notation is generally unambiguous. I have endeavored to use notation consistently.

This appendix is not intended to be a comprehensive listing of definitions. The Index, beginning on page 519, is a more reliable set of pointers to definitions, except for symbols that are not words.

A.1 General Notation

Uppercase italic Latin and Greek letters, such as A, B, E, Λ, etc., are generally used to represent either matrices or random variables. Random variables are usually denoted by letters nearer the end of the Latin alphabet, such X, Y, and Z, and by the Greek letter E. Parameters in models (that is, unobservables in the models), whether or not they are considered to be random variables, are generally represented by lowercase Greek letters. Uppercase Latin and Greek letters are also used to represent cumulative distribution functions. Also, uppercase Latin letters are used to denote sets.

Lowercase Latin and Greek letters are used to represent ordinary scalar or vector variables and functions. **No distinction in the notation is made between scalars and vectors**; thus, β may represent a vector and β_i may represent the ith element of the vector β. In another context, however, β may represent a scalar. All vectors are considered to be column vectors, although we may write a vector as $x = (x_1, x_2, \ldots, x_n)$. Transposition of a vector or a matrix is denoted by the superscript "T".

Uppercase calligraphic Latin letters, such $\mathcal{D}$, $\mathcal{V}$, and $\mathcal{W}$, are generally used to represent either vector spaces or transforms (functionals).

Subscripts generally represent indexes to a larger structure; for example, x_{ij} may represent the $(i, j)^{\text{th}}$ element of a matrix, X. A subscript in parentheses represents an order statistic. A superscript in parentheses represents an iteration; for example, $x_i^{(k)}$ may represent the value of x_i at the kth step of an iterative process.

x_i	The ith element of a structure (including a sample, which is a multiset).
$x_{(i)}$	The ith order statistic.
$x^{(i)}$	The value of x at the ith iteration.

Realizations of random variables and placeholders in functions associated with random variables are usually represented by lowercase letters corresponding to the uppercase letters; thus, ϵ may represent a realization of the random variable E.

A single symbol in an italic font is used to represent a single variable. A Roman font or a special font is often used to represent a standard operator or a standard mathematical structure. Sometimes a string of symbols in a Roman font is used to represent an operator (or a standard function); for example, $\exp(\cdot)$ represents the exponential function. But a string of symbols in an italic font on the same baseline should be interpreted as representing a composition (probably by multiplication) of separate objects; for example, exp represents the product of e, x, and p. Likewise a string of symbols in a Roman font (usually a single symbol) is used to represent a fundamental constant; for example, e represents the base of the natural logarithm, while e represents a variable.

A fixed-width font is used to represent computer input or output, for example,

```
a = bx + sin(c).
```

In computer text, a string of letters or numerals with no intervening spaces or other characters, such as `bx` above, represents a single object, and there is no distinction in the font to indicate the type of object.

Some important mathematical structures and other objects are:

$\mathbb{R}$	The field of reals or the set over which that field is defined.
$\mathbb{R}^d$	The usual d-dimensional vector space over the reals or the set of all d-tuples with elements in $\mathbb{R}$.
$\mathbb{Z}$	The ring of integers or the set over which that ring is defined.

$\mathcal{G}L(n)$ — The general linear group; that is, the group of $n \times n$ full rank (real) matrices with Cayley multiplication.

$\mathcal{O}(n)$ — The orthogonal group; that is, the group of $n \times n$ orthogonal (orthonormal) matrices with Cayley multiplication.

e — The base of the natural logarithm. This is a constant; e may be used to represent a variable. (Note the difference in the font.)

i — The imaginary unit, $\sqrt{-1}$. This is a constant; i may be used to represent a variable. (Note the difference in the font.)

A.2 Computer Number Systems

Computer number systems are used to simulate the more commonly used number systems. It is important to realize that they have different properties, however. Some notation for computer number systems follows.

$\mathbb{F}$ — The set of floating-point numbers with a given precision, on a given computer system, or this set together with the four operators +, -, *, and /. ($\mathbb{F}$ is similar to $\mathbb{R}$ in some useful ways; see Section 10.1.1.)

$\mathbb{I}$ — The set of fixed-point numbers with a given length, on a given computer system, or this set together with the four operators +, -, *, and /. ($\mathbb{I}$ is similar to $\mathbb{Z}$ in some useful ways; see Section 10.1.2 and Table 10.3 on page 400.)

$e_{\min}$ and $e_{\max}$ — The minimum and maximum values of the exponent in the set of floating-point numbers with a given length (see page 381).

$\epsilon_{\min}$ and $\epsilon_{\max}$ — The minimum and maximum spacings around 1 in the set of floating-point numbers with a given length (see page 383).

ϵ or ϵ_{mach} — The machine epsilon, the same as $\epsilon_{\min}$ (see page 383).

$[\cdot]_c$ — The computer version of the object $\cdot$ (see page 393).

NA   Not available; a missing-value indicator.

NaN   Not-a-number (see page 386).

A.3 General Mathematical Functions and Operators

Functions such as sin, max, span, and so on that are commonly associated with groups of Latin letters are generally represented by those letters in a Roman font.

Operators such as d (the differential operator) that are commonly associated with a Latin letter are generally represented by that letter in a Roman font.

Note that some symbols, such as $|\cdot|$, are overloaded; such symbols are generally listed together below.

$\times$ The Cartesian or cross product of sets, or multiplication of elements of a field or ring.

$|x|$ The modulus of the real or complex number x; if x is real, $|x|$ is the absolute value of x.

$\lceil x \rceil$ The ceiling function evaluated at the real number x: $\lceil x \rceil$ is the largest integer less than or equal to x.

$\lfloor x \rfloor$ The floor function evaluated at the real number x: $\lfloor x \rfloor$ is the smallest integer greater than or equal to x.

$x!$ The factorial of x. If x is a positive integer, $x! = x(x-1)\cdots 2\cdot 1$.

$O(f(n))$ Big O; $g(n) = O(f(n))$ means $g(n)/f(n) \to c$ as $n \to \infty$, where c is a nonzero finite constant. In particular, $g(n) = O(1)$ means $g(n)$ is bounded.

$o(f(n))$ Little o; $g(n) = o(f(n))$ means $g(n)/f(n) \to 0$ as $n \to \infty$. In particular, $g(n) = o(1)$ means $g(n) \to 0$.

$o_P(f(n))$ Convergent in probability; $X(n) = o_P(f(n))$ means that for any positive ϵ, $\Pr(|X(n) - f(n)| > \epsilon) \to 0$ as $n \to \infty$.

d   The differential operator.

Δ A perturbation operator; Δx represents a perturbation of x and not a multiplication of x by Δ, even if x is a type of object for which a multiplication is defined.

$\Delta(\cdot,\cdot)$ A real-valued difference function; $\Delta(x,y)$ is a measure of the difference of x and y. For simple objects, $\Delta(x,y) = |x - y|$. For more complicated objects, a subtraction operator may not be defined, and Δ is a generalized difference.

$\tilde{x}$ A perturbation of the object x; $\Delta(x,\tilde{x}) = \Delta x$.

$\tilde{x}$ An average of a sample of objects generically denoted by x.

$\bar{x}$ The mean of a sample of objects generically denoted by x.

$\bar{x}$ The complex conjugate of the complex number x; that is, if $x = r + ic$, then $\bar{x} = r - ic$.

$\mathrm{sign}(x)$ For the vector x, a vector of units corresponding to the signs:
$$
\begin{aligned}
\mathrm{sign}(x)_i &= \ \ \ 1 && \text{if } x_i > 0,\\
&= \ \ \ 0 && \text{if } x_i = 0,\\
&= -1 && \text{if } x_i < 0,
\end{aligned}
$$
with a similar meaning for a scalar.

Special Functions

A good general reference on special functions in mathematics is the venerable book edited by Abramowitz and Stegun (1964), which has been kept in print by Dover Publications.

$\log x$ The natural logarithm evaluated at x.

$\sin x$ The sine evaluated at x (in radians) and similarly for other trigonometric functions.

$\Gamma(x)$ The complete gamma function: $\Gamma(x) = \int_0^\infty t^{x-1}\mathrm{e}^{-t}\mathrm{d}t$. (This is called Euler's integral.) Integration by parts immediately gives the replication formula $\Gamma(x+1) = x\Gamma(x)$, and so if x is a positive integer, $\Gamma(x+1) = x!$, and more generally, $\Gamma(x+1)$ defines $x!$. Direct evaluation of the integral yields $\Gamma(1/2) = \sqrt{\pi}$. Using this and the replication formula, with some manipulation we get for the positive integer j

$$\Gamma(j+1/2) = \frac{1 \cdot 2 \cdots (2j-1)}{2^j}\sqrt{\pi}.$$

The notation $\Gamma_d(x)$ denotes the multivariate gamma function (page 169), although in other literature this notation denotes the incomplete univariate gamma function.

A.4 Linear Spaces and Matrices

$\mathcal{V}(G)$ For the set of vectors (all of the same order) G, the vector space generated by that set.

$\mathcal{V}(X)$ For the matrix X, the vector space generated by the columns of X.

$\dim(\mathcal{V})$ The dimension of the vector space $\mathcal{V}$; that is, the maximum number of linearly independent vectors in the vector space.

$\mathrm{span}(Y)$ For Y either a set of vectors or a matrix, the vector. space $\mathcal{V}(Y)$

$\perp$ Orthogonality relationship (vectors, see page 22; vector spaces, see page 23).

$\mathcal{V}^\perp$ The orthogonal complement of the vector space $\mathcal{V}$ (see page 23).

$\mathcal{N}(A)$ The null space of the matrix A; that is, the set of vectors generated by all solutions, z, of the homogeneous system $Az = 0$; $\mathcal{N}(A)$ is the orthogonal complement of $\mathcal{V}(A^{\mathrm{T}})$.

$\text{tr}(A)$	The trace of the square matrix A, that is, the sum of the diagonal elements.
$\text{rank}(A)$	The rank of the matrix A, that is, the maximum number of independent rows (or columns) of A.
$\rho(A)$	The spectral radius of the matrix A (the maximum absolute value of its eigenvalues).
$A > 0$ $A \geq 0$	If A is a matrix, this notation means, respectively, that each element of A is positive or nonnegative.
$A \succ 0$ $A \succeq 0$	This notation means that A is a symmetric matrix and that it is, respectively, positive definite or nonnegative definite.
A^{T}	For the matrix A, its transpose (also used for a vector to represent the corresponding row vector).
A^{H}	The conjugate transpose, also called the adjoint, of the matrix A; $\quad A^{\mathrm{H}} = \bar{A}^{\mathrm{T}} = \overline{A^{\mathrm{T}}}$.
A^{-1}	The inverse of the square, nonsingular matrix A.
$A^{-\mathrm{T}}$	The inverse of the transpose of the square, nonsingular matrix A.
A^{+}	The g_4 inverse, the Moore-Penrose inverse, or the pseudoinverse of the matrix A (see page 102).
A^{-}	A g_1, or generalized, inverse of the matrix A (see page 102).
$A^{\frac{1}{2}}$	The square root of a nonnegative definite or positive definite matrix A; $(A^{\frac{1}{2}})^2 = A$.
$A^{-\frac{1}{2}}$	The square root of the inverse of a positive definite matrix A; $(A^{-\frac{1}{2}})^2 = A^{-1}$.
$\otimes$	Kronecker multiplication (see page 72).
$\oplus$	The direct sum of two matrices; $A \oplus B = \text{diag}(A, B)$ (see page 47).
$\oplus$	Direct sum of vector spaces (see page 13).

Norms and Inner Products

L_p For real $p \geq 1$, a norm formed by accumulating the pth powers of the moduli of individual elements in an object and then taking the $(1/p)^{\text{th}}$ power of the result (see page 17).

$\|\cdot\|$ In general, the norm of the object $\cdot$.

$\|\cdot\|_p$ In general, the L_p norm of the object $\cdot$.

$\|x\|_p$ For the vector x, the L_p norm

$$\|x\|_p = \left(\sum |x_i|^p \right)^{\frac{1}{p}}$$

(see page 17).

$\|X\|_p$ For the matrix X, the L_p norm

$$\|X\|_p = \max_{\|v\|_p = 1} \|Xv\|_p$$

(see page 130).

$\|X\|_F$ For the matrix X, the Frobenius norm

$$\|X\|_F = \sqrt{\sum_{i,j} x_{ij}^2}$$

(see page 131).

$\langle x, y \rangle$ The inner product or dot product of x and y (see page 15; and see page 74 for matrices).

$\kappa_p(A)$ The L_p condition number of the nonsingular square matrix A with respect to inversion (see page 203).

Matrix Shaping Notation

$\mathrm{diag}(v)$

For the vector v, the diagonal matrix whose nonzero elements are those of v; that is, the square matrix, A, such that $A_{ii} = v_i$ and for $i \neq j$, $A_{ij} = 0$.

$\mathrm{diag}(A_1, A_2, \ldots, A_k)$

The block diagonal matrix whose submatrices along the diagonal are $A_1, A_2, \ldots, A_k$.

$\mathrm{vec}(A)$

The vector consisting of the columns of the matrix A all strung into one vector; if the column vectors of A are $a_1, a_2, \ldots, a_m$, then

$$\mathrm{vec}(A) = (a_1^{\mathrm{T}}, a_2^{\mathrm{T}}, \ldots, a_m^{\mathrm{T}}).$$

$\mathrm{vech}(A)$

For the $m \times m$ symmetric matrix A, the vector consisting of the lower triangular elements all strung into one vector:

$$\mathrm{vech}(A) = (a_{11}, a_{21}, \ldots, a_{m1}, a_{22}, \ldots, a_{m2}, \ldots, a_{mm}).$$

$A_{(i_1,\ldots,i_k)}$

The matrix formed from rows $i_1, \ldots, i_k$ and columns $i_1, \ldots, i_k$ from a given matrix A. This kind of submatrix and the ones below occur often when working with determinants (for square matrices). If A is square, the determinants of these submatrices are called *minors* (see page 51). Because the principal diagonal elements of this matrix are principal diagonal elements of A, it is called a principal submatrix of A. Generally, but not necessarily, $i_j < i_{j+1}$.

$A_{(i_1,\ldots,i_k)(j_1,\ldots,j_l)}$

The submatrix of a given matrix A formed from rows $i_1, \ldots, i_k$ and columns $j_1, \ldots, j_l$ from A.

$A_{(i_1,\ldots,i_k)(*)}$
or
$A_{(*)(j_1,\ldots,j_l)}$

The submatrix of a given matrix A formed from rows $i_1, \ldots, i_k$ and all columns or else all rows and columns $j_1, \ldots, j_l$ from A.

$A_{-(i_1,\ldots,i_k)(j_1,\ldots,j_l)}$

The submatrix formed from a given matrix A by deleting rows $i_1, \ldots, i_k$ and columns $j_1, \ldots, j_l$.

$A_{-(i_1,\dots,i_k)()}$
or
$A_{-()(j_1,\dots,j_l)}$

The submatrix formed from a given matrix A by deleting rows $i_1,\dots,i_k$ (and keeping all columns) or else by deleting columns $j_1,\dots,j_l$ from A.

Notation for Rows or Columns of Matrices

a_{i*}

The vector that corresponds to the ith row of the matrix A. As with all vectors, this is a column vector, so it often appears in the form a_{i*}^{T}.

a_{*j}

The vector that corresponds to the jth column of the matrix A.

Notation Relating to Matrix Determinants

$|A|$

The determinant of the square matrix A, $\quad |A| = \det(A)$.

$\det(A)$

The determinant of the square matrix A, $\quad \det(A) = |A|$.

$|A_{(i_1,\dots,i_k)}|$

A principal minor of a square matrix A; in this case, it is the minor corresponding to the matrix formed from rows $i_1,\dots,i_k$ and columns $i_1,\dots,i_k$ from a given matrix A.

$|A_{-(i)(j)}|$

The minor associated with the $(i,j)^{\text{th}}$ element of a square matrix A.

$a_{(ij)}$

The cofactor associated with the $(i,j)^{\text{th}}$ element of a square matrix A; that is, $a_{(ij)} = (-1)^{i+j}|A_{-(i)(j)}|$.

$\text{adj}(A)$

The adjugate, also called the classical adjoint, of the square matrix A: $\text{adj}(A) = (a_{(ji)})$; that is, the matrix of the same size as A formed from the cofactors of the elements of A^{T}.

Matrix-Vector Differentiation

$\mathrm{d}t$

The differential operator on the scalar, vector, or matrix t. This is an operator; d may be used to represent a variable. (Note the difference in the font.)

g_f
or ∇f

For the scalar-valued function f of a vector variable, the vector whose i^{th} element is $\partial f/\partial x_i$. This is the gradient, also often denoted as g_f.

∇f

For the vector-valued function f of a vector variable, the matrix whose element in position (i, j) is

$$\frac{\partial f_j(x)}{\partial x_i}.$$

This is also written as $\partial f^{\mathrm{T}}/\partial x$ or just as $\partial f/\partial x$. This is the transpose of the Jacobian of f.

J_f

For the vector-valued function f of a vector variable, the Jacobian of f denoted as J_f. The element in position (i, j) is

$$\frac{\partial f_i(x)}{\partial x_j}.$$

This is the transpose of (∇f): $\mathrm{J}_f = (\nabla f)^{\mathrm{T}}$.

H_f
or $\nabla\nabla f$
or $\nabla^2 f$

The Hessian of the scalar-valued function f of a vector variable. The Hessian is the transpose of the Jacobian of the gradient. Except in pathological cases, it is symmetric. The element in position (i, j) is

$$\frac{\partial^2 f(x)}{\partial x_i \partial x_j}.$$

The symbol $\nabla^2 f$ is sometimes also used to denote the diagonal of the Hessian, in which case it is called the Laplacian.

Special Vectors and Matrices

1 or 1_n

A vector (of length n) whose elements are all 1s.

0 or 0_n A vector (of length n) whose elements are all 0s.

I or I_n The $(n \times n)$ identity matrix.

e_i The ith unit vector (with implied length) (see page 12).

Elementary Operator Matrices

E_{pq} The $(p, q)^{\text{th}}$ elementary permutation matrix (see page 63).

E_π The permutation matrix that permutes the rows according to the permutation π.

$E_p(a)$ The pth elementary scalar multiplication matrix (see page 64).

$E_{pq}(a)$ The $(p, q)^{\text{th}}$ elementary axpy matrix (see page 65).

A.5 Models and Data

A form of model used often in statistics and applied mathematics has three parts: a left-hand side representing an object of primary interest; a function of another variable and a parameter, each of which is likely to be a vector; and an adjustment term to make the right-hand side equal the left-hand side. The notation varies depending on the meaning of the terms. One of the most common models used in statistics, the linear regression model with normal errors, is written as

$$Y = \beta^{\text{T}} x + E. \tag{A.1}$$

The adjustment term is a random variable, denoted by an uppercase epsilon. The term on the left-hand side is also a random variable. This model does not represent observations or data. A slightly more general form is

$$Y = f(x; \theta) + E. \tag{A.2}$$

A single observation or a single data item that corresponds to model (A.1) may be written as

$$y = \beta^{\text{T}} x + \epsilon,$$

or, if it is one of several,

$$y_i = \beta^{\mathrm{T}} x_i + \epsilon_i.$$

Similar expressions are used for a single data item that corresponds to model (A.2).

In these cases, rather than being a random variable, ϵ or ϵ_i may be a realization of a random variable, or it may just be an adjustment factor with no assumptions about its origin.

A set of n such observations is usually represented in an n-vector y, a matrix X with n rows, and an n-vector ϵ:

$$y = X\beta + \epsilon$$

or

$$y = f(X;\theta) + \epsilon.$$

B

Solutions and Hints for Selected Exercises

Exercises Beginning on Page 37

2.2. Let one vector space consist of all vectors of the form $(a, 0)$ and the other consist of all vectors of the form $(0, b)$. The vector (a, b) is not in the union if $a \neq 0$ and $b \neq 0$.

2.4. Give a counterexample to the triangle inequality; for example, let $x = (9, 25)$ and $y = (16, 144)$.

2.6a. We first observe that if $\|x\|_p = 0$ or $\|y\|_q = 0$, we have $x = 0$ or $y = 0$, and so the inequality is satisfied because both sides are 0; hence, we need only consider the case $\|x\|_p > 0$ and $\|y\|_q > 0$. We also observe that if $p = 1$ or $q = 1$, we have the Manhattan and Chebyshev norms and the inequality is satisfied; hence we need only consider the case $1 < p < \infty$. Now, for p and q as given, for any numbers a_i and b_i, there are numbers s_i and t_i such that $|a_i| = e^{s_i/p}$ and $|b_i| = e^{t_i/q}$. Because e^x is a convex function, we have $e^{s_i/p + t_i/q} \leq \frac{1}{p} e_i^s + \frac{1}{q} e_i^t$, or

$$a_i b_i \leq |a_i||b_i| \leq |a_i|^p/p + |b_i|^q/q.$$

Now let

$$a_i = \frac{x_i}{\|x\|_p} \quad \text{and} \quad b_i = \frac{y_i}{\|y\|_q},$$

and so

$$\frac{x_i}{\|x\|_p} \frac{y_i}{\|y\|_q} \leq \frac{1}{p} \frac{|x_i|^p}{\|x\|_p^p} + \frac{1}{q} \frac{|y_i|^q}{\|y\|_q^q}.$$

Now, summing these equations over i, we have

$$\frac{\langle x, y \rangle}{\|x\|_p \|y\|_q} \leq \frac{1}{p} \frac{\|x\|_p^p}{\|x\|_p^p} + \frac{1}{q} \frac{\|y\|_q^q}{\|y\|_q^q}$$
$$= 1.$$

Hence, we have the desired result.

As we see from this proof, the inequality is actually a little stronger than stated. If we define u and v by $u_i = |x_i|$ and $v_i = |y_i|$, we have

$$\langle x, y \rangle \leq \langle u, v \rangle \leq \|x\|_p \|y\|_q.$$

We observe that equality occurs if and only if

$$\left(\frac{|x_i|}{\|x\|_p} \right)^{\frac{1}{q}} = \left(\frac{|y_i|}{\|y\|_q} \right)^{\frac{1}{p}}$$

and

$$\operatorname{sign}(x_i) = \operatorname{sign}(y_i)$$

for all i.

We note a special case by letting $y = 1$:

$$\bar{x} \leq \|x\|_p,$$

and with $p = 2$, we have a special case of the Cauchy-Schwarz inequality,

$$n\bar{x}^2 \leq \|x\|_2^2,$$

which guarantees that $V(x) \geq 0$.

2.6b. Using the triangle inequality for the absolute value, we have $|x_i + y_i| \leq |x_i| + |y_i|$. This yields the result for $p = 1$ and $p = \infty$ (in the limit). Now assume $1 < p < \infty$. We have

$$\|x + y\|_p^p \leq \sum_{i=1}^{n} |x_i + y_i|^{p-1} |x_i| + \sum_{i=1}^{n} |x_i + y_i|^{p-1} |y_i|.$$

Now, letting $q = p/(p-1)$, we apply Hölder's inequality to each of the terms on the right:

$$\sum_{i=1}^{n} |x_i + y_i|^{p-1} |x_i| \leq \left(\sum_{i=1}^{n} |x_i + y_i|^{(p-1)q} \right)^{\frac{1}{q}} \left(\sum_{i=1}^{n} |x_i|^p \right)^{\frac{1}{p}}$$

and

$$\sum_{i=1}^{n} |x_i + y_i|^{p-1} |x_i| \leq \left(\sum_{i=1}^{n} |x_i + y_i|^{(p-1)q} \right)^{\frac{1}{q}} \left(\sum_{i=1}^{n} |y_i|^p \right)^{\frac{1}{p}},$$

so

$$\sum_{i=1}^{n} |x_i + y_i|^p \leq \left(\sum_{i=1}^{n} |x_i + y_i|^{(p-1)q} \right)^{\frac{1}{q}} \left(\left(\sum_{i=1}^{n} |x_i|^p \right)^{\frac{1}{p}} + \left(\sum_{i=1}^{n} |y_i|^p \right)^{\frac{1}{p}} \right)$$

or, because $(p-1)q = p$ and $1 - \frac{1}{q} = \frac{1}{p}$,

$$\left(\sum_{i=1}^{n} |x_i + y_i|^p\right)^{\frac{1}{p}} \le \left(\sum_{i=1}^{n} |x_i|^p\right)^{\frac{1}{p}} + \left(\sum_{i=1}^{n} |y_i|^p\right)^{\frac{1}{p}},$$

which is the same as

$$\|x + y\|_p \le \|x\|_p + \|y\|_p,$$

the triangle inequality.

2.13e. In $\mathbb{R}^3$,

$$\text{angle}(x, y) = \sin^{-1}\left(\frac{\|x \times y\|}{\|x\|\,\|y\|}\right).$$

Because $x \times y = -y \times x$, this allows us to determine the angle from x to y; that is, the *direction* within $(-\pi, \pi]$ in which x would be rotated to y.

2.15. Just consider the orthogonal vectors $x = (1,0)$ and $y = (0,1)$. The centered vectors are $x_c = (\frac{1}{2}, -\frac{1}{2})$ and $y_c = (-\frac{1}{2}, \frac{1}{2})$. The angle between the uncentered vectors is $\pi/2$, while that between the centered vectors is π.

Exercises Beginning on Page 140

3.16. For property 7, let c be a nonzero eigenvalue of AB. Then there exists v ($\ne 0$) such that $ABv = cv$, that is, $BABv = Bcv$. But this means $BAw = cw$, where $w = Bv \ne 0$ (because $ABv \ne 0$) and so c is an eigenvalue of BA. We use the same argument starting with an eigenvalue of BA. For square matrices, there are no other eigenvalues, so the set of eigenvalues is the same.

For property 8, see the discussion of similarity transformations on page 114.

3.27. Let A and B be such that AB is defined.

$$\|AB\|_F^2 = \sum_{ij} \left|\sum_k a_{ik} b_{kj}\right|^2$$

$$\le \sum_{ij} \left(\sum_k a_{ik}^2\right)\left(\sum_k b_{kj}^2\right) \qquad \text{(Cauchy-Schwarz)}$$

$$= \left(\sum_{i,k} a_{ik}^2\right)\left(\sum_{k,j} b_{kj}^2\right)$$

$$= \|A\|_F^2 \|B\|_F^2.$$

Exercises Beginning on Page 169

4.5b. The first step is to use the trick of equation (3.63), $x^{\mathrm{T}}Ax = \mathrm{tr}(Axx^{\mathrm{T}})$, again to undo the earlier expression, and write the last term in equation (4.36) as

$$-\frac{n}{2}\mathrm{tr}\left(\Sigma^{-1}(\bar{y}-\mu)(\bar{y}-\mu)^{\mathrm{T}}\right) = -\frac{n}{2}(\bar{y}-\mu)\Sigma^{-1}(\bar{y}-\mu)^{\mathrm{T}}.$$

Now Σ^{-1} is positive definite, so $(\bar{y}-\mu)\Sigma^{-1}(\bar{y}-\mu)^{\mathrm{T}} \geq 0$ and hence is minimized for $\hat{\mu} = \bar{y}$. Decreasing this term increases the value of $l(\mu, \Sigma; y)$, and so $l(\hat{\mu}, \Sigma; y) \geq l(\mu, \Sigma; y)$ for all positive definite Σ^{-1}. Now, we consider the other term. Let $A = \sum_{i=1}^{n}(y_i - \bar{y})(y_i - \bar{y})^{\mathrm{T}}$. The first question is whether A is positive definite. We will refer to a text on multivariate statistics for the proof that A is positive definite with probability 1 (see Muirhead, 1982, for example). We have

$$l(\hat{\mu}, \Sigma; y) = c - \frac{n}{2}\log|\Sigma| - \frac{1}{2}\mathrm{tr}\left(\Sigma^{-1}A\right)$$
$$= c - \frac{n}{2}\left(\log|\Sigma| + \mathrm{tr}\left(\Sigma^{-1}A/n\right)\right).$$

Because c is constant, the function is maximized at the minimum of the latter term subject to Σ being positive definite, which, as shown for expression (4.32), occurs at $\widehat{\Sigma} = A/n$.

4.8. $2^{dn/2}\Gamma_d(n/2)|\Sigma|^{n/2}$.

Make the change of variables $W = 2\Sigma^{\frac{1}{2}}Y\Sigma^{\frac{1}{2}}$, determine the Jacobian, and integrate.

Exercises Beginning on Page 198

5.2. The R code that will produce the graph is

```
x<-c(0,1)
y<-c(0,1)
z<-matrix(c(0,0,1,1),nrow=2)
persp(x, y, z, theta = 45, phi = 30)
bottom<-c(.5,0,0,1)%*%trans
top<-c(.5,1,1,1)%*%trans
xends<-c(top[,1]/top[,4],bottom[,1]/bottom[,4])
yends<-c(top[,2]/top[,4],bottom[,2]/bottom[,4])
lines(xends,yends,lwd=2)
```

Exercises Beginning on Page 238

6.1. First, show that

$$\max_{x\neq 0}\frac{\|Ax\|}{\|x\|} = \left(\min_{x\neq 0}\frac{\|A^{-1}x\|}{\|x\|}\right)^{-1}$$

and

$$\max_{x\neq0}\frac{\|A^{-1}x\|}{\|x\|}=\left(\min_{x\neq0}\frac{\|Ax\|}{\|x\|}\right)^{-1}.$$

6.2a. The matrix at the first elimination is

$$\begin{bmatrix} 2\,5\,3 & 19 \\ 1\,4\,1 & 12 \\ 1\,2\,2 & 9 \end{bmatrix}.$$

The solution is $(3,2,1)$.

6.2b. The matrix at the first elimination is

$$\begin{bmatrix} 5\,2\,3 & 19 \\ 4\,1\,1 & 12 \\ 2\,1\,2 & 9 \end{bmatrix},$$

and x_1 and x_2 have been interchanged.

6.2c.

$$D = \begin{bmatrix} 2\,0\,0 \\ 0\,4\,0 \\ 0\,0\,2 \end{bmatrix},$$

$$L = \begin{bmatrix} 0\,0\,0 \\ 1\,0\,0 \\ 1\,2\,0 \end{bmatrix},$$

$$U = \begin{bmatrix} 0 & -5 & -3 \\ 0 & 0 & -1 \\ 0 & 0 & -2 \end{bmatrix},$$

$$\rho((D+L)^{-1}U) = 0.9045.$$

6.4a. $nm(m+1) - m(m+1)/2$. (Remember $A^{\mathrm{T}}A$ is symmetric.)

6.4g. Using the normal equations with the Cholesky decomposition requires only about half as many flops as the QR, when n is much larger than m. The QR method oftens yields better accuracy, however.

Exercises Beginning on Page 256

7.1a. 1.

7.1b. 1.

7.1d. 1. (All that was left was to determine the probability that $c_n \neq 0$ and $c_{n-1} \neq 0$.)

7.2a. 11.6315.

7.3.

$$\begin{bmatrix} 3.08 & -0.66 & 0 & 0 \\ -0.66 & 4.92 & -3.27 & 0 \\ 0 & -3.27 & 7.00 & -3.74 \\ 0 & 0 & -3.74 & 7.00 \end{bmatrix}.$$

Exercises Beginning on Page 317

8.10a.

$$p(c) = c^m - \alpha_1 c^{m-1} - \alpha_2 \sigma_1 c^{m-2} - \alpha_3 \sigma_1 \sigma_2 c^{m-3} - \cdots - \alpha_m \sigma_1 \sigma_2 \cdots \sigma_{m-1}.$$

8.10b. Define

$$f(c) = 1 - \frac{p(c)}{c^m}.$$

This is a monotone decreasing continuous function in c, with $f(c) \to \infty$ as $c \to 0_+$ and $f(c) \to 0$ as $c \to \infty$. Therefore, there is a unique value c_* for which $f(c_*) = 1$. The uniqueness also follows from Descartes' rule of signs, which states that the maximum number of positive roots of a polynomial is the number of sign changes of the coefficients, and in the case of the polynomial $p(c)$, this is one.

8.12. $(-1)^{\lfloor n/2 \rfloor} n^n$, where $\lfloor \cdot \rfloor$ is the floor function (the greatest integer function). For $n = 1, 2, 3, 4$, the determinants are $1, -4, -27, 256$.

Exercises Beginning on Page 365

9.1. 1. This is because the subspace that generates a singular matrix is a lower dimensional space than the full sample space, and so its measure is 0.

9.4d. Assuming W is positive definite, we have

$$\widehat{\beta}_{W,C} = (X^\mathrm{T} W X)^{-1} X^\mathrm{T} W y +$$

$$(X^\mathrm{T} W X)^{-1} L^\mathrm{T} (L(X^\mathrm{T} W X)^+ L^\mathrm{T})^{-1} (c - L(X^\mathrm{T} W X)^+ X^\mathrm{T} W y).$$

9.11. Let $X = [X_\mathrm{i} \mid X_\mathrm{o}]$ and $Z = X_\mathrm{o}^\mathrm{T} X_\mathrm{o} - X_\mathrm{o}^\mathrm{T} X_\mathrm{i} (X_\mathrm{i}^\mathrm{T} X_\mathrm{i})^{-1} X_\mathrm{i}^\mathrm{T} X_\mathrm{o}$. Note that $X_\mathrm{o}^\mathrm{T} X_\mathrm{i} = X_\mathrm{i}^\mathrm{T} X_\mathrm{o}$. We have

$$X_\mathrm{i}^\mathrm{T} X (X^\mathrm{T} X)^{-1} X^\mathrm{T}$$

$$= X_\mathrm{i}^\mathrm{T} [X_\mathrm{i} \mid X_\mathrm{o}] \begin{bmatrix} X_\mathrm{i}^\mathrm{T} X_\mathrm{i} & X_\mathrm{i}^\mathrm{T} X_\mathrm{o} \\ X_\mathrm{o}^\mathrm{T} X_\mathrm{i} & X_\mathrm{o}^\mathrm{T} X_\mathrm{o} \end{bmatrix}^{-1} [X_\mathrm{i} \mid X_\mathrm{o}]^\mathrm{T}$$

$$= \begin{bmatrix} X_\mathrm{i}^\mathrm{T} X_\mathrm{i} \mid X_\mathrm{i}^\mathrm{T} X_\mathrm{o} \end{bmatrix} \begin{bmatrix} (X_\mathrm{i}^\mathrm{T} X_\mathrm{i})^{-1} - (X_\mathrm{i}^\mathrm{T} X_\mathrm{i})^{-1}(X_\mathrm{o}^\mathrm{T} X_\mathrm{i}) Z^{-1}(X_\mathrm{i}^\mathrm{T} X_\mathrm{o})(X_\mathrm{i}^\mathrm{T} X_\mathrm{i})^{-1} & -(X_\mathrm{i}^\mathrm{T} X_\mathrm{i})^{-1}(X_\mathrm{i}^\mathrm{T} X_\mathrm{o}) Z^{-1} \\ -Z^{-1}(X_\mathrm{o}^\mathrm{T} X_\mathrm{i})(X_\mathrm{i}^\mathrm{T} X_\mathrm{i})^{-1} & Z^{-1} \end{bmatrix}$$

$$\begin{bmatrix} X_\mathrm{i}^\mathrm{T} \\ X_\mathrm{o}^\mathrm{T} \end{bmatrix}$$

$$= \begin{bmatrix} I - (X_\mathrm{o}^\mathrm{T} X_\mathrm{i}) Z^{-1}(X_\mathrm{i}^\mathrm{T} X_\mathrm{o})(X_\mathrm{i}^\mathrm{T} X_\mathrm{i})^{-1} - X_\mathrm{i}^\mathrm{T} X_\mathrm{o} Z^{-1}(X_\mathrm{o}^\mathrm{T} X_\mathrm{i})(X_\mathrm{i}^\mathrm{T} X_\mathrm{i})^{-1} & -X_\mathrm{i}^\mathrm{T} X_\mathrm{o} Z^{-1} + X_\mathrm{i}^\mathrm{T} X_\mathrm{o} Z^{-1} \end{bmatrix}$$

$$\begin{bmatrix} X_\mathrm{i}^\mathrm{T} \\ X_\mathrm{o}^\mathrm{T} \end{bmatrix}$$

$$= X_\mathrm{i}^\mathrm{T},$$

9.13. One possibility is

$$\begin{bmatrix} 20 & 100 \\ 5 & 25 \\ 5 & 25 \\ 10 & \text{NA} \\ 10 & \text{NA} \\ 10 & \text{NA} \\ \text{NA} & 10 \\ \text{NA} & 10 \\ \text{NA} & 10 \end{bmatrix}.$$

The variance-covariance matrix computed from all pairwise complete observations is

$$\begin{bmatrix} 30 & 375 \\ 375 & 1230 \end{bmatrix},$$

while that computed only from complete cases is

$$\begin{bmatrix} 75 & 375 \\ 375 & 1875 \end{bmatrix}.$$

The correlation matrix computed from all pairwise complete observations is

$$\begin{bmatrix} 1.00 & 1.95 \\ 1.95 & 1.00 \end{bmatrix}.$$

Note that this example is not a pseudo-correlation matrix.

In the R software system, the `cov` and `cor` functions have an argument called "use", which can take the values "all.obs", "complete.obs", or "pairwise.complete.obs". The value "all.obs" yields an error if the data matrix contains any missing values. In `cov`, the values "complete.obs" and "pairwise.complete.obs" yield the variance-covariances shown above. The function `cor` with `use="pairwise.complete.obs"` yields

$$\begin{bmatrix} 1.00 & 1.00 \\ 1.00 & 1.00 \end{bmatrix}.$$

However, if `cov` is invoked with `use="pairwise.complete.obs"` and the function `cov2cor` is applied to the result, the correlations are 1.95, as in the first correlation matrix above.

9.16. This is an open question. If you get a proof of convergence, submit it for publication. You may wish to try several examples and observe the performance of the intermediate steps. I know of no case in which the method has not converged.

9.18b. Starting with the correlation matrix given above as a possible solution for Exercise 9.13, four iterations of equation (9.52) using $\delta = 0.05$ and $f(x) = \tanh(x)$ yield

$$\begin{bmatrix} 1.00 & 0.997 \\ 0.997 & 1.00 \end{bmatrix}.$$

9.20. We can develop a recursion for p_{11}^t based on p_{11}^{t-1} and p_{12}^{t-1},

$$p_{11}^t = p_{11}^{t-1}(1-\alpha) + p_{12}^{t-1}\beta,$$

and because $p_{11} + p_{12} = 1$, we have $p_{11}^t = p_{11}^{t-1}(1-\alpha-\beta) + \beta$. Putting this together, we have

$$\lim_{t\to\infty} P = \begin{bmatrix} \beta/(\alpha+\beta) & \alpha/(\alpha+\beta) \\ \beta/(\alpha+\beta) & \alpha/(\alpha+\beta) \end{bmatrix},$$

and so the limiting (and invariant) distribution is $\pi_s = (\beta/(\alpha+\beta), \alpha/(\alpha+\beta))$.

9.21c. From the exponential growth, we have $N^{(T)} = N^{(0)}e^{rT}$; hence,

$$r = \frac{1}{T} \log\left(N^{(T)}/N^{(0)}\right) = \frac{1}{T}\log(r_0).$$

Exercises Beginning on Page 422

10.1a. The computations do not overflow. The first floating-point number x such that $x+1 = x$ is

$$0.10\cdots0 \times b^{p+1}.$$

Therefore, the series converges at the value of i such that $i(i+1)/2 = x$. Now solve for i.

10.2. The function is $\log(n)$, and Euler's constant is $0.57721\ldots$.

10.5. 2^{-56}. (The standard has 53 bits normalized, so the last bit is 2^{-55}, and half of that is 2^{-56}.)

10.6a. Normalized: $2b^{p-1}(b-1)(e_{\max} - e_{\min} + 1) + 1$.
  Nonnormalized: $2b^{p-1}(b-1)(e_{\max} - e_{\min} + 1) + 1 + 2b^{p-1}$.

10.6b. Normalized: $b^{e_{\min}-1}$.
  Nonnormalized: $b^{e_{\min}-p}$.

10.6c. $1 + b^{-p+1}$ or $1 + b^{-p}$ when $b = 2$ and the first bit is hidden.

10.6d. b^p.

10.6e. 22.

10.11. First of all, we recognize that the full sum in each case is 1. We therefore accumulate the sum from the direction in which there are fewer terms. After computing the first term from the appropriate direction, take a logarithm to determine a scaling factor, say s^k. (This term will be the smallest in the sum.) Next, proceed to accumulate terms until the sum is of a different order of magnitude than the next term. At that point, perform a scale adjustment by dividing by s. Resume summing, making similar scale adjustments as necessary, until the limit of the summation is reached.

10.13. The result is close to 1.
  What is relevant here is that numbers close to 1 have only a very few digits of accuracy; therefore, it would be better to design this program so that it returns $1 - \Pr(X \le x)$ (the "significance level"). The purpose and the anticipated use of a program determine how it should be designed.

10.16a. 2.

10.16b. 0.

10.16c. No (because the operations in the "for" loop are not chained).

10.17c.

$$a = x_1$$
$$b = y_1$$
$$s = 0$$
$$\text{for } i = 2, n$$
$$\{$$
$$d = (x_i - a)/i$$
$$e = (y_i - b)/i$$
$$a = d + a$$
$$b = e + b$$
$$s = i(i - 1)de + s$$
$$\}.$$

10.20. 1. No; 2. Yes; 3. No; 4. No.

10.21. A very simple example is

$$\begin{bmatrix} 1 & 1 + \epsilon \\ 1 & 1 \end{bmatrix},$$

where $\epsilon < b^{-p}$, because in this case the matrix stored in the computer would be singular. Another example is

$$\begin{bmatrix} 1 & a(1 + \epsilon) \\ a(1 + \epsilon) & a^2(1 + 2\epsilon) \end{bmatrix},$$

where ϵ is the machine epsilon.

Exercises Beginning on Page 441

11.2a. O(nk).

11.2c. At each successive stage in the fan-in, the number of processors doing the additions goes down by approximately one-half.

If $p \approx k$, then O$(n \log k)$ (fan-in on one element of c at a time)

If $p \approx nk$, then O$(\log k)$ (fan-in on all elements of c simultaneously)

If p is a fixed constant smaller than k, the order of time does not change; only the multiplicative constant changes.

Notice the difference in the order of time and the order of the number of computations. Often there is very little that can be done about the order of computations.

11.2d. Because in a serial algorithm the magnitudes of the summands become more and more different. In the fan-in, they are more likely to remain relatively equal. Adding magnitudes of different quantities results in benign roundoff, but many benign roundoffs become bad. (This is not catastrophic cancellation.) Clearly, if all elements are nonnegative, this argument would hold. Even if the elements are randomly distributed, there is likely to be a drift in the sum (this can be thought of as a random walk). There is *no difference* in the number of computations.

11.2e. Case 1: $p \approx n$. Give each c_i a processor – do an outer loop on each. This would likely be more efficient because all processors are active at once.

Case 2: $p \approx nk$. Give each $a_{ij}b_j$ a processor – fan-in for each. This would be the same as the other.

If p is a fixed constant smaller than n, set it up as in Case 1, using n/p groups of c_i's.

11.2f. If $p \approx n$, then $O(k)$.

If $p \approx nk$, then $O(\log k)$.

If p is some small fixed constant, the order of time does not change; only the multiplicative constant changes.

Exercises Beginning on Page 475

12.1. Here is a recursive Matlab function for the Strassen algorithm due to Coleman and Van Loan. When it uses the Strassen algorithm, it requires the matrices to have even dimension.

```
    function C = strass(A,B,nmin)
%
% Strassen matrix multiplication C=AB
%       A, B must be square and of even dimension
% From Coleman and Van Loan
% If n <= nmin, the multiplication is done conventionally
%
    [n n ] = size(A);
    if n <= nmin
       C = A * B;    % n is small, get C conventionally
    else
       m = n/2; u = 1:m; v = m+1:n;
       P1 = strass(A(u,u)+A(v,v), B(u,u)+B(v,v), nmin);
       P2 = strass(A(v,u)+A(v,v), B(u,u),          nmin);
       P3 = strass(A(u,u),        B(u,v)-B(v,v), nmin);
       P4 = strass(A(v,v),        B(v,u)-B(u,u), nmin);
       P5 = strass(A(u,u)+A(u,v), B(v,v),          nmin);
       P6 = strass(A(v,u)-A(u,u), B(u,u)+B(u,v), nmin);
       P7 = strass(A(u,v)-A(v,v), B(v,u)+B(v,v), nmin);
       C = [P1+P4-P5+P7 P3+P5; P2+P4 P1+P3-P2+P6];
    end
```

12.3a.

```
real a(4,3)
data a/3.,6.,8.,2.,5.,1.,6.,3.,6.,2.,7.,1./
n = 4
m = 3
x1 = a(2,2) ! Temporary variables must be used because of
x2 = a(4,2) ! the side effects of srotg.
call srotg(x1, x2,, c, s)
call srot(m, a(2,1), n, a(4,1), n, c, s)
print *, c, s
print *, a
end
```

This yields 0.3162278 and 0.9486833 for c and s. The transformed matrix
is

$$\begin{bmatrix} 3.000000 & 5.000000 & 6.000000 \\ 3.794733 & 3.162278 & 1.581139 \\ 8.000000 & 6.000000 & 7.000000 \\ -5.059644 & -0.00000002980232 & -1.581139 \end{bmatrix}.$$

12.5. 10.7461941829033 and 10.7461941829034.

Bibliography

The references that I have cited in this text are generally traditional books, journal articles, or compact discs. This usually means that the material has been reviewed by someone other than the author. It also means that the author possibly has newer thoughts on the same material. The Internet provides a mechanism for the dissemination of large volumes of information that can be updated readily. The ease of providing material electronically is also the source of the major problem with the material: it is often half-baked and has not been reviewed critically. Another reason that I have refrained from making frequent reference to material available over the Internet is the unreliability of some sites. The average life of a Web site is measured in weeks.

For statistics, one of the most useful sites on the Internet is the electronic repository `statlib`, maintained at Carnegie Mellon University, which contains programs, datasets, and other items of interest. The URL is

> `http://lib.stat.cmu.edu.`

The collection of algorithms published in *Applied Statistics* is available in `statlib`. These algorithms are sometimes called the *ApStat* algorithms.

Another very useful site for scientific computing is `netlib`, which was established by research workers at AT&T (now Alcatel-Lucent) Bell Laboratories and national laboratories, primarily Oak Ridge National Laboratories. The URL is

> `http://www.netlib.org`

The *Collected Algorithms of the ACM (CALGO)*, which are the Fortran, C, and Algol programs published in *ACM Transactions on Mathematical Software* (or in *Communications of the ACM* prior to 1975), are available in `netlib` under the TOMS link.

A wide range of software is used in the computational sciences. Some of the software is produced by a single individual who is happy to share the software, sometimes for a fee, but who has no interest in maintaining it. At the other extreme is software produced by large commercial companies whose continued existence depends on a process of production, distribution, and maintenance of the software. Information on much of the software can be obtained from GAMS, as we mentioned at the begining of Chapter 12. Some of the free software can be obtained from `statlib` or `netlib`.

The following bibliography obviously covers a wide range of topics in statistical computing and computational statistics. Except for a few of the general references, all of these entries have been cited in the text.

The purpose of this bibliography is to help the reader get more information; hence I eschew "personal communications" and references to technical reports that may or may not exist. Those kinds of references are generally for the author rather than for the reader.

Abramowitz, Milton, and Irene A. Stegun (Editors) (1964), *Handbook of Mathematical Functions with Formulas, Graphs, and Mathematical Tables*, National Bureau of Standards (NIST), Washington. (Reprinted in 1965 by Dover Publications, Inc., New York.)

Alefeld, Göltz, and Jürgen Herzberger (1983), *Introduction to Interval Computation*, Academic Press, New York.

Amdahl, G. M. (1967), Validity of the single processor approach to achieving large-scale computing capabilities, *Proceedings of the American Federation of Information Processing Societies* **30**, Washington, D.C., 483–485.

Ammann, Larry, and John Van Ness (1988), A routine for converting regression algorithms into corresponding orthogonal regression algorithms, *ACM Transactions on Mathematical Software* **14**, 76–87.

Ammann, Larry, and John Van Ness (1989), Standard and robust orthogonal regression, *Communications in Statistics — Simulation and Computation* **18**, 145–162.

Anda, Andrew A., and Haesun Park (1994), Fast plane rotations with dynamic scaling, *SIAM Journal of Matrix Analysis and Applications* **15**, 162–174.

Anda, Andrew A., and Haesun Park (1996), Self-scaling fast rotations for stiff least squares problems, *Linear Algebra and Its Applications* **234**, 137–162.

Anderson, E.; Z. Bai; C. Bischof; L. S. Blackford; J. Demmel; J. Dongarra; J. Du Croz; A. Greenhaum; S. Hammarling; A. McKenney; and D. Sorensen (2000), *LAPACK Users' Guide*, third edition, Society for Industrial and Applied Mathematics, Philadelphia.

Anderson, T. W. (2003), *An Introduction to Multivariate Statistical Analysis*, third edition, John Wiley and Sons, New York.

ANSI (1978), *American National Standard for Information Systems — Programming Language FORTRAN*, Document X3.9-1978, American National Standards Institute, New York.

ANSI (1989), *American National Standard for Information Systems — Programming Language C*, Document X3.159-1989, American National Standards Institute, New York.

ANSI (1992), *American National Standard for Information Systems — Programming Language Fortran-90*, Document X3.9-1992, American National Standards Institute, New York.

ANSI (1998), *American National Standard for Information Systems — Programming Language C++*, Document ISO/IEC 14882-1998, American National Standards Institute, New York.

Atkinson, A. C., and A. N. Donev (1992), *Optimum Experimental Designs*, Oxford University Press, Oxford, United Kingdom.

Bailey, David H. (1993), Algorithm 719: Multiprecision translation and execution of FORTRAN programs, *ACM Transactions on Mathematical Software* **19**, 288–319.

Bailey, David H. (1995), A Fortran 90-based multiprecision system, *ACM Transactions on Mathematical Software* **21**, 379–387.

Bailey, David H.; King Lee; and Horst D. Simon (1990), Using Strassen's algorithm to accelerate the solution of linear systems, *Journal of Supercomputing* **4**, 358–371.

Barker, V. A.; L. S. Blackford; J. Dongarra; J. Du Croz; S. Hammarling; M. Marinova; J. Wasniewsk; and P. Yalamov (2001), *LAPACK95 Users' Guide*, Society for Industrial and Applied Mathematics, Philadelphia.

Barrett, R.; M. Berry; T. F. Chan; J. Demmel; J. Donato; J. Dongarra; V. Eijkhout; R. Pozo; C. Romine; and H. Van der Vorst (1994), *Templates for the Solution of Linear Systems: Building Blocks for Iterative Methods*, second edition, Society for Industrial and Applied Mathematics, Philadelphia.

Basilevsky, Alexander (1983), *Applied Matrix Algebra in the Statistical Sciences*, North Holland, New York.

Beaton, Albert E.; Donald B. Rubin; and John L. Barone (1976), The acceptability of regression solutions: Another look at computational accuracy, *Journal of the American Statistical Association* **71**, 158–168.

Benzi, Michele (2002), Preconditioning techniques for large linear systems: A survey, *Journal of Computational Physics* **182**, 418–477.

Bickel, Peter J., and Joseph A. Yahav (1988), Richardson extrapolation and the bootstrap, *Journal of the American Statistical Association* **83**, 387–393.

Bindel, David; James Demmel; William Kahan; and Osni Marques (2002), On computing Givens rotations reliably and efficiently, *ACM Transactions on Mathematical Software* **28**, 206–238.

Birkhoff, Garrett, and Surender Gulati (1979), Isotropic distributions of test matrices, *Journal of Applied Mathematics and Physics (ZAMP)* **30**, 148–158.

Bischof, Christian H. (1990), Incremental condition estimation, *SIAM Journal of Matrix Analysis and Applications* **11**, 312–322.

Bischof, Christian H., and Gregorio Quintana-Ortí (1998a), Computing rank-revealing QR factorizations, *ACM Transactions on Mathematical Software* **24**, 226–253.

Bischof, Christian H., and Gregorio Quintana-Ortí (1998b), Algorithm 782: Codes for rank-revealing QR factorizations of dense matrices, *ACM Transactions on Mathematical Software* **24**, 254–257.

Björck, Åke (1967), Solving least squares problems by Gram-Schmidt orthogonalization, *BIT* **7**, 1–21.

Björck, Åke (1996), *Numerical Methods for Least Squares Problems*, Society for Industrial and Applied Mathematics, Philadelphia.

Blackford, L. S.; J. Choi; A. Cleary; E. D'Azevedo; J. Demmel; I. Dhillon; J. Dongarra; S. Hammarling; G. Henry; A. Petitet; K. Stanley; D. Walker; and R. C. Whaley (1997a), *ScaLAPACK Users' Guide*, Society for Industrial and Applied Mathematics, Philadelphia.

Blackford, L. S.; A. Cleary; A. Petitet; R. C. Whaley; J. Demmel; I. Dhillon; H. Ren; K. Stanley; J. Dongarra; and S. Hammarling (1997b), Practical experience in the numerical dangers of heterogeneous computing, *ACM Transactions on Mathematical Software* **23**, 133–147.

Blackford, L. Susan; Antoine Petitet; Roldan Pozo; Karin Remington; R. Clint Whaley; James Demmel; Jack Dongarra; Iain Duff; Sven Hammarling; Greg Henry; Michael Heroux; Linda Kaufman; and Andrew Lumsdaine (2002), An updated set of basic linear algebra subprograms (BLAS), *ACM Transactions on Mathematical Software* **28**, 135–151.

Brent, Richard P. (1978), A FORTRAN multiple-precision arithmetic package, *ACM Transactions on Mathematical Software* **4**, 57–70.

Brown, Peter N., and Homer F. Walker (1997), GMRES on (nearly) singular systems, *SIAM Journal of Matrix Analysis and Applications* **18**, 37–51.

Bunch, James R., and Linda Kaufman (1977), Some stable methods for calculating inertia and solving symmetric linear systems, *Mathematics of Computation* **31**, 163–179.

Calvetti, Daniela (1991), Roundoff error for floating point representation of real data, *Communications in Statistics* **20**, 2687–2695.

Campbell, S. L., and C. D. Meyer, Jr. (1991), *Generalized Inverses of Linear Transformations*, Dover Publications, Inc., New York.

Carmeli, Moshe (1983), *Statistical Theory and Random Matrices*, Marcel Dekker, Inc., New York.

Carrig, James J., Jr., and Gerard G. L. Meyer (1997), Efficient Householder QR factorization for superscalar processors, *ACM Transactions on Mathematical Software* **23**, 362–378.

Chaitin-Chatelin, Françoise, and Valérie Frayssé (1996), *Lectures on Finite Precision Computations*, Society for Industrial and Applied Mathematics, Philadelphia.

Chambers, John M. (1998), *Programming with Data: A Guide to the S Language*, Springer-Verlag, New York.

Chan, T. F. (1982a), An improved algorithm for computing the singular value decomposition, *ACM Transactions on Mathematical Software* **8**, 72–83.

Chan, T. F. (1982b), Algorithm 581: An improved algorithm for computing the singular value decomposition, *ACM Transactions on Mathematical Software* **8**, 84–88.

Chan, T. F.; G. H. Golub; and R. J. LeVeque (1982), Updating formulae and a pairwise algorithm for computing sample variances, in *Compstat 1982: Proceedings in Computational Statistics* (edited by H. Caussinus, P. Ettinger, and R. Tomassone), Physica-Verlag, Vienna, 30–41.

Chan, Tony F.; Gene H. Golub; and Randall J. LeVeque (1983), Algorithms for computing the sample variance: Analysis and recommendations, *The American Statistician* **37**, 242–247.

Chan, Tony F., and John Gregg Lewis (1979), Computing standard deviations: Accuracy, *Communications of the ACM* **22**, 526–531.

Chu, Moody T. (1991), Least squares approximation by real normal matrices with specified spectrum, *SIAM Journal on Matrix Analysis and Applications* **12**, 115–127.

Chung, Fan R. K. (1997), *Spectral Graph Theory*, American Mathematical Society, Providence, Rhode Island.

Čížková, Lenka, and Pavel Čížek (2004), Numerical linear algebra, in *Handbook of Computational Statistics: Concepts and Methods* (edited by James E. Gentle, Wolfgang Härdle, and Yuichi Mori), Springer, Berlin, 103–136.

Cline, Alan K.; Andrew R. Conn; and Charles F. Van Loan (1982), Generalizing the LINPACK condition estimator, in *Numerical Analysis, Mexico, 1981* (edited by J. P. Hennart), Springer-Verlag, Berlin, 73–83.

Cline, A. K.; C. B. Moler; G. W. Stewart; and J. H. Wilkinson (1979), An estimate for the condition number of a matrix, *SIAM Journal of Numerical Analysis* **16**, 368–375.

Cline, A. K., and R. K. Rew (1983), A set of counter-examples to three condition number estimators, *SIAM Journal on Scientific and Statistical Computing* **4**, 602–611.

Cody, W. J. (1988), Algorithm 665: MACHAR: A subroutine to dynamically determine machine parameters, *ACM Transactions on Mathematical Software* **14**, 303–329.

Cody, W. J., and Jerome T. Coonen (1993), Algorithm 722: Functions to support the IEEE standard for binary floating-point arithmetic, *ACM Transactions on Mathematical Software* **19**, 443–451.

Coleman, Thomas F., and Charles Van Loan (1988), *Handbook for Matrix Computations*, Society for Industrial and Applied Mathematics, Philadelphia.

Cullen, M. R. (1985), *Linear Models in Biology*, Halsted Press, New York.

Dauger, Dean E., and Viktor K. Decyk (2005), Plug-and-play cluster computing: High-performance computing for the mainstream, *Computing in Science and Engineering* **07**(2), 27–33.

Davies, Philip I., and Nicholas J. Higham (2000), Numerically stable generation of correlation matrices and their factors, *BIT* **40**, 640–651.

Dempster, Arthur P., and Donald B. Rubin (1983), Rounding error in regression: The appropriateness of Sheppard's corrections, *Journal of the Royal Statistical Society, Series B* **39**, 1–38.

Devlin, Susan J.; R. Gnanadesikan; and J. R. Kettenring (1975), Robust estimation and outlier detection with correlation coefficients, *Biometrika* **62**, 531–546.

Dey, Aloke, and Rahul Mukerjee (1999), *Fractional Factorial Plans*, John Wiley and Sons, New York.

Dodson, David S.; Roger G. Grimes; and John G. Lewis (1991), Sparse extensions to the FORTRAN basic linear algebra subprograms, *ACM Transactions on Mathematical Software* **17**, 253–263.

Dongarra, J. J.; J. R. Bunch; C. B. Moler; and G. W. Stewart (1979), *LINPACK Users' Guide*, Society for Industrial and Applied Mathematics, Philadelphia.

Dongarra, J. J.; J. DuCroz; S. Hammarling; and I. Duff (1990), A set of level 3 basic linear algebra subprograms, *ACM Transactions on Mathematical Software* **16**, 1–17.

Dongarra, J. J.; J. DuCroz; S. Hammarling; and R. J. Hanson (1988), An extended set of Fortran basic linear algebra subprograms, *ACM Transactions on Mathematical Software* **14**, 1–17.

Dongarra, Jack J.; Ian S. Duff; Danny C. Sorensen; and Henk A. van der Vorst (1998), *Numerical Linear Algebra for High-Performance Computers*, Society for Industrial and Applied Mathematics, Philadelphia.

Dongarra, Jack J., and Victor Eijkhout (2000), Numerical linear algebra algorithms and software, *Journal of Computational and Applied Mathematics* **123**, 489–514.

Dongarra, Jack J.; Ian Foster; Geoffrey C. Fox; William Gropp; Ken Kennedy; Linda Torczon; and Andy White (2002), *The Sourcebook of Parallel Computing*, Morgan Kaufmann, San Francisco.

Dongarra, Jack J., and David W. Walker (1995), Software libraries for linear algebra computations on high performance computers, *SIAM Review* **37**, 151–180. (Also published as Libraries for linear algebra, in *High Performance Computing*, edited by Gary W. Sabot, 1995, Addison-Wesley Publishing Company, Reading, Massachusetts, 93–134.)

Draper, Norman R., and Harry Smith (1998), *Applied Regression Analysis*, third edition, John Wiley and Sons, New York.

Duff, Iain S.; Michael A. Heroux; and Roldan Pozo (2002), An overview of the sparse basic linear algebra subprograms: the new standard from the BLAS technical forum, *ACM Transactions on Mathematical Software* **28**, 239–267.

Duff, Iain S.; Michele Marrone; Guideppe Radicati; and Carlo Vittoli (1997), Level 3 basic linear algebra subprograms for sparse matrices: A user-level interface, *ACM Transactions on Mathematical Software* **23**, 379–401.

Duff, Iain S., and Christof Vömel (2002), Algorithm 818: A reference model implementation of the sparse BLAS in Fortran 95, *ACM Transactions on Mathematical Software* **28**, 268–283.

Eckart, Carl, and Gale Young (1936), The approximation of one matrix by another of lower rank, *Psychometrika* **1**, 211–218.

Ericksen, Wilhelm S. (1985), Inverse pairs of matrices, *ACM Transactions on Mathematical Software* **11**, 302–304.

Efron, Bradley; Trevor Hastie; Iain Johnstone; and Robert Tibshirani (2004), Least angle regression, *The Annals of Statistics* **32**, 407–499.

Escobar, Luis A., and E. Barry Moser (1993), A note on the updating of regression estimates, *The American Statistician* **47**, 192–194.

Eskow, Elizabeth, and Robert B. Schnabel (1991), Algorithm 695: Software for a new modified Cholesky factorization, *ACM Transactions on Mathematical Software* **17**, 306–312.

Everitt, Brian S., and Torsten Nothorn (2006), *A Handbook of Statistical Analyses Using R*, Chapman and Hall, New York.

Fasino, Dario, and Luca Gemignani (2003), A Lanczos-type algorithm for the QR factorization of Cauchy-like matrices, in *Fast Algorithms for Structured Matrices: Theory and Applications* (edited by Vadim Olshevsky), American Mathematical Society, Providence, Rhode Island, 91–104.

Filippone, Salvatore, and Michele Colajanni (2000), PSBLAS: A library for parallel linear algebra computation on sparse matrices, *ACM Transactions on Mathematical Software* **26**, 527–550.

Forsythe, George E., and Cleve B. Moler (1967), *Computer Solution of Linear Algebraic Systems*, Prentice-Hall, Englewood Cliffs, New Jersey.

Fuller, Wayne A. (1995), *Introduction to Statistical Time Series*, second edition, John Wiley and Sons, New York.

Galassi, Mark; Jim Davies; James Theiler; Brian Gough; Gerard Jungman; Michael Booth; and Fabrice Rossi (2002), *GNU Scientific Library Reference Manual*, second edition, Network Theory Limited, Bristol, United Kingdom.

Gantmacher, F. R. (1959) *The Theory of Matrices*, Volumes I and II, translated by K. A. Hirsch, Chelsea, New York.

Geist, Al; Adam Beguelin; Jack Dongarra; Weicheng Jiang; Robert Manchek; and Vaidy Sunderam (1994), *PVM. Parallel Virtual Machine. A Users' Guide and Tutorial for Networked Parallel Computing*, The MIT Press, Cambridge, Massachusetts.

Gentle, James E. (2002), *Elements of Computational Statistics*, Springer-Verlag, New York.

Gentle, James E. (2003), *Random Number Generation and Monte Carlo Methods*, second edition, Springer-Verlag, New York.

Gentle, James E. (2007), *Optimization Methods for Applications in Statistics*, Springer-Verlag, New York.

Gentleman, W. M. (1974), Algorithm AS 75: Basic procedures for large, sparse or weighted linear least squares problems, *Applied Statistics* **23**, 448–454.

Gill, Len, and Arthur Lewbel (1992), Testing the rank and definiteness of estimated matrices with applications to factor, state-space and ARMA models, *Journal of the American Statistical Association* **87**, 766–776.

Goedecker, Stefan, and Adolfy Hoisie (2001), *Performance Optimization of Numerically Intensive Code*, Society for Industrial and Applied Mathematics, Philadelphia.

Golub, G., and W. Kahan (1965), Calculating the singular values and pseudo-inverse of a matrix, *SIAM Journal of Numerical Analysis, Series B* **2**, 205–224.

Golub, G. H., and C. Reinsch (1970), Singular value decomposition and least squares solutions, *Numerische Mathematik* **14**, 403–420.

Golub, G. H., and C. F. Van Loan (1980), An analysis of the total least squares problem, *SIAM Journal of Numerical Analysis* **17**, 883–893.

Golub, Gene H., and Charles F. Van Loan (1996), *Matrix Computations*, third edition, The Johns Hopkins Press, Baltimore.

Graybill, Franklin A. (1983), *Introduction to Matrices with Applications in Statistics*, second edition, Wadsworth Publishing Company, Belmont, California.

Greenbaum, Anne, and Zdeněk Strakoš (1992), Predicting the behavior of finite precision Lanczos and conjugate gradient computations, *SIAM Journal for Matrix Analysis and Applications* **13**, 121–137.

Gregory, Robert T., and David L. Karney (1969), *A Collection of Matrices for Testing Computational Algorithms*, John Wiley and Sons, New York.

Gregory, R. T., and E. V. Krishnamurthy (1984), *Methods and Applications of Error-Free Computation*, Springer-Verlag, New York.

Grewal, Mohinder S., and Angus P. Andrews (1993), *Kalman Filtering Theory and Practice*, Prentice-Hall, Englewood Cliffs, New Jersey.

Griffiths, P., and I. D. Hill (Editors) (1985), *Applied Statistics Algorithms*, Ellis Horwood Limited, Chichester, United Kingdom.

Gropp, William D. (2005), Issues in accurate and reliable use of parallel computing in numerical programs, in *Accuracy and Reliability in Scientific Computing* (edited by Bo Einarsson), Society for Industrial and Applied Mathematics, Philadelphia, 253–263.

Gropp, William; Ewing Lusk; and Anthony Skjellum (1999), *Using MPI*, second edition, The MIT Press, Cambridge, Massachusetts.

Gropp, William; Ewing Lusk; and Thomas Sterling (Editors) (2003), *Beowulf Cluster Computing with Linux*, second edition, The MIT Press, Cambridge, Massachusetts.

Haag, J. B., and D. S. Watkins (1993), *QR*-like algorithms for the nonsymmetric eigenvalue problem, *ACM Transactions on Mathematical Software* **19**, 407–418.

Hager, W. W. (1984), Condition estimates, *SIAM Journal on Scientific and Statistical Computing* **5**, 311–316.

Hanselman, Duane C., and Bruce L. Littlefield (2004), *Mastering MATLAB 7*, Prentice-Hall, Englewood Cliffs, New Jersey.

Hansen, Per Christian (1998), *Rank-Deficient and Discrete Ill-Posed Problems: Numerical Aspects of Linear Inversion*, Society for Industrial and Applied Mathematics, Philadelphia.

Harville, David A. (1997), *Matrix Algebra from a Statistician's Point of View*, Springer-Verlag, New York.

Heath, M. T.; E. Ng; and B. W. Peyton (1991), Parallel algorithms for sparse linear systems, *SIAM Review* **33**, 420–460.

Hedayat, A. S.; N. J. A. Sloane; and John Stufken (1999), *Orthogonal Arrays: Theory and Applications*, Springer-Verlag, New York.

Heiberger, Richard M. (1978), Algorithm AS127: Generation of random orthogonal matrices, *Applied Statistics* **27**, 199–205.

Higham, Nicholas J. (1987), A survey of condition number estimation for triangular matrices, *SIAM Review* **29**, 575–596.

Higham, Nicholas J. (1988), FORTRAN codes for estimating the one-norm of a real or complex matrix, with applications to condition estimation, *ACM Transactions on Mathematical Software* **14**, 381–386.

Higham, Nicholas J. (1990), Experience with a matrix norm estimator, *SIAM Journal on Scientific and Statistical Computing* **11**, 804–809.

Higham, Nicholas J. (1991), Algorithm 694: A collection of test matrices in Matlab, *ACM Transactions on Mathematical Software* **17**, 289–305.

Higham, Nicholas J. (1997), Stability of the diagonal pivoting method with partial pivoting, *SIAM Journal of Matrix Analysis and Applications* **18**, 52–65.

Higham, Nicholas J. (2002), *Accuracy and Stability of Numerical Algorithms*, second edition, Society for Industrial and Applied Mathematics, Philadelphia.

Hoffman, A. J., and H. W. Wielandt (1953), The variation of the spectrum of a normal matrix, *Duke Mathematical Journal* **20**, 37–39.

Hong, H. P., and C. T. Pan (1992), Rank-revealing QR factorization and SVD, *Mathematics of Computation* **58**, 213–232.

Horn, Roger A., and Charles R. Johnson (1991), *Topics in Matrix Analysis*, Cambridge University Press, Cambridge, United Kingdom.

IEEE (1985), *IEEE Standard for Binary Floating-Point Arithmetic, Std 754-1985*, IEEE, Inc. New York.

Jansen, Paul, and Peter Weidner (1986), High-accuracy arithmetic software — some tests of the ACRITH problem-solving routines, *ACM Transactions on Mathematical Software* **12**, 62–70.

Jolliffe, I. T. (2002), *Principal Component Analysis*, second edition, Springer-Verlag, New York.

Kearfott, R. Baker (1996), INTERVAL_ARITHMETIC: A Fortran 90 module for an interval data type. *ACM Transactions on Mathematical Software* **22**, 385–392.

Kearfott, R. Baker, and Vladik Kreinovich (Editors) (1996), *Applications of Interval Computations*, Kluwer, Dordrecht, Netherlands.

Kearfott, R. B.; M. Dawande; K. Du; and C. Hu (1994), Algorithm 737: INTLIB: A portable Fortran 77 interval standard-function library, *ACM Transactions on Mathematical Software* **20**, 447–459.

Keller-McNulty, Sallie, and W. J. Kennedy (1986), An error-free generalized matrix inversion and linear least squares method based on bordering, *Communications in Statistics — Simulation and Computation* **15**, 769–785.

Kendall, M. G. (1961), *A Course in the Geometry of n Dimensions*, Charles Griffin and Company Limited, London.

Kennedy, William J., and James E. Gentle (1980), *Statistical Computing*, Marcel Dekker, Inc., New York.

Kenney, C. S., and A. J. Laub (1994), Small-sample statistical condition estimates for general matrix functions, *SIAM Journal on Scientific Computing* **15**, 191–209.

Kenney, C. S.; A. J. Laub; and M. S. Reese (1998), Statistical condition estimation for linear systems, *SIAM Journal on Scientific Computing* **19**, 566–583.

Lawson, C. L.; R. J. Hanson; D. R. Kincaid; and F. T. Krogh (1979), Basic linear algebra subprograms for Fortran usage, *ACM Transactions on Mathematical Software* **5**, 308–323.

Lehoucq, R. B.; D. C. Sorensen; and C. Yang (1998), *ARPACK Users' Guide: Solution of Large-Scale Eigenvalue Problems with Implicitly Restarted Arnoldi Methods*, Society for Industrial and Applied Mathematics, Philadelphia.

Lemmon, David R., and Joseph L. Schafer (2005), *Developing Statistical Software in Fortran 95*, Springer-Verlag, New York.

Liem, C. B.; T. Lü; and T. M. Shih (1995), *The Splitting Extrapolation Method*, World Scientific, Singapore.

Linnainmaa, Seppo (1975), Towards accurate statistical estimation of rounding errors in floating-point computations, *BIT* **15** 165–173.

Liu, Shuangzhe and Heinz Neudecker (1996), Several matrix Kantorovich-type inequalities, *Journal of Mathematical Analysis and Applications* **197**, 23–26.

Loader, Catherine (2004), Smoothing: Local regression techniques, in *Handbook of Computational Statistics: Concepts and Methods* (edited by James E. Gentle, Wolfgang Härdle, and Yuichi Mori), Springer, Berlin, 539–563.

Longley, James W. (1967), An appraisal of least squares problems for the electronic computer from the point of view of the user, *Journal of the American Statistical Association* **62**, 819–841.

Luk, F. T., and H. Park (1989), On parallel Jacobi orderings, *SIAM Journal on Scientific and Statistical Computing* **10**, 18–26.

Magnus, Jan R., and Heinz Neudecker (1999), *Matrix Differential Calculus with Applications in Statistics and Econometrics*, revised edition, John Wiley and Sons, New York.

Maindonald, John, and John Braun (2003), *Data Analysis and Graphics Using R*, Cambridge University Press, Cambridge, United Kingdom.

Marshall, A. W., and I. Olkin (1990), Matrix versions of the Cauchy and Kantorovich inequalities, *Aequationes Mathematicae* **40**, 89–93.

Metcalf, Michael, John Reid, and Malcolm Cohen (2004), *Fortran 95/2003 Explained*, third edition, Oxford University Press, Oxford, United Kingdom.

Miller, Alan J. (1992), Algorithm AS 274: Least squares routines to supplement those of Gentleman, *Applied Statistics* **41**, 458–478 (Corrections, 1994, *ibid.* **43**, 678).

Miller, Alan (2002), *Subset Selection in Regression*, second edition, Chapman and Hall/CRC, Boca Raton.

Miller, Alan J., and Nam-Ky Nguyen (1994), A Fedorov exchange algorithm for D-optimal design, *Applied Statistics* **43**, 669–678.

Mizuta, Masahiro (2004), Dimension reduction methods, in *Handbook of Computational Statistics: Concepts and Methods* (edited by James E. Gentle, Wolfgang Härdle, and Yuichi Mori), Springer, Berlin, 565–589.

Moore, E. H. (1920), On the reciprocal of the general algebraic matrix, *Bulletin of the American Mathematical Society*, **26**, 394–395.

Moore, Ramon E. (1979), *Methods and Applications of Interval Analysis*, Society for Industrial and Applied Mathematics, Philadelphia.

Mortenson, Michael E. (1997), *Geometric Modeling*, second edition, John Wiley and Sons, New York.

Muirhead, Robb J. (1982), *Aspects of Multivariate Statistical Theory*, John Wiley and Sons, New York.

Mullet, Gary M., and Tracy W. Murray (1971), A new method for examining rounding error in least-squares regression computer programs, *Journal of the American Statistical Association* **66**, 496–498.

Nachbin, Leopoldo (1965), *The Haar Integral*, translated by Lulu Bechtolsheim, D. Van Nostrand Co Inc, Princeton, New Jersey.

Nakano, Junji (2004), Parallel computing techniques, in *Handbook of Computational Statistics: Concepts and Methods* (edited by James E. Gentle, Wolfgang Härdle, and Yuichi Mori), Springer, Berlin, 237–266.

Nash, Stephen G., and Ariela Sofer (1996), *Linear and Nonlinear Programming*, McGraw-Hill, New York.

Nguyen, Nam-Ky, and Alan J. Miller (1992), A review of some exchange algorithms for constructing D-optimal designs, *Computational Statistics and Data Analysis* **14**, 489–498.

Norris, J. R. (1997), *Markov Chains*, Cambridge University Press, Cambridge, United Kingdom.

Olshevsky, Vadim (Editor) (2003), *Fast Algorithms for Structured Matrices: Theory and Applications*, American Mathematical Society, Providence, Rhode Island.

Overton, Michael L. (2001), *Numerical Computing with IEEE Floating Point Arithmetic*, Society for Industrial and Applied Mathematics, Philadelphia.

Penrose, R. (1955), A generalized inverse for matrices, *Proceedings of the Cambridge Philosophical Society*, **51**, 406–413.

Press, William H.; Saul A. Teukolsky; William T. Vetterling; and Brian P. Flannery (1996), *Numerical Recipes in Fortran 90*, Cambridge University Press, Cambridge, United Kingdom. (Also called *Fortran Numerical Recipes, Volume 2*, second edition.)

Quinn, Michael J. (2003), *Parallel Programming in C with MPI and OpenMP*, McGraw-Hill, New York.

Rice, John R. (1966), Experiments on Gram-Schmidt orthogonalization, *Mathematics of Computation* **20**, 325–328.

Rice, John R. (1993), *Numerical Methods, Software, and Analysis*, second edition, McGraw-Hill Book Company, New York.

Riesenfeld, R. F. (1981), Homogeneous coordinates and projective planes in computer graphics, *IEEE Computer Graphics and Applications* **1**, 50–55.

Rogers, Gerald S. (1980), *Matrix Derivatives*, Marcel Dekker, Inc., New York.

Roosta, Seyed H. (2000), *Parallel Processing and Parallel Algorithms: Theory and Computation*, Springer-Verlag, New York.

Rousseeuw, Peter J., and Geert Molenberghs (1993), Transformation of non-positive semidefinite correlation matrices, *Communications in Statistics — Theory and Methods* **22**, 965–984.

Rust, Bert W. (1994), Perturbation bounds for linear regression problems, *Computing Science and Statistics* **26**, 528–532.

Saad, Y., and M. H. Schultz (1986), GMRES: A generalized minimal residual algorithm for solving nonsymmetric linear systems, *SIAM Journal on Scientific and Statistical Computing* **7**, 856–869.

Schott, James R. (2004), *Matrix Analysis for Statistics*, second edition, John Wiley and Sons, New York.

Searle, S. R. (1971), *Linear Models*, John Wiley and Sons, New York.

Searle, Shayle R. (1982), *Matrix Algebra Useful for Statistics*, John Wiley and Sons, New York.

Shao, Jun (2003), *Mathematical Statistics*, second edition, Springer-Verlag, New York.

Sherman, J., and W. J. Morrison (1950), Adjustment of an inverse matrix corresponding to a change in one element of a given matrix, *Annals of Mathematical Statistics* **21**, 124–127.

Siek, Jeremy, and Andrew Lumsdaine (2000), A modern framework for portable high-performance numerical linear algebra, in *Advances in Software Tools for Scientific Computing* (edited by Are Bruaset, H. Langtangen, and E. Quak), Springer-Verlag, New York, 1–56.

Skeel, R. D. (1980), Iterative refinement implies numerical stability for Gaussian elimination, *Mathematics of Computation* **35**, 817–832.

Smith, B. T.; J. M. Boyle; J. J. Dongarra; B. S. Garbow; Y. Ikebe; V. C. Klema; and C. B. Moler (1976), *Matrix Eigensystem Routines — EISPACK Guide*, Springer-Verlag, Berlin.

Smith, David M. (1991), Algorithm 693: A FORTRAN package for floating-point multiple-precision arithmetic, *ACM Transactions on Mathematical Software* **17**, 273–283.

Stallings, W. T., and T. L. Boullion (1972), Computation of pseudo-inverse using residue arithmetic, *SIAM Review* **14**, 152–163.

Stewart, G. W. (1980), The efficient generation of random orthogonal matrices with an application to condition estimators, *SIAM Journal of Numerical Analysis* **17**, 403–409.

Stewart, G. W. (1990), Stochastic perturbation theory, *SIAM Review* **32**, 579–610.

Strang, Gilbert, and Tri Nguyen (2004), The interplay of ranks of submatrices, *SIAM Review* **46**, 637–646.

Strassen, V. 1969, Gaussian elimination is not optimal, *Numerische Mathematik* **13**, 354–356.

Szabó, S., and R. Tanaka (1967), *Residue Arithmetic and Its Application to Computer Technology*, McGraw-Hill, New York.

Tanner, M. A., and R. A. Thisted (1982), A remark on AS127. Generation of random orthogonal matrices, *Applied Statistics* **31**, 190–192.

Titterington, D. M. (1975), Optimal design: Some geometrical aspects of D-optimality, *Biometrika* **62**, 313–320.

Trosset, Michael W. (2002), Extensions of classical multidimensional scaling via variable reduction, *Computational Statistics* **17**, 147–163.

Unicode Consortium (1990), *The Unicode Standard, Worldwide Character Encoding, Version 1.0, Volume 1*, Addison-Wesley Publishing Company, Reading, Massachusetts.

Unicode Consortium (1992), *The Unicode Standard, Worldwide Character Encoding, Version 1.0, Volume 2*, Addison-Wesley Publishing Company, Reading, Massachusetts.

Van de Geijn, Robert (1997), *Using PLAPACK: Parallel Linear Algebra Package*, The MIT Press, Cambridge, Massachusetts.

Van Loan, Charles F. (1987), On estimating the condition of eigenvalues and eigenvectors, *Linear Algebra and Its Applications* **88**, 715–732.

Van Loan, Charles F. (1997), *Introduction to Scientific Computing: A Matrix-Vector Approach Using MATLAB*, Prentice-Hall, Englewood Cliffs, New Jersey.

Venables, W. N., and B. D. Ripley (2003), *Modern Applied Statistics with S*, fourth edition, Springer-Verlag, New York.

Walker, Homer F. (1988), Implementation of the GMRES method using Householder transformations, *SIAM Journal on Scientific and Statistical Computing* **9**, 152–163.

Walker, Homer F., and Lu Zhou (1994), A simpler GMRES, *Numerical Linear Algebra with Applications* **1**, 571–581.

Walster, G. William (1996), Stimulating hardware and software support for interval arithmetic, in *Applications of Interval Computations*, (edited by R. Baker Kearfott and Vladik Kreinovich), Kluwer, Dordrecht, Netherlands, 405–416.

Walster, G. William (2005), The use and implementation of interval data types, in *Accuracy and Reliability in Scientific Computing* (edited by Bo Einarsson), 173–194.

Watkins, David S. (2002), *Fundamentals of Matrix Computations*, second edition, John Wiley and Sons, New York.

Wilkinson, J. H. (1959), The evaluation of the zeros of ill-conditioned polynomials, *Numerische Mathematik* **1**, 150–180.

Wilkinson, J. H. (1963), *Rounding Errors in Algebraic Processes*, Prentice-Hall, Englewood Cliffs, New Jersey. (Reprinted by Dover Publications, Inc., New York, 1994).

Wilkinson, J. H. (1965), *The Algebraic Eigenvalue Problem*, Oxford University Press, New York.

Woodbury, M. A. (1950), "Inverting Modified Matrices", Memorandum Report 42, Statistical Research Group, Princeton University.

Zhou, Bing Bing, and Richard P. Brent (2003), An efficient method for computing eigenvalues of a real normal matrix, *Journal of Parallel and Distributed Computing* **63**, 638–648.

Index

CPSIA information can be obtained
at www.ICGtesting.com
Printed in the USA
BVOW06s1325220817
492739BV00020B/39/P